Differential Equations

Linear, Nonlinear, Ordinary, Partial

When mathematical modelling is used to describe physical, biological or chemical phenomena, one of the most common results of the modelling process is a system of ordinary or partial differential equations. Finding and interpreting the solutions of these differential equations is therefore a central part of applied mathematics, and a thorough understanding of differential equations is essential for any applied mathematician. The aim of this book is to develop the required skills on the part of the reader.

The authors focus on the business of constructing solutions analytically and interpreting their meaning, although they do use rigorous analysis where needed. The reader is assumed to have some basic knowledge of linear, constant coefficient ordinary differential equations, real analysis and linear algebra. The book will thus appeal to undergraduates in mathematics, but would also be of use to physicists and engineers. MATLAB is used extensively to illustrate the material. There are many worked examples based on interesting real-world problems. A large selection of exercises is provided, including several lengthier projects, some of which involve the use of MATLAB. The coverage is broad, ranging from basic second-order ODEs including the method of Frobenius, Sturm-Liouville theory, Fourier and Laplace transforms, and existence and uniqueness, through to techniques for nonlinear differential equations including phase plane methods, bifurcation theory and chaos, asymptotic methods, and control theory. This broad coverage, the authors' clear presentation and the fact that the book has been thoroughly class-tested will increase its appeal to undergraduates at each stage of their studies.

Differential Equations

Linear, Nonlinear, Ordinary, Partial

A.C. King, J. Billingham and S.R. Otto

PUBLISHED BY THE PRESS SYNDICATE OF THE UNIVERSITY OF CAMBRIDGE
The Pitt Building, Trumpington Street, Cambridge, United Kingdom

CAMBRIDGE UNIVERSITY PRESS
The Edinburgh Building, Cambridge CB2 2RU, UK
40 West 20th Street, New York, NY 10011–4211, USA
477 Williamstown Road, Port Melbourne, Vic 3207, Australia
Ruiz de Alarcón 13, 28014 Madrid, Spain
Dock House, The Waterfront, Cape Town 8001, South Africa

http://www.cambridge.org

© Cambridge University Press 2003

This book is in copyright. Subject to statutory exception
and to the provisions of relevant collective licensing agreements,
no reproduction of any part may take place without
the written permission of Cambridge University Press.

First published 2003
First South Asian edition 2005

Printed in India by Replika Press Pvt. Ltd.

Typeface Computer Modern 10/12.5pt   *System* LATEX   [TB]

A catalogue record for this book is available from the British Library

Library of Congress Cataloguing in Publication data

ISBN 0 521 67045 4 paperback

This edition for sale in South Asia only, not for export elsewhere.

Contents

Preface   page ix

    Part One: Linear Equations   1

1   **Variable Coefficient, Second Order, Linear, Ordinary Differential Equations**   3
    1.1   The Method of Reduction of Order   5
    1.2   The Method of Variation of Parameters   7
    1.3   Solution by Power Series: The Method of Frobenius   11

2   **Legendre Functions**   31
    2.1   Definition of the Legendre Polynomials, $P_n(x)$ 31
    2.2   The Generating Function for $P_n(x)$ 35
    2.3   Differential and Recurrence Relations Between Legendre Polynomials   38
    2.4   Rodrigues' Formula   39
    2.5   Orthogonality of the Legendre Polynomials   41
    2.6   Physical Applications of the Legendre Polynomials   44
    2.7   The Associated Legendre Equation   52

3   **Bessel Functions**   58
    3.1   The Gamma Function and the Pockhammer Symbol   58
    3.2   Series Solutions of Bessel's Equation   60
    3.3   The Generating Function for $J_n(x)$, n an integer   64
    3.4   Differential and Recurrence Relations Between Bessel Functions   69
    3.5   Modified Bessel Functions   71
    3.6   Orthogonality of the Bessel Functions   71
    3.7   Inhomogeneous Terms in Bessel's Equation   77
    3.8   Solutions Expressible as Bessel Functions   79
    3.9   Physical Applications of the Bessel Functions   80

4   **Boundary Value Problems, Green's Functions and Sturm–Liouville Theory**   93
    4.1   Inhomogeneous Linear Boundary Value Problems   96
    4.2   The Solution of Boundary Value Problems by Eigenfunction Expansions   100
    4.3   Sturm–Liouville Systems   107

5 Fourier Series and the Fourier Transform — 123
- 5.1 General Fourier Series — 127
- 5.2 The Fourier Transform — 133
- 5.3 Green's Functions Revisited — 141
- 5.4 Solution of Laplace's Equation Using Fourier Transforms — 143
- 5.5 Generalization to Higher Dimensions — 145

6 Laplace Transforms — 152
- 6.1 Definition and Examples — 152
- 6.2 Properties of the Laplace Transform — 154
- 6.3 The Solution of Ordinary Differential Equations using Laplace Transforms — 157
- 6.4 The Inversion Formula for Laplace Transforms — 162

7 Classification, Properties and Complex Variable Methods for Second Order Partial Differential Equations — 175
- 7.1 Classification and Properties of Linear, Second Order Partial Differential Equations in Two Independent Variables — 175
- 7.2 Complex Variable Methods for Solving Laplace's Equation — 186

Part Two: Nonlinear Equations and Advanced Techniques — 201

8 Existence, Uniqueness, Continuity and Comparison of Solutions of Ordinary Differential Equations — 203
- 8.1 Local Existence of Solutions — 204
- 8.2 Uniqueness of Solutions — 210
- 8.3 Dependence of the Solution on the Initial Conditions — 211
- 8.4 Comparison Theorems — 212

9 Nonlinear Ordinary Differential Equations: Phase Plane Methods — 217
- 9.1 Introduction: The Simple Pendulum — 217
- 9.2 First Order Autonomous Nonlinear Ordinary Differential Equations — 222
- 9.3 Second Order Autonomous Nonlinear Ordinary Differential Equations — 224
- 9.4 Third Order Autonomous Nonlinear Ordinary Differential Equations — 249

10 Group Theoretical Methods — 256
- 10.1 Lie Groups — 257
- 10.2 Invariants Under Group Action — 261
- 10.3 The Extended Group — 262
- 10.4 Integration of a First Order Equation with a Known Group Invariant — 263

	10.5	Towards the Systematic Determination of Groups Under Which a First Order Equation is Invariant	265
	10.6	Invariants for Second Order Differential Equations	266
	10.7	Partial Differential Equations	270
11	**Asymptotic Methods: Basic Ideas**		274
	11.1	Asymptotic Expansions	275
	11.2	The Asymptotic Evaluation of Integrals	280
12	**Asymptotic Methods: Differential Equations**		303
	12.1	An Instructive Analogy: Algebraic Equations	303
	12.2	Ordinary Differential Equations	306
	12.3	Partial Differential Equations	351
13	**Stability, Instability and Bifurcations**		372
	13.1	Zero Eigenvalues and the Centre Manifold Theorem	372
	13.2	Lyapunov's Theorems	381
	13.3	Bifurcation Theory	388
14	**Time-Optimal Control in the Phase Plane**		417
	14.1	Definitions	418
	14.2	First Order Equations	418
	14.3	Second Order Equations	422
	14.4	Examples of Second Order Control Problems	426
	14.5	Properties of the Controllable Set	429
	14.6	The Controllability Matrix	433
	14.7	The Time-Optimal Maximum Principle (TOMP)	436
15	**An Introduction to Chaotic Systems**		447
	15.1	Three Simple Chaotic Systems	447
	15.2	Mappings	452
	15.3	The Poincaré Return Map	467
	15.4	Homoclinic Tangles	472
	15.5	Quantifying Chaos: Lyapunov Exponents and the Lyapunov Spectrum	484

Appendix 1 **Linear Algebra** — 495

Appendix 2 **Continuity and Differentiability** — 502

Appendix 3 **Power Series** — 505

Appendix 4 **Sequences of Functions** — 509

Appendix 5 **Ordinary Differential Equations** — 511

Appendix 6 **Complex Variables** — 517

Appendix 7 **A Short Introduction to MATLAB** — 526

Bibliography — 534

Index — 536

Preface

When mathematical modelling is used to describe physical, biological or chemical phenomena, one of the most common results is either a differential equation or a system of differential equations, together with appropriate boundary and initial conditions. These differential equations may be ordinary or partial, and finding and interpreting their solution is at the heart of applied mathematics. A thorough introduction to differential equations is therefore a necessary part of the education of any applied mathematician, and this book is aimed at building up skills in this area. For similar reasons, the book should also be of use to mathematically-inclined physicists and engineers.

Although the importance of studying differential equations is not generally in question, exactly how the theory of differential equations should be taught, and what aspects should be emphasized, is more controversial. In our experience, textbooks on differential equations usually fall into one of two categories. Firstly, there is the type of textbook that emphasizes the importance of abstract mathematical results, proving each of its theorems with full mathematical rigour. Such textbooks are usually aimed at graduate students, and are inappropriate for the average undergraduate. Secondly, there is the type of textbook that shows the student how to construct solutions of differential equations, with particular emphasis on algorithmic methods. These textbooks often tackle only linear equations, and have no pretension to mathematical rigour. However, they are usually well-stocked with interesting examples, and often include sections on numerical solution methods.

In this textbook, we steer a course between these two extremes, starting at the level of preparedness of a typical, but well-motivated, second year undergraduate at a British university. As such, the book begins in an unsophisticated style with the clear objective of obtaining quantitative results for a particular linear ordinary differential equation. The text is, however, written in a progressive manner, with the aim of developing a deeper understanding of ordinary and partial differential equations, including conditions for the existence and uniqueness of solutions, solutions by group theoretical and asymptotic methods, the basic ideas of control theory, and nonlinear systems, including bifurcation theory and chaos. The emphasis of the book is on analytical and asymptotic solution methods. However, where appropriate, we have supplemented the text by including numerical solutions and graphs produced using MATLAB†, version 6. We assume some knowledge of

† MATLAB is a registered trademark of The MathWorks, Inc.

MATLAB (summarized in Appendix 7), but explain any nontrivial aspects as they arise. Where mathematical rigour is required, we have presented the appropriate analysis, on the basis that the student has taken first courses in analysis and linear algebra. We have, however, avoided any functional analysis. Most of the material in the book has been taught by us in courses for undergraduates at the University of Birmingham. This has given us some insight into what students find difficult, and, as a consequence, what needs to be emphasized and re-iterated.

The book is divided into two parts. In the first of these, we tackle linear differential equations. The first three chapters are concerned with variable coefficient, linear, second order ordinary differential equations, emphasizing the methods of reduction of order and variation of parameters, and series solution by the method of Frobenius. In particular, we discuss Legendre functions (Chapter 2) and Bessel functions (Chapter 3) in detail, and motivate this by giving examples of how they arise in real modelling problems. These examples lead to partial differential equations, and we use separation of variables to obtain Legendre's and Bessel's equations. In Chapter 4, the emphasis is on boundary value problems, and we show how these differ from initial value problems. We introduce Sturm–Liouville theory in this chapter, and prove various results on eigenvalue problems. The next two chapters of the first part of the book are concerned with Fourier series, and Fourier and Laplace transforms. We discuss in detail the convergence of Fourier series, since the analysis involved is far more straightforward than that associated with other basis functions. Our approach to Fourier transforms involves a short introduction to the theory of generalized functions. The advantage of this approach is that a discussion of what types of function possess a Fourier transform is straightforward, since *all* generalized functions possess a Fourier transform. We show how Fourier transforms can be used to construct the free space Green's function for both ordinary and partial differential equations. We also use Fourier transforms to derive the solutions of the Dirichlet and Neumann problems for Laplace's equation. Our discussion of the Laplace transform includes an outline proof of the inversion theorem, and several examples of physical problems, for example involving diffusion, that can be solved by this method. In Chapter 7 we discuss the classification of linear, second order partial differential equations, emphasizing the reasons why the canonical examples of elliptic, parabolic and hyperbolic equations, namely Laplace's equation, the diffusion equation and the wave equation, have the properties that they do. We also consider complex variable methods for solving Laplace's equation, emphasizing their application to problems in fluid mechanics.

The second part of the book is concerned with nonlinear problems and more advanced techniques. Although we have used a lot of the material in Chapters 9 and 14 (phase plane techniques and control theory) in a course for second year undergraduates, the bulk of the material here is aimed at third year students. We begin in Chapter 8 with a brief introduction to the rigorous analysis of ordinary differential equations. Here the emphasis is on existence, uniqueness and comparison theorems. In Chapter 9 we introduce the phase plane and its associated techniques. This is the first of three chapters (the others being Chapters 13 and 15) that form an introduction to the theory of nonlinear ordinary differential equations, often known as dynamical systems. In Chapter 10, we show how the ideas of group theory can be used to find exact solutions of ordinary and partial differential equations. In Chapters 11 and 12 we discuss the theory and practice of asymptotic analysis. After discussing the basic ideas at the beginning of Chapter 11, we move on to study the three most important techniques for the asymptotic evaluation of integrals: Laplace's method, the method of stationary phase and the method of steepest descents. Chapter 12 is devoted to the asymptotic solution of differential equations, and we introduce the method of matched asymptotic expansions, and the associated idea of asymptotic matching, the method of multiple scales, including Kuzmak's method for analysing the slow damping of nonlinear oscillators, and the WKB expansion. We illustrate each of these methods with a wide variety of examples, for both nonlinear ordinary differential equations and partial differential equations. In Chapter 13 we cover the centre manifold theorem, Lyapunov functions and an introduction to bifurcation theory. Chapter 14 is about time-optimal control theory in the phase plane, and includes a discussion of the controllability matrix and the time-optimal maximum principle for second order linear systems of ordinary differential equations. Chapter 15 is on chaotic systems, and, after some illustrative examples, emphasizes the theory of homoclinic tangles and Mel'nikov theory.

There is a set of exercises at the end of each chapter. Harder exercises are marked with a star, and many chapters include a project, which is rather longer than the average exercise, and whose solution involves searches in the library or on the Internet, and deeper study. Bona fide teachers and instructors can obtain full worked solutions to many of the exercises by emailing solutions@cambridge.org.

In order to follow many of the ideas and calculations that we describe in this book, and to fully appreciate the more advanced material, the reader may need to acquire (or refresh) some basic skills. These are covered in the appendices, and fall into six basic areas: linear algebra, continuity and differentiability, power series, sequences and series of functions, ordinary differential equations and complex variables.

We would like to thank our friends and colleagues, Adam Burbidge (Nestlé Research Centre, Lausanne), Norrie Everitt (Birmingham), Chris Good (Birmingham), Ray Jones (Birmingham), John King (Nottingham), Dave Needham (Reading), Nigel Scott (East Anglia) and Warren Smith (Birmingham), who read and commented on one or more chapters of the book before it was published. Any nonsense remaining is, of course, our fault and not theirs.

ACK, JB and SRO, Birmingham 2002

Part One
Linear Equations

CHAPTER ONE

Variable Coefficient, Second Order, Linear, Ordinary Differential Equations

Many physical, chemical and biological systems can be described using mathematical models. Once the model is formulated, we usually need to solve a differential equation in order to predict and quantify the features of the system being modelled. As a precursor to this, we consider linear, second order ordinary differential equations of the form

$$P(x)\frac{d^2y}{dx^2} + Q(x)\frac{dy}{dx} + R(x)y = F(x),$$

with $P(x)$, $Q(x)$ and $R(x)$ finite polynomials that contain no common factor. This equation is inhomogeneous and has variable coefficients. The form of these polynomials varies according to the underlying physical problem that we are studying. However, we will postpone any discussion of the physical origin of such equations until we have considered some classical mathematical models in Chapters 2 and 3.

After dividing through by $P(x)$, we obtain the more convenient, equivalent form,

$$\frac{d^2y}{dx^2} + a_1(x)\frac{dy}{dx} + a_0(x)y = f(x). \tag{1.1}$$

This process is mathematically legitimate, provided that $P(x) \neq 0$. If $P(x_0) = 0$ at some point $x = x_0$, it is *not* legitimate, and we call x_0 a **singular** point of the equation. If $P(x_0) \neq 0$, x_0 is a **regular** or **ordinary** point of the equation. If $P(x) \neq 0$ for all points x in the interval where we want to solve the equation, we say that the equation is **nonsingular**, or **regular**, on the interval.

We usually need to solve (1.1) subject to either **initial conditions** of the form $y(a) = \alpha$, $y'(a) = \beta$ or **boundary conditions**, of which $y(a) = \alpha$ and $y(b) = \beta$ are typical examples. It is worth reminding ourselves that, given the ordinary differential equation and initial conditions (an **initial value problem**), the objective is to determine the solution for other values of x, typically, $x > a$, as illustrated in Figure 1.1. As an example, consider a projectile. The initial conditions are the position of the projectile and the speed and angle to the horizontal at which it is fired. We then want to know the path of the projectile, given these initial conditions.

For initial value problems of this form, it is possible to show that:

(i) If $a_1(x)$, $a_0(x)$ and $f(x)$ are continuous on some open interval I that contains the initial point a, a unique solution of the initial value problem exists on the interval I, as we shall demonstrate in Chapter 8.

4   VARIABLE COEFFICIENT, SECOND ORDER DIFFERENTIAL EQUATIONS

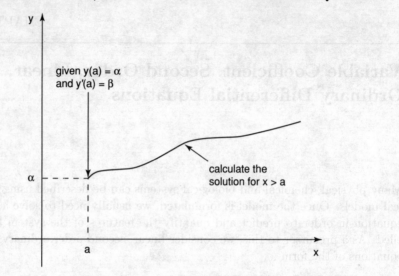

Fig. 1.1. An initial value problem.

(ii) The structure of the solution of the initial value problem is of the form

$$y = \underbrace{A u_1(x) + B u_2(x)}_{\text{Complementary function}} + \underbrace{G(x)}_{\text{Particular integral}},$$

where A, B are constants that are fixed by the initial conditions and $u_1(x)$ and $u_2(x)$ are linearly independent solutions of the corresponding homogeneous problem $y'' + a_1(x)y' + a_0(x)y = 0$.

These results can be proved rigorously, but nonconstructively, by studying the operator

$$Ly \equiv \frac{d^2y}{dx^2} + a_1(x)\frac{dy}{dx} + a_0(x)y,$$

and regarding $L : C^2(I) \to C^0(I)$ as a linear transformation from the space of twice-differentiable functions defined on the interval I to the space of continuous functions defined on I. The solutions of the homogeneous equation are elements of the null space of L. This subspace is completely determined once its basis is known. The solution of the inhomogeneous problem, $Ly = f$, is then given formally as $y = L^{-1}f$. Unfortunately, if we actually want to construct the solution of a particular equation, there is a lot more work to do.

Before we try to construct the general solution of the inhomogeneous initial value problem, we will outline a series of subproblems that are more tractable.

1.1 The Method of Reduction of Order

As a first simplification we discuss the solution of the homogeneous differential equation

$$\frac{d^2y}{dx^2} + a_1(x)\frac{dy}{dx} + a_0(x)y = 0, \qquad (1.2)$$

on the assumption that we know one solution, say $y(x) = u_1(x)$, and only need to find the second solution. We will look for a solution of the form $y(x) = U(x)u_1(x)$. Differentiating $y(x)$ using the product rule gives

$$\frac{dy}{dx} = \frac{dU}{dx}u_1 + U\frac{du_1}{dx},$$

$$\frac{d^2y}{dx^2} = \frac{d^2U}{dx^2}u_1 + 2\frac{dU}{dx}\frac{du_1}{dx} + U\frac{d^2u_1}{dx^2}.$$

If we substitute these expressions into (1.2) we obtain

$$\frac{d^2U}{dx^2}u_1 + 2\frac{dU}{dx}\frac{du_1}{dx} + U\frac{d^2u_1}{dx^2} + a_1(x)\left(\frac{dU}{dx}u_1 + U\frac{du_1}{dx}\right) + a_0(x)Uu_1 = 0.$$

We can now collect terms to get

$$U\left(\frac{d^2u_1}{dx^2} + a_1(x)\frac{du_1}{dx} + a_0(x)u_1\right) + u_1\frac{d^2U}{dx^2} + \frac{dU}{dx}\left(2\frac{du_1}{dx} + a_1u_1\right) = 0.$$

Now, since $u_1(x)$ is a solution of (1.2), the term multiplying U is zero. We have therefore obtained a differential equation for dU/dx, and, by defining $Z = dU/dx$, have

$$u_1\frac{dZ}{dx} + Z\left(2\frac{du_1}{dx} + a_1u_1\right) = 0.$$

Dividing through by Zu_1 we have

$$\frac{1}{Z}\frac{dZ}{dx} + \frac{2}{u_1}\frac{du_1}{dx} + a_1 = 0,$$

which can be integrated directly to yield

$$\log|Z| + 2\log|u_1| + \int^x a_1(s)\,ds = C,$$

where s is a dummy variable, for some constant C. Thus

$$Z = \frac{c}{u_1^2}\exp\left\{-\int^x a_1(s)\,ds\right\} = \frac{dU}{dx}$$

where $c = e^C$. This can then be integrated to give

$$U(x) = \int^x \frac{c}{u_1^2(t)}\exp\left\{-\int^t a_1(s)\,ds\right\}dt + \tilde{c},$$

for some constant $\tilde{c}$. The solution is therefore

VARIABLE COEFFICIENT, SECOND ORDER DIFFERENTIAL EQUATIONS

$$y(x) = u_1(x) \int^x \frac{c}{u_1^2(t)} \exp\left\{-\int^t a_1(s)\,ds\right\} dt + \tilde{c}u_1(x).$$

We can recognize $\tilde{c}u_1(x)$ as the part of the complementary function that we knew to start with, and

$$u_2(x) = u_1(x) \int^x \frac{1}{u_1^2(t)} \exp\left\{-\int^t a_1(s)\,ds\right\} dt \qquad (1.3)$$

as the second part of the complementary function. This result is called **the reduction of order formula**.

Example

Let's try to determine the full solution of the differential equation

$$(1 - x^2)\frac{d^2y}{dx^2} - 2x\frac{dy}{dx} + 2y = 0,$$

given that $y = u_1(x) = x$ is a solution. We firstly write the equation in standard form as

$$\frac{d^2y}{dx^2} - \frac{2x}{1-x^2}\frac{dy}{dx} + \frac{2}{1-x^2}y = 0.$$

Comparing this with (1.2), we have $a_1(x) = -2x/(1-x^2)$. After noting that

$$\int^t a_1(s)\,ds = \int^t -\frac{2s}{1-s^2}\,ds = \log(1 - t^2),$$

the reduction of order formula gives

$$u_2(x) = x \int^x \frac{1}{t^2} \exp\{-\log(1-t^2)\}\,dt = x \int^x \frac{dt}{t^2(1-t^2)}.$$

We can express the integrand in terms of its partial fractions as

$$\frac{1}{t^2(1-t^2)} = \frac{1}{t^2} + \frac{1}{1-t^2} = \frac{1}{t^2} + \frac{1}{2(1+t)} + \frac{1}{2(1-t)}.$$

This gives the second solution of (1.2) as

$$u_2(x) = x \int^x \left\{\frac{1}{t^2} + \frac{1}{2(1+t)} + \frac{1}{2(1-t)}\right\} dt$$

$$= x\left[-\frac{1}{t} + \frac{1}{2}\log\left(\frac{1+t}{1-t}\right)\right]^x = \frac{x}{2}\log\left(\frac{1+x}{1-x}\right) - 1,$$

and hence the general solution is

$$y = Ax + B\left\{\frac{x}{2}\log\left(\frac{1+x}{1-x}\right) - 1\right\}.$$

1.2 The Method of Variation of Parameters

Let's now consider how to find the particular integral *given* the complementary function, comprising $u_1(x)$ and $u_2(x)$. As the name of this technique suggests, we take the constants in the complementary function to be variable, and assume that

$$y = c_1(x)u_1(x) + c_2(x)u_2(x).$$

Differentiating, we find that

$$\frac{dy}{dx} = c_1 \frac{du_1}{dx} + u_1 \frac{dc_1}{dx} + c_2 \frac{du_2}{dx} + u_2 \frac{dc_2}{dx}.$$

We will choose to impose the condition

$$u_1 \frac{dc_1}{dx} + u_2 \frac{dc_2}{dx} = 0, \qquad (1.4)$$

and thus have

$$\frac{dy}{dx} = c_1 \frac{du_1}{dx} + c_2 \frac{du_2}{dx},$$

which, when differentiated again, yields

$$\frac{d^2y}{dx^2} = c_1 \frac{d^2u_1}{dx^2} + \frac{du_1}{dx}\frac{dc_1}{dx} + c_2 \frac{d^2u_2}{dx^2} + \frac{du_2}{dx}\frac{dc_2}{dx}.$$

This form can then be substituted into the original differential equation to give

$$c_1 \frac{d^2u_1}{dx^2} + \frac{du_1}{dx}\frac{dc_1}{dx} + c_2 \frac{d^2u_2}{dx^2} + \frac{du_2}{dx}\frac{dc_2}{dx} + a_1\left(c_1 \frac{du_1}{dx} + c_2 \frac{du_2}{dx}\right) + a_0(c_1 u_1 + c_2 u_2) = f.$$

This can be rearranged to show that

$$c_1\left(\frac{d^2u_1}{dx^2} + a_1 \frac{du_1}{dx} + a_0 u_1\right) + c_2\left(\frac{d^2u_2}{dx^2} + a_1 \frac{du_2}{dx} + a_0 u_2\right) + \frac{du_1}{dx}\frac{dc_1}{dx} + \frac{du_2}{dx}\frac{dc_2}{dx} = f.$$

Since u_1 and u_2 are solutions of the homogeneous equation, the first two terms are zero, which gives us

$$\frac{du_1}{dx}\frac{dc_1}{dx} + \frac{du_2}{dx}\frac{dc_2}{dx} = f. \qquad (1.5)$$

We now have two simultaneous equations, (1.4) and (1.5), for $c_1' = dc_1/dx$ and $c_2' = dc_2/dx$, which can be written in matrix form as

$$\begin{pmatrix} u_1 & u_2 \\ u_1' & u_2' \end{pmatrix} \begin{pmatrix} c_1' \\ c_2' \end{pmatrix} = \begin{pmatrix} 0 \\ f \end{pmatrix}$$

These can easily be solved to give

$$c_1' = -\frac{fu_2}{W}, \quad c_2' = \frac{fu_1}{W},$$

where

$$W = u_1 u_2' - u_2 u_1' = \begin{vmatrix} u_1 & u_2 \\ u_1' & u_2' \end{vmatrix}$$

is called the **Wronskian**. These expressions can be integrated to give

$$c_1 = \int^x -\frac{f(s)u_2(s)}{W(s)}\,ds + A, \quad c_2 = \int^x \frac{f(s)u_1(s)}{W(s)}\,ds + B.$$

We can now write down the solution of the entire problem as

$$y(x) = u_1(x)\int^x -\frac{f(s)u_2(s)}{W(s)}\,ds + u_2(x)\int^x \frac{f(s)u_1(s)}{W(s)}\,ds + Au_1(x) + Bu_2(x).$$

The particular integral is therefore

$$y(x) = \int^x f(s)\left\{\frac{u_1(s)u_2(x) - u_1(x)u_2(s)}{W(s)}\right\}\,ds. \tag{1.6}$$

This is called the **variation of parameters formula**.

Example

Consider the equation

$$\frac{d^2y}{dx^2} + y = x\sin x.$$

The homogeneous form of this equation has constant coefficients, with solutions

$$u_1(x) = \cos x, \quad u_2(x) = \sin x.$$

The variation of parameters formula then gives the particular integral as

$$y = \int^x s\sin s\left\{\frac{\cos s \sin x - \cos x \sin s}{1}\right\}\,ds,$$

since

$$W = \begin{vmatrix} \cos x & \sin x \\ -\sin x & \cos x \end{vmatrix} = \cos^2 x + \sin^2 x = 1.$$

We can split the particular integral into two integrals as

$$y(x) = \sin x \int^x s\sin s \cos s\,ds - \cos x \int^x s\sin^2 s\,ds$$

$$= \frac{1}{2}\sin x \int^x s\sin 2s\,ds - \frac{1}{2}\cos x \int^x s(1 - \cos 2s)\,ds.$$

Using integration by parts, we can evaluate this, and find that

$$y(x) = -\frac{1}{4}x^2\cos x + \frac{1}{4}x\sin x + \frac{1}{8}\cos x$$

is the required particular integral. The general solution is therefore

$$y = c_1\cos x + c_2\sin x - \frac{1}{4}x^2\cos x + \frac{1}{4}x\sin x.$$

Although we have given a rational derivation of the reduction of order and variation of parameters formulae, we have made no comment so far about why the procedures we used in the derivation should work at all! It turns out that this has a close connection with the theory of continuous groups, which we will investigate in Chapter 10.

1.2 THE METHOD OF VARIATION OF PARAMETERS

1.2.1 The Wronskian

Before we carry on, let's pause to discuss some further properties of the Wronskian. Recall that if V is a vector space over $\mathbb{R}$, then two elements $\mathbf{v}_1, \mathbf{v}_2 \in V$ are linearly dependent if $\exists\, \alpha_1, \alpha_2 \in \mathbb{R}$, with α_1 and α_2 not both zero, such that $\alpha_1 \mathbf{v}_1 + \alpha_2 \mathbf{v}_2 = \mathbf{0}$.

Now let $V = C^1(a,b)$ be the set of once-differentiable functions over the interval $a < x < b$. If $u_1, u_2 \in C^1(a,b)$ are linearly dependent, $\exists\, \alpha_1, \alpha_2 \in \mathbb{R}$ such that $\alpha_1 u_1(x) + \alpha_2 u_2(x) = 0\ \forall x \in (a,b)$. Notice that, by direct differentiation, this also gives $\alpha_1 u_1'(x) + \alpha_2 u_2'(x) = 0$ or, in matrix form,

$$\begin{pmatrix} u_1(x) & u_2(x) \\ u_1'(x) & u_2'(x) \end{pmatrix} \begin{pmatrix} \alpha_1 \\ \alpha_2 \end{pmatrix} = \begin{pmatrix} 0 \\ 0 \end{pmatrix}.$$

These are homogeneous equations of the form

$$\mathbf{A}\mathbf{x} = \mathbf{0},$$

which only have nontrivial solutions if $\det(\mathbf{A}) = 0$, that is

$$W = \begin{vmatrix} u_1(x) & u_2(x) \\ u_1'(x) & u_2'(x) \end{vmatrix} = u_1 u_2' - u_1' u_2 = 0.$$

In other words, the Wronskian of two linearly dependent functions is identically zero on (a,b). The contrapositive of this result is that if $W \neq 0$ on (a,b), then u_1 and u_2 are linearly independent on (a,b).

Example

The functions $u_1(x) = x^2$ and $u_2(x) = x^3$ are linearly independent on the interval $(-1, 1)$. To see this, note that, since $u_1(x) = x^2$, $u_2(x) = x^3$, $u_1'(x) = 2x$, and $u_2'(x) = 3x^2$, the Wronskian of these two functions is

$$W = \begin{vmatrix} x^2 & x^3 \\ 2x & 3x^2 \end{vmatrix} = 3x^4 - 2x^4 = x^4.$$

This quantity is not identically zero, and hence x^2 and x^3 are linearly independent on $(-1, 1)$.

Example

The functions $u_1(x) = f(x)$ and $u_2(x) = kf(x)$, with k a constant, are linearly dependent on any interval, since their Wronskian is

$$W = \begin{vmatrix} f & kf \\ f' & kf' \end{vmatrix} = 0.$$

If the functions u_1 and u_2 are solutions of (1.2), we can show by differentiating $W = u_1 u_2' - u_1' u_2$ directly that

$$\frac{dW}{dx} + a_1(x) W = 0.$$

VARIABLE COEFFICIENT, SECOND ORDER DIFFERENTIAL EQUATIONS

This first order differential equation has solution

$$W(x) = W(x_0) \exp\left\{-\int_{x_0}^{x} a_1(t)dt\right\}, \qquad (1.7)$$

which is known as **Abel's formula**. This gives us an easy way of finding the Wronskian of the solutions of any second order differential equation without having to construct the solutions themselves.

Example

Consider the equation

$$y'' + \frac{1}{x}y' + \left(1 - \frac{1}{x^2}\right)y = 0.$$

Using Abel's formula, this has Wronskian

$$W(x) = W(x_0)\exp\left\{-\int_{x_0}^{x}\frac{dt}{t}\right\} = \frac{x_0 W(x_0)}{x} = \frac{A}{x}$$

for some constant A. To find this constant, it is usually necessary to know more about the solutions $u_1(x)$ and $u_2(x)$. We will describe a technique for doing this in Section 1.3.

We end this section with a couple of useful theorems.

Theorem 1.1 *If u_1 and u_2 are linearly independent solutions of the homogeneous, nonsingular ordinary differential equation (1.2), then the Wronskian is either strictly positive or strictly negative.*

Proof From Abel's formula, and since the exponential function does not change sign, the Wronskian is identically positive, identically negative or identically zero. We just need to exclude the possibility that W is ever zero. Suppose that $W(x_1) = 0$. The vectors $\begin{pmatrix} u_1(x_1) \\ u_1'(x_1) \end{pmatrix}$ and $\begin{pmatrix} u_2(x_1) \\ u_2'(x_1) \end{pmatrix}$ are then linearly dependent, and hence $u_1(x_1) = ku_2(x_1)$ and $u_1'(x) = ku_2'(x)$ for some constant k. The function $u(x) = u_1(x) - ku_2(x)$ is also a solution of (1.2) by linearity, and satisfies the initial conditions $u(x_1) = 0$, $u'(x_1) = 0$. Since (1.2) has a unique solution, the obvious solution, $u \equiv 0$, is the only solution. This means that $u_1 \equiv ku_2$. Hence u_1 and u_2 are linearly dependent – a contradiction. □

The nonsingularity of the differential equation is crucial here. If we consider the equation $x^2 y'' - 2xy' + 2y = 0$, which has $u_1(x) = x^2$ and $u_2(x) = x$ as its linearly independent solutions, the Wronksian is $-x^2$, which vanishes at $x = 0$. This is because the coefficient of y'' also vanishes at $x = 0$.

Theorem 1.2 (The Sturm separation theorem) *If $u_1(x)$ and $u_2(x)$ are the linearly independent solutions of a nonsingular, homogeneous equation, (1.2), then* the zeros of $u_1(x)$ and $u_2(x)$ occur alternately. In other words, successive zeros of $u_1(x)$ are separated by successive zeros of $u_2(x)$ and vice versa.

Proof Suppose that x_1 and x_2 are successive zeros of $u_2(x)$, so that $W(x_i) = u_1(x_i)u_2'(x_i)$ for $i = 1$ or 2. We also know that $W(x)$ is of one sign on $[x_1, x_2]$, since $u_1(x)$ and $u_2(x)$ are linearly independent. This means that $u_1(x_i)$ and $u_2'(x_i)$ are nonzero. Now if $u_2'(x_1)$ is positive then $u_2'(x_2)$ is negative (or vice versa), since $u_2(x_2)$ is zero. Since the Wronskian cannot change sign between x_1 and x_2, $u_1(x)$ must change sign, and hence u_1 has a zero in $[x_1, x_2]$, as we claimed. □

As an example of this, consider the equation $y'' + \omega^2 y = 0$, which has solution $y = A\sin\omega x + B\cos\omega x$. If we consider any two of the zeros of $\sin\omega x$, it is immediately clear that $\cos\omega x$ has a zero between them.

1.3 Solution by Power Series: The Method of Frobenius

Up to this point, we have considered ordinary differential equations for which we know at least one solution of the homogeneous problem. From this we have seen that we can easily construct the second independent solution and, in the inhomogeneous case, the particular integral. We now turn our attention to the more difficult case, in which we cannot determine a solution of the homogeneous problem by inspection. We must devise a method that is capable of solving variable coefficient ordinary differential equations in general. As we noted at the start of the chapter, we will restrict our attention to the case where the variable coefficients are simple polynomials. This suggests that we can look for a solution of the form

$$y = x^c \sum_{n=0}^{\infty} a_n x^n = \sum_{n=0}^{\infty} a_n x^{n+c}, \tag{1.8}$$

and hence

$$\frac{dy}{dx} = \sum_{n=0}^{\infty} a_n (n+c) x^{n+c-1}, \tag{1.9}$$

$$\frac{d^2 y}{dx^2} = \sum_{n=0}^{\infty} a_n (n+c)(n+c-1) x^{n+c-2}, \tag{1.10}$$

where the constants $c, a_0, a_1, \ldots$, are as yet undetermined. This is known as the **method of Frobenius**. Later on, we will give some idea of why and when this method can be used. For the moment, we will just try to make it work. We proceed by example, with the simplest case first.

1.3.1 The Roots of the Indicial Equation Differ by an Integer
Consider the equation

$$x^2 \frac{d^2 y}{dx^2} + x \frac{dy}{dx} + \left(x^2 - \frac{1}{4}\right) y = 0. \tag{1.11}$$

12 VARIABLE COEFFICIENT, SECOND ORDER DIFFERENTIAL EQUATIONS

We substitute (1.8) to (1.10) into (1.11), which gives

$$x^2 \sum_{n=0}^{\infty} a_n (n+c)(n+c-1) x^{n+c-2} + x \sum_{n=0}^{\infty} a_n (n+c) x^{n+c-1}$$

$$+ \left(x^2 - \frac{1}{4} \right) \sum_{n=0}^{\infty} a_n x^{n+c} = 0.$$

We can rearrange this slightly to obtain

$$\sum_{n=0}^{\infty} a_n \left\{ (n+c)(n+c-1) + (n+c) - \frac{1}{4} \right\} x^{n+c} + \sum_{n=0}^{\infty} a_n x^{n+c+2} = 0,$$

and hence, after simplifying the terms in the first summation,

$$\sum_{n=0}^{\infty} a_n \left\{ (n+c)^2 - \frac{1}{4} \right\} x^{n+c} + \sum_{n=0}^{\infty} a_n x^{n+c+2} = 0.$$

We now extract the first two terms from the first summation to give

$$a_0 \left(c^2 - \frac{1}{4} \right) x^c + a_1 \left\{ (c+1)^2 - \frac{1}{4} \right\} x^{c+1}$$

$$+ \sum_{n=2}^{\infty} a_n \left\{ (n+c)^2 - \frac{1}{4} \right\} x^{n+c} + \sum_{n=0}^{\infty} a_n x^{n+c+2} = 0. \qquad (1.12)$$

Notice that the first term is the only one containing x^c and similarly for the second term in x^{c+1}

The two summations in (1.12) begin at the same power of x, namely x^{2+c}. If we let $m = n + 2$ in the last summation (notice that if $n = 0$ then $m = 2$, and $n = \infty$ implies that $m = \infty$), (1.12) becomes

$$a_0 \left(c^2 - \frac{1}{4} \right) x^c + a_1 \left\{ (c+1)^2 - \frac{1}{4} \right\} x^{c+1}$$

$$+ \sum_{n=2}^{\infty} a_n \left\{ (n+c)^2 - \frac{1}{4} \right\} x^{n+c} + \sum_{m=2}^{\infty} a_{m-2} x^{m+c} = 0.$$

Since the variables in the summations are merely dummy variables,

$$\sum_{m=2}^{\infty} a_{m-2} x^{m+c} = \sum_{n=2}^{\infty} a_{n-2} x^{n+c},$$

and hence

$$a_0 \left(c^2 - \frac{1}{4} \right) x^c + a_1 \left\{ (c+1)^2 - \frac{1}{4} \right\} x^{c+1}$$

$$+ \sum_{n=2}^{\infty} a_n \left\{ (n+c)^2 - \frac{1}{4} \right\} x^{n+c} + \sum_{n=2}^{\infty} a_{n-2} x^{n+c} = 0.$$

1.3 SOLUTION BY POWER SERIES: THE METHOD OF FROBENIUS

Since the last two summations involve identical powers of x, we can combine them to obtain

$$a_0 \left(c^2 - \frac{1}{4}\right) x^c + a_1 \left\{(c+1)^2 - \frac{1}{4}\right\} x^{c+1}$$

$$+ \sum_{n=2}^{\infty} \left[a_n \left\{(n+c)^2 - \frac{1}{4}\right\} + a_{n-2}\right] x^{n+c} = 0. \tag{1.13}$$

Although the operations above are straightforward, we need to take some care to avoid simple slips.

Since (1.13) must hold for *all* values of x, the coefficient of each power of x must be zero. The coefficient of x^c is therefore

$$a_0 \left(c^2 - \frac{1}{4}\right) = 0.$$

Up to this point, most Frobenius analysis is very similar. It is here that the different structures come into play. If we were to use the solution $a_0 = 0$, the series (1.8) would have $a_1 x^{c+1}$ as its first term. This is just equivalent to increasing c by 1. We therefore assume that $a_0 \neq 0$, which means that c must satisfy $c^2 - \frac{1}{4} = 0$. This is called the **indicial equation**, and implies that $c = \pm \frac{1}{2}$. Now, progressing to the next term, proportional to x^{c+1}, we find that

$$a_1 \left\{(c+1)^2 - \frac{1}{4}\right\} = 0.$$

Choosing $c = \frac{1}{2}$ gives $a_1 = 0$, and, if we were to do this, we would find that we had constructed a solution with one arbitrary constant. However, if we choose $c = -\frac{1}{2}$ the indicial equation is satisfied for arbitrary values of a_1, and a_1 will act as the second arbitrary constant for the solution. In order to generate this more general solution, we therefore let $c = -\frac{1}{2}$.

We now progress to the individual terms in the summation. The general term yields

$$a_n \left\{\left(n - \frac{1}{2}\right)^2 - \frac{1}{4}\right\} + a_{n-2} = 0 \quad \text{for } n = 2, 3, \ldots.$$

This is called a **recurrence relation**. We solve it by *observation* as follows. We start by rearranging to give

$$a_n = -\frac{a_{n-2}}{n(n-1)}. \tag{1.14}$$

By putting $n = 2$ in (1.14) we obtain

$$a_2 = -\frac{a_0}{2 \cdot 1}.$$

For $n = 3$,

$$a_3 = -\frac{a_1}{3 \cdot 2}.$$

For $n = 4$,
$$a_4 = -\frac{a_2}{4 \cdot 3},$$
and substituting for a_2 in terms of a_0 gives
$$a_4 = -\frac{1}{4 \cdot 3}\left(-\frac{a_0}{2 \cdot 1}\right) = \frac{a_0}{4 \cdot 3 \cdot 2 \cdot 1} = \frac{a_0}{4!}.$$
Similarly for $n = 5$, using the expression for a_3 in terms of a_1,
$$a_5 = -\frac{a_3}{5 \cdot 4} = -\frac{1}{5 \cdot 4}\left(-\frac{a_1}{3 \cdot 2}\right) = \frac{a_1}{5!}.$$
A pattern is emerging here and we propose that
$$a_{2n} = (-1)^n \frac{a_0}{(2n)!}, \quad a_{2n+1} = (-1)^n \frac{a_1}{(2n+1)!}. \tag{1.15}$$
This can be proved in a straightforward manner by induction, although we will not dwell upon the details here.†

We can now deduce the full solution. Starting from (1.8), we substitute $c = -\frac{1}{2}$, and write out the first few terms in the summation
$$y = x^{-1/2}(a_0 + a_1 x + a_2 x^2 + \cdots).$$
Now, using the forms of the even and odd coefficients given in (1.15),
$$y = x^{-1/2}\left(a_0 + a_1 x - \frac{a_0 x^2}{2!} - \frac{a_1 x^3}{3!} + \frac{a_0 x^4}{4!} + \frac{a_1 x^5}{5!} + \cdots\right).$$
This series splits naturally into two proportional to a_0 and a_1, namely
$$y = x^{-1/2} a_0 \left(1 - \frac{x^2}{2!} + \frac{x^4}{4!} - \cdots\right) + x^{-1/2} a_1 \left(x - \frac{x^3}{3!} + \frac{x^5}{5!} - \cdots\right).$$
The solution is therefore
$$y(x) = a_0 \frac{\cos x}{x^{1/2}} + a_1 \frac{\sin x}{x^{1/2}},$$
since we can recognize the Taylor series expansions for sine and cosine.

This particular differential equation is an example of the use of the method of Frobenius, formalized by

> **Frobenius General Rule I**
> If the indicial equation has **two distinct roots**, $c = \alpha, \beta$ ($\alpha < \beta$), **whose difference is an integer**, and one of the coefficients of x^k becomes indeterminate on putting $c = \alpha$, both solutions can be generated by putting $c = \alpha$ in the recurrence relation.

† In the usual way, we must show that (1.15) is true for $n = 0$ and that, when the value of a_{2n+1} is substituted into the recurrence relation, we obtain $a_{2(n+1)+1}$, as given by substituting $n+1$ for n in (1.15).

1.3 SOLUTION BY POWER SERIES: THE METHOD OF FROBENIUS

In the above example the indicial equation was $c^2 - \frac{1}{4} = 0$, which has solutions $c = \pm\frac{1}{2}$, whose difference is an integer. The coefficient of x^{c+1} was $a_1\{(c+1)^2 - \frac{1}{4}\} = 0$. When we choose the lower of the two values ($c = -\frac{1}{2}$) this expression does not give us any information about the constant a_1, in other words a_1 is indeterminate.

1.3.2 The Roots of the Indicial Equation Differ by a Noninteger Quantity

We now consider the differential equation

$$2x(1-x)\frac{d^2y}{dx^2} + (1-x)\frac{dy}{dx} + 3y = 0. \tag{1.16}$$

As before, let's assume that the solution can be written as the power series (1.8). As in the previous example, this can be differentiated and substituted into the equation to yield

$$2x(1-x)\sum_{n=0}^{\infty} a_n(n+c)(n+c-1)x^{n+c-2} + (1-x)\sum_{n=0}^{\infty} a_n(n+c)x^{n+c-1}$$

$$+3\sum_{n=0}^{\infty} a_n x^{n+c} = 0.$$

The various terms can be multiplied out, which gives us

$$\sum_{n=0}^{\infty} a_n(n+c)(n+c-1)2x^{n+c-1} - \sum_{n=0}^{\infty} a_n(n+c)(n+c-1)2x^{n+c}$$

$$+\sum_{n=0}^{\infty} a_n(n+c)x^{n+c-1} - \sum_{n=0}^{\infty} a_n(n+c)x^{n+c} + 3\sum_{n=0}^{\infty} a_n x^{n+c} = 0.$$

Collecting similar terms gives

$$\sum_{n=0}^{\infty} a_n\{2(n+c)(n+c-1) + (n+c)\}x^{n+c-1}$$

$$+\sum_{n=0}^{\infty} a_n\{3 - 2(n+c)(n+c-1) - (n+c)\}x^{n+c} = 0,$$

and hence

$$\sum_{n=0}^{\infty} a_n(n+c)(2n+2c-1)x^{n+c-1} + \sum_{n=0}^{\infty} a_n\{3 - (n+c)(2n+2c-1)\}x^{n+c} = 0.$$

We now extract the first term from the left hand summation so that both summations start with a term proportional to x^c. This gives

$$a_0 c(2c-1)x^{c-1} + \sum_{n=1}^{\infty} a_n(n+c)(2n+2c-1)x^{n+c-1}$$

VARIABLE COEFFICIENT, SECOND ORDER DIFFERENTIAL EQUATIONS

$$+\sum_{n=0}^{\infty} a_n\{3 - (n+c)(2n+2c-1)\}x^{n+c} = 0.$$

We now let $m = n+1$ in the second summation, which then becomes

$$\sum_{m=1}^{\infty} a_{m-1}\{3 - (m-1+c)(2(m-1)+2c-1)\}x^{m+c-1}.$$

We again note that m is merely a dummy variable which for ease we rewrite as n, which gives

$$a_0 c(2c-1)x^{c-1} + \sum_{n=1}^{\infty} a_n(n+c)(2n+2c-1)x^{n+c-1}$$

$$+ \sum_{n=1}^{\infty} a_{n-1}\{3 - (n-1+c)(2n+2c-3)\}x^{n+c-1} = 0.$$

Finally, we can combine the two summations to give

$$a_0 c(2c-1)x^{c-1}$$

$$+ \sum_{n=1}^{\infty} \{a_n(n+c)(2n+2c-1) + a_{n-1}\{3-(n-1+c)(2n+2c-3)\}\}x^{n+c-1} = 0.$$

As in the previous example we can now consider the coefficients of successive powers of x. We start with the coefficient of x^{c-1}, which gives the indicial equation, $a_0 c(2c-1) = 0$. Since $a_0 \neq 0$, this implies that $c = 0$ or $c = \frac{1}{2}$. Notice that these roots do not differ by an integer. The general term in the summation shows that

$$a_n = a_{n-1}\left\{\frac{(n+c-1)(2n+2c-3)-3}{(n+c)(2n+2c-1)}\right\} \quad \text{for } n = 1, 2, \ldots. \tag{1.17}$$

We now need to solve this recurrence relation, considering each root of the indicial equation separately.

Case I: $c = 0$

In this case, we can rewrite the recurrence relation (1.17) as

$$a_n = a_{n-1}\left\{\frac{(n-1)(2n-3)-3}{n(2n-1)}\right\} = a_{n-1}\left\{\frac{2n^2 - 5n}{n(2n-1)}\right\} = a_{n-1}\left(\frac{2n-5}{2n-1}\right).$$

We recall that this holds for $n \geqslant 1$, so we start with $n = 1$, which yields

$$a_1 = a_0\left(-\frac{3}{1}\right) = -3a_0.$$

For $n = 2$

$$a_2 = a_1\left(-\frac{1}{3}\right) = -3a_0\left(-\frac{1}{3}\right) = a_0,$$

1.3 SOLUTION BY POWER SERIES: THE METHOD OF FROBENIUS

where we have used the expression for a_1 in terms of a_0. Now progressing to $n = 3$, we have

$$a_3 = a_2\left(\frac{1}{5}\right) = a_0\frac{3}{5 \cdot 3},$$

and for $n = 4$,

$$a_4 = a_3\left(\frac{3}{7}\right) = a_0\frac{3}{7 \cdot 5}.$$

Finally, for $n = 5$ we have

$$a_5 = a_4\left(\frac{5}{9}\right) = a_0\frac{3}{9 \cdot 7}.$$

In general,

$$a_n = \frac{3a_0}{(2n-1)(2n-3)},$$

which again can be proved by induction. We conclude that one solution of the differential equation is

$$y = x^c\sum_{n=0}^{\infty}a_n x^n = x^0\sum_{n=0}^{\infty}\frac{3a_0}{(2n-1)(2n-3)}x^n.$$

This can be tidied up by putting $3a_0 = A$, so that the solution is

$$y = A\sum_{n=0}^{\infty}\frac{x^n}{(2n-1)(2n-3)}. \tag{1.18}$$

Note that there is no obvious way of writing this solution in terms of elementary functions. In addition, a simple application of the ratio test shows that this power series is only convergent for $|x| \leqslant 1$, for reasons that we discuss below.

A simple MATLAB† function that evaluates (1.18) is

```
function frob = frob(x)
n = 100:-1:0; a = 1./(2*n-1)./(2*n-3);
frob = polyval(a,x);
```

which sums the first 100 terms of the series. The function `polyval` evaluates the polynomial formed by the first 100 terms in the sum (1.18) in an efficient manner. Figure 1.2 can then be produced using the command `ezplot(@frob,[-1,1])`.

Although we could now use the method of reduction of order, since we have constructed a solution, this would be very complicated. It is easier to consider the second root of the indicial equation.

† See Appendix 7 for a short introduction to MATLAB.

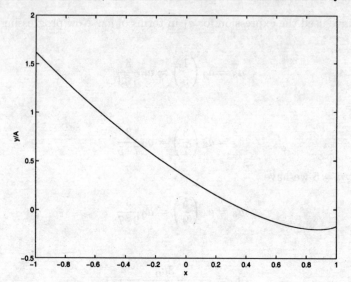

Fig. 1.2. The solution of (1.16) given by (1.18).

Case II: $c = \frac{1}{2}$

In this case, we simplify the recurrence relation (1.17) to give

$$a_n = a_{n-1} \left\{ \frac{\left(n - \frac{1}{2}\right)(2n - 2) - 3}{\left(n + \frac{1}{2}\right) 2n} \right\} = a_{n-1} \left(\frac{2n^2 - 3n - 2}{2n^2 + n} \right)$$

$$= a_{n-1} \left\{ \frac{(2n+1)(n-2)}{n(2n+1)} \right\} = a_{n-1} \left(\frac{n-2}{n} \right).$$

We again recall that this relation holds for $n \geqslant 1$ and start with $n = 1$, which gives $a_1 = a_0(-1)$. Substituting $n = 2$ gives $a_2 = 0$ and, since all successive a_i will be written in terms of a_2, $a_i = 0$ for $i = 2, 3, \ldots$. The second solution of the equation is therefore $y = Bx^{1/2}(1-x)$. We can now use this simple solution in the reduction of order formula, (1.3), to determine an analytical formula for the first solution, (1.18). For example, for $0 \leqslant x \leqslant 1$, we find that (1.18) can be written as

$$y = -\frac{1}{6} A \left[3x - 2 + 3x^{1/2}(1-x) \log \left\{ \frac{1 + x^{1/2}}{(1-x)^{1/2}} \right\} \right]$$

This expression has a logarithmic singularity in its derivative at $x = 1$, which explains why the radius of convergence of the power series solution (1.18) is $|x| \leqslant 1$.

This differential equation is an example of the second major case of the method of Frobenius, formalized by

1.3 SOLUTION BY POWER SERIES: THE METHOD OF FROBENIUS

> **Frobenius General Rule II**
> If the indicial equation has **two distinct roots**, $c = \alpha, \beta$ $(\alpha < \beta)$, **whose difference is *not* an integer**, the general solution of the equation is found by successively substituting $c = \alpha$ then $c = \beta$ into the general recurrence relation.

1.3.3 The Roots of the Indicial Equation are Equal

Let's try to determine the two solutions of the differential equation

$$x\frac{d^2y}{dx^2} + (1+x)\frac{dy}{dx} + 2y = 0.$$

We substitute in the standard power series, (1.8), which gives

$$x\sum_{n=0}^{\infty} a_n(n+c)(n+c-1)x^{n+c-2} + (1+x)\sum_{n=0}^{\infty} a_n(n+c)x^{n+c-1}$$

$$+2\sum_{n=0}^{\infty} a_n x^{n+c} = 0.$$

This can be simplified to give

$$\sum_{n=0}^{\infty} a_n(n+c)^2 x^{n+c-1} + \sum_{n=0}^{\infty} a_n(n+c+2)x^{n+c} = 0.$$

We can extract the first term from the left hand summation to give

$$a_0 c^2 x^{c-1} + \sum_{n=1}^{\infty} a_n(n+c)^2 x^{n+c-1} + \sum_{n=0}^{\infty} a_n(n+c+2)x^{n+c} = 0.$$

Now shifting the series using $m = n+1$ (and subsequently changing dummy variables from m to n) we have

$$a_0 c^2 x^{c-1} + \sum_{n=1}^{\infty} \{a_n(n+c)^2 + a_{n-1}(n+c+1)\}x^{n+c} = 0, \qquad (1.19)$$

where we have combined the two summations. The indicial equation is $c^2 = 0$ which has a **double root** at $c = 0$. We know that there must be two solutions, but it appears that there is only one available to us. For the moment let's see how far we can get by setting $c = 0$. The recurrence relation is then

$$a_n = -a_{n-1}\frac{n+1}{n^2} \quad \text{for } n = 1, 2, \ldots.$$

When $n = 1$ we find that

$$a_1 = -a_0 \frac{2}{1^2},$$

and with $n = 2$,
$$a_2 = -a_1 \frac{3}{2^2} = a_0 \frac{3 \cdot 2}{1^2 \cdot 2^2}.$$

Using $n = 3$ gives
$$a_3 = -a_2 \frac{4}{3^2} = -a_0 \frac{4 \cdot 3 \cdot 2}{1^2 \cdot 2^2 \cdot 3^2},$$

and we conclude that
$$a_n = (-1)^n \frac{(n+1)!}{(n!)^2} a_0 = (-1)^n \frac{n+1}{n!} a_0.$$

One solution is therefore
$$y = a_0 \sum_{n=0}^{\infty} (-1)^n \frac{(n+1)}{n!} x^n,$$

which can also be written as
$$y = a_0 \left\{ x \sum_{n=1}^{\infty} \frac{(-1)^n x^{n-1}}{(n-1)!} + \sum_{n=0}^{\infty} \frac{(-1)^n x^n}{n!} \right\}$$
$$= a_0 \left\{ -x \sum_{m=0}^{\infty} \frac{(-1)^m x^m}{m!} + e^{-x} \right\} = a_0 (1-x) e^{-x}.$$

This solution is one that we could not have readily determined simply by inspection. We could now use the method of reduction of order to find the second solution, but we will proceed with the method of Frobenius so that we can see how it works in this case.

Consider (1.19), which we write out more fully as
$$x \frac{d^2 y}{dx^2} + (1+x) \frac{dy}{dx} + 2y =$$
$$a_0 c^2 x^{c-1} + \sum_{n=1}^{\infty} \{a_n (n+c)^2 + a_{n-1}(n+c+1)\} x^{n+c} = 0.$$

The best we can do at this stage is to set $a_n(n+c)^2 + a_{n-1}(n+c+1) = 0$ for $n \geqslant 1$, as this gets rid of most of the terms. This gives us a_n as a function of c for $n \geqslant 1$, and leaves us with
$$x \frac{d^2 y}{dx^2} + (1+x) \frac{dy}{dx} + 2y = a_0 c^2 x^{c-1}. \tag{1.20}$$

Let's now take a partial derivative with respect to c, where we regard y as a function of both x and c, making use of
$$\frac{d}{dx} = \frac{\partial}{\partial x}, \quad \frac{\partial}{\partial c}\left(\frac{\partial y}{\partial x}\right) = \frac{\partial}{\partial x}\left(\frac{\partial y}{\partial c}\right).$$

This gives
$$x \frac{\partial^2}{\partial x^2}\left(\frac{\partial y}{\partial c}\right) + (1+x) \frac{\partial}{\partial x}\left(\frac{\partial y}{\partial c}\right) + 2 \left(\frac{\partial y}{\partial c}\right) = a_0 \frac{\partial}{\partial c}(c^2 x^{c-1}).$$

1.3 SOLUTION BY POWER SERIES: THE METHOD OF FROBENIUS

Notice that we have used the fact that a_0 is independent of c. We need to be careful when evaluating the right hand side of this expression. Differentiating using the product rule we have

$$\frac{\partial}{\partial c}(c^2 x^{c-1}) = c^2 \frac{\partial}{\partial c}(x^{c-1}) + x^{c-1} \frac{\partial}{\partial c}(c^2).$$

We rewrite x^{c-1} as $x^c x^{-1} = e^{c \log x} x^{-1}$, so that we have

$$\frac{\partial}{\partial c}(c^2 x^{c-1}) = c^2 \frac{\partial}{\partial c}(e^{c \log x} x^{-1}) + x^{c-1} \frac{\partial}{\partial c}(c^2).$$

Differentiating the exponential gives

$$\frac{\partial}{\partial c}(c^2 x^{c-1}) = c^2 (\log x \, e^{c \log x}) x^{-1} + x^{c-1} 2c,$$

which we can tidy up to give

$$\frac{\partial}{\partial c}(c^2 x^{c-1}) = c^2 x^{c-1} \log x + x^{c-1} 2c.$$

Substituting this form back into the differential equation gives

$$x \frac{\partial^2}{\partial x^2}\left(\frac{\partial y}{\partial c}\right) + (1+x)\frac{\partial}{\partial x}\left(\frac{\partial y}{\partial c}\right) + 2 \frac{\partial y}{\partial c} = a_0 \{c^2 x^{c-1} \log x + x^{c-1} 2c\}.$$

Now letting $c \to 0$ gives

$$x \frac{\partial^2}{\partial x^2} \frac{\partial y}{\partial c}\bigg|_{c=0} + (1+x)\frac{\partial}{\partial x}\frac{\partial y}{\partial c}\bigg|_{c=0} + 2 \frac{\partial y}{\partial c}\bigg|_{c=0} = 0.$$

Notice that this procedure only works because (1.20) has a repeated root at $c = 0$. We conclude that $\dfrac{\partial y}{\partial c}\bigg|_{c=0}$ is a second solution of our ordinary differential equation.

To construct this solution, we differentiate the power series (1.8) (carefully!) to give

$$\frac{\partial y}{\partial c} = x^c \sum_{n=0}^{\infty} \frac{da_n}{dc} x^n + \sum_{n=0}^{\infty} a_n x^n x^c \log x,$$

using a similar technique as before to deal with the differentiation of x^c with respect to c. Note that, although a_0 is not a function of c, the other coefficients are. Putting $c = 0$ gives

$$\frac{\partial y}{\partial c}\bigg|_{c=0} = \sum_{n=0}^{\infty} \frac{da_n}{dc}\bigg|_{c=0} x^n + \log x \sum_{n=0}^{\infty} a_n|_{c=0} x^n.$$

We therefore need to determine $\dfrac{da_n}{dc}\bigg|_{c=0}$. We begin with the recurrence relation, which is

$$a_n = -\frac{a_{n-1}(n+c+1)}{(n+c)^2}.$$

Starting with $n = 1$ we find that
$$a_1 = \frac{-a_0(c+2)}{(c+1)^2},$$
whilst for $n = 2$,
$$a_2 = \frac{-a_1(c+3)}{(c+2)^2},$$
and substituting for a_1 in terms of a_0 gives us
$$a_2 = \frac{a_0(c+2)(c+3)}{(c+1)^2(c+2)^2}.$$
This process can be continued to give
$$a_n = (-1)^n a_0 \frac{(c+2)(c+3)\ldots(c+n+1)}{(c+1)^2(c+2)^2\ldots(c+n)^2},$$
which we can write as
$$a_n = (-1)^n a_0 \frac{\prod_{j=1}^n (c+j+1)}{\left\{\prod_{j=1}^n (c+j)\right\}^2}.$$
We now take the logarithm of this expression, recalling that the logarithm of a product is the sum of the terms, which leads to
$$\log(a_n) = \log((-1)^n a_0) + \log\left(\prod_{j=1}^n (c+j+1)\right) - 2\log\left(\prod_{j=1}^n (c+j)\right)$$
$$= \log((-1)^n a_0) + \sum_{j=1}^n \log(c+j+1) - 2\sum_{j=1}^n \log(c+j).$$
Now differentiating with respect to c gives
$$\frac{1}{a_n}\frac{da_n}{dc} = \sum_{j=1}^n \frac{1}{c+j+1} - 2\sum_{j=1}^n \frac{1}{c+j},$$
and setting $c = 0$ we have
$$\left.\left(\frac{1}{a_n}\frac{da_n}{dc}\right)\right|_{c=0} = \sum_{j=1}^n \frac{1}{j+1} - 2\sum_{j=1}^n \frac{1}{j}.$$
Since we know a_n when $c = 0$, we can write
$$\left.\frac{da_n}{dc}\right|_{c=0} = (-1)^n a_0 \frac{\prod_{j=1}^n (j+1)}{\left(\prod_{j=1}^n j\right)^2}\left(\sum_{j=1}^n \frac{1}{j+1} - 2\sum_{j=1}^n \frac{1}{j}\right),$$
$$= (-1)^n a_0 \frac{(n+1)!}{(n!)^2}\left(\sum_{j=1}^{n+1} \frac{1}{j} - 1 - 2\sum_{j=1}^n \frac{1}{j}\right)$$

In this expression, we have manipulated the products and written them as factorials, changed the first summation and removed the extra term that this incurs.

1.3 SOLUTION BY POWER SERIES: THE METHOD OF FROBENIUS

Simplifying, we obtain
$$\left.\frac{da_n}{dc}\right|_{c=0} = (-1)^n a_0 \frac{(n+1)}{n!} \left(\phi(n+1) - 2\phi(n) - 1\right),$$
where we have introduced the notation
$$\phi(n) \equiv \sum_{m=1}^{n} \frac{1}{m}. \tag{1.21}$$

The second solution is therefore
$$y = a_0 \left[\sum_{n=0}^{\infty} (-1)^n \frac{(n+1)}{n!} \{\phi(n+1) - 2\phi(n) - 1\} x^n + \sum_{n=0}^{\infty} (-1)^n \frac{(n+1)}{n!} x^n \log x \right].$$

This methodology is formalized in

> **Frobenius General Rule III**
> If the indicial equation has **a double root**, $c = \alpha$, one solution is obtained by putting $c = \alpha$ into the recurrence relation.
> The second independent solution is $(\partial y / \partial c)_{c=\alpha}$ where $a_n = a_n(c)$ for the calculation.

There are several other cases that can occur in the method of Frobenius, which, due to their complexity, we will not go into here. One method of dealing with these is to notice that the method outlined in this chapter always produces a solution of the form $y = u_1(x) = \sum_{n=0}^{\infty} a_n x^{n+c}$. This can be used in the reduction of order formula, (1.3), to find the second linearly independent solution. Of course, it is rather difficult to get all of $u_2(x)$ this way, but the first few terms are usually easy enough to compute by expanding for small x. Having got these, we can assume a general series form for $u_2(x)$, and find the coefficients in the usual way.

Example

Let's try to solve
$$x^2 y'' + xy' + (x^2 - 1)y = 0, \tag{1.22}$$
using the method of Frobenius. If we look for a solution in the usual form, $y = \sum_{n=0}^{\infty} a_n x^{n+c}$, we find that
$$a_0(c^2 - 1)x^c + a_1\left\{(1+c)^2 - 1\right\} x^{c+1} + \sum_{k=2}^{\infty} \left[a_k \left\{(k+c)^2 - 1\right\} + a_{k-2} \right] x^{k+c} = 0.$$
The indicial equation has roots $c = \pm 1$, and, by choosing either of these, we find that $a_1 = 0$. If we now look for the general solution of
$$a_k = -\frac{a_{k-2}}{(k+c)^2 - 1},$$
we find that
$$a_2 = -\frac{a_0}{(2+c)^2 - 1} = -\frac{a_0}{(1+c)(3+c)},$$

VARIABLE COEFFICIENT, SECOND ORDER DIFFERENTIAL EQUATIONS

$$a_4 = -\frac{a_2}{(4+c)^2-1} = \frac{a_0}{(1+c)(3+c)^2(5+c)},$$

and so on. This gives us a solution of the form

$$y(x,c) = a_0 x^c \left\{ 1 - \frac{x^2}{(1+c)(3+c)} + \frac{x^4}{(1+c)(3+c)^2(5+c)} - \cdots \right\}.$$

Now, by choosing $c = 1$, we obtain one of the linearly independent solutions of (1.22),

$$u_1(x) = y(x,1) = x\left(1 - \frac{x^2}{2\cdot 4} + \frac{x^4}{2\cdot 4^2 \cdot 6} - \cdots\right).$$

However, if $c = -1$, the coefficients a_{2n} for $n = 1, 2, \ldots$ are singular.

In order to find the structure of the second linearly independent solution, we use the reduction of order formula, (1.3). Substituting for $u_1(x)$ gives

$$\begin{aligned}
u_2(x) &= x\left(1 - \frac{x^2}{8} + \cdots\right)\int^x \frac{1}{t^2\left(1 - \frac{t^2}{8} + \cdots\right)^2} \exp\left(-\int^t \frac{1}{s}\,ds\right) dt \\
&= x\left(1 - \frac{x^2}{8} + \cdots\right)\int^x \frac{1}{t^2}\left(1 + \frac{t^2}{4} + \cdots\right)\frac{1}{t}\,dt \\
&= x\left(1 - \frac{x^2}{8} + \cdots\right)\left(-\frac{1}{2x^2} + \frac{1}{4}\log x + \cdots\right).
\end{aligned}$$

The second linearly independent solution of (1.22) therefore has the *structure*

$$u_2(x) = \frac{1}{4}u_1(x)\log x - \frac{1}{2x}v(x),$$

where $v(x) = 1 + b_2 x^2 + b_4 x^4 + \cdots$. If we assume a solution structure of this form and substitute it into (1.22), it is straightforward to pick off the coefficients b_{2n}.

Finally, note that we showed in Section 1.2.1 that the Wronskian of (1.22) is $W = A/x$ for some constant A. Now, since we know that $u_1 = x + \cdots$ and $u_2 = -1/2x + \cdots$, we must have $W = x(1/2x^2) + 1/2x + \cdots = 1/x + \cdots$, and hence $A = 1$.

1.3.4 Singular Points of Differential Equations

In this section, we give some definitions and a statement of a theorem that tells us when the method of Frobenius can be used, and for what values of x the infinite series will converge. We consider a second order, variable coefficient differential equation of the form

$$P(x)\frac{d^2y}{dx^2} + Q(x)\frac{dy}{dx} + R(x)y = 0. \qquad (1.23)$$

Before we proceed, we need to further refine our ideas about singular points. If x_0 is a singular point and $(x-x_0)Q(x)/P(x)$ and $(x-x_0)^2 R(x)/P(x)$ have convergent

Taylor series expansions about x_0, then $x = x_0$ is called a **regular singular** point; otherwise, x_0 is called an **irregular singular** point.

Example

Consider the equation

$$x(x-1)\frac{d^2y}{dx^2} + (1+x)\frac{dy}{dx} + y = 0. \qquad (1.24)$$

There are singular points where $x^2 - x = 0$, at $x = 0$ and $x = 1$. Let's start by looking at the singular point $x = 0$. Consider the expression

$$\frac{xQ}{P} = \frac{x(1+x)}{x^2-x} = \frac{(1+x)}{(x-1)} = -(1+x)(1-x)^{-1}.$$

Upon expanding $(1-x)^{-1}$ using the binomial expansion we have

$$\frac{xQ}{P} = -(1+x)(1+x+x^2+\cdots+x^n+\cdots),$$

which can be multiplied out to give

$$\frac{xQ}{P} = -1 - 2x + \cdots.$$

This power series is convergent provided $|x| < 1$ (by considering the binomial expansion used above). Now

$$\frac{x^2 R}{P} = \frac{x^2}{x^2-x} = \frac{x}{x-1} = -x(1-x)^{-1}.$$

Again, using the binomial expansion, which is convergent provided $|x| < 1$,

$$\frac{x^2 R}{P} = -x(1+x+x^2+\cdots) = -x - x^2 - x^3 - \cdots.$$

Since xQ/P and x^2R/P have convergent Taylor series about $x = 0$, this is a regular singular point.

Now consider the other singular point, $x = 1$. We note that

$$(x-1)\frac{Q}{P} = \frac{(x-1)(1+x)}{(x^2-x)} = \frac{(x-1)(1+x)}{x(x-1)} = \frac{1+x}{x}.$$

At this point we need to recall that we want information near $x = 1$, so we rewrite x as $1 - (1-x)$, and hence

$$(x-1)\frac{Q}{P} = \frac{2-(1-x)}{\{1-(1-x)\}}.$$

Expanding in powers of $(1-x)$ using the binomial expansion gives

$$(x-1)\frac{Q}{P} = 2\left(1 - \frac{1-x}{2}\right)\{1 + (1-x) + (1-x)^2 + \cdots\},$$

which is a power series in $(x-1)$ that is convergent provided $|x-1| < 1$. We also need to consider

$$(x-1)^2 \frac{R}{P} = \frac{(x-1)^2}{x^2 - x} = \frac{(x-1)^2}{x(x-1)} = \frac{x-1}{x} = \frac{x-1}{\{1-(1-x)\}}$$

$$= (x-1)\{1-(1-x)\}^{-1} = (x-1)\{1+(x-1)+(x-1)^2+\cdots\},$$

which is again convergent provided $|x-1| < 1$. Therefore $x = 1$ is a regular singular point.

Theorem 1.3 *If x_0 is an ordinary point of the ordinary differential equation (1.23), then there exists a unique series solution in the neighbourhood of x_0 which converges for $|x-x_0| < \rho$, where ρ is the smaller of the radii of convergence of the series for $Q(x)/P(x)$ and $R(x)/P(x)$.*

If x_0 is a regular singular point, then there exists a unique series solution in the neighbourhood of x_0, which converges for $|x-x_0| < \rho$, where ρ is the smaller of the radii of convergence of the series $(x-x_0)Q(x)/P(x)$ and $(x-x_0)^2 R(x)/P(x)$.

Proof This can be found in Kreider, Kuller, Ostberg and Perkins (1966) and is due to Fuchs. We give an outline of why the result should hold in Section 1.3.5. We are more concerned here with using the result to tell us when a series solution will converge. □

Example

Consider the differential equation (1.24). We have already seen that $x = 0$ is a regular singular point. The radii of convergence of $xQ(x)/P(x)$ and $x^2 R(x)/P(x)$ are both unity, and hence the series solution $\sum_{n=0}^{\infty} a_n x^{n+c}$ exists, is unique, and will converge for $|x| < 1$.

Example

Consider the differential equation

$$x^4 \frac{d^2 y}{dx^2} - \frac{dy}{dx} + y = 0.$$

For this equation, $x = 0$ is a singular point but it is **not** regular. The series solution $\sum_{n=0}^{\infty} a_n x^{n+c}$ is not guaranteed to exist, since $xQ/P = -1/x^3$, which cannot be expanded about $x = 0$.

1.3.5 An outline proof of Theorem 1.3

We will now give a sketch of why Theorem 1.3 holds. Consider the equation

$$P(x)y'' + Q(x)y' + R(x)y = 0.$$

When $x = 0$ is an ordinary point, assuming that

$$P(x) = P_0 + xP_1 + \cdots, \quad Q(x) = Q_0 + xQ_1 + \cdots, \quad R(x) = R_0 + xR_1 + \cdots,$$

1.3 SOLUTION BY POWER SERIES: THE METHOD OF FROBENIUS

we can look for a solution using the method of Frobenius. When the terms are ordered, we find that the first two terms in the expansion are

$$P_0 a_0 c(c-1) x^{c-2} + \{P_0 a_1 c(c+1) + P_1 a_1 c(c-1) + Q_0 a_1 c\} x^{c-1} + \cdots = 0.$$

The indicial equation, $c(c-1) = 0$, has two distinct roots that differ by an integer and, following Frobenius General Rule I, we can choose $c = 0$ and find a solution of the form $y = a_0 y_0(x) + a_1 y_1(x)$.

When $x = 0$ is a singular point, the simplest case to consider is when

$$P(x) = P_1 x + P_2 x^2 + \cdots .$$

We can then ask what form $Q(x)$ and $R(x)$ must take to ensure that a series solution exists. When $x = 0$ is an ordinary point, the indicial equation was formed from the y'' term in the equation alone. Let's now try to include the y' and y terms as well, by making the assumption that

$$Q(x) = Q_0 + Q_1 x + \cdots, \quad R(x) = \frac{R_{-1}}{x} + R_0 + \cdots .$$

Then, after substitution of the Frobenius series into the equation, the coefficient of x^{c-1} gives the indicial equation as

$$P_1 c(c-1) + Q_0 c + R_{-1} = 0.$$

This is a quadratic equation in c, with the usual possibilities for its types of roots. As $x \to 0$

$$\frac{xQ(x)}{P(x)} \to \frac{Q_0}{P_1}, \quad \frac{x^2 R(x)}{P(x)} \to \frac{R_{-1}}{P_1},$$

so that both of these quantities have a Taylor series expansion about $x = 0$. This makes it clear that, when $P(x) = P_1 x + \cdots$, these choices of expressions for the behaviour of $Q(x)$ and $R(x)$ close to $x = 0$ are what is required to make it a regular singular point. That the series converges, as claimed by the theorem, is most easily shown using the theory of complex variables; this is done in Section A6.5.

1.3.6 The point at infinity

Our discussion of singular points can be extended to include the point at infinity by defining $s = 1/x$ and considering the properties of the point $s = 0$. In particular, after using

$$\frac{dy}{dx} = -s^2 \frac{dy}{ds}, \quad \frac{d^2 y}{dx^2} = s^2 \frac{d}{ds}\left(s^2 \frac{dy}{ds}\right)$$

in (1.23), we find that

$$\hat{P}(s) \frac{d^2 y}{ds^2} + \hat{Q}(s) \frac{dy}{ds} + \hat{R}(s) y = 0,$$

where

$$\hat{P}(s) = s^4 P\left(\frac{1}{s}\right), \quad \hat{Q}(s) = 2s^3 P\left(\frac{1}{s}\right) - s^2 Q\left(\frac{1}{s}\right), \quad \hat{R}(s) = R\left(\frac{1}{s}\right).$$

VARIABLE COEFFICIENT, SECOND ORDER DIFFERENTIAL EQUATIONS

For example, Bessel's equation, which we will study in Chapter 3, has $P(x) = x^2$, $Q(x) = x$ and $R(x) = x^2 - \nu^2$, where ν is a constant, and hence has a regular singular point at $x = 0$. In this case, $\hat{P}(s) = s^2$, $\hat{Q}(s) = s$ and $\hat{R}(s) = 1/s^2 - \nu^2$. Since $\hat{P}(0) = 0$, Bessel's equation also has a singular point at infinity. In addition, $s^2\hat{R}(s)/\hat{P}(s) = (1 - s^2\nu)/s$, which is singular at $s = 0$, and we conclude that the point at infinity is an irregular singular point. We will study the behaviour of solutions of Bessel's equation in the neighbourhood of the point at infinity in Section 12.2.7.

Exercises

1.1 Show that the functions u_1 are solutions of the differential equations given below. Use the reduction of order method to find the second independent solution, u_2.

(a) $u_1 = e^x$, $(x-1)\dfrac{d^2y}{dx^2} - x\dfrac{dy}{dx} + y = 0$,

(b) $u_1 = x^{-1}\sin x$, $x\dfrac{d^2y}{dx^2} + 2\dfrac{dy}{dx} + xy = 0$.

1.2 Find the Wronskian of

(a) x, x^2,
(b) e^x, e^{-x},
(c) $x\cos(\log|x|)$, $x\sin(\log|x|)$.

Which of these pairs of functions are linearly independent on the interval $[-1, 1]$?

1.3 Find the general solution of

(a) $\dfrac{d^2y}{dx^2} - 2\dfrac{dy}{dx} + y = x^{3/2}e^x$,

(b) $\dfrac{d^2y}{dx^2} + 4y = 2\sec 2x$,

(c) $\dfrac{d^2y}{dx^2} + \dfrac{1}{x}\dfrac{dy}{dx} + \left(1 - \dfrac{1}{4x^2}\right)y = x$,

(d) $\dfrac{d^2y}{dx^2} + y = f(x)$, subject to $y(0) = y'(0) = 0$.

1.4 If u_1 and u_2 are linearly independent solutions of

$$y'' + p(x)y' + q(x)y = 0$$

and y is any other solution, show that the Wronskian of $\{y, u_1, u_2\}$,

$$W(x) = \begin{vmatrix} y & u_1 & u_2 \\ y' & u_1' & u_2' \\ y'' & u_1'' & u_2'' \end{vmatrix},$$

is zero everywhere. Hence find a second order differential equation which has solutions $y = x$ and $y = \log x$.

1.5 Find the Wronskian of the solutions of the differential equation $(1-x^2)y'' - 2xy' + 2y = 0$ to within a constant. Use the method of Frobenius to determine this constant.

1.6 Find the two linearly independent solutions of each of the differential equations

(a) $x^2 \dfrac{d^2y}{dx^2} + x\left(x - \dfrac{1}{2}\right)\dfrac{dy}{dx} + \dfrac{1}{2}y = 0$,

(b) $x^2 \dfrac{d^2y}{dx^2} + x(x+1)\dfrac{dy}{dx} - y = 0$,

using the method of Frobenius.

1.7 Show that the indicial equation for

$$x(1-x)\dfrac{d^2y}{dx^2} + (1-5x)\dfrac{dy}{dx} - 4y = 0$$

has a double root. Obtain one series solution of the equation in the form

$$y = A\sum_{n=1}^{\infty} n^2 x^{n-1} = Au_1(x).$$

What is the radius of convergence of this series? Obtain the second solution of the equation in the form

$$u_2(x) = u_1(x)\log x + u_1(x)(-4x + \cdots).$$

1.8 (a) The points $x = \pm 1$ are singular points of the differential equation

$$(x^2-1)^2 \dfrac{d^2y}{dx^2} + (x+1)\dfrac{dy}{dx} - y = 0.$$

Show that one of them is a regular singular point and that the other is an irregular singular point.

(b) Find two linearly independent Frobenius solutions of

$$x\dfrac{d^2y}{dx^2} + 4\dfrac{dy}{dx} - xy = 0,$$

which are valid for $x > 0$.

1.9 Find the general solution of the differential equation

$$2x\dfrac{d^2y}{dx^2} + (1+x)\dfrac{dy}{dx} - ky = 0$$

(where k is a real constant) in power series form. For which values of k is there a polynomial solution?

1.10 Let α, β, γ denote real numbers and consider the hypergeometric equation

$$x(1-x)\dfrac{d^2y}{dx^2} + \{\gamma - (\alpha + \beta + 1)x\}\dfrac{dy}{dx} - \alpha\beta y = 0.$$

Show that $x = 0$ and $x = 1$ are regular singular points and obtain the roots of the indicial equation at the point $x = 0$. Show that if γ is not an integer, there are two linearly independent solutions of Frobenius form. Express a_1

and a_2 in terms of a_0 for each of the solutions.

1.11 Show that each of the equations

(a) $x^3 \dfrac{d^2y}{dx^2} + x^2 \dfrac{dy}{dx} + y = 0,$

(b) $x^2 \dfrac{d^2y}{dx^2} + \dfrac{dy}{dx} - 2y = 0,$

has an irregular singular point at $x = 0$. Show that equation (a) has no solution of Frobenius type but that equation (b) does. Obtain this solution and hence find the general solution of equation (b).

1.12 Show that $x = 0$ is a regular singular point of the differential equation

$$2x^2 \dfrac{d^2y}{dx^2} + x(1-x) \dfrac{dy}{dx} - y = 0.$$

Find two linearly independent Frobenius solutions and show that one of these solutions can be expressed in terms of elementary functions. Verify directly that this function satisfies the differential equation.

1.13 Find the two linearly independent solutions of the ordinary differential equation

$$x(x-1)y'' + 3xy' + y = 0$$

in the form of a power series. *Hint:* It is straightforward to find one solution, but you will need to use the reduction of order formula to determine the structure of the second solution.

1.14 * Show that, if $f(x)$ and $g(x)$ are nontrivial solutions of the differential equations $u'' + p(x)u = 0$ and $v'' + q(x)v = 0$, and $p(x) \geqslant q(x)$, $f(x)$ vanishes at least once between any two zeros of $g(x)$ unless $p \equiv q$ and $f = \mu g$, $\mu \in \mathbb{R}$ (this is known as the **Sturm comparison theorem**).

1.15 * Show that, if $q(x) \leqslant 0$, no nontrivial solution of $u'' + q(x)u = 0$ can have more than one zero.

CHAPTER TWO

Legendre Functions

Legendre's equation occurs in many areas of applied mathematics, physics and chemistry in physical situations with a spherical geometry. We are now in a position to examine Legendre's equation in some detail using the ideas that we developed in the previous chapter. The **Legendre functions** are the solutions of Legendre's equation, a second order linear differential equation with variable coefficients. This equation was introduced by Legendre in the late 18th century, and takes the form

$$(1-x^2)\frac{d^2y}{dx^2} - 2x\frac{dy}{dx} + n(n+1)y = 0, \qquad (2.1)$$

where n is a parameter, called the **order** of the equation. The equation is usually defined for $-1 < x < 1$ for reasons that will become clear in Section 2.6. We can see immediately that $x = 0$ is an ordinary point of the equation, and, by Theorem 1.3, a series solution will be convergent for $|x| < 1$.

2.1 Definition of the Legendre Polynomials, $P_n(x)$

We will use the method of Frobenius, and seek a power series solution of (2.1) in the form

$$y = \sum_{i=0}^{\infty} a_i x^{i+c}.$$

Substitution of this series into (2.1) leads to

$$(1-x^2)\sum_{i=0}^{\infty} a_i(i+c)(i+c-1)x^{i+c-2} - 2x\sum_{i=0}^{\infty} a_i(i+c)x^{i+c-1}$$

$$+ n(n+1)\sum_{i=0}^{\infty} a_i x^{i+c} = 0.$$

Tidying this up gives

$$\sum_{i=0}^{\infty} a_i(i+c)(i+c-1)x^{i+c-2} + \sum_{i=0}^{\infty} a_i\{n(n+1) - (i+c)(i+c+1)\}x^{i+c} = 0,$$

and collecting like powers of x we get

$$a_0 c(c-1)x^{c-2} + a_1 c(c+1)x^{c-1} + \sum_{i=2}^{\infty} a_i(i+c)(i+c-1)x^{i+c-2}$$

$$+ \sum_{i=0}^{\infty} a_i \{n(n+1) - (i+c)(i+c+1)\} x^{i+c} = 0.$$

Rearranging the series gives us

$$a_0 c(c-1) x^{c-2} + a_1 c(c+1) x^{c-1}$$

$$+ \sum_{i=2}^{\infty} \{a_i(i+c)(i+c-1) + a_{i-2}(n(n+1) - (i+c-2)(i+c-1))\} x^{i+c-2} = 0.$$

The indicial equation is therefore

$$c(c-1) = 0,$$

which gives $c = 0$ or 1. Following Frobenius General Rule I, we choose $c = 0$, so that a_1 is arbitrary and acts as a second constant. In general we have,

$$a_i = \frac{(i-1)(i-2) - n(n+1)}{i(i-1)} a_{i-2} \quad \text{for } i = 2, 3, \ldots.$$

For $i = 2$,

$$a_2 = \frac{-n(n+1)}{2} a_0.$$

For $i = 3$,

$$a_3 = \frac{\{2 \cdot 1 - n(n+1)\}}{3 \cdot 2} a_1.$$

For $i = 4$,

$$a_4 = \frac{\{3 \cdot 2 - n(n+1)\}}{4 \cdot 3} a_2 = \frac{-\{3 \cdot 2 - n(n+1)\} n(n+1)}{4!} a_0.$$

For $i = 5$,

$$a_5 = \frac{\{4 \cdot 3 - n(n+1)\}}{5 \cdot 4} a_3 = \frac{\{4 \cdot 3 - n(n+1)\}\{2 \cdot 1 - n(n+1)\}}{5!} a_1.$$

The solution of Legendre's equation is therefore

$$y = a_0 \left[1 - \frac{n(n+1)}{2!} x^2 - \frac{\{3 \cdot 2 - n(n+1)\} n(n+1)}{4!} x^4 + \cdots \right]$$

$$+ a_1 \left[x + \frac{\{2 \cdot 1 - n(n+1)\}}{3!} x^3 + \frac{\{4 \cdot 3 - n(n+1)\}(2 \cdot 1 - n(n+1))}{5!} x^5 + \cdots \right],$$

(2.2)

where a_0 and a_1 are arbitrary constants. If n is not a positive integer, we have two infinite series solutions, convergent for $|x| < 1$. If n is a positive integer, one of the infinite series terminates to give a simple **polynomial solution**.

If we write the solution (2.2) as $y = a_0 u_n(x) + a_1 v_n(x)$, when n is an integer,

2.1 DEFINITION OF THE LEGENDRE POLYNOMIALS, $P_n(x)$

then $u_0(x)$, $v_1(x)$, $u_2(x)$ and $v_3(x)$ are the first four polynomial solutions. It is convenient to write the solution in the form

$$y = \underset{\substack{\uparrow \\ \text{Polynomial} \\ \text{of degree } n}}{AP_n(x)} + \underset{\substack{\uparrow \\ \text{Infinite series,} \\ \text{converges for } |x| < 1}}{BQ_n(x),}$$

where we have defined $P_n(x) = u_n(x)/u_n(1)$ for n even and $P_n(x) = v_n(x)/v_n(1)$ for n odd. The polynomials $P_n(x)$ are called the **Legendre polynomials** and can be written as

$$P_n(x) = \sum_{r=0}^{m} (-1)^r \frac{(2n-2r)! x^{n-2r}}{2^n r! (n-r)! (n-2r)!},$$

where m is the integer part of $n/2$. Note that by definition $P_n(1) = 1$. The first five of these are

$$P_0(x) = 1, \quad P_1(x) = x, \quad P_2(x) = \frac{1}{2}(3x^2 - 1),$$

$$P_3(x) = \frac{1}{2}(5x^3 - 3x), \quad P_4(x) = \frac{35}{8}x^4 - \frac{15}{4}x^2 + \frac{3}{8}.$$

Graphs of these Legendre polynomials are shown in Figure 2.1, which was generated using the MATLAB script

```
x = linspace(-1,1,500); pout = [];
for k = 1:4
    p = legendre(k,x); p=p(1,:);
    pout = [pout; p];
end
plot(x,pout(1,:),x,pout(2,:),'--',...
                x,pout(3,:),'-.',x,pout(4,:),':')
legend('P_1(x)','P_2(x)','P_3(x)','P_4(x)',4),xlabel('x')
```

Note that the MATLAB function `legendre(n,x)` generates both the Legendre polynomial of order n *and* the associated Legendre functions of orders 1 to n, which we will meet later, so we have to pick off the Legendre polynomial as the first row of the matrix p.

Simple expressions for the $Q_n(x)$ are available for $n = 0, 1, 2, 3$ using the reduction of order formula, (1.3). In particular

$$Q_0(x) = \frac{1}{2} \log\left(\frac{1+x}{1-x}\right), \quad Q_1(x) = \frac{1}{2} x \log\left(\frac{1+x}{1-x}\right) - 1,$$

$$Q_2(x) = \frac{1}{4}(3x^2 - 1) \log\left(\frac{1+x}{1-x}\right) - \frac{3}{2}x, \quad Q_3(x) = \frac{1}{4}(5x^3 - 3x) \log\left(\frac{1+x}{1-x}\right) - \frac{5}{2}x^2 + \frac{2}{3}.$$

These functions are singular at $x = \pm 1$. Notice that part of the infinite series has

LEGENDRE FUNCTIONS

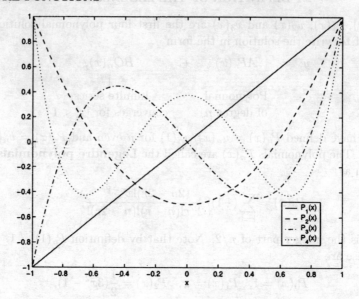

Fig. 2.1. The Legendre polynomials $P_1(x)$, $P_2(x)$, $P_3(x)$ and $P_4(x)$.

been summed to give us the logarithmic terms. Graphs of these functions are shown in Figure 2.2.

Example

Let's try to find the general solution of

$$(1-x^2)y'' - 2xy' + 2y = \frac{1}{1-x^2},$$

for $-1 < x < 1$. This is just an inhomogeneous version of Legendre's equation of order one. The complementary function is

$$y_h = AP_1(x) + BQ_1(x).$$

The variation of parameters formula, (1.6), then shows that the particular integral is

$$y_p = \int^x \frac{P_1(s)Q_1(x) - P_1(x)Q_1(s)}{(1-s^2)\{P_1(s)Q_1'(s) - P_1'(s)Q_1(s)\}} ds.$$

We can considerably simplify this rather complicated looking result. Firstly, Abel's formula, (1.7), shows that the Wronskian is

$$W = P_1(s)Q_1'(s) - P_1'(s)Q_1(s) = \frac{W_0}{1-s^2}.$$

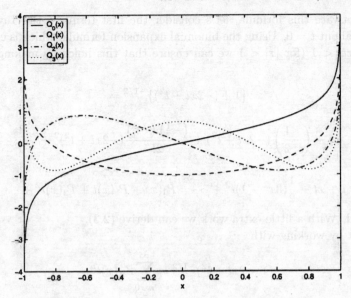

Fig. 2.2. The first four Legendre functions, $Q_0(x)$, $Q_1(x)$, $Q_2(x)$ and $Q_3(x)$.

We can determine the constant W_0 by considering the behaviour of W as $s \to 0$. Since

$$P_1(s) = s, \quad Q_1(s) = \frac{1}{2} s \log\left(\frac{1+s}{1-s}\right) - 1,$$

$W = 1 + s^2 + \cdots$ for $s \ll 1$. From the binomial theorem, $1/(1-s^2) = 1 + s^2 + \cdots$ for $s \ll 1$. We conclude that $W_0 = 1$. This means that

$$y_p = Q_1(x) \int^x P_1(s)\, ds - P_1(x) \int^x Q_1(s)\, ds$$

$$= \frac{1}{2} x^2 Q_1(x) - x \left\{ \frac{1}{4}(x^2 - 1) \log\left(\frac{1+x}{1-x}\right) - \frac{1}{2} x \right\}.$$

The general solution is this particular integral plus the complementary function $(y_p(x) + y_h(x))$.

2.2 The Generating Function for $P_n(x)$

In order to make a more systematic study of the Legendre polynomials, it is helpful to introduce a **generating function**, $G(x,t)$. This function is defined in such a way that the coefficients of the Taylor series of $G(x,t)$ around $t = 0$ are $P_n(x)$.

We start with the assertion that

$$G(x,t) = (1 - 2xt + t^2)^{-1/2} = \sum_{n=0}^{\infty} P_n(x) t^n. \qquad (2.3)$$

LEGENDRE FUNCTIONS

Just to motivate this formula, let's consider the first terms in the Taylor series expansion about $t = 0$. Using the binomial expansion formula, which is convergent for $|t^2 - 2xt| < 1$ (for $|x| < 1$ we can ensure that this holds by making $|t|$ small enough),

$$\{1 + (-2xt + t^2)\}^{-1/2} =$$

$$1 + \left(-\frac{1}{2}\right)(-2xt + t^2) + \frac{\left(-\frac{1}{2}\right)\left(-\frac{3}{2}\right)}{2!}(-2xt + t^2)^2 + \cdots$$

$$= 1 + xt + \frac{1}{2}(3x^2 - 1)t^2 + \cdots = P_0(x) + P_1(x)t + P_2(x)t^2 + \cdots,$$

as expected. With a little extra work we can derive (2.3).

We start by working with

$$(1 - 2xt + t^2)^{-1/2} = \sum_{n=0}^{\infty} Z_n(x) t^n, \qquad (2.4)$$

where, using the binomial expansion, we know that $Z_n(x)$ is a polynomial of degree n. We first differentiate with respect to x, which gives us

$$t(1 - 2xt + t^2)^{-3/2} = \sum_{n=0}^{\infty} Z'_n(x) t^n,$$

and again gives

$$3t^2(1 - 2xt + t^2)^{-5/2} = \sum_{n=0}^{\infty} Z''_n(x) t^n.$$

Now we differentiate (2.4) with respect to t, which leads to

$$(x - t)(1 - 2xt + t^2)^{-3/2} = \sum_{n=0}^{\infty} Z_n(x) n t^{n-1}.$$

Multiplying this last result by t^2 and differentiating with respect to t gives

$$\sum_{n=0}^{\infty} Z_n(x) n(n+1) t^n = \frac{\partial}{\partial t}\{t^2(x - t)(1 - 2xt + t^2)^{-3/2}\},$$

$$= t^2\{(x - t)(1 - 2xt + t^2)^{-5/2} 3(x - t) + (1 - 2xt + t^2)^{-3/2} - 1\}$$

$$+ (x - t)(1 - 2xt + t^2)^{-3/2} 2t,$$

which we can simplify to get

$$(1 - 2xt + t^2)^{-3/2}\{3t^2(x - t)^2(1 - 2xt + t^2)^{-1} - 1 + 2t(x - t)\}$$

$$= \sum_{n=0}^{\infty} Z_n(x) n(n+1) t^n.$$

Combining all of these results gives

$$(1-x^2)\sum_{n=0}^{\infty} Z_n''(x)t^n - 2x\sum_{n=0}^{\infty} Z_n'(x)t^n + \sum_{n=0}^{\infty} n(n+1)Z_n(x)t^n$$

$$= (1-x^2)3t^2(1-2xt+t^2)^{-5/2} - 2xt(1-2xt+t^2)^{-3/2}$$

$$+(1-2xt+t^2)^{-3/2}\{3t^2(x-t)^2(1-2xt+t^2)^{-1} - 1 + 2t(x-t)\} = 0,$$

for any t. Therefore

$$(1-x^2)Z_n''(x) - 2xZ_n'(x) + n(n+1)Z_n(x) = 0,$$

which is just Legendre's equation. This means that $Z_n(x) = \alpha P_n(x) + \beta Q_n(x)$ where α, β are constants. As $Z_n(x)$ is a polynomial of degree n, β must be zero.

Finally, we need to show that $Z_n(1) = 1$. This is done by putting $x = 1$ in the generating function relationship, (2.3), to obtain

$$(1-2t+t^2)^{-1/2} = \sum_{n=0}^{\infty} Z_n(1)t^n.$$

Since

$$(1-2t+t^2)^{-1/2} = \{(1-t)^2\}^{-1/2} = (1-t)^{-1} = \sum_{n=0}^{\infty} t^n,$$

at least for $|t| < 1$, we have $Z_n(1) = 1$. Since we know that $P_n(1) = 1$, we conclude that $Z_n(x) \equiv P_n(x)$, as required.

The generating function, $G(x,t) = (1-2xt+t^2)^{-1/2}$, can be used to prove a number of interesting properties of Legendre polynomials, as well as some recurrence formulae. We will give a few examples of its application.

Special Values

The generating function is useful for determining the values of the Legendre polynomials for certain special values of x. For example, substituting $x = -1$ in (2.3) gives

$$(1+2t+t^2)^{-1/2} = \sum_{n=0}^{\infty} P_n(-1)t^n.$$

By the binomial expansion we have that

$$(1+2t+t^2)^{-1/2} = \{(1+t)^2\}^{-1/2} = (1+t)^{-1}$$

$$= 1 - t + t^2 - \cdots + (-t)^n + \cdots = \sum_{n=0}^{\infty} (-1)^n t^n.$$

We conclude that

$$\sum_{n=0}^{\infty} (-1)^n t^n = \sum_{n=0}^{\infty} P_n(-1)t^n,$$

and therefore $P_n(-1) = (-1)^n$ for $n = 1, 2, \ldots$.

LEGENDRE FUNCTIONS

2.3 Differential and Recurrence Relations Between Legendre Polynomials

The generating function can also be used to derive recurrence relations between the various $P_n(x)$. Starting with (2.3), we differentiate with respect to t, and find that

$$(x-t)(1-2xt+t^2)^{-3/2} = \sum_{n=0}^{\infty} P_n(x)nt^{n-1}.$$

We now multiply through by $(1-2xt+t^2)$ to obtain

$$(x-t)(1-2xt+t^2)^{-1/2} = (1-2xt+t^2)\sum_{n=0}^{\infty} P_n(x)nt^{n-1},$$

which leads to

$$x\sum_{n=0}^{\infty} P_n(x)t^n - \sum_{n=0}^{\infty} P_n(x)t^{n+1}$$

$$= n\sum_{n=0}^{\infty} P_n(x)t^{n-1} - 2xn\sum_{n=0}^{\infty} P_n(x)t^n + n\sum_{n=0}^{\infty} P_n(x)t^{n+1}.$$

Equating coefficients of t^n on both sides shows that

$$xP_n(x) - P_{n-1}(x) = (n+1)P_{n+1}(x) - 2xnP_n(x) + (n-1)P_{n-1}(x),$$

and hence

$$(n+1)P_{n+1}(x) - (2n+1)xP_n(x) + nP_{n-1}(x) = 0. \qquad (2.5)$$

This is a recurrence relation between $P_{n+1}(x)$, $P_n(x)$ and $P_{n-1}(x)$, which can be used to compute the polynomials $P_n(x)$. Starting with $P_0(x) = 1$ and $P_1(x) = x$, we substitute $n = 1$ into (2.5), which gives

$$2P_2(x) - 3x^2 + 1 = 0,$$

and hence

$$P_2(x) = \frac{1}{2}(3x^2 - 1).$$

By iterating this procedure, we can generate the Legendre polynomials $P_n(x)$ for any n.

In a rather similar manner we can generate a recurrence relation that involves the derivatives of the Legendre polynomials. Firstly, we differentiate the generating function, (2.3), with respect to x to get

$$t(1-2xt+t^2)^{-3/2} = \sum_{n=0}^{\infty} t^n P_n'(x).$$

Differentiation of (2.3) with respect to t gives

$$(x-t)(1-2xt+t^2)^{-3/2} = \sum_{n=0}^{\infty} nt^{n-1}P_n(x).$$

Combining these gives

$$\sum_{n=0}^{\infty} n t^n P_n(x) = (x-t) \sum_{n=0}^{\infty} t^n P'_n(x),$$

and by equating coefficients of t^n we obtain the recurrence relation

$$n P_n(x) = x P'_n(x) - P'_{n-1}(x).$$

An Example From Electrostatics

In electrostatics, the potential due to a unit point charge at $\mathbf{r} = \mathbf{r}_0$ is

$$V = \frac{1}{|\mathbf{r} - \mathbf{r}_0|}.$$

If this unit charge lies on the z-axis, at $x = y = 0$, $z = a$, this becomes

$$V = \frac{1}{\sqrt{x^2 + y^2 + (z-a)^2}}.$$

In terms of spherical polar coordinates, (r, θ, ϕ),

$$x = r \sin\theta \cos\phi, \quad y = r \sin\theta \sin\phi, \quad z = r \cos\theta.$$

This means that

$$x^2 + y^2 + (z-a)^2 = x^2 + y^2 + z^2 - 2az + a^2 = r^2 + a^2 - 2az,$$

and hence

$$V = \frac{1}{\sqrt{r^2 + a^2 - 2ar\cos\theta}} = \frac{1}{a}\left(1 - 2\cos\theta\frac{r}{a} + \frac{r^2}{a^2}\right)^{-1/2}.$$

As we would expect from the symmetry of the problem, there is no dependence upon the azimuthal angle, ϕ. We can now use the generating function to write this as a power series,

$$V = \frac{1}{a} \sum_{n=0}^{\infty} P_n(\cos\theta) \left(\frac{r}{a}\right)^n.$$

2.4 Rodrigues' Formula

There are other methods of generating the Legendre polynomials, and the most useful of these is **Rodrigues' formula**,

$$P_n(x) = \frac{1}{2^n n!} \frac{d^n}{dx^n}\{(x^2 - 1)^n\}.$$

For example, for $n = 1$,

$$P_1(x) = \frac{1}{2^1 1!} \frac{d^1}{dx^1}\{(x^2 - 1)\} = x,$$

LEGENDRE FUNCTIONS

whilst for $n = 2$,

$$P_2(x) = \frac{1}{2^2 2!} \frac{d^2}{dx^2}\{(x^2 - 1)^2\} = \frac{1}{4 \cdot 2} \frac{d}{dx}\{4x(x^2 - 1)\} = \frac{1}{2}(3x^2 - 1).$$

The general proof of this result is by induction on n, which we leave as an exercise.

Rodrigues' formula can also be used to develop an integral representation of the Legendre polynomials. In order to show how this is done, it is convenient to switch from the real variable x to the complex variable $z = x + iy$. We define the finite complex Legendre polynomials in the same way as for a real variable. In particular Rodrigues' formula,

$$P_n(z) = \frac{1}{2^n n!} \frac{d^n}{dz^n}(z^2 - 1)^n,$$

will be useful to us here. Recall from complex variable theory (see Appendix 6) that, if $f(z)$ is analytic and single-valued inside and on a simple closed curve $\mathcal{C}$,

$$f^{(n)}(z) = \frac{n!}{2\pi i} \int_{\mathcal{C}} \frac{f(\xi)}{(\xi - z)^{n+1}} d\xi,$$

for $n \geq 0$, when z is an interior point of $\mathcal{C}$. Now, using Rodrigues' formula, we have

$$P_n(z) = \frac{1}{2^{n+1} \pi i} \int_{\mathcal{C}} \frac{(\xi^2 - 1)^n}{(\xi - z)^{n+1}} d\xi, \qquad (2.6)$$

which is known as **Schläfli's representation**. The contour $\mathcal{C}$ must, of course, enclose the point $\xi = z$ and be traversed in an anticlockwise sense. To simplify matters, we now choose $\mathcal{C}$ to be a circle, centred on $\xi = z$ with radius $|\sqrt{z^2 - 1}|$, with $z \neq 1$. Putting $\xi = z + \sqrt{z^2 - 1} e^{i\theta}$ gives, after some simple manipulation,

$$P_n(z) = \frac{1}{2\pi} \int_{\theta=0}^{2\pi} \left(z + \sqrt{z^2 - 1} \cos \theta\right)^n d\theta.$$

This is known as **Laplace's representation**. In fact it is also valid when $z = 1$, since

$$P_n(1) = \frac{1}{2\pi} \int_0^{2\pi} 1 d\theta = 1.$$

Laplace's representation is useful, amongst other things, for providing a bound on the size of the Legendre polynomials of real argument. For $z \in [-1, 1]$, we can write $z = \cos \phi$ and use Laplace's representation to show that

$$|P_n(\cos \phi)| \leq \frac{1}{2\pi} \int_0^{2\pi} |\cos \phi + i \sin \phi \cos \theta|^n d\theta.$$

Now, since

$$|\cos \phi + i \sin \phi \cos \theta| = \sqrt{\cos^2 \phi + \sin^2 \phi \cos^2 \theta} \leq \sqrt{\cos^2 \phi + \sin^2 \phi} = 1,$$

we have $|P_n(\cos \phi)| \leq 1$.

2.5 Orthogonality of the Legendre Polynomials

Legendre polynomials have the very important property of **orthogonality** on $[-1, 1]$, that is

$$\int_{-1}^{1} P_n(x) P_m(x) dx = \frac{2}{2n+1} \delta_{mn}, \tag{2.7}$$

where the **Kronecker delta** is defined by

$$\delta_{mn} = \begin{cases} 1 & \text{for } m = n, \\ 0 & \text{for } m \neq n. \end{cases}$$

To show this, note that if $f \in C[-1, 1]$, Rodrigues' formula shows that

$$\int_{-1}^{1} f(x) P_n(x) dx = \frac{1}{2^n n!} \int_{-1}^{1} f(x) \frac{d^n}{dx^n} \{(x^2 - 1)^n\} dx$$

$$= \frac{1}{2^n n!} \left\{ \left[f(x) \frac{d^{n-1}}{dx^{n-1}} \{(x^2 - 1)^n\} \right]_{-1}^{1} - \int_{-1}^{1} f'(x) \frac{d^{n-1}}{dx^{n-1}} \{(x^2 - 1)^n\} dx \right\}$$

$$= -\frac{1}{2^n n!} \int_{-1}^{1} f'(x) \frac{d^{n-1}}{dx^{n-1}} \{(x^2 - 1)^n\} dx.$$

Repeating this integration by parts $(n-1)$ more times gives

$$\int_{-1}^{1} f(x) P_n(x) dx = \frac{(-1)^n}{2^n n!} \int_{-1}^{1} (x^2 - 1)^n f^{(n)}(x) dx, \tag{2.8}$$

where $f^{(n)}(x)$ is the nth derivative of $f(x)$. This result is interesting in its own right, but for the time being consider the case $f(x) = P_m(x)$ with $m < n$, so that

$$\int_{-1}^{1} P_m(x) P_n(x) dx = \frac{(-1)^n}{2^n n!} \int_{-1}^{1} (x^2 - 1)^n P_m^{(n)}(x) dx.$$

Since the nth derivative of an mth order polynomial is zero for $m < n$, we have

$$\int_{-1}^{1} P_m(x) P_n(x) dx = 0 \text{ for } m < n.$$

By symmetry, this must also hold for $m > n$.

Let's now consider the case $n = m$, for which

$$\int_{-1}^{1} P_n(x) P_n(x) dx = \frac{(-1)^n}{2^n n!} \int_{-1}^{1} (x^2 - 1)^n P_n^{(n)}(x) dx$$

$$= \frac{(-1)^n}{2^n n!} \int_{-1}^{1} (x^2 - 1)^n \frac{1}{2^n n!} \frac{d^{2n}}{dx^{2n}} \{(x^2 - 1)^n\} dx.$$

Noting the fact that

$$(x^2 - 1)^n = x^{2n} + \cdots + (-1)^n,$$

LEGENDRE FUNCTIONS

and hence that

$$\frac{d^{2n}}{dx^{2n}}(x^2-1)^n = 2n(2n-1)\ldots 3\cdot 2\cdot 1 = (2n)!,$$

we can see that

$$\int_{-1}^{1} P_n(x)P_n(x)\,dx = \frac{(-1)^n}{2^n n!}\int_{-1}^{1}(x^2-1)^n \frac{1}{2^n n!}(2n)!\,dx$$

$$= \frac{(-1)^n(2n)!}{2^{2n}(n!)^2}\int_{-1}^{1}(x^2-1)^n\,dx.$$

To evaluate the remaining integral, we use a reduction formula to show that

$$\int_{-1}^{1}(x^2-1)^n\,dx = \frac{2^{2n+2}n!(n+1)!(-1)^n}{(2n+2)!},$$

and hence

$$\int_{-1}^{1} P_n^2(x)\,dx = \frac{2}{2n+1}.$$

This completes the derivation of (2.7). There is an easier proof of the first part of this, using the idea of a self-adjoint linear operator, which we will discuss in Chapter 4.

We have now shown that the Legendre polynomials are orthogonal on $[-1, 1]$. It is also possible to show that these polynomials are **complete** in the function space $C[-1, 1]$. This means that a continuous function can be expanded as a linear combination of the Legendre polynomials. The proof of the completeness property is rather difficult, and will be omitted here.† What we can present here is the procedure for obtaining the coefficients of such an expansion for a given function $f(x)$ belonging to $C[-1, 1]$. To do this we write

$$f(x) = a_0 P_0(x) + a_1 P_1(x) + \cdots + a_n P_n(x) + \cdots = \sum_{n=0}^{\infty} a_n P_n(x).$$

Multiplying by $P_m(x)$ and integrating over $[-1, 1]$ (more precisely, forming the inner product with $P_m(x)$) gives

$$\int_{-1}^{1} f(x)P_m(x)\,dx = \int_{-1}^{1} P_m(x)\sum_{n=0}^{\infty} a_n P_n(x)\,dx.$$

---

† Just to give a flavour of the completeness proof for the case of Legendre polynomials, we note that, because of their polynomial form, we can deduce that *any* polynomial can be written as a linear combination of Legendre polynomials. However, according to a fundamental result due to Weierstrass (the Weierstrass polynomial approximation theorem) *any* function which is continuous on some interval can be approximated as closely as we wish by a polynomial. The completeness then follows from the application of this theorem. The treatment of completeness of other solutions of Sturm–Liouville problems may be more complicated, for example for Bessel functions. A complete proof of this can be found in Kreider, Kuller, Ostberg and Perkins (1966). We will return to this topic in Chapter 4.

2.5 ORTHOGONALITY OF THE LEGENDRE POLYNOMIALS

Interchanging the order of summation and integration leads to

$$\int_{-1}^{1} f(x) P_m(x) dx = \sum_{n=0}^{\infty} a_n \left\{ \int_{-1}^{1} P_n(x) P_m(x) dx \right\}$$

$$= \sum_{n=0}^{\infty} a_n \frac{2}{2m+1} \delta_{mn} = \frac{2 a_m}{2m+1},$$

using the orthogonality property, (2.7). This means that

$$a_m = \frac{2m+1}{2} \int_{-1}^{1} f(x) P_m(x) dx, \qquad (2.9)$$

and we write the series as

$$f(x) = \sum_{n=0}^{\infty} \left\{ \frac{2n+1}{2} \int_{-1}^{1} f(x) P_n(x) dx \right\} P_n(x).$$

This is called a **Fourier–Legendre series**.

Let's consider a couple of examples. Firstly, when $f(x) = x^2$,

$$a_m = \frac{2m+1}{2} \int_{-1}^{1} x^2 P_m(x) dx,$$

so that

$$a_0 = \frac{1}{2} \int_{-1}^{1} x^2 \cdot 1 \, dx = \frac{1}{2} \left[\frac{x^3}{3} \right]_{-1}^{1} = \frac{1}{2} \frac{2}{3} = \frac{1}{3},$$

$$a_1 = \frac{(2 \cdot 1) + 1}{2} \int_{-1}^{1} x^2 \cdot x \, dx = \frac{3}{2} \left[\frac{x^4}{4} \right]_{-1}^{1} = 0,$$

$$a_2 = \frac{(2 \cdot 2) + 1}{2} \int_{-1}^{1} x^2 \frac{1}{2}(3x^2 - 1) dx = \frac{5}{2} \left[\frac{3x^5}{2 \cdot 5} - \frac{x^3}{3 \cdot 2} \right]_{-1}^{1} = \frac{2}{3}.$$

Also, (2.8) shows that

$$a_m = 0 \text{ for } m = 3, 4, \ldots,$$

and therefore,

$$x^2 = \frac{1}{3} P_0(x) + \frac{2}{3} P_2(x).$$

A finite polynomial clearly has a finite Fourier–Legendre series.

Secondly, consider $f(x) = e^x$. In this case

$$a_m = \frac{2m+1}{2} \int_{-1}^{1} e^x P_m(x) \, dx,$$

and hence

$$a_0 = \frac{1}{2} \int_{-1}^{1} e^x \, dx = \frac{1}{2} \left(e - e^{-1} \right),$$

$$a_1 = \frac{3}{2}\int_{-1}^{1} x\,e^x\,dx = 3\,e^{-1}.$$

To proceed with this calculation it is necessary to find a recurrence relation between the a_n. This is best done by using Rodrigues' formula, which gives

$$a_n = (2n+1)\left(\frac{a_{n-2}}{2n-3} - a_{n-1}\right) \quad \text{for } n = 2, 3, \ldots, \tag{2.10}$$

from which the values of $a_4, a_5, \ldots$ are easily computed.

We will not examine the convergence of Fourier–Legendre series here as the details are rather technical. Instead we content ourselves with a statement that the Fourier–Legendre series converges uniformly on any closed subinterval of $(-1, 1)$ in which f is continuous and differentiable. An extension of this result to the space of piecewise continuous functions is that the series converges to the value $\frac{1}{2}\{f(x_0^+) + f(x_0^-)\}$ at each point $x_0 \in (-1, 1)$ where f has a right and left derivative. We will prove a related theorem in Chapter 5.

2.6 Physical Applications of the Legendre Polynomials

In this section we present some examples of Legendre polynomials as they arise in mathematical models of heat conduction and fluid flow in spherical geometries. In general, we will encounter the Legendre equation in situations where we have to solve partial differential equations containing the Laplacian in spherical polar coordinates.

2.6.1 Heat Conduction

Let's derive the equation that governs the evolution of an initial distribution of heat in a solid body with temperature T, density ρ, specific heat capacity c and thermal conductivity k. Recall that the specific heat capacity, c, is the amount of heat required to raise the temperature of a unit mass of a substance by one degree. The thermal conductivity, k, of a body appears in Fourier's law, which states that the heat flux per unit area, per unit time, $\mathbf{Q} = (Q_x, Q_y, Q_z)$, is related to the temperature gradient, ∇T, by the simple linear relationship $\mathbf{Q} = -k\nabla T$. If we now consider a small element of our solid body at (x, y, z) with sides of length δx, δy and δz, the temperature change in this element over a time interval δt is determined by the difference between the amount of heat that flows into the element and the amount of heat that flows out, which gives

$$\rho c\left\{T\left(x,y,z,t+\delta t\right) - T\left(x,y,z,t\right)\right\}\delta x \delta y \delta z$$

$$= \{Q_x(x,y,z,t) - Q_x(x+\delta x, y, z, t)\}\delta t \delta y \delta z$$

$$+ \{Q_y(x,y,z,t) - Q_y(x, y+\delta y, z, t)\}\delta t \delta x \delta z \tag{2.11}$$

$$+ \{Q_z(x,y,z,t) - Q_z(x, y, z+\delta z, t)\}\delta t \delta x \delta y.$$

2.6 PHYSICAL APPLICATIONS OF THE LEGENDRE POLYNOMIALS

	ρ kg m^{-3}	k J m^{-1} s^{-1} K^{-1}	c J kg^{-1} K^{-1}	K m^2 s^{-1}
copper	8920	385	386	1.1×10^{-4}
water	1000	254	4186	6.1×10^{-5}
glass	2800	0.8	840	3.4×10^{-7}

Table 2.1. *Some typical physical properties of copper, water (at room temperature and pressure) and glass.*

Note that a typical term on the right hand side of this, for example,
$$\{Q_x(x, y, z, t) - Q_x(x + \delta x, y, z, t)\} \delta t \delta y \delta z,$$
is the amount of heat crossing the x-orientated faces of the element, each with area $\delta y \delta z$, during the time interval $(t, t + \delta t)$. Taking the limit $\delta t, \delta x, \delta y, \delta z \to 0$, we obtain
$$\rho c \frac{\partial T}{\partial t} = -\left\{\frac{\partial Q_x}{\partial x} + \frac{\partial Q_y}{\partial y} + \frac{\partial Q_z}{\partial z}\right\} = -\nabla \cdot \mathbf{Q}.$$

Substituting in Fourier's law, $\mathbf{Q} = -k\nabla T$, gives the **diffusion equation**,
$$\frac{\partial T}{\partial t} = K\nabla^2 T, \tag{2.12}$$

where $K = k/\rho c$ is called the **thermal diffusivity**. Table 2.1 contains the values of relevant properties for three everyday materials.

When the temperature reaches a steady state $(\partial T/\partial t = 0)$, this equation takes the simple form
$$\nabla^2 T = 0, \tag{2.13}$$
which is known as **Laplace's equation**. It must be solved in conjunction with appropriate boundary conditions, which drive the temperature gradients in the body.

Example

Let's try to find the steady state temperature distribution in a solid, uniform sphere of unit radius, when the surface temperature is held at $f(\theta) = T_0 \sin^4 \theta$ in spherical polar coordinates, (r, θ, ϕ). This temperature distribution will satisfy Laplace's equation, (2.13). Since the equation and boundary conditions do not depend upon the azimuthal angle, ϕ, neither does the solution, and hence Laplace's equation takes the form
$$\frac{1}{r^2}\frac{\partial}{\partial r}\left(r^2 \frac{\partial T}{\partial r}\right) + \frac{1}{r^2 \sin\theta}\frac{\partial}{\partial \theta}\left(\sin\theta \frac{\partial T}{\partial \theta}\right) = 0.$$

Let's look for a **separable solution**, $T(r, \theta) = R(r)\Theta(\theta)$. This gives
$$\Theta \frac{d}{dr}\left(r^2 \frac{dR}{dr}\right) + \frac{R}{\sin\theta}\frac{d}{d\theta}\left(\sin\theta \frac{d\Theta}{d\theta}\right) = 0,$$

LEGENDRE FUNCTIONS

and hence
$$\frac{1}{R}\frac{d}{dr}\left(r^2\frac{dR}{dr}\right) = -\frac{1}{\Theta\sin\theta}\frac{d}{d\theta}\left(\sin\theta\frac{d\Theta}{d\theta}\right).$$

Since the left hand side is a function of r only and the right hand side is a function of θ only, this equality can only be valid if both sides are equal to some constant, with
$$\frac{1}{R}\frac{d}{dr}\left(r^2\frac{dR}{dr}\right) = -\frac{1}{\Theta\sin\theta}\frac{d}{d\theta}\left(\sin\theta\frac{d\Theta}{d\theta}\right) = \text{constant} = n(n+1).$$

This choice of constant may seem rather ad hoc at this point, but all will become clear later.

We now have an ordinary differential equation for Θ, namely
$$\frac{1}{\sin\theta}\frac{d}{d\theta}\left(\sin\theta\frac{d\Theta}{d\theta}\right) + n(n+1)\Theta = 0.$$

Changing variables to
$$\mu = \cos\theta, \quad y(\mu) = \Theta(\theta),$$

and using
$$\sin^2\theta = 1 - \mu^2, \quad \frac{d}{d\mu} = \frac{d\theta}{d\mu}\frac{d}{d\theta} = \frac{-1}{\sin\theta}\frac{d}{d\theta},$$

leads to
$$\frac{d}{d\mu}\left[(1-\mu^2)\frac{dy}{d\mu}\right] + n(n+1)y = 0,$$

or, equivalently
$$(1-\mu^2)y'' - 2\mu y' + n(n+1)y = 0,$$

which is Legendre's equation. Since a physically meaningful solution must in this case be finite at $\mu = \pm 1$ (the north and south poles of the sphere), the solution, to within an arbitrary multiplicative constant, is the Legendre polynomial, $y(\mu) = P_n(\mu)$, and hence $\Theta_n(\theta) = P_n(\cos\theta)$. We have introduced the subscript n for Θ so that we can specify that this is the solution corresponding to a particular choice of n. Note that we have just solved our first boundary value problem. Specifically, we have found the solution of Legendre's equation that is bounded at $\mu = \pm 1$.

We must now consider the solution of the equation for $R(r)$. This is
$$\frac{d}{dr}\left(r^2\frac{dR}{dr}\right) = n(n+1)R.$$

By inspection, by the method of Frobenius or by noting that the equation is invariant under the transformation group $(r, R) \mapsto (\lambda r, \lambda R)$ (see Chapter 10), we can find the solution of this equation as
$$R_n(r) = A_n r^n + B_n r^{-1-n},$$

where again the subscript n denotes the dependence of the solution on n and A_n and B_n are arbitrary constants. At the centre of the sphere, the temperature must

2.6 PHYSICAL APPLICATIONS OF THE LEGENDRE POLYNOMIALS

be finite, so we set $B_n = 0$. Our solution of Laplace's equation therefore takes the form

$$T = A_n r^n P_n(\cos\theta).$$

As this is a solution for arbitrary n, the general solution for the temperature will be a **linear combination** of these solutions,

$$T = \sum_{n=0}^{\infty} A_n r^n P_n(\cos\theta).$$

The remaining task is to evaluate the coefficients A_n. This can be done using the specified temperature on the surface of the sphere, $T = T_0 \sin^4\theta$ at $r = 1$. We substitute $r = 1$ into the general expression for the temperature to get

$$f(\theta) = T_0 \sin^4\theta = \sum_{n=0}^{\infty} A_n P_n(\cos\theta).$$

The A_n will therefore be the coefficients in the Fourier–Legendre expansion of the function $f(\theta) = T_0 \sin^4\theta$.

It is best to work in terms of the variable $\mu = \cos\theta$. We then have

$$T_0 (1 - \mu^2)^2 = \sum_{n=0}^{\infty} A_n P_n(\mu).$$

From (2.9),

$$A_n = \frac{2n+1}{2} \int_{-1}^{1} T_0 (1 - \mu^2)^2 P_n(\mu) d\mu.$$

Since the function that we want to expand is a finite polynomial in μ, we expect to obtain a finite Fourier–Legendre series. A straightforward calculation of the integral gives us

$$A_0 = \frac{1}{2} \cdot \frac{16}{15} T_0, \quad A_1 = 0, \quad A_2 = -\frac{5}{2} \cdot \frac{32}{105} T_0, \quad A_3 = 0, \quad A_4 = \frac{9}{2} \cdot \frac{16}{315} T_0,$$

$$A_m = 0 \text{ for } m = 5, 6, \ldots.$$

The solution is therefore

$$T = T_0 \left\{ \frac{8}{15} P_0(\cos\theta) - \frac{16}{21} r^2 P_2(\cos\theta) + \frac{8}{35} r^4 P_4(\cos\theta) \right\}. \tag{2.14}$$

This solution when $T_0 = 1$ is shown in Figure 2.3, which is a polar plot in a plane of constant ϕ. Note that the temperature at the centre of the sphere is $8T_0/15$. We produced Figure 2.3 using the MATLAB script

```
ezmesh('r*cos(t)','r*sin(t)','8/15-8*r^2*(3*cos(t)^2-1)/21+...
8*r^4*(35*cos(t)^4/8-15*cos(t)^2/4+3/8)/35', [0 1 0 pi])
```

LEGENDRE FUNCTIONS

The function `ezmesh` gives an easy way of plotting parametric surfaces like this. The first three arguments give the x, y and z coordinates as parametric functions of r and $t = \theta$, whilst the fourth specifies the ranges of r and t.

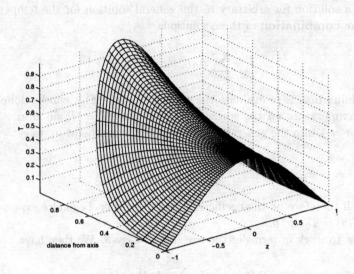

Fig. 2.3. The steady state temperature in a uniform sphere with surface temperature $\sin^4\theta$, given by (2.14).

2.6.2 Fluid Flow

Consider a fixed volume V bounded by a surface S within an incompressible fluid. Although fluid flows into and out of V, the mass of fluid within V remains constant, since the fluid is incompressible, so any flux out of V at one place is balanced by a flux into V at another place. Mathematically, we can express this as

$$\int_S \mathbf{u} \cdot \mathbf{n}\, dS = 0,$$

where $\mathbf{u}$ is the velocity of the fluid, $\mathbf{n}$ is the outward unit normal to S, and hence $\mathbf{u} \cdot \mathbf{n}$ is the normal component of the fluid velocity out through S. If the fluid velocity field, $\mathbf{u}$, is smooth, we can use the divergence theorem to rewrite this statement of conservation of mass as

$$\int_V \nabla \cdot \mathbf{u}\, dV = 0.$$

As this applies to an arbitrary volume V within the fluid, we must have $\nabla \cdot \mathbf{u} = 0$ throughout the flow. If we also suppose that the fluid is inviscid (there is no friction as one fluid element flows past another), it is possible to make the further assumption that the flow is irrotational ($\nabla \times \mathbf{u} = \mathbf{0}$). Physically, this means that

2.6 PHYSICAL APPLICATIONS OF THE LEGENDRE POLYNOMIALS

there is no spin in any fluid element as it flows around. Inviscid, irrotational, incompressible flows are therefore governed by the two equations $\nabla \cdot \mathbf{u} = 0$ and $\nabla \times \mathbf{u} = \mathbf{0}$.† For simply connected flow domains, $\nabla \times \mathbf{u} = \mathbf{0}$ if and only if $\mathbf{u} = \nabla \phi$ for some scalar potential function ϕ, known as the **velocity potential**. Substituting into $\nabla \cdot \mathbf{u} = 0$ gives $\nabla^2 \phi = 0$. In other words, the velocity potential satisfies Laplace's equation. As for boundary conditions, there can be no flux of fluid into a solid body so that $\mathbf{u} \cdot \mathbf{n} = \mathbf{n} \cdot \nabla \phi = \partial \phi / \partial n = 0$ where the fluid is in contact with a solid body.

As an example of such a flow, let's consider what happens when a sphere of radius $r = a$ is placed in a uniform stream, assuming that the flow is steady, inviscid and irrotational. The flow at infinity must be uniform, with $\mathbf{u} = U\mathbf{i}$ where $\mathbf{i}$ is the unit vector in the x-direction. First of all, it is clear that the problem will be best treated in spherical polar coordinates. We know from the previous example that the bounded, axisymmetric solution of Laplace's equation is

$$\phi = \sum_{n=0}^{\infty} \left(A_n r^n + B_n r^{-1-n} \right) P_n(\cos \theta). \tag{2.15}$$

The flow at infinity has potential $\phi = Ux = Ur\cos\theta$. Since $P_1(\cos\theta) = \cos\theta$, we see that we must take $A_1 = 1$ and $A_n = 0$ for $n > 1$. To fix the constants B_n, notice that there can be no flow through the surface of the sphere. This gives us the boundary condition on the radial velocity as

$$u_r = \frac{\partial \phi}{\partial r} = 0 \quad \text{at } r = a.$$

On substituting (2.15) into this boundary condition, we find that $B_1 = \frac{1}{2}a^3$ and $B_n = 0$ for $n > 1$. The solution is therefore

$$\phi = U\left(r + \frac{a^3}{2r^2}\right)\cos\theta. \tag{2.16}$$

The **streamlines** (continuous curves that are tangent to the velocity vector) are shown in Figure 2.4 when $a = U = 1$. In order to obtain this figure, we note that (see Section 7.2.2) the streamlines are given by

$$\psi = U\left(r - \frac{a^3}{r^2}\right)\sin\theta = Uy\left(1 - \frac{a^3}{(x^2 + y^2)^{3/2}}\right) = \text{constant}.$$

We can then use the MATLAB script

```
x = linspace(-2,2,400); y = linspace(0,2,200);
[X Y] = meshgrid(x,y);
Z = Y.*(1-1./(X.^2+Y.^2).^(3/2));
v = linspace(0,2,15); contour(X,Y,Z,v)
```

† We will consider an example of viscous flow in Section 6.4.

50 LEGENDRE FUNCTIONS

The command `meshgrid` creates a grid suitable for use with the plotting command `contour` out of the two vectors, x and y. The vector v specifies the values of ψ for which a contour is to be plotted.

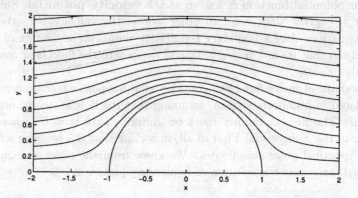

Fig. 2.4. The streamlines for inviscid, irrotational flow about a unit sphere.

In order to complete this example, we must consider the **pressure** in our ideal fluid. The force exerted by the fluid on a surface S by the fluid outside S is purely a pressure, p, which acts normally to S. In other words, the force on a surface element with area dS and outward unit normal $\mathbf{n}$ is $-p\mathbf{n}\,dS$. We would now like to apply Newton's second law to the motion of the fluid within V, the volume enclosed by S. In order to do this, we need an expression for the acceleration of the fluid. Let's consider the change in the velocity of a fluid particle between times t and $t + \delta t$, which, for small δt, we can Taylor expand to give

$$\mathbf{u}(\mathbf{x}(t+\delta t), t+\delta t) - \mathbf{u}(\mathbf{x}(t),t) = \delta t \left\{ \frac{\partial \mathbf{u}}{\partial t}(\mathbf{x},t) + \frac{d\mathbf{x}}{dt} \cdot \nabla \mathbf{u}(\mathbf{x},t) \right\} + \cdots$$

$$= \delta t \left\{ \frac{\partial \mathbf{u}}{\partial t}(\mathbf{x},t) + (\mathbf{u}(\mathbf{x},t) \cdot \nabla)\, \mathbf{u}(\mathbf{x},t) \right\} + \cdots,$$

since $\mathbf{u} = d\mathbf{x}/dt$. This means that the fluid acceleration is

$$\frac{D\mathbf{u}}{Dt} = \frac{\partial \mathbf{u}}{\partial t} + (\mathbf{u} \cdot \nabla)\,\mathbf{u},$$

where D/Dt is the usual notation for the **material derivative**, or time derivative following the fluid particle.

We can now use Newton's second law on the fluid within S to obtain

$$\int_S -p\mathbf{n}\,dS = \int_V \rho \frac{D\mathbf{u}}{Dt}\,dV.$$

After noting that

$$\int_S p\mathbf{n}\,dS = \mathbf{i}\int_S (p\mathbf{i}) \cdot \mathbf{n}\,dS + \mathbf{j}\int_S (p\mathbf{j}) \cdot \mathbf{n}\,dS + \mathbf{k}\int_S (p\mathbf{k}) \cdot \mathbf{n}\,dS,$$

2.6 PHYSICAL APPLICATIONS OF THE LEGENDRE POLYNOMIALS

where **i**, **j**, **k** are unit vectors in the coordinate directions, the divergence theorem, applied to each of these integrals, shows that

$$\int_S p\mathbf{n}\, dS = \mathbf{i}\int_V \frac{\partial p}{\partial x} dV + \mathbf{j}\int_V \frac{\partial p}{\partial y} dV + \mathbf{k}\int_V \frac{\partial p}{\partial z} dV = \int_V \nabla p\, dV,$$

and hence that

$$\int_V \left\{ \rho \frac{D\mathbf{u}}{Dt} + \nabla p \right\} dV = 0.$$

Since V is arbitrary,

$$\frac{D\mathbf{u}}{Dt} = -\frac{1}{\rho}\nabla p,$$

and for steady flows

$$(\mathbf{u}.\nabla)\mathbf{u} = -\frac{1}{\rho}\nabla p. \tag{2.17}$$

We earlier used the irrotational nature of the flow to write the velocity field as the gradient of a scalar potential, $\mathbf{u} = \nabla \phi$. Since $(\mathbf{u}\cdot\nabla)\mathbf{u} \equiv \nabla\left(\frac{1}{2}\mathbf{u}\cdot\mathbf{u}\right) - \mathbf{u}\times(\nabla\times\mathbf{u})$, for irrotational flow $(\mathbf{u}\cdot\nabla)\mathbf{u} = \nabla\left(\frac{1}{2}\mathbf{u}\cdot\mathbf{u}\right)$ and we can write (2.17) as

$$\nabla\left(\frac{1}{2}|\nabla\phi|^2\right) = -\frac{1}{\rho}\nabla p,$$

and hence

$$\nabla\left(\frac{p}{\rho} + \frac{1}{2}|\nabla\phi|^2\right) = 0,$$

which we can integrate to give

$$\frac{p}{\rho} + \frac{1}{2}|\nabla\phi|^2 = C, \tag{2.18}$$

which is known as **Bernoulli's equation**. Its implication for inviscid, irrotational flow is that the pressure can be calculated once we know the velocity potential, ϕ. In our example, ϕ is given by (2.16), so that

$$|\nabla\phi|^2 = \phi_r^2 + \frac{1}{r^2}\phi_\theta^2 = U^2\left\{\left(1 - \frac{a^3}{r^3}\right)^2 \cos^2\theta + \frac{1}{r^2}\left(r + \frac{a^3}{2r^2}\right)^2 \sin^2\theta\right\},$$

and Bernoulli's equation gives

$$p = p_\infty + \frac{1}{2}\rho U^2 \left[1 - \left\{\left(1 - \frac{a^3}{r^3}\right)^2 \cos^2\theta + \frac{1}{r^2}\left(r + \frac{a^3}{2r^2}\right)^2 \sin^2\theta\right\}\right],$$

where we have written p_∞ for the pressure at infinity. This expression simplifies on the surface of the sphere to give

$$p|_{r=a} = p_\infty + \frac{1}{2}\rho U^2 \left(1 - \frac{9}{4}\sin^2\theta\right).$$

This shows that the pressure is highest at $\theta = 0$ and π, the stream-facing poles of the sphere, with $p = p_\infty + \frac{1}{2}\rho U^2$, and drops below p_∞ over a portion of the rest of the boundary, with the lowest pressure, $p = p_\infty - \frac{5}{8}\rho U^2$, on the equator, $\theta = \pi/2$. The implications of this and the modelling assumptions made are discussed more fully by Acheson (1990).

2.7 The Associated Legendre Equation

We can also look for a separable solution of Laplace's equation in spherical polar coordinates (r, θ, ϕ), that is not necessarily axisymmetric. If we seek a solution of the form $y(\theta)\Phi(\phi)R(r)$, we find that $\Phi = e^{im\phi}$ with m an integer. The equation for y is, after using the change of variable $x = \cos\theta$,

$$(1-x^2)\frac{d^2y}{dx^2} - 2x\frac{dy}{dx} + \left\{n(n+1) - \frac{m^2}{1-x^2}\right\}y = 0. \qquad (2.19)$$

This equation is known as the **associated Legendre equation**. It trivially reduces to Legendre's equation when $m = 0$, corresponding to separable axisymmetric solutions of Laplace's equation. However, the connection is more profound than this.

If we define

$$Y = (1-x^2)^{-m/2}y,$$

Y is the solution of

$$(1-x^2)Y'' - 2(m+1)xY' + (n-m)(n+m+1)Y = 0. \qquad (2.20)$$

If we write Legendre's equation in the form

$$(1-x^2)Z'' - 2xZ' + n(n+1)Z = 0,$$

which has solution $Z = AP_n(x) + BQ_n(x)$, and differentiate m times, we get

$$(1-x^2)[Z^{(m)}]'' - 2(m+1)x[Z^{(m)}]' + (n-m)(n+m+1)Z^{(m)} = 0, \qquad (2.21)$$

where we have written $Z^{(m)} = d^m Z/dx^m$. A comparison of (2.20) and (2.21) shows that

$$Y = \frac{d^m}{dx^m}\left[AP_n(x) + BQ_n(x)\right],$$

and therefore that

$$y = (1-x^2)^{m/2}\left\{A\frac{d^m}{dx^m}P_n(x) + B\frac{d^m}{dx^m}Q_n(x)\right\}.$$

We have now shown that the solutions of the differential equation (2.19) are

$$y = AP_n^m(x) + BQ_n^m(x),$$

where

$$P_n^m(x) = (1-x^2)^{m/2}\frac{d^m}{dx^m}P_n(x), \quad Q_n^m(x) = (1-x^2)^{m/2}\frac{d^m}{dx^m}Q_n(x).$$

The functions $P_n^m(x)$ and $Q_n^m(x)$ are called the **associated Legendre functions**. Clearly, from what we know about the Legendre polynomials, $P_n^m(x) = 0$ if $m > n$

and $Q_n^m(x)$ is singular at $x = \pm 1$. It is straightforward to show from their definitions that

$$P_1^1(x) = (1-x^2)^{1/2} = \sin\theta, \quad P_2^1(x) = 3x(1-x^2)^{1/2} = 3\sin\theta\cos\theta = \frac{3}{2}\sin 2\theta,$$

$$P_2^2(x) = 3(1-x^2) = 3\sin^2\theta = \frac{3}{2}(1-\cos 2\theta),$$

$$P_3^1(x) = \frac{3}{2}(5x^2-1)(1-x^2)^{1/2} = \frac{3}{8}(\sin\theta + 5\sin 3\theta),$$

where $x = \cos\theta$.

There are various recurrence formulae and orthogonality relations between the associated Legendre functions, which can be derived in much the same way as those for the ordinary Legendre functions.

Example: Spherical harmonics

Let's try to find a representation of the solutions of Laplace's equation in three dimensions,

$$\frac{\partial^2 u}{\partial x^2} + \frac{\partial^2 u}{\partial y^2} + \frac{\partial^2 u}{\partial z^2} = 0,$$

that are homogeneous in x, y and z. By homogeneous we mean of the form $x^i y^j z^k$. It is simplest to work in spherical polar coordinates, in terms of which Laplace's equation takes the form

$$\frac{\partial}{\partial r}\left(r^2 \frac{\partial u}{\partial r}\right) + \frac{1}{\sin\theta}\frac{\partial}{\partial \theta}\left(\sin\theta\frac{\partial u}{\partial \theta}\right) + \frac{1}{\sin^2\theta}\frac{\partial^2 u}{\partial \phi^2} = 0.$$

A homogeneous solution of order $n = i + j + k$ in (x, y, z) will look, in the new coordinates, like $u = r^n S_n(\theta, \phi)$. Substituting this expression for u in Laplace's equation, we find that

$$\frac{1}{\sin\theta}\frac{\partial}{\partial \theta}\left(\sin\theta\frac{\partial S_n}{\partial \theta}\right) + \frac{1}{\sin^2\theta}\frac{\partial^2 S_n}{\partial \phi^2} + n(n+1)S_n = 0.$$

Separable solutions take the form

$$S_n = (A_m \cos m\phi + B_m \sin m\phi)F(\theta),$$

with m an integer. The function $F(\theta)$ satisfies

$$\frac{1}{\sin\theta}\frac{d}{d\theta}\left(\sin\theta\frac{dF}{d\theta}\right) + \left\{n(n+1) - \frac{m^2}{\sin^2\theta}\right\}F = 0.$$

If we now make the usual transformation, $\mu = \cos\theta$, $F(\theta) = y(\mu)$, we get

$$(1-\mu^2)y'' - 2\mu y' + \left\{n(n+1) - \frac{m^2}{1-\mu^2}\right\}y = 0,$$

54 **LEGENDRE FUNCTIONS**

which is the associated Legendre equation. Our solutions can therefore be written in the form

$$u = r^n(A_m \cos m\phi + B_m \sin m\phi)\{C_{n,m}P_n^m(\cos\theta) + D_{n,m}Q_n^m(\cos\theta)\}.$$

The quantities $r^n \cos m\phi \, P_n^m(\cos\theta)$, which appear as typical terms in the solution, are called **spherical harmonics** and appear widely in the solution of linear boundary value problems in spherical geometry. We will see how they arise in a quantum mechanical description of the hydrogen atom in Section 4.3.6.

Exercises

2.1 Solve Legendre's equation of order two, $(1-x^2)y'' - 2xy' + 6y = 0$, by the method of Frobenius. What is the simplest solution of this equation? By using this simple solution and the reduction of order method find a closed form expression for $Q_1(x)$.

2.2 Use the generating function to evaluate (a) $P_n'(1)$, (b) $P_n(0)$.

2.3 Prove that

(a) $\quad P_{n+1}'(x) - P_{n-1}'(x) = (2n+1)P_n(x),$

(b) $\quad (1-x^2)P_n'(x) = nP_{n-1}(x) - nxP_n(x).$

2.4 Find the first four nonzero terms in the Fourier–Legendre expansion of the function

$$f(x) = \begin{cases} 0 & \text{for } -1 < x < 0, \\ 1 & \text{for } 0 < x < 1. \end{cases}$$

What value will this series have at $x = 0$?

2.5 Establish the results

(a) $\quad \displaystyle\int_{-1}^{1} x P_n(x) P_{n-1}(x)\,dx = \frac{2n}{4n^2 - 1} \quad \text{for } n = 1, 2, \ldots,$

(b) $\quad \displaystyle\int_{-1}^{1} P_n(x) P_{n+1}'(x)\,dx = 2 \quad \text{for } n = 0, 1, \ldots,$

(c) $\quad \displaystyle\int_{-1}^{1} x P_n'(x) P_n(x)\,dx = \frac{2n}{2n+1} \quad \text{for } n = 0, 1, \ldots.$

2.6 Determine the Wronskian of P_n and Q_n for $n = 0, 1, 2, \ldots$.

2.7 Solve the axisymmetric boundary value problem for Laplace's equation,

$$\nabla^2 T = 0 \quad \text{for } 0 < r < a,\ 0 < \theta < \pi,$$
$$T(a, \theta) = 2\cos^5\theta.$$

2.8 Show that

(a) $P_3^2(x) = 15x(1-x^2),$

(b) $P_4^1(x) = \dfrac{5}{2}(7x^3 - 3x)(1-x^2)^{1/2}.$

2.9 * Prove that $|P_n'(x)| < n^2$ and $|P_n''(x)| < n^4$ for $-1 < x < 1$.

2.10 Derive Equation (2.10).

2.11 * Find the solution of the Dirichlet problem, $\nabla^2 \Phi = 0$ in $r > 2$ subject to $\Phi \to 0$ as $r \to \infty$ and $\Phi(2, \theta, \phi) = \sin^2 \theta \cos 2\phi$.

2.12 * The self-adjoint form of the associated Legendre equation is

$$\frac{d}{dx}\{(1-x^2)P_n^m{}'(x)\} + \left\{n(n+1) - \frac{m^2}{1-x^2}\right\} P_n^m(x) = 0.$$

Using this directly, prove the orthogonality property

$$\int_{-1}^{1} P_l^m(x) P_n^m(x)\, dx = 0 \quad \text{for } l \neq n.$$

Evaluate

$$\int_{-1}^{1} [P_m^m(x)]^2\, dx.$$

2.13 (a) Suppose $P_n(x_0) = 0$ for some $x_0 \in (-1, 1)$. Show that x_0 is a simple zero.

(b) Show that P_n with $n \geq 1$ has n distinct zeros in $(-1, 1)$.

2.14 **Project** A simplified model for the left ventricle of a human heart is provided by a sphere of time-dependent radius $R = R(t)$ with a circular aortic opening of constant area A, as shown in Figure 2.5. During contraction we suppose that the opening remains fixed whilst the centre of the sphere moves directly toward the centre of the opening and the radius $R(t)$ decreases accordingly. As a result, some of the blood filling the ventricle cavity is ejected though the opening with mean speed $U = U(t)$ into the attached cylindrical aorta. This occurs sufficiently rapidly that we can assume that the flow is inviscid, irrotational, incompressible, and symmetric with respect to the aortal axis.

(a) State a partial differential equation appropriate to the fluid flow for this situation.

(b) During contraction, show that a point on the surface of the ventricle has velocity

$$\dot R \mathbf{n} - (\dot R a)\mathbf{i},$$

where $\mathbf{n}$ is the outward unit normal at time t, $\mathbf{i}$ is the unit vector in the aortic flow direction and $a = \cos \alpha(t)$. Show that having $(R \sin \alpha)^2 = R^2(1 - a^2)$ constant gives

$$\frac{\partial \phi}{\partial n} = f(s) = \begin{cases} Us & \text{for } a < s < 1, \\ \dot R(1 - s/a) & \text{for } -1 < s < a, \end{cases}$$

where $s = \cos \theta$ with θ the usual spherical polar coordinate.

56 LEGENDRE FUNCTIONS

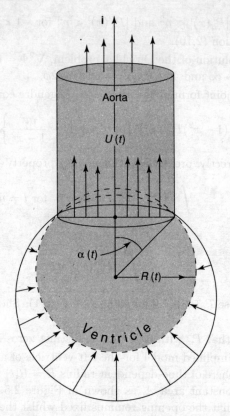

Fig. 2.5. A simple model for the left ventricle of the human heart.

(c) Show that, for a solution to exist, $\int_{-1}^{1} f(s)\,ds = 0$. Deduce that $\dot{R} = [a(a-1)/(a+1)]U$, which relates the geometry to the mean aortal speed.

(d) Let $V = V(t)$ denote both the interior of the sphere at time t and its volume. The total momentum in the direction of $\mathbf{i}$ is the blood-density times the integral

$$I = \int_V \nabla \phi \cdot \mathbf{i}\, dV = \int_S (s\phi)|_{r=R}\, dS,$$

where S is the surface of V. Hence show that

$$I = 2\pi R^2 \sum_{n=0}^{\infty} c_n \int_{-1}^{1} s P_n(s)\,ds = \frac{4}{3}\pi R^2 c_1 = \frac{3}{2} V \int_{-1}^{1} s f(s)\, ds.$$

(e) Use the answer to part (c) to show that $4I = (1-a)(a^2+4+3a)VU$.

(f) Explain how this model could be made more realistic in terms of the fluid mechanics and the physiology. You may like to refer to Pedley (1980) for some ideas on this.

CHAPTER THREE

Bessel Functions

In this chapter, we will discuss a class of functions known as **Bessel functions**. These are named after the German mathematician and astronomer Friedrich Bessel, who first used them to analyze planetary orbits, as we shall discuss later. Bessel functions occur in many other physical problems, usually in a cylindrical geometry, and we will discuss some examples of these at the end of this chapter.

Bessel's equation can be written in the form

$$x^2 \frac{d^2 y}{dx^2} + x \frac{dy}{dx} + \left(x^2 - \nu^2\right) y = 0, \tag{3.1}$$

with ν real and positive. Note that (3.1) has a regular singular point at $x = 0$. Using the notation of Chapter 1,

$$\frac{xQ}{P} = \frac{x^2}{x^2} = 1, \quad \frac{x^2 R}{P} = \frac{x^2 \left(x^2 - \nu^2\right)}{x^2} = x^2 - \nu^2,$$

both of which are polynomials and have Taylor expansions with infinite radii of convergence. Any series solution will therefore also have an infinite radius of convergence.

3.1 The Gamma Function and the Pockhammer Symbol

Before we use the method of Frobenius to construct the solutions of Bessel's equation, it will be useful for us to make a couple of definitions. The **gamma function** is defined by

$$\Gamma(x) = \int_0^\infty e^{-q} q^{x-1} \, dq, \quad \text{for } x > 0. \tag{3.2}$$

Note that the integration is over the dummy variable q and x is treated as constant during the integration. We will start by considering the function evaluated at $x = 1$. By definition,

$$\Gamma(1) = \int_0^\infty e^{-q} \, dq = 1.$$

We also note that

$$\Gamma(x+1) = \int_0^\infty e^{-q} q^x \, dq,$$

3.1 THE GAMMA FUNCTION AND THE POCKHAMMER SYMBOL

which can be integrated by parts to give

$$\Gamma(x+1) = \left[-q^x e^{-q}\right]_0^\infty + \int_0^\infty e^{-q} x q^{x-1}\, dq = x \int_0^\infty e^{-q} q^{x-1}\, dq = x\Gamma(x).$$

We therefore have the recursion formula

$$\Gamma(x+1) = x\Gamma(x). \tag{3.3}$$

Suppose that $x = n$ is a positive integer. Then

$$\Gamma(n+1) = n\Gamma(n) = n(n-1)\Gamma(n-1) = \cdots = n(n-1)\ldots 2\cdot 1 = n!. \tag{3.4}$$

We therefore have the useful result, $\Gamma(n+1) = n!$ for n a positive integer.

We will often need to know $\Gamma(1/2)$. Firstly, consider the definition,

$$\Gamma\left(\frac{1}{2}\right) = \int_0^\infty e^{-q} q^{-1/2}\, dq.$$

If we introduce the new variable $Q = \sqrt{q}$, so that $dQ = \frac{1}{2}q^{-1/2}dq$, this integral becomes

$$\Gamma\left(\frac{1}{2}\right) = 2\int_0^\infty e^{-Q^2}\, dQ.$$

We can also write this integral in terms of another new variable, $Q = \tilde{Q}$, to obtain

$$\left\{\Gamma\left(\frac{1}{2}\right)\right\}^2 = \left(2\int_0^\infty e^{-Q^2}\, dQ\right)\left(2\int_0^\infty e^{-\tilde{Q}^2}\, d\tilde{Q}\right).$$

Since the limits are independent, we can combine the integrals as

$$\left\{\Gamma\left(\frac{1}{2}\right)\right\}^2 = 4\int_0^\infty \int_0^\infty e^{-(Q^2+\tilde{Q}^2)}\, dQ\, d\tilde{Q}.$$

If we now change to standard polar coordinates we have $dQ\, d\tilde{Q} = r\, dr\, d\theta$, where $Q = r\cos\theta$ and $\tilde{Q} = r\sin\theta$, and hence

$$\left\{\Gamma\left(\frac{1}{2}\right)\right\}^2 = 4\int_{\theta=0}^{\pi/2} \int_{r=0}^\infty e^{-r^2} r\, dr\, d\theta.$$

The limits of integration give us the positive quadrant of the $(Q,\tilde{Q})$-plane, as required. Performing the integration over θ we have

$$\left\{\Gamma\left(\frac{1}{2}\right)\right\}^2 = 2\pi \int_0^\infty re^{-r^2}\, dr,$$

and integrating with respect to r gives

$$\left\{\Gamma\left(\frac{1}{2}\right)\right\}^2 = 2\pi \left[-\frac{1}{2}e^{-r^2}\right]_0^\infty = \pi.$$

Finally, we have

$$\Gamma\left(\frac{1}{2}\right) = 2\int_0^\infty e^{-Q^2}\, dQ = \sqrt{\pi}. \tag{3.5}$$

60 BESSEL FUNCTIONS

We can use $\Gamma(x) = \Gamma(x+1)/x$ to define $\Gamma(x)$ for negative values of x. For example,

$$\Gamma\left(-\frac{1}{2}\right) = \frac{\Gamma\left(-\frac{1}{2}+1\right)}{-\frac{1}{2}} = -2\Gamma\left(\frac{1}{2}\right) = -2\sqrt{\pi}.$$

We also find that $\Gamma(x)$ is singular at $x = 0$. From the definition, (3.2), the integrand diverges like $1/q$ as $q \to 0$, which is not integrable. Alternatively, $\Gamma(x) = \Gamma(x+1)/x$ shows that $\Gamma(x) \sim 1/x$ as $x \to 0$. Note that the gamma function is available in MATLAB as the function **gamma**.

The **Pockhammer symbol** is a simple way of writing down long products. It is defined as

$$(\alpha)_r = \alpha(\alpha+1)(\alpha+2)\ldots(\alpha+r-1),$$

so that, for example, $(\alpha)_1 = \alpha$ and $(\alpha)_2 = \alpha(\alpha+1)$, and, in general, $(\alpha)_r$ is a product of r terms. We also choose to define $(\alpha)_0 = 1$. Note that $(1)_n = n!$. A relationship between the gamma function and the Pockhammer symbol that we will need later is

$$\Gamma(x)\,(x)_n = \Gamma(x+n) \qquad (3.6)$$

for x real and n a positive integer. To derive this, we start with the definition of the Pockhammer symbol,

$$(x)_n = x(x+1)(x+2)\ldots(x+n-1).$$

Now

$$\Gamma(x)\,(x)_n = \Gamma(x)\,\{x(x+1)(x+2)\ldots(x+n-1)\}$$

$$= \{\Gamma(x)x\}\,\{(x+1)(x+2)\ldots(x+n-1)\}.$$

Using the recursion relation (3.3),

$$\Gamma(x)\,(x)_n = \Gamma(x+1)\,\{(x+1)(x+2)\ldots(x+n-1)\}.$$

We can repeat this to give

$$\Gamma(x)\,(x)_n = \Gamma(x+n-1)\,(x+n-1) = \Gamma(x+n).$$

3.2 Series Solutions of Bessel's Equation

We can now proceed to consider a Frobenius solution,

$$y(x) = \sum_{n=0}^{\infty} a_n x^{n+c}.$$

When substituted into (3.1), this yields

$$x^2 \sum_{n=0}^{\infty} a_n(n+c)(n+c-1)x^{n+c-2} + x\sum_{n=0}^{\infty} a_n(n+c)x^{n+c-1}$$

3.2 SERIES SOLUTIONS OF BESSEL'S EQUATION

$$+(x^2 - \nu^2) \sum_{n=0}^{\infty} a_n x^{n+c} = 0.$$

We can now rearrange this equation and combine corresponding terms to obtain

$$\sum_{n=0}^{\infty} a_n \{(n+c)^2 - \nu^2\} x^{n+c} + \sum_{n=0}^{\infty} a_n x^{n+c+2} = 0.$$

We can extract two terms from the first summation and then shift the second one to obtain

$$a_0(c^2 - \nu^2)x^c + a_1\{(1+c)^2 - \nu^2\}x^{c+1}$$

$$+ \sum_{n=2}^{\infty} [a_n\{(n+c)^2 - \nu^2\} + a_{n-2}] x^{n+c} = 0. \tag{3.7}$$

The indicial equation is therefore $c^2 - \nu^2 = 0$, so that $c = \pm \nu$.

We can now distinguish various cases. We start with the case for which the difference between the two roots of the indicial equation, 2ν, is not an integer. Using Frobenius General Rule II, we can consider both roots of the indicial equation at once. From the second term of (3.7), which is proportional to x^{c+1}, we have

$$a_1\{(1 \pm \nu)^2 - \nu^2\} = a_1(1 \pm 2\nu) = 0,$$

which implies that $a_1 = 0$ since 2ν is not an integer. The recurrence relation that we obtain from the general term of (3.7) is

$$a_n \{(n \pm \nu)^2 - \nu^2\} + a_{n-2} = 0 \text{ for } n = 2, 3, \ldots,$$

and hence $a_n = 0$ for n odd. Note that it is not possible for the series solution to terminate and give a polynomial solution, which is what happens to give the Legendre polynomials, which we studied in the previous chapter. This makes the Bessel functions rather more difficult to study.

We will now determine an expression for the value of a_n for general values of the parameter ν. The recurrence relation gives us

$$a_n = -\frac{a_{n-2}}{n(n \pm 2\nu)}.$$

Let's start with $n = 2$, which yields

$$a_2 = -\frac{a_0}{2(2 \pm 2\nu)},$$

and now with $n = 4$,

$$a_4 = -\frac{a_2}{4(4 \pm 2\nu)}.$$

Substituting for a_2 in terms of a_0 gives

$$a_4 = \frac{a_0}{(4 \pm 2\nu)(2 \pm 2\nu) \cdot 4 \cdot 2} = \frac{a_0}{2^2(2 \pm \nu)(1 \pm \nu)2^2(2 \cdot 1)}.$$

BESSEL FUNCTIONS

We can continue this process, and we find that

$$a_{2n} = (-1)^n \frac{a_0}{2^{2n} (1 \pm \nu)_n \, n!}, \tag{3.8}$$

where we have used the Pockhammer symbol to simplify the expression. From this expression for a_{2n},

$$y(x) = a_0 x^{\pm \nu} \sum_{n=0}^{\infty} (-1)^n \frac{x^{2n}}{2^{2n} (1 \pm \nu)_n \, n!}.$$

With a suitable choice of a_0 we can write this as

$$y(x) = A \frac{x^{\pm \nu}}{2^{\pm \nu} \Gamma(1 \pm \nu)} \sum_{n=0}^{\infty} (-1)^n \frac{(x^2/4)^n}{(1 \pm \nu)_n \, n!} = A J_{\pm \nu}(x).$$

These are the Bessel functions of order $\pm \nu$. The general solution of Bessel's equation, (3.1), is therefore

$$y(x) = A J_\nu(x) + B J_{-\nu}(x),$$

for arbitrary constants A and B, with

$$J_{\pm \nu}(x) = \frac{x^{\pm \nu}}{2^{\pm \nu} \Gamma(1 \pm \nu)} \sum_{n=0}^{\infty} (-1)^n \frac{(x^2/4)^n}{(1 \pm \nu)_n \, n!}. \tag{3.9}$$

Remember that 2ν is not an integer.

Let's now consider what happens when $\nu = 0$, in which case the indicial equation has a repeated root and we need to apply Frobenius General Rule III. By setting $\nu = 0$ in the expression (3.8) for a_n and exploiting the fact that $(1)_n = n!$, one solution is

$$J_0(x) = \sum_{n=0}^{\infty} (-1)^n \frac{1}{(n!)^2} \left(\frac{x^2}{4} \right)^n.$$

Using Frobenius General Rule III, we can show that the other solution is

$$Y_0(x) = J_0(x) \log x - \sum_{n=0}^{\infty} (-1)^n \frac{\phi(n)}{(n!)^2} \left(\frac{x^2}{4} \right)^n,$$

which is called **Weber's Bessel function** of order zero. This expression can be derived by evaluating $\partial y / \partial c$ at $c = 0$. Note that we have made use of the function $\phi(n)$ defined in (1.21).

We now consider the case for which 2ν is a nonzero integer, beginning with 2ν an odd integer, $2n + 1$. In this case, the solution takes the form

$$y(x) = A J_{n+1/2}(x) + B J_{-n-1/2}(x).$$

As an example, let's consider the case $\nu = \frac{1}{2}$ so that Bessel's equation is

$$\frac{d^2 y}{dx^2} + \frac{1}{x} \frac{dy}{dx} + \left(1 - \frac{1}{4x^2} \right) y = 0.$$

We considered this example in detail in Section 1.3.1, and found that

$$y(x) = a_0 \frac{\cos x}{x^{1/2}} + a_1 \frac{\sin x}{x^{1/2}}.$$

This means that (see Exercise 3.2)

$$J_{1/2}(x) = \sqrt{\frac{2}{\pi x}} \sin x, \quad J_{-1/2}(x) = \sqrt{\frac{2}{\pi x}} \cos x.$$

The recurrence relations (3.21) and (3.22), which we will derive later, then show that $J_{n+1/2}(x)$ and $J_{-n-1/2}(x)$ are products of finite polynomials with $\sin x$ and $\cos x$.

Finally we consider what happens when 2ν is an even integer, and hence ν is an integer. A rather lengthy calculation allows us to write the solution in the form

$$y = AJ_\nu(x) + BY_\nu(x),$$

where Y_ν is Weber's Bessel function of order ν defined as

$$Y_\nu(x) = \frac{J_\nu(x) \cos \nu\pi - J_{-\nu}(x)}{\sin \nu\pi}. \tag{3.10}$$

Notice that the denominator of this expression is obviously zero when ν is an integer, so this case requires careful treatment. We note that the second solution of Bessel's equation can also be determined using the method of reduction of order as

$$y(x) = AJ_\nu(x) + BJ_\nu(x) \int^x \frac{1}{q\, J_\nu(q)^2}\, dq.$$

In Figure 3.1 we show $J_i(x)$ for $i = 0$ to 3. Note that $J_0(0) = 1$, but that $J_i(0) = 0$ for $i > 0$, and that $J_i^{(j)}(0) = 0$ for $j < i$, $i > 1$. We generated Figure 3.1 using the MATLAB script

```
x=0:0.02:20;
subplot(2,2,1), plot(x,besselj(0,x)), title('J_0(x)')
subplot(2,2,2), plot(x,besselj(1,x)), title('J_1(x)')
subplot(2,2,3), plot(x,besselj(2,x)), title('J_2(x)')
subplot(2,2,4), plot(x,besselj(3,x)), title('J_3(x)')
```

We produced Figures 3.2, 3.5 and 3.6 in a similar way, using the MATLAB functions bessely, besseli and besselk.

In Figure 3.2 we show the first two Weber's Bessel functions of integer order. Notice that as $x \to 0$, $Y_n(x) \to -\infty$. As you can see, all of these Bessel functions are oscillatory. The first three zeros of $J_0(x)$ are 2.4048, 5.5201 and 8.6537, whilst the first three nontrivial zeros of $J_1(x)$ are 3.8317, 7.0156 and 10.1735, all to four decimal places.

BESSEL FUNCTIONS

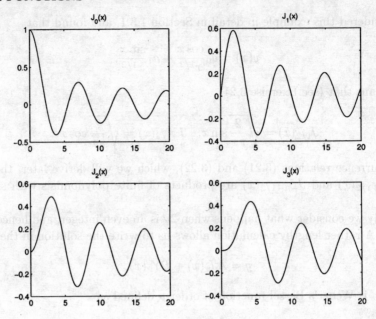

Fig. 3.1. The functions $J_0(x)$, $J_1(x)$, $J_2(x)$ and $J_3(x)$.

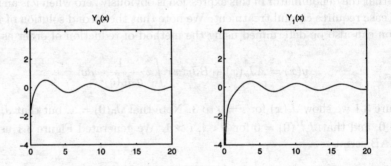

Fig. 3.2. The functions $Y_0(x)$ and $Y_1(x)$.

3.3 The Generating Function for $J_n(x)$, n an integer

Rather like the Legendre polynomials, there is a simple function that will generate all of the Bessel functions of integer order. In order to establish this, it is useful to manipulate the definition of the $J_\nu(x)$. From (3.9), we have, for $\nu = n$,

$$J_n(x) = \frac{x^n}{2^n \Gamma(1+n)} \sum_{i=0}^{\infty} \frac{(-x^2/4)^i}{i!(1+n)_i} = \sum_{i=0}^{\infty} \frac{(-1)^i x^{2i+n}}{2^{2i+n} i!(1+n)_i \Gamma(1+n)}.$$

3.3 THE GENERATING FUNCTION FOR $J_n(x)$, n AN INTEGER

Using (3.6) we note that $\Gamma(1+n)(1+n)_i = \Gamma(1+n+i)$, and using (3.4) we find that this is equal to $(n+i)!$. For n an integer, we can therefore write

$$J_n(x) = \sum_{i=0}^{\infty} \frac{(-1)^i x^{2i+n}}{2^{2i+n} i!(n+i)!}. \tag{3.11}$$

Let's now consider the generating function

$$g(x,t) = \exp\left\{\frac{1}{2}x\left(t - \frac{1}{t}\right)\right\}. \tag{3.12}$$

The series expansions of each of the constituents of this are

$$\exp(xt/2) = \sum_{j=0}^{\infty} \frac{(xt/2)^j}{j!}, \quad \exp\left(-\frac{x}{2t}\right) = \sum_{i=0}^{\infty} \frac{(-x/2t)^i}{i!},$$

both from the Taylor series for e^x. These summations can be combined to produce

$$g(x,t) = \sum_{j=0}^{\infty} \sum_{i=0}^{\infty} \frac{(-1)^i x^{i+j} t^{j-i}}{2^{j+i} i! j!}.$$

Now, putting $j = i + n$ so that $-\infty \leqslant n \leqslant \infty$, this becomes

$$g(x,t) = \sum_{n=-\infty}^{\infty} \left\{ \sum_{i=0}^{\infty} \frac{(-1)^i x^{2i+n}}{2^{2i+n} i!(n+i)!} \right\} t^n.$$

Comparing the coefficients in this series with the expression (3.11) we find that

$$g(x,t) = \exp\left\{\frac{1}{2}x\left(t - \frac{1}{t}\right)\right\} = \sum_{n=-\infty}^{\infty} J_n(x) t^n. \tag{3.13}$$

We can now exploit this relation to study Bessel functions.

Using the fact that the generating function is invariant under $t \mapsto -1/t$, we have

$$\sum_{n=-\infty}^{\infty} J_n(x) t^n = \sum_{n=-\infty}^{\infty} J_n(x) \left(-\frac{1}{t}\right)^n = \sum_{n=-\infty}^{\infty} J_n(x)(-1)^n t^{-n}.$$

Now putting $m = -n$, this is equal to

$$\sum_{m=\infty}^{-\infty} J_{-m}(x)(-1)^m t^m.$$

Now let $n = m$ in the series on the right hand side, which gives

$$\sum_{n=-\infty}^{\infty} J_n(x) t^n = \sum_{n=-\infty}^{\infty} J_{-n}(x)(-1)^n t^n.$$

Comparing like terms in the series, we find that $J_n(x) = (-1)^n J_{-n}(x)$, and hence that $J_n(x)$ and $J_{-n}(x)$ are linearly dependent over $\mathbb{R}$ (see Section 1.2.1). This explains why the solution of Bessel's equation proceeds rather differently when ν is an integer, since $J_\nu(x)$ and $J_{-\nu}(x)$ cannot then be independent solutions.

The generating function can also be used to derive an integral representation of $J_n(x)$. The first step is to put $t = e^{\pm i\theta}$ in (3.13), which yields

$$e^{\pm ix\sin\theta} = J_0(x) + \sum_{n=1}^{\infty}(\pm 1)^n \left\{e^{ni\theta} + (-1)^n e^{-ni\theta}\right\} J_n(x),$$

or, in terms of sine and cosine,

$$e^{\pm ix\sin\theta} = J_0(x) + 2\sum_{n=1}^{\infty} J_{2n}(x)\cos 2n\theta \pm i\sum_{n=0}^{\infty} J_{2n+1}(x)\sin(2n+1)\theta.$$

Appropriate combinations of these two cases show that

$$\cos(x\sin\theta) = J_0(x) + 2\sum_{n=1}^{\infty} J_{2n}(x)\cos 2n\theta, \tag{3.14}$$

$$\sin(x\sin\theta) = 2\sum_{n=0}^{\infty} J_{2n+1}(x)\sin(2n+1)\theta, \tag{3.15}$$

and, substituting $\eta = \frac{\pi}{2} - \theta$, we obtain

$$\cos(x\cos\eta) = J_0(x) + 2\sum_{n=1}^{\infty}(-1)^n J_{2n}(x)\cos 2n\eta, \tag{3.16}$$

$$\sin(x\cos\eta) = 2\sum_{n=0}^{\infty}(-1)^n J_{2n+1}(x)\cos(2n+1)\eta. \tag{3.17}$$

Multiplying (3.16) by $\cos m\eta$ and integrating from zero to π, we find that

$$\int_{\eta=0}^{\pi} \cos m\eta \cos(x\cos\eta)\, d\eta = \begin{cases} \pi J_m(x) & \text{for } m \text{ even,} \\ 0 & \text{for } m \text{ odd.} \end{cases}$$

Similarly we find that

$$\int_{\eta=0}^{\pi} \sin m\eta \sin(x\cos\eta)\, d\eta = \begin{cases} 0 & \text{for } m \text{ even,} \\ \pi J_m(x) & \text{for } m \text{ odd,} \end{cases}$$

from (3.17). Adding these expressions gives us the integral representation

$$J_n(x) = \frac{1}{\pi}\int_0^{\pi} \cos(n\theta - x\sin\theta)\, d\theta. \tag{3.18}$$

We shall now describe a problem concerning planetary motion that makes use of (3.18). This is the context in which Bessel originally studied these functions.

Example: Planetary motion

We consider the motion of a planet under the action of the gravitational force exerted by the Sun. In doing this, we neglect the effect of the gravitational attraction of the other bodies in the solar system, which are all much less massive than the Sun. Under this approximation, it is straightforward to prove Kepler's first and second laws, that the planet moves on an ellipse, with the Sun at one of its foci,

3.3 THE GENERATING FUNCTION FOR $J_n(x)$, n AN INTEGER

and that the line from the planet to the Sun (PS in Figure 3.3) sweeps out equal areas of the ellipse in equal times (see, for example, Lunn, 1990). Our aim now is to use Kepler's first and second laws to obtain a measure of how the passage of the planet around its orbit depends upon time. We will denote the length of the

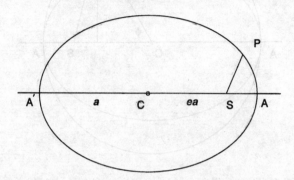

Fig. 3.3. The elliptical orbit of a planet, P, around the Sun at a focus, S. Here, C is the centre of the ellipse and A and A' the extrema of the orbit on the major axis of the ellipse.

semi-major axis, $A'C$, by a and the eccentricity by e. Note that distance of the Sun from the centre of the ellipse, SC, is ea. We also define the **mean anomaly** to be

$$\mu = 2\pi \frac{\text{Area of the elliptic sector } ASP}{\text{Area of the ellipse}},$$

which Kepler's second law tells us is proportional to the time of passage from A to P.

Let's now consider the **auxiliary circle**, which has centre C and passes through A and A', as shown in Figure 3.4. We label the projection of the point P onto the auxiliary circle as Q. We will also need to introduce the **eccentric anomaly** of P, which is defined to be the angle ACQ, and can be written as

$$\phi = 2\pi \frac{\text{Area of the sector } ACQ}{\text{Area of the auxiliary circle}}.$$

We now note that, by orthogonal projection, the ratio of the area of ASP to that of the ellipse is the same as the ratio of the area of ASQ to that of the auxiliary circle. The area of ASQ is given by the area of the sector ACQ ($\frac{1}{2}\phi a^2$) minus the area of the triangle CSQ ($\frac{1}{2}ea^2 \sin\phi$), so that

$$\frac{\mu}{\pi} = \frac{\frac{1}{2}a^2\phi - \frac{1}{2}ea^2 \sin\phi}{\frac{1}{2}\pi a^2},$$

and hence

$$\mu = \phi - e \sin\phi. \tag{3.19}$$

Now, in order to determine ϕ as a function of μ, we note that $\phi - \mu$ is a periodic

BESSEL FUNCTIONS

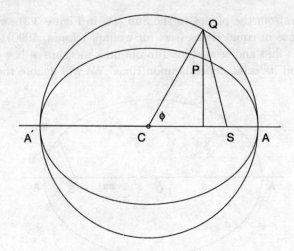

Fig. 3.4. The auxiliary circle and the projection of P onto Q.

function of μ, which vanishes when P and Q are coincident with A or A', that is when μ is an integer multiple of π. Hence we must be able to write

$$\phi - \mu = \sum_{n=1}^{\infty} A_n \sin n\mu. \tag{3.20}$$

As we shall see in Chapter 5, this is a Fourier series. In order to determine the constant coefficients A_n, we differentiate (3.20) with respect to μ to yield

$$\frac{d\phi}{d\mu} - 1 = \sum_{n=1}^{\infty} nA_n \cos n\mu.$$

We can now exploit the orthogonality of the functions $\cos n\mu$ to determine A_n. We multiply through by $\cos m\mu$ and integrate from zero to π to obtain

$$\int_{\mu=0}^{\pi} \left(\frac{d\phi}{d\mu} - 1\right) \cos m\mu \, d\mu = \int_{\mu=0}^{\pi} \cos m\mu \frac{d\phi}{d\mu} d\mu = \frac{\pi m}{2} A_m.$$

Since $\phi = 0$ when $\mu = 0$ and $\phi = \pi$ when $\mu = \pi$, we can change the independent variable to give

$$\frac{\pi m}{2} A_m = \int_{\phi=0}^{\pi} \cos m\mu \, d\phi.$$

Substituting for μ from (3.19), we have that

$$A_m = \frac{2}{m\pi} \int_{\phi=0}^{\pi} \cos m \left(\phi - e \sin \phi\right) d\phi.$$

Finally, by direct comparison with (3.18), $A_m = \frac{2}{m} J_m(me)$, so that

$$\phi = \mu + 2 \left\{ J_1(e) \sin \mu + \frac{1}{2} J_2(2e) \sin 2\mu + \frac{1}{3} J_3(3e) \sin 3\mu + \cdots \right\}.$$

3.4 DIFFERENTIAL AND RECURRENCE RELATIONS

Since μ is proportional to the time the planet takes to travel from A to P, this expression gives us the variation of the angle ϕ with time.

3.4 Differential and Recurrence Relations Between Bessel Functions

It is often useful to find relationships between Bessel functions with different indices. We will derive two such relationships. We start with (3.9), multiply by x^ν and differentiate to obtain

$$\frac{d}{dx}[x^\nu J_\nu(x)] = \frac{d}{dx}\left\{\sum_{n=0}^{\infty}\frac{(-1)^n\, x^{2n+2\nu}}{2^{2n+\nu}\, n!\,\Gamma(1+\nu+n)}\right\} = \sum_{n=0}^{\infty}\frac{(-1)^n\,(2n+2\nu)\,x^{2n+2\nu-1}}{2^{2n+\nu}\, n!\,\Gamma(1+\nu+n)}.$$

Since $\Gamma(1+\nu+n) = (n+\nu)\Gamma(n+\nu)$, this gives a factor that cancels with the term $2(n+\nu)$ in the numerator to give

$$\frac{d}{dx}[x^\nu J_\nu(x)] = \sum_{n=0}^{\infty}\frac{(-1)^n x^{2n}\, x^\nu\, x^{\nu-1}}{2^{2n+\nu-1}\, n!\,\Gamma(\nu+n)}.$$

We can rewrite this so that we have the series expansion for $J_{\nu-1}(x)$, as

$$\frac{d}{dx}[x^\nu J_\nu(x)] = \frac{x^\nu x^{\nu-1}}{2^{\nu-1}}\sum_{n=0}^{\infty}\frac{(-1)^n\, x^{2n}}{2^{2n}\, n!\,\Gamma(\nu)(\nu)_n},$$

so that

$$\frac{d}{dx}\{x^\nu J_\nu(x)\} = x^\nu J_{\nu-1}(x). \tag{3.21}$$

Later, we will use this expression to develop relations between general Bessel functions. Note that by putting ν equal to zero

$$\frac{d}{dx}\{J_0(x)\} = J_{-1}(x).$$

However, we recall that $J_n(x) = (-1)^n J_{-n}(x)$ for n an integer, so that $J_1(x) = -J_{-1}(x)$ and hence

$$J_0'(x) = -J_1(x),$$

where we have used a prime to denote the derivative with respect to x.

In the same vein as the derivation of (3.21),

$$\frac{d}{dx}(x^{-\nu}J_\nu(x)) = \frac{d}{dx}\left(\frac{x^{-\nu}x^\nu}{2^\nu\Gamma(1+\nu)}\sum_{n=0}^{\infty}\frac{(-x^2/4)^n}{n!\,(1+\nu)_n}\right)$$

$$= \frac{d}{dx}\left(\sum_{n=0}^{\infty}\frac{(-1)^n\, x^{2n}}{2^{2n+\nu}\, n!\,\Gamma(1+\nu+n)}\right) = \sum_{n=0}^{\infty}\frac{(-1)^n\, x^{2n-1}\, n}{2^{2n+\nu-1}\, n!\,\Gamma(1+\nu+n)}.$$

Notice that the first term in this series is zero (due to the factor of n in the numerator), so we can start the series at $n=1$ and cancel the factors of n. The series

BESSEL FUNCTIONS

can then be expressed in terms of the dummy variable $m = n - 1$ as

$$\frac{d}{dx}(x^{-\nu}J_\nu(x)) = \frac{1}{2^{\nu-1}} \sum_{m=0}^{\infty} \frac{(-1)^{m+1} x^{2m+1}}{2^{2m+2} \, m! \, \Gamma(\nu + m + 2)}$$

$$= -\frac{1}{2^{\nu+1}} \sum_{m=0}^{\infty} \frac{(-1)^m x^{2m+1}}{2^{2m} \, m! \, \Gamma(\nu + m + 2)}.$$

Using the fact that $\Gamma(\nu + m + 2) = \Gamma(\nu + 2)(\nu + 2)_m$ and the series expansion of $J_{\nu+1}(x)$, (3.11), we have

$$\frac{d}{dx}\{x^{-\nu}J_\nu(x)\} = -\frac{x}{2^{\nu+1}\Gamma(\nu+1)} \sum_{m=0}^{\infty} \frac{(-x^2/4)^m}{m!\,(2+\nu)_m},$$

and consequently

$$\frac{d}{dx}\{x^{-\nu}J_\nu(x)\} = -x^{-\nu}J_{\nu+1}(x). \qquad (3.22)$$

Notice that (3.21) and (3.22) both hold for $Y_\nu(x)$ as well.

We can use these relationships to derive **recurrence relations** between the Bessel functions. We expand the differentials in each expression to give the equations

$$J'_\nu(x) + \frac{\nu}{x}J_\nu(x) = J_{\nu-1}(x),$$

where we have divided through by x^ν, and

$$J'_\nu(x) - \frac{\nu}{x}J_\nu(x) = -J_{\nu+1}(x),$$

where this time we have multiplied by x^ν. By adding these expressions we find that

$$J'_\nu(x) = \frac{1}{2}\{J_{\nu-1}(x) - J_{\nu+1}(x)\},$$

and by subtracting them

$$\frac{2\nu}{x}J_\nu(x) = J_{\nu-1}(x) + J_{\nu+1}(x),$$

which is a pure recurrence relationship.

These results can also be used when integrating Bessel functions. For example, consider the integral

$$I = \int x J_0(x)\, dx.$$

This can be integrated using (3.21) with $\nu = 1$, since

$$I = \int x J_0(x)\, dx = \int (xJ_1(x))' \, dx = xJ_1(x).$$

3.5 Modified Bessel Functions

We will now consider the solutions of the **modified Bessel equation**,

$$x^2 \frac{d^2y}{dx^2} + x\frac{dy}{dx} - (x^2 + \nu^2)y = 0, \qquad (3.23)$$

which can be derived from Bessel's equation using $x \mapsto ix$, so that the solutions could be written down as conventional Bessel functions with purely imaginary arguments. However, it is more transparent to introduce the **modified Bessel function of first kind** of order ν, $I_\nu(x)$, so that the complete solution of (3.23) is

$$y(x) = AI_\nu(x) + BI_{-\nu}(x),$$

provided ν is not an integer. As was the case when we introduced the function $Y_\nu(x)$, there is a corresponding function here $K_\nu(x)$, the **modified Bessel function of second kind** of order ν. This is defined as

$$K_\nu(x) = \frac{\pi}{2\sin\nu\pi}\{I_{-\nu}(x) - I_\nu(x)\}.$$

Note that the slight difference between the definition of this Bessel function and that of Weber's Bessel function of order ν, (3.10), occurs because these functions must agree with the definition of the ordinary Bessel functions when ν is an integer. Most of the results we have derived in this chapter can be modified by changing the argument from x to ix in deriving, for example, the recurrence relations. The first few modified Bessel functions are shown in Figures 3.5 and 3.6. Note the contrast in behaviour between these and the Bessel functions $J_\nu(x)$ and $Y_\nu(x)$.

Equations (3.21) and (3.22) also hold for the modified Bessel functions and are given by

$$\frac{d}{dx}\{x^\nu I_\nu(x)\} = x^\nu I_{\nu-1}(x), \quad \frac{d}{dx}\{x^{-\nu}I_\nu(x)\} = x^{-\nu}I_{\nu+1}(x)$$

and

$$\frac{d}{dx}\{x^\nu K_\nu(x)\} = -x^\nu K_{\nu-1}(x), \quad \frac{d}{dx}\{x^{-\nu}K_\nu(x)\} = -x^{-\nu}K_{\nu+1}(x).$$

3.6 Orthogonality of the Bessel Functions

In this section we will show that the Bessel functions are orthogonal, and hence can be used as a basis over a certain interval. This will then allow us to develop the Fourier–Bessel series, which can be used to represent functions in much the same way as the Fourier–Legendre series.

We will consider the interval $[0, a]$ where, at this stage, a remains arbitrary. We start from the fact that the function $J_\nu(x)$ satisfies the Bessel equation, (3.1), and make a simple transformation, replacing x by λx, where λ is a real constant, to give

$$x^2\frac{d^2}{dx^2}J_\nu(\lambda x) + x\frac{d}{dx}J_\nu(\lambda x) + (\lambda^2 x^2 - \nu^2)J_\nu(\lambda x) = 0. \qquad (3.24)$$

We choose λ so that $J_\nu(\lambda a)$ is equal to zero. There is a countably infinite number

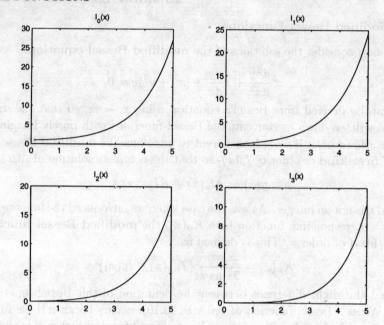

Fig. 3.5. The modified Bessel functions of first kind of order zero to three. Note that they are bounded at $x = 0$ and monotone increasing for $x > 0$, with $I_n(x) \sim e^x/\sqrt{2\pi x}$ as $x \to \infty$.

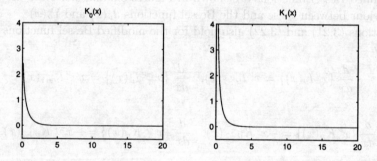

Fig. 3.6. The modified Bessel functions of the second kind of order zero and order one. Note that they are singular, with $K_0(x) \sim -\log x$ and $K_n(x) \sim 2^{n-1}(n-1)!/x^n$ for n a positive integer, as $x \to 0$. These functions are monotone decreasing for $x > 0$, tending to zero exponentially fast as $x \to \infty$.

of values of λ for which this is true, as we can deduce from Figure 3.1. Now choose $\mu \neq \lambda$ so that $J_\nu(\mu a) = 0$. Of course $J_\nu(\mu x)$ also satisfies (3.24) with λ replaced by μ. We now multiply (3.24) through by $J_\nu(\mu x)/x$ and integrate between zero and a, which yields

$$\int_0^a J_\nu(\mu x) \left\{ x \frac{d^2}{dx^2} J_\nu(\lambda x) + \frac{d}{dx} J_\nu(\lambda x) + \frac{1}{x}(\lambda^2 x^2 - \nu^2) J_\nu(\lambda x) \right\} dx = 0. \quad (3.25)$$

3.6 ORTHOGONALITY OF THE BESSEL FUNCTIONS

Notice that we could have also multiplied the differential equation for $J_\nu(\mu x)$ by $J_\nu(\lambda x)/x$ and integrated to give

$$\int_0^a J_\nu(\lambda x) \left\{ x \frac{d^2}{dx^2} J_\nu(\mu x) + \frac{d}{dx} J_\nu(\mu x) + \frac{1}{x}(\mu^2 x^2 - \nu^2) J_\nu(\mu x) \right\} dx = 0. \quad (3.26)$$

We now subtract (3.25) from (3.26) to give

$$\int_0^a \left\{ J_\nu(\mu x) \frac{d}{dx}\left(x \frac{d}{dx} J_\nu(\lambda x) \right) - J_\nu(\lambda x) \frac{d}{dx}\left(x \frac{d}{dx} J_\nu(\mu x) \right) \right.$$

$$\left. + x(\lambda^2 - \mu^2) J_\nu(\lambda x) J_\nu(\mu x) \right\} dx = 0. \quad (3.27)$$

We have simplified these expressions slightly by observing that $xJ_\nu'' + J_\nu' = (xJ_\nu')'$. Now consider the first term of the integrand and note that it can be integrated by parts to give

$$\int_0^a J_\nu(\mu x) \frac{d}{dx}\left(x \frac{d}{dx} J_\nu(\lambda x) \right) dx = \left[J_\nu(\mu x) x \frac{d}{dx} J_\nu(\lambda x) \right]_0^a$$

$$- \int_0^a x \frac{d}{dx} J_\nu(\lambda x) \frac{d}{dx} J_\nu(\mu x) dx.$$

Similarly for the second term, which is effectively the same with μ and λ interchanged,

$$\int_0^a J_\nu(\lambda x) \frac{d}{dx}\left(x \frac{d}{dx} J_\nu(\mu x) \right) dx = \left[J_\nu(\lambda x) x \frac{d}{dx} J_\nu(\mu x) \right]_0^a$$

$$- \int_0^a x \frac{d}{dx} J_\nu(\mu x) \frac{d}{dx} J_\nu(\lambda x) dx.$$

Using these expressions in (3.27), we find that

$$(\lambda^2 - \mu^2) \int_0^a x J_\nu(\lambda x) J_\nu(\mu x) dx = J_\nu(\lambda a) a\mu J_\nu'(\mu a) - J_\nu(\mu a) a\lambda J_\nu'(\lambda a). \quad (3.28)$$

Finally, since we chose λ and μ so that $J_\nu(\lambda a) = J_\nu(\mu a) = 0$ and $\mu \ne \lambda$,

$$\int_0^a x J_\nu(\mu x) J_\nu(\lambda x) dx = 0.$$

This is an orthogonality relation for the Bessel functions, with weighting function $w(x) = x$. We now need to calculate the value of $\int_0^a x J_\nu(\mu x)^2 \, dx$. To this end, we substitute $\lambda = \mu + \epsilon$ into (3.28). For $\epsilon \ll 1$, neglecting terms in ϵ^2 and smaller, we find that

$$-2\mu\epsilon \int_0^a x J_\nu(\mu x) J_\nu(\mu x + \epsilon x) \, dx$$

$$= a\left[\mu J_\nu(\mu a + \epsilon a) J_\nu'(\mu a) - (\mu + \epsilon) J_\nu(\mu a) J_\nu'(\mu a + \epsilon a) \right]. \quad (3.29)$$

In order to deal with the terms evaluated at $x = a(\mu + \epsilon)$ we consider Taylor series

74   BESSEL FUNCTIONS

expansions, $J_\nu(a(\mu+\epsilon)) = J_\nu(a\mu) + \epsilon a J'_\nu(a\mu) + \cdots$, $J'_\nu(a(\mu+\epsilon)) = J'_\nu(a\mu) + \epsilon a J''_\nu(a\mu) + \cdots$. These expansions can then be substituted into (3.29). On dividing through by ϵ and considering the limit $\epsilon \to 0$, we find that

$$\int_0^a x[J_\nu(\mu x)]^2\, dx = \frac{a^2}{2}[J'_\nu(\mu a)]^2 - \frac{1}{2}a^2 J_\nu(\mu a) J''_\nu(\mu a) - \frac{a}{2\mu} J_\nu(\mu a) J'_\nu(\mu a).$$

We now suppose that $J_\nu(\mu a) = 0$, which gives

$$\int_0^a x[J_\nu(\mu x)]^2\, dx = \frac{a^2}{2}[J'_\nu(\mu a)]^2.$$

In general,

$$\int_0^a x J_\nu(\mu x) J_\nu(\lambda x)\, dx = \frac{a^2}{2}[J'_\nu(\mu a)]^2 \delta_{\lambda\mu}, \qquad (3.30)$$

where $J_\nu(\mu a) = J_\nu(\lambda a) = 0$ and $\delta_{\lambda\mu}$ is the Kronecker delta function.

We can now construct a series expansion of the form

$$f(x) = \sum_{i=1}^\infty C_i J_\nu(\lambda_i x). \qquad (3.31)$$

This is known as a **Fourier–Bessel series**, and the λ_i are chosen such that $J_\nu(\lambda_i a) = 0$ for $i = 1, 2, \ldots$, $\lambda_1 < \lambda_2 < \cdots$. As we shall see later, both $f(x)$ and $f'(x)$ must be piecewise continuous for this series to converge. After multiplying both sides of (3.31) by $xJ_\nu(\lambda_j x)$ and integrating over the interval $[0,a]$ we find that

$$\int_0^a x J_\nu(\lambda_j x) f(x)\, dx = \int_0^a x J_\nu(\lambda_j x) \sum_{i=1}^\infty C_i J_\nu(\lambda_i x)\, dx.$$

Assuming that the series converges, we can interchange the integral and the summation to obtain

$$\int_0^a x J_\nu(\lambda_j x) f(x)\, dx = \sum_{i=1}^\infty C_i \int_0^a x J_\nu(\lambda_j x) J_\nu(\lambda_i x)\, dx.$$

We can now use (3.30) to give

$$\int_0^a x J_\nu(\lambda_j x) f(x)\, dx = \sum_{i=1}^\infty C_i \frac{a^2}{2}[J'_\nu(\lambda_j a)]^2 \delta_{\lambda_j \lambda_i} = C_j \frac{\lambda_j a^2}{2}[J'_\nu(\lambda_j a)]^2,$$

and hence

$$C_j = \frac{2}{a^2[J'_\nu(\lambda_j a)]^2} \int_0^a x J_\nu(\lambda_j x) f(x)\, dx. \qquad (3.32)$$

Example

Let's try to expand the function $f(x) = 1$ on the interval $0 \leqslant x \leqslant 1$, as a Fourier–Bessel series. Since $J_0(0) = 1$ but $J_i(0) = 0$ for $i > 0$, we will choose ν to be zero for our expansion. We rely on the existence of a set of values λ_j such that $J_0(\lambda_j) = 0$

for $j = 1, 2, \ldots$ (see Figure 3.1). We will need to determine these values either from tables or numerically.

Using (3.32), we have

$$C_j = \frac{2}{[J_0'(\lambda_j)]^2} \int_0^1 x J_0(\lambda_j x) \, dx.$$

If we introduce the variable $y = \lambda_j x$ (so that $dy = \lambda_j dx$), the integral becomes

$$C_j = \frac{2}{[J_0'(\lambda_j)]^2} \frac{1}{\lambda_j^2} \int_{y=0}^{\lambda_j} y J_0(y) \, dy.$$

Using the expression (3.21) with $\nu = 1$, we have

$$C_j = \frac{2}{[J_0'(\lambda_j)]^2} \frac{1}{\lambda_j^2} \int_{y=0}^{\lambda_j} \frac{d}{dy} (y J_1(y)) \, dy$$

$$= \frac{2}{\lambda_j^2 [J_0'(\lambda_j)]^2} [y J_1(y)]_0^{\lambda_j} = \frac{2 J_1(\lambda_j)}{\lambda_j [J_0'(\lambda_j)]^2} = \frac{2}{\lambda_j J_1(\lambda_j)},$$

and hence

$$\sum_{i=1}^{\infty} \frac{2}{\lambda_i J_1(\lambda_i)} J_0(\lambda_i x) = 1 \quad \text{for } 0 \leqslant x < 1,$$

where $J_0(\lambda_i) = 0$ for $i = 1, 2, \ldots$. In Figure 3.7 we show the sum of the first fifteen terms of the Fourier–Bessel series. Notice the oscillatory nature of the solution, which is more pronounced in the neighbourhood of the discontinuity at $x = 1$. This phenomenon always occurs in series expansions relative to sequences of orthogonal functions, and is called **Gibbs' phenomenon**.

Before we can give a MATLAB script that produces Figure 3.7, we need to be able to calculate the zeros of Bessel functions. Here we merely state a couple of simple results and explain how these are helpful. The interested reader is referred to Watson (1922, Chapter 15) for a full discussion of this problem. The **Bessel–Lommel theorem** on the location of the zeros of the Bessel functions $J_\nu(x)$ states that when $-\frac{1}{2} < \nu \leqslant \frac{1}{2}$ and $m\pi < x < (m + \frac{1}{2})\pi$, $J_\nu(x)$ is positive for even m and negative for odd m. This implies that $J_\nu(x)$ has an odd number of zeros in the intervals $(2n-1)\pi/2 < x < 2n\pi$ for n an integer. In fact, it can be shown that the positive zeros of $J_0(x)$ lie in the intervals $n\pi + 3\pi/4 < x < n\pi + 7\pi/8$. This allows us to use the ends of these intervals as an initial bracketing interval for the roots of $J_0(x)$. A simple MATLAB script that uses this result is

```
global nu ze
nu=0;
for ir=1:20
        start_int = (ir-1)*pi+3*pi/4;
        end_int   = (ir-1)*pi+7*pi/8;
        ze(ir) = fzero('bessel',[start_int end_int]);
end
```

BESSEL FUNCTIONS

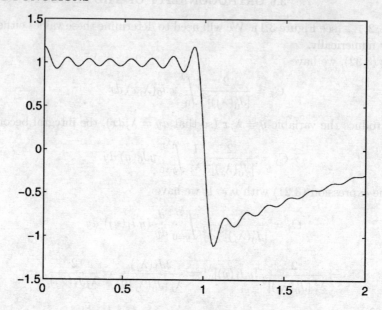

Fig. 3.7. Fourier–Bessel series representation of the function $f(x) = 1$ on the interval $[0, 1)$ (truncated after fifteen terms). The expansion is only valid in the interval $x \in [0, 1)$. The series is shown for $x > 1$ only to demonstrate that convergence is only guaranteed for the associated interval.

together with the function

```
function bessel = bessel(x);
global nu
bessel = besselj(nu,x);
```

The MATLAB function fzero finds a zero of functions of a single variable in a given interval. By defining the variables nu and ze as global, we make them available to other functions. In particular, this allows us to use the computed positions of the zeros, ze, in the script below.

The zeros of $J_1(x)$ interlace those of $J_0(x)$. We can see this by noting that $J_1(x) = -J_0'(x)$ and that both functions are continuous. Consequently the zeros of $J_0(x)$ can be used as the bracketing intervals for the determination of the zeros of $J_1(x)$. A MATLAB script for this is

3.7 INHOMOGENEOUS TERMS IN BESSEL'S EQUATION

```
global nu ze ze1
nu=1;
for ir=1:19
        start_int = ze(ir);
        end_int   = ze(ir+1);
        ze1(ir)   = fzero('bessel',[start_int end_int]);
end
```

The zeros of the other Bessel functions can be found by exploiting similar analytical results.

We can now define a MATLAB function

```
function fourierbessel = fourierbessel(x)
global ze
n=15; a = 2./ze(1:n)./besselj(1,ze(1:n));
for k = 1:n
    X(:,k) = besselj(0,ze(k)*x(:));
end
fourierbessel = X*a';
```

which can be plotted with ezplot(@fourierbessel, [0 2]) to produce Figure 3.7.

3.7 Inhomogeneous Terms in Bessel's Equation

So far we have only concerned ourselves with homogeneous forms of Bessel's equation. The inhomogeneous version of Bessel's equation,

$$x^2 \frac{d^2y}{dx^2} + x\frac{dy}{dx} + (x^2 - \nu^2)y = f(x),$$

can be dealt with by using the technique of variation of parameters (see Section 1.2). The solution can be written as

$$y(x) = \int^x \frac{s \sin \nu \pi}{2\nu} f(s) \left(J_\nu(s)Y_\nu(x) - J_\nu(x)Y_\nu(s)\right) ds + AJ_\nu(x) + BY_\nu(x). \quad (3.33)$$

Here we have made use of the fact that the Wronskian associated with $J_\nu(x)$ and $Y_\nu(x)$ is $2\nu/x \sin \nu\pi$, which can be derived using Abel's formula, (1.7). The constant can then be found by considering the behaviour of the functions close to $x = 0$ (see Exercise 3.5).

Example

Let's find the general solution of

$$x^2 \frac{d^2y}{dx^2} + x\frac{dy}{dx} + x^2 y = x^2.$$

BESSEL FUNCTIONS

This can be determined by using (3.33) with $f(x) = x^2$ and $\nu = 0$, so that the general solution is

$$y(x) = \int^x \frac{\pi s}{2} \left(J_0(s) Y_0(x) - J_0(x) Y_0(s) \right) ds + A J_0(x) + B Y_0(x).$$

In order to integrate $s J_0(s)$ we note that this can be written as $(s J_1(s))'$ and similarly for $s Y_0(s)$, which gives

$$y(x) = \frac{\pi x}{2} \left(J_1(x) Y_0(x) - J_0(x) Y_1(x) \right) + A J_0(x) + B Y_0(x).$$

But we note that $J_1(x) = -J_0'(x)$ and $Y_1(x) = -Y_0'(x)$, so that the expression in the brackets is merely the Wronskian, and hence

$$y(x) = 1 + A J_0(x) + B Y_0(x).$$

Although it is clear, with hindsight, that $y(x) = 1$ is the particular integral solution, simple solutions are not always easy to spot *a priori*.

Example

Let's find the particular integral of

$$x^2 \frac{d^2 y}{dx^2} + x \frac{dy}{dx} + (x^2 - \nu^2) y = x. \tag{3.34}$$

We will look for a series solution as used in the method of Frobenius, namely

$$y(x) = \sum_{n=0}^{\infty} a_n x^{n+c}.$$

Substituting this into (3.34), we obtain an expression similar to (3.7),

$$a_0 (c^2 - \nu^2) x^c + a_1 \{(1+c)^2 - \nu^2\} x^{c+1}$$

$$+ \sum_{n=2}^{\infty} [a_n \{(n+c)^2 - \nu^2\} + a_{n-2}] x^{n+c} = x.$$

Note that x^c needs to match with the x on the right hand side so that $c = 1$ and $a_0 = 1/(1 - \nu^2)$. We will defer discussion of the case $\nu = 1$. At next order we find that $a_1 (2^2 - \nu^2) = 0$, and consequently, unless $\nu = 2$, we have $a_1 = 0$. For the general terms in the summation we have

$$a_n = -\frac{a_{n-2}}{\{(n+1)^2 - \nu^2\}}.$$

Note that since $a_1 = 0$, $a_n = 0$ for n odd. It now remains to determine the general term in the sequence. For $n = 2$ we find that

$$a_2 = -\frac{a_0}{3^2 - \nu^2} = -\frac{1}{(1^2 - \nu^2)(3^2 - \nu^2)}$$

and then with $n = 4$ and using the form of a_2 we have

$$a_4 = \frac{1}{(1^2 - \nu^2)(3^2 - \nu^2)(5^2 - \nu^2)}.$$

The general expression is

$$a_{2n} = (-1)^n \prod_{i=1}^{n+1} \frac{1}{(2i-1)^2 - \nu^2}.$$

This can be manipulated by factorizing the denominator and extracting a factor of 2^2 from each term in the product (of which there are $n+1$). This gives

$$a_{2n} = \frac{(-1)^n}{2^{2n+2}} \prod_{i=1}^{n+1} \frac{1}{i - \frac{1}{2} + \frac{1}{2}\nu} \prod_{i=1}^{n+1} \frac{1}{i - \frac{1}{2} - \frac{1}{2}\nu} = \frac{(-1)^n}{2^{2n+2} \left(\frac{1}{2} + \frac{1}{2}\nu\right)_{n+1} \left(\frac{1}{2} - \frac{1}{2}\nu\right)_{n+1}}.$$

Hence the particular integral of the differential equation is

$$y(x) = x \sum_{n=0}^{\infty} \frac{(-1)^n}{2^{2n+2} \left(\frac{1}{2} + \frac{1}{2}\nu\right)_{n+1} \left(\frac{1}{2} - \frac{1}{2}\nu\right)_{n+1}} x^{2n}. \qquad (3.35)$$

In fact, solutions of the equation

$$x^2 \frac{d^2 y}{dx^2} + x \frac{dy}{dx} + (x^2 - \nu^2) y = x^{\mu+1} \qquad (3.36)$$

are commonly referred to as $s_{\mu,\nu}$ and are called **Lommel's functions**. They are undefined when $\mu \pm \nu$ is an odd negative integer, a case that is discussed in depth by Watson (1922). The series expansion of the solution of (3.36) is

$$y(x) = x^{\mu-1} \sum_{m=0}^{\infty} \frac{(-1)^m \left(\frac{1}{2}x\right)^{2m+2} \Gamma\left(\frac{1}{2}\mu - \frac{1}{2}\nu + \frac{1}{2}\right) \Gamma\left(\frac{1}{2}\mu + \frac{1}{2}\nu + \frac{1}{2}\right)}{\Gamma\left(\frac{1}{2}\mu - \frac{1}{2}\nu + m + \frac{3}{2}\right) \Gamma\left(\frac{1}{2}\mu + \frac{1}{2}\nu + m + \frac{3}{2}\right)}.$$

We can use this to check that (3.35) is correct. Note that we need to use (3.6).

3.8 Solutions Expressible as Bessel Functions

There are many other differential equations whose solutions can be written in terms of Bessel functions. In order to determine some of these, we consider the transformation

$$y(x) = x^\alpha \tilde{y}(x^\beta).$$

Since α and β could be fractional, we will restrict our attention to $x \geqslant 0$. We substitute this expression into the differential equation and seek values of α and β which give Bessel's equation for $\tilde{y}$.

Example

Let's try to express the solutions of the differential equation

$$\frac{d^2 y}{dx^2} - xy = 0$$

in terms of Bessel functions. This is called **Airy's equation** and has solutions $\text{Ai}(x)$ and $\text{Bi}(x)$, the **Airy functions**. We start by introducing the function $\tilde{y}$.

Differentiating with respect to x we have

$$\frac{dy}{dx} = \alpha x^{\alpha-1}\tilde{y} + \beta x^{\alpha+\beta-1}\tilde{y}'.$$

Differentiating again we obtain

$$\frac{d^2y}{dx^2} = \alpha(\alpha-1)x^{\alpha-2}\tilde{y} + (2\alpha\beta + \beta^2 - \beta)x^{\alpha+\beta-2}\tilde{y}' + \beta^2 x^{\alpha+2\beta-2}\tilde{y}''.$$

These expressions can now be substituted into Airy's equation to give

$$\alpha(\alpha-1)x^{\alpha-2}\tilde{y} + (2\alpha\beta + \beta^2 - \beta)x^{\alpha+\beta-2}\tilde{y}' + \beta^2 x^{\alpha+2\beta-2}\tilde{y}'' - x^{\alpha+1}\tilde{y} = 0.$$

It is now convenient to multiply the entire equation by $x^{-\alpha+2}/\beta^2$ (this means that the coefficient of $\tilde{y}''$ is $x^{2\beta}$), which gives

$$x^{2\beta}\tilde{y}'' + \frac{(2\alpha + \beta - 1)}{\beta}x^{\beta}\tilde{y}' + \left\{-\frac{1}{\beta^2}x^3 + \frac{\alpha(\alpha-1)}{\beta^2}\right\}\tilde{y} = 0.$$

Considering the coefficient of $\tilde{y}$ we note that we require $x^3 \propto x^{2\beta}$ which gives $\beta = \frac{3}{2}$. The coefficient of $\tilde{y}'$ gives us that $\alpha = \frac{1}{2}$. The equation is now

$$x^{2\beta}\tilde{y}'' + x^{\beta}\tilde{y}' + \left\{-\left(\frac{2x^{\beta}}{3}\right)^2 - \frac{1}{9}\right\}\tilde{y} = 0,$$

which has solutions $K_{1/3}(2x^{3/2}/3)$ and $I_{1/3}(2x^{3/2}/3)$. The general solution of Airy's equation in terms of Bessel's functions is therefore

$$y(x) = x^{1/2}\left\{AK_{1/3}\left(\frac{2}{3}x^{3/2}\right) + BI_{1/3}\left(\frac{2}{3}x^{3/2}\right)\right\}.$$

In fact, $Ai(x) = \sqrt{x/3}K_{1/3}(2x^{3/2}/3)/\pi$. A graph of this Airy function is shown in Figure 11.12. The Airy functions Ai and Bi are available in MATLAB through the function airy.

3.9 Physical Applications of the Bessel Functions

3.9.1 Vibrations of an Elastic Membrane

We will now derive the equation that governs small displacements, $z = z(x, y, t)$, of an elastic membrane. We start by considering a small membrane with sides of length δS_x in the x-direction and δS_y in the y-direction, which makes angles $\psi_x, \psi_x + \delta\psi_x$ and $\psi_y, \psi_y + \delta\psi_y$ with the horizontal, as shown in Figure 3.8. Newton's second law of motion in the vertical, z-direction gives

$$\frac{\partial^2 z}{\partial t^2}\rho\delta S_x\delta S_y = \delta S_y\{T\sin(\psi_x + \delta\psi_x) - T\sin(\psi_x)\} + \delta S_x\{T\sin(\psi_y + \delta\psi_y) - T\sin(\psi_y)\},$$

where ρ is the constant density (mass per unit area) of the membrane and T is the tension, assumed constant for small vibrations of the membrane. We will eventually consider the angles ψ_x and ψ_y to be small, but at the outset we will consider the

3.9 PHYSICAL APPLICATIONS OF THE BESSEL FUNCTIONS   81

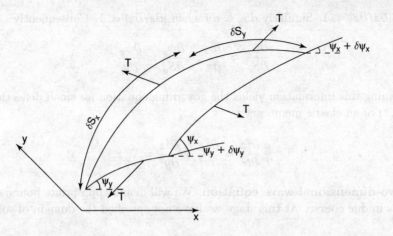

Fig. 3.8. A small section of an elastic membrane.

changes in these angles, $\delta\psi_x$ and $\delta\psi_y$, to be smaller. Accordingly we expand the trigonometric functions and find that

$$\frac{\rho}{T}\frac{\partial^2 z}{\partial t^2} = \cos\psi_x \frac{\delta\psi_x}{\delta S_x} + \cos\psi_y \frac{\delta\psi_y}{\delta S_y} + \cdots,$$

where we have divided through by the area of the element, $\delta S_x \delta S_y$. We now consider the limit as the size of the element shrinks to zero, and therefore let δS_x and δS_y tend to zero. Consequently we find that

$$\frac{\rho}{T}\frac{\partial^2 z}{\partial t^2} = \cos\psi_x \frac{\partial\psi_x}{\partial S_x} + \cos\psi_y \frac{\partial\psi_y}{\partial S_y}. \qquad (3.37)$$

We can now use the definition of the partial derivatives,

$$\tan\psi_x = \frac{\partial z}{\partial x}, \quad \tan\psi_y = \frac{\partial z}{\partial y}.$$

By differentiating these expressions with respect to x and y respectively we find that

$$\sec^2\psi_x \frac{\partial\psi_x}{\partial x} = \frac{\partial^2 z}{\partial x^2}, \quad \sec^2\psi_y \frac{\partial\psi_y}{\partial y} = \frac{\partial^2 z}{\partial y^2}.$$

For small slopes, $\cos\psi_x$ and $\sec^2\psi_x$ are both approximately unity (and similarly for variation in the y direction). Also, using the formula for arc length we have

$$dS_x = \sqrt{1 + \left(\frac{\partial z}{\partial x}\right)^2}\, dx \approx dx,$$

when $|\partial z/\partial x| \ll 1$. Similarly $dS_y \approx dy$ when $|\partial z/\partial y| \ll 1$. Consequently

$$\frac{\partial \psi_x}{\partial S_x} \approx \frac{\partial \psi_x}{\partial x}, \quad \frac{\partial \psi_y}{\partial S_y} \approx \frac{\partial \psi_y}{\partial y}.$$

Combining this information yields the governing equation for small deflections $z = z(x, y, t)$ of an elastic membrane,

$$\frac{\rho}{T}\frac{\partial^2 z}{\partial t^2} = \frac{\partial^2 z}{\partial x^2} + \frac{\partial^2 z}{\partial y^2} = \nabla^2 z, \qquad (3.38)$$

the **two-dimensional wave equation**. We will define appropriate boundary conditions in due course. At this stage we have not specified the domain of solution.

One-Dimensional Solutions of the Wave Equation

We will start by considering the solution of this equation in one dimension. If we look for solutions of the form $z \equiv z(x, t)$, independent of y, as illustrated in Figure 3.9, we need to solve the **one-dimensional wave equation**,

$$\frac{\partial^2 z}{\partial t^2} = c^2 \frac{\partial^2 z}{\partial x^2}, \qquad (3.39)$$

where $c = \sqrt{T/\rho}$. This equation also governs the propagation of small-amplitude waves on a stretched string (for further details, see Billingham and King, 2001). The easiest way to solve (3.39) is to define new variables, known as **characteristic**

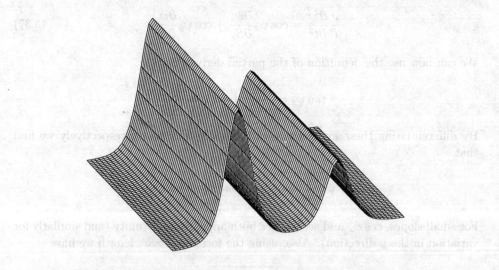

Fig. 3.9. A one-dimensional solution of the two-dimensional wave equation.

variables, $\xi = x - ct$ and $\eta = x + ct$ (see Section 7.1 for an explanation of where

3.9 PHYSICAL APPLICATIONS OF THE BESSEL FUNCTIONS

these come from). In terms of these variables, (3.39) becomes

$$\frac{\partial^2 z}{\partial \xi \partial \eta} = 0.$$

Integrating this with respect to η gives

$$\frac{\partial z}{\partial \xi} = F(\xi),$$

and with respect to ξ

$$z = \int^\xi F(s)\,ds + g(\eta) = f(\xi) + f(\eta),$$

and hence we have that

$$z(x,t) = f(x - ct) + g(x + ct). \tag{3.40}$$

This represents the sum of two waves, one, represented by $f(x - ct)$, propagating from left to right without change of form at speed c, and one, represented by $g(x+ct)$, from right to left at speed c. To see this, consider, for example, the solution $y = f(x - ct)$, and simply note that on the paths $x = ct +$ constant, $f(x - ct)$ is constant. Similarly, $g(x + ct)$ is constant on the paths $x = -ct +$ constant.

The functions f and g can be determined from the initial displacement and velocity. If

$$z(x,0) = z_0(x), \quad \frac{\partial z}{\partial t}(x,0) = u_0(x),$$

then (3.40) with $t = 0$ gives us

$$f(x) + g(x) = z_0(x). \tag{3.41}$$

If we now differentiate (3.40) with respect to time, we have

$$\frac{\partial z}{\partial t} = -cf'(x - ct) + cg'(x + ct),$$

and hence when $t = 0$,

$$-cf'(x) + cg'(x) = u_0(x).$$

This can be integrated to yield

$$-cf(x) + cg(x) = \int_a^x u_0(s)\,ds, \tag{3.42}$$

where a is an arbitrary constant. Solving the simultaneous equations (3.41) and (3.42), we find that

$$f(x) = \frac{1}{2}z_0(x) - \frac{1}{2c}\int_a^x u_0(s)\,ds, \quad g(x) = \frac{1}{2}z_0(x) + \frac{1}{2c}\int_a^x u_0(s)\,ds.$$

On substituting these into (3.40), we obtain **d'Alembert's solution** of the one-dimensional wave equation,

$$z(x,t) = \frac{1}{2}\{z_0(x - ct) + z_0(x + ct)\} + \frac{1}{2c}\int_{x-ct}^{x+ct} u_0(s)\,ds. \tag{3.43}$$

In particular, if $u_0 = 0$, a string released from rest, the solution consists of a left-travelling and a right-travelling wave, each with the same shape but half the amplitude of the initial displacement. The solution when $z_0 = 0$ for $|x| > a$ and $z_0 = 1$ for $|x| < a$, a **top hat** initial displacement, is shown in Figure 3.10.

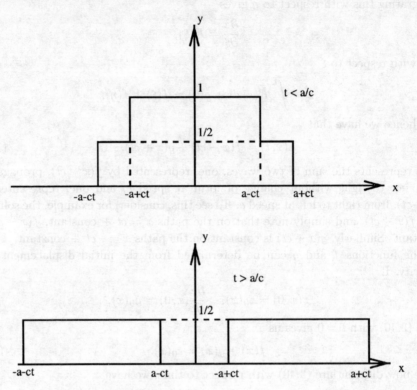

Fig. 3.10. D'Alembert's solution for an initially stationary top hat displacement.

Two-Dimensional Solutions of the Wave Equation

Let's now consider the solution of the two-dimensional wave equation, (3.38), for an elastic, disc-shaped membrane fixed to a circular support. In cylindrical polar coordinates, the two-dimensional wave equation becomes

$$\frac{1}{c^2}\frac{\partial^2 z}{\partial t^2} = \frac{\partial^2 z}{\partial r^2} + \frac{1}{r}\frac{\partial z}{\partial r} + \frac{1}{r^2}\frac{\partial^2 z}{\partial \theta^2}. \qquad (3.44)$$

We will look for solutions in the circular domain $0 \leqslant r \leqslant a$ and $0 \leqslant \theta < 2\pi$. Such solutions must be periodic in θ with period 2π. The boundary condition is $z = 0$ at $r = a$. We seek a separable solution, $z = R(r)\tau(t)\Theta(\theta)$. On substituting this into (3.44), we find that

$$\frac{1}{c^2}\frac{\tau''}{\tau} = \frac{rR'' + R'}{rR} + \frac{1}{r^2}\frac{\Theta''}{\Theta} = -\frac{\omega^2}{c^2},$$

3.9 PHYSICAL APPLICATIONS OF THE BESSEL FUNCTIONS

where $-\omega^2/c^2$ is the separation constant. An appropriate solution is $\tau = e^{i\omega t}$, which represents a time-periodic solution. This is what we would expect for the vibrations of an elastic membrane, which is, after all, a drum. The **angular frequency**, ω, is yet to be determined. We can now write

$$\frac{r^2 R'' + rR'}{R} + \frac{\omega^2 r^2}{c^2} = -\frac{\Theta''}{\Theta} = n^2,$$

where n^2 is the next separation constant. This gives us $\Theta = A\cos n\theta + B\sin n\theta$, with n a positive integer for 2π-periodicity. Finally, $R(r)$ satisfies

$$r^2 \frac{d^2R}{dr^2} + r\frac{dR}{dr} + \left(\frac{\omega^2}{c^2}r^2 - n^2\right) R = 0.$$

We can simplify this by introducing a new coordinate $s = \lambda r$ where $\lambda = \omega/c$, which gives

$$s^2 \frac{d^2R}{ds^2} + s\frac{dR}{ds} + \left(s^2 - n^2\right) R = 0,$$

which is Bessel's equation with $\nu = n$. Consequently the solutions can be written as

$$R(s) = A J_n(s) + B Y_n(s).$$

We need a solution that is bounded at the origin so, since $Y_n(s)$ is unbounded at $s = 0$, we require $B = 0$. The other boundary condition is that the membrane is constrained not to move at the edge of the domain, so that $R(s) = 0$ at $s = \lambda a$. This gives the condition

$$J_n(\lambda a) = 0, \qquad (3.45)$$

which has an infinite number of solutions $\lambda = \lambda_{ni}$. Specifying the value of $\lambda = \omega/c$ prescribes the frequency at which the membrane will oscillate. Consequently the functional form of the natural modes of oscillation of a circular membrane is

$$z = J_n(\lambda_{ni} r)\left(A\cos n\theta + B\sin n\theta\right) e^{i\omega_i t}, \qquad (3.46)$$

where $\omega_i = c\lambda_{ni}$ and the values of λ_{ni} are solutions of (3.45). Figure 3.11 shows a few of these natural modes when $a = 1$, which we created using the MATLAB function ezmesh (see Section 2.6.1).

Here we have considered the natural modes of oscillation. We could however have tackled an initial value problem. Let's consider an example where the initial displacement of the membrane is specified to be $z(r, \theta, 0) = G(r, \theta)$ and the membrane is released from rest, so that $\partial z/\partial t = 0$, when $t = 0$. The fact that the membrane is released from rest implies that the temporal variation will be even, so we need only consider a solution proportional to $\cos\omega_i t$. Consequently, using the linearity of the wave equation to add all the possible solutions of the form (3.46), the general form of the displacement of the membrane is

$$z(r, \theta, t) = \sum_{i=0}^{\infty}\sum_{n=0}^{\infty} J_n(\lambda_{ni} r)\left(A_{ni}\cos n\theta + B_{ni}\sin n\theta\right)\cos\omega_i t.$$

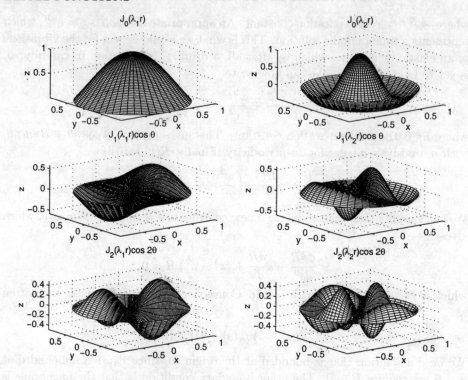

Fig. 3.11. Six different modes of oscillation of an elastic membrane.

Using the condition that $z(r,\theta,0) = G(r,\theta)$ we have the equations

$$G(r,\theta) = \sum_{i=0}^{\infty}\sum_{n=0}^{\infty} J_n(\lambda_{ni}r)\left(A_{ni}\cos n\theta + B_{ni}\sin n\theta\right). \tag{3.47}$$

In order to find the coefficients A_{ni} and B_{ni} we need exploit the orthogonality of the Bessel functions, given by (3.30), and of the trigonometric functions, using

$$\int_0^{2\pi} \cos m\theta \cos n\theta \, d\theta = \int_0^{2\pi} \sin m\theta \sin n\theta \, d\theta = \pi \delta_{mn},$$

for m and n integers. Multiplying (3.47) through by $\cos m\theta$ and integrating from 0 to 2π we have

$$\frac{1}{\pi}\int_0^{2\pi} \cos m\theta \, G(r,\theta) \, d\theta = \sum_{i=0}^{\infty} A_{mi} J_n(\lambda_{ni}r),$$

and with $\sin m\theta$ we have

$$\frac{1}{\pi}\int_0^{2\pi} \sin m\theta \, G(r,\theta) \, d\theta = \sum_{i=0}^{\infty} B_{mi} J_n(\lambda_{ni}r).$$

3.9 PHYSICAL APPLICATIONS OF THE BESSEL FUNCTIONS

Now, using (3.32), we have

$$A_{mj} = \frac{2}{a^2 \pi \left[J_n'(\lambda_j a)\right]^2} \int_{r=0}^{a} r J_n(\lambda_j r) \int_0^{2\pi} \cos m\theta \, G(r,\theta) \, d\theta \, dr$$

and

$$B_{mj} = \frac{2}{a^2 \pi \left[J_n'(\lambda_j a)\right]^2} \int_{r=0}^{a} r J_n(\lambda_j r) \int_0^{2\pi} \sin m\theta \, G(r,\theta) \, d\theta \, dr.$$

Note that if the function $G(r, \theta)$ is even, $B_{mj} = 0$, and similarly if it is odd, $A_{mj} = 0$. The expansion in θ is an example of a **Fourier series expansion**, about which we will have more to say in Chapter 5.

3.9.2 Frequency Modulation (FM)

We will now discuss an application in which Bessel functions occur within the description of a modulated wave (for further information, see Dunlop and Smith, 1977). The expression for a frequency carrier comprises a carrier frequency f_c and the modulating signal with frequency f_m. The phase shift of the carrier is related to the time integral of the modulating wave $F_m(t) = \cos 2\pi f_m t$, so that the actual signal is

$$F_c(t) = \cos\left\{2\pi f_c t + 2\pi K_2 \int_0^t a \cos(2\pi f_m t) dt\right\},$$

which we can integrate to obtain

$$F_c(t) = \cos\left\{2\pi f_c t + \beta \sin(2\pi f_m t)\right\}, \tag{3.48}$$

where $\beta = K_2 a / f_m$, a constant called the **modulation index**. We can expand (3.48) to give

$$F_c(t) = \cos(2\pi f_c t) \cos\left\{\beta \sin(2\pi f_m t)\right\} - \sin(2\pi f_c t) \sin\left\{\beta \sin(2\pi f_m t)\right\}. \tag{3.49}$$

An example of this signal is shown in Figure 3.12.

We would now like to split the signal (3.49) into its various frequency components. We can do this by exploiting (3.16) and (3.17) to give

$$F_c(t) = \cos(2\pi f_c t) \left\{ J_0(\beta) + 2 \sum_{n=1}^{\infty} (-1)^n J_{2n}(\beta) \cos 2n(2\pi f_m t) \right\}$$

$$+ \sin(2\pi f_c t) \left\{ 2 \sum_{n=0}^{\infty} (-1)^n J_{2n+1}(\beta) \cos(2n+1)(2\pi f_m t) \right\}. \tag{3.50}$$

Using simple trigonometry,

$$2 \cos(2\pi f_c t) \cos\left\{(2n) 2\pi f_m t\right\} =$$

$$\cos\left\{2\pi f_c t + (2n) 2\pi f_m t\right\} + \cos\left\{2\pi f_c t - (2n) 2\pi f_m t\right\},$$

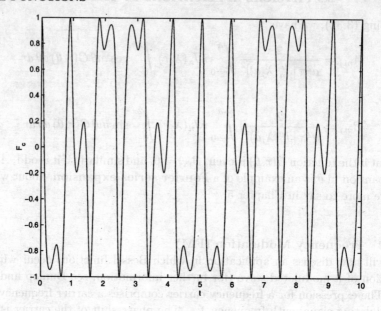

Fig. 3.12. An example of frequency modulation, with $\beta = 1.2$, $f_m = 1.2$ and $f_c = 1$.

and

$$2\sin(2\pi f_c t)\sin\{(2n+1)2\pi f_m t\} =$$
$$-\cos\{2\pi f_c t + (2n+1)2\pi f_m t\} + \cos\{2\pi f_c t - (2n+1)2\pi f_m t\}.$$

Substituting these expressions back into (3.50), we find that $F_c(t)$ can be written as

$$F_c(t) = J_0(\beta)\cos(2\pi f_c t)$$
$$+ \sum_{n=1}^{\infty} J_{2n}(\beta)\left[\cos\{2\pi t(f_c + 2nf_m)\} + \cos\{2\pi t(f_c - 2nf_m)\}\right]$$
$$- \sum_{n=0}^{\infty} J_{2n+1}(\beta)\left[\cos\{2\pi t(f_c - (2n+1)f_m)\} - \cos\{2\pi t(f_c + (2n+1)f_m)\}\right],$$

the first few terms of which are given by

$$F_c(t) = J_0(\beta)\cos(2\pi f_c t) - J_1(\beta)\left[\cos\{2\pi(f_c - f_m)t\} - \cos\{2\pi(f_c + f_m)t\}\right]$$
$$+ J_2(\beta)\left[\cos\{2\pi(f_c - 2f_m)t\} + \cos\{2\pi(f_c + 2f_m)t\}\right]$$
$$- J_3(\beta)\left[\cos\{2\pi(f_c - 3f_m)t\} - \cos\{2\pi(f_c + 3f_m)t\}\right] + \cdots.$$

This means that the main carrier signal has frequency f_c and amplitude $J_0(\beta)$, and that the other components, known as **sidebands**, have frequencies $f_c \pm nf_m$ for

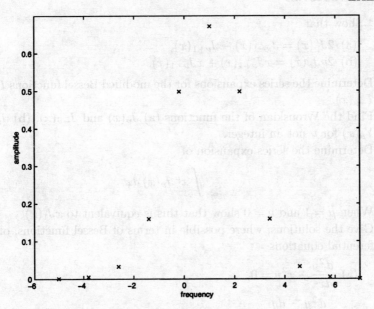

Fig. 3.13. The frequency spectrum of a signal with $\beta = 0.2$, $f_m = 1.2$ and $f_c = 1$.

n an integer, and amplitudes $J_n(\beta)$. These amplitudes, known as the **frequency spectrum** of the signal, are shown in Figure 3.13.

Exercises

3.1 The functions $J_0(x)$ and $Y_0(x)$ are solutions of the equation

$$x\frac{d^2y}{dx^2} + \frac{dy}{dx} + xy = 0.$$

Show that if $Y_0(x)$ is the *singular* solution then

$$Y_0(x) = J_0(x)\log x - \sum_{i=1}^{\infty} \frac{\phi(i)}{(i!)^2}\left(-\frac{x^2}{4}\right)^i,$$

where

$$\phi(i) = \sum_{k=1}^{i} \frac{1}{k}.$$

3.2 Using the definition

$$J_\nu(x) = \frac{x^\nu}{2^\nu \Gamma(1+\nu)} \sum_{i=0}^{\infty} \frac{(-x^2/4)^i}{i!(1+\nu)_i},$$

show that $J_{1/2}(x) = \sqrt{\frac{2}{\pi x}} \sin x$.

BESSEL FUNCTIONS

3.3 * Show that
  (a) $2J'_\nu(x) = J_{\nu-1}(x) - J_{\nu+1}(x)$,
  (b) $2\nu J_\nu(x) = xJ_{\nu+1}(x) + xJ_{\nu-1}(x)$.

3.4 Determine the series expansions for the modified Bessel functions $I_\nu(x)$ and $I_{-\nu}(x)$.

3.5 Find the Wronskian of the functions (a) $J_\nu(x)$ and $J_{-\nu}(x)$, (b) $J_\nu(x)$ and $Y_\nu(x)$ for ν not an integer.

3.6 Determine the series expansion of
$$\int x^\mu J_\nu(x)\,dx.$$
When $\mu = 1$ and $\nu = 0$ show that this is equivalent to $xJ_1(x)$.

3.7 Give the solutions, where possible in terms of Bessel functions, of the differential equations

  (a) $\dfrac{d^2y}{dx^2} - x^2 y = 0$,

  (b) $x\dfrac{d^2y}{dx^2} + \dfrac{dy}{dx} + y = 0$,

  (c) $x\dfrac{d^2y}{dx^2} + (x+1)^2 y = 0$,

  (d) $\dfrac{d^2y}{dx^2} + \alpha^2 y = 0$,

  (e) $\dfrac{d^2y}{dx^2} - \alpha^2 y = 0$,

  (f) $\dfrac{d^2y}{dx^2} + \beta\dfrac{dy}{dx} + \gamma y = 0$,

  (g) $(1-x^2)\dfrac{d^2y}{dx^2} - 2x\dfrac{dy}{dx} + n(n+1)y = 0$.

3.8 Using the expression
$$J_\nu(z) = \frac{1}{2\pi}\int_0^{2\pi} \cos(\nu\theta - z\sin\theta)\,d\theta,$$
show that $J_\nu(0) = 0$ for ν a nonzero integer and $J_0(0) = 1$.

3.9 Determine the coefficients of the Fourier–Bessel series for the function
$$f(x) = \begin{cases} 1 & \text{for } 0 \leq x < 1, \\ -1 & \text{for } 1 \leq x \leq 2, \end{cases}$$
in terms of the Bessel function $J_0(x)$.

3.10 Determine the coefficients of the Fourier–Bessel series for the function $f(x) = x$ on the interval $0 \leq x \leq 1$ in terms of the Bessel function $J_1(x)$ (and repeat the exercise for $J_2(x)$). Modify the MATLAB code used to generate Figure 3.7 to check your answers.

3.11 Calculate the Fourier–Bessel expansion for the functions

(a)
$$f(x) = \begin{cases} x & \text{for } 0 \leq x < 1, \\ 2-x & \text{for } 1 \leq x \leq 2, \end{cases}$$

(b)
$$f(x) = \begin{cases} x^2 + 2 & \text{for } 0 \leq x < 1, \\ 3 & \text{for } 1 \leq x \leq 3, \end{cases}$$

in terms of the Bessel function $J_1(x)$.

3.12 Construct the general solution of the differential equation

$$x^2 \frac{d^2 y}{dx^2} + x \frac{dy}{dx} + (x^2 - \nu^2)y = \sin x.$$

3.13 **Project** This project arises from attempts to model the baking of foodstuffs, and thereby improve the quality of mass-produced baked foods. A key element of such modelling is the temperature distribution in the food, which, as we showed in Chapter 2, is governed by the diffusion equation,

$$\frac{\partial T}{\partial t} = \nabla \cdot (D \nabla T).$$

The diffusivity, D, is a function of the properties of the food.

We will consider some problems associated with the baking of infinitely-long, cylindrical, axisymmetric foodstuffs under axisymmetric conditions (a first approximation to, for example, the baking of a loaf of bread). In this case, the diffusion equation becomes

$$\frac{\partial T}{\partial t} = \frac{1}{r} \frac{\partial}{\partial r}\left(rD(r) \frac{\partial T}{\partial r}\right). \tag{E3.1}$$

(a) Look for a separable solution, $T(r,t) = f(r)e^{\omega t}$, of (E3.1) with $D(r) = D_0$, a constant, subject to the boundary conditions $\partial T/\partial r = 0$ at $r = 0$ and $\partial T/\partial r = h(T_a - T)$ at $r = r_0$. Here, $h > 0$ is a heat transfer coefficient and T_a is the ambient temperature. Determine the possible values of ω.

(b) If the initial temperature profile within the foodstuff is $T(r,0) = T_0(r)$, determine the temperature for $t > 0$.

(c) In many baking problems, there are two distinct layers of food. When

$$D(r) = \begin{cases} D_0 & \text{for } 0 \leq r \leq r_1, \\ D_1 & \text{for } r_1 < r \leq r_0, \end{cases}$$

and the heat flux and the temperature are continuous at $r = r_1$, determine the effect of changing r_1/r_0 and D_0 and D_1 on the possible values of ω.

(d) Solve the initial–boundary value problem when the initial temperature is uniform in each layer, with

$$T(r,0) = \begin{cases} T_0 & \text{for } 0 \leq r \leq r_1, \\ T_1 & \text{for } r_1 < r \leq r_0. \end{cases}$$

(e) Find some realistic values of D_0 and h for bread dough, either from your library or the Internet. How does the temperature at the centre of baking bread vary with time, if this simple model is to be believed?

CHAPTER FOUR

Boundary Value Problems, Green's Functions and Sturm–Liouville Theory

We now turn our attention to **boundary value problems** for ordinary differential equations, for which the boundary conditions are specified at two different points, $x = a$ and $x = b$. The solutions of boundary value problems have some rather different properties to those of solutions of initial value problems. In order to ground our discussion in terms of a real physical problem, we will consider the dynamics of a string fixed at $x = y = 0$ and $x = l$, $y = 0$, and rotating steadily about the x-axis at constant angular velocity, as shown in Figure 4.1. This is rather like a skipping rope, but one whose ends are motionless. In order to be able

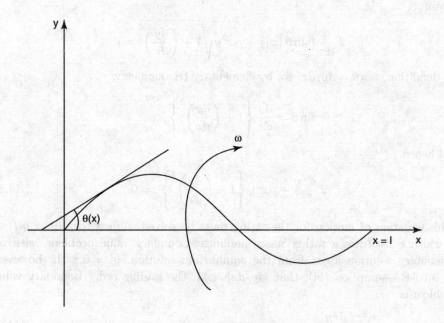

Fig. 4.1. A string rotating about the x-axis, fixed at $x = 0$ and $x = l$.

to formulate a differential equation that captures the dynamics of this string, we will make several assumptions.

BOUNDARY VALUE PROBLEMS

(i) The tension in the string is large enough that any additional forces introduced by the bending of the string are negligible in comparison with the tension force.

(ii) The tension force acts along the local tangent to the string, and is of constant magnitude, T.

(iii) The slope of the string, and hence its displacement from the x-axis, is small.

(iv) The effect of gravity is negligible.

(v) There is no friction, either due to air resistance on the string or to the fixing of the string at each of its ends.

(vi) The thickness of the string is negligible, but it has a constant line density, ρ, a mass per unit length.

We denote the constant angular velocity of the string about the x-axis by ω, and the angle that the tangent to the string makes with the x-axis by the function $\theta(x)$. Working in a frame of reference that rotates with the string, in which the string is stationary, Newton's first law shows that the forces that act on the string, including the centrifugal force, must be in balance, and hence that

$$T \sin \theta(x + \delta x) - T \sin \theta(x) = -\rho \omega^2 y \sqrt{\delta x^2 + \delta y^2},$$

as shown in Figure 4.2. If we divide through by δx and take the limit $\delta x \to 0$ we obtain

$$T \frac{d}{dx} \{\sin \theta(x)\} + \rho \omega^2 y \sqrt{1 + \left(\frac{dy}{dx}\right)^2} = 0.$$

By definition, $\tan \theta = dy/dx$, so, by elementary trigonometry,

$$\sin \theta = \frac{dy}{dx} \left\{ 1 + \left(\frac{dy}{dx}\right)^2 \right\}^{-1/2},$$

and hence

$$T \frac{d^2 y}{dx^2} + \rho \omega^2 y \left\{ 1 + \left(\frac{dy}{dx}\right)^2 \right\}^2 = 0. \tag{4.1}$$

This equation of motion for the string must be solved subject to $y(0) = y(l) = 0$, which constitutes a rather nasty nonlinear boundary value problem, with no elementary solution apart from the equilibrium solution, $y = 0$.† If, however, we invoke assumption (iii), that $|dy/dx| \ll 1$, the leading order boundary value problem is

$$\frac{d^2 y}{dx^2} + \lambda y = 0, \quad \text{subject to } y(0) = y(l) = 0, \tag{4.2}$$

where $\lambda = \rho \omega^2 / T$. This is now a linear boundary value problem, which we can solve by elementary methods.

---
† This nonlinear boundary value problem can, however, easily be studied in the phase plane (see Chapter 9).

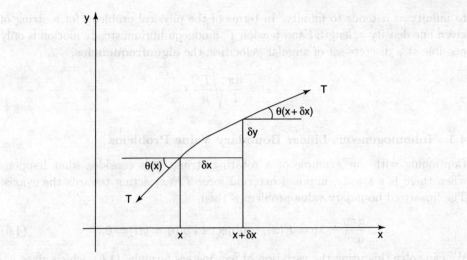

Fig. 4.2. A small element of the rotating string.

If we look for a solution of the form $y = Ae^{mx}$, we obtain $m^2 + \lambda = 0$, and hence $m = \pm i\lambda^{1/2}$, so that the solution is $y = Ae^{i\lambda^{1/2}x} + Be^{-i\lambda^{1/2}x}$. Since $y(0) = 0$, $B = -A$, and $y(l) = 0$ gives

$$e^{i\lambda^{1/2}l} - e^{-i\lambda^{1/2}l} = 0. \tag{4.3}$$

At this stage, we do not know whether λ is real, although, since it represents a ratio of real quantities, it should be. If we write $\lambda^{1/2} = \alpha + i\beta$ and equate real and imaginary parts in (4.3), we find that

$$\left(e^{-\beta l} - e^{\beta l}\right)\cos\alpha l = 0, \quad \left(e^{-\beta l} + e^{\beta l}\right)\sin\alpha l = 0. \tag{4.4}$$

From the first of these, we have either $e^{-\beta l} - e^{\beta l} = 0$ or $\cos\alpha l = 0$, and hence either $\beta = 0$ or $\alpha l = \left(n + \frac{1}{2}\right)\pi$ for n an integer. The latter leaves the second of equations (4.4) with no solution, so we must have $\beta = 0$, and hence $\sin\alpha l = 0$. This gives $\alpha l = n\pi$ with n an integer, which means that $\lambda^{1/2}$ is real, as expected, with $\lambda^{1/2} = n\pi/l$ and $y = A_n\left(e^{in\pi x/l} - e^{-in\pi x/l}\right) = 2iA_n\sin(n\pi x/l)$. To ensure that y, the displacement of the string, is real, we write $A_n = a_n/2i$ for a_n real, which gives

$$y = a_n \sin\left(\frac{n\pi x}{l}\right) \quad \text{for } n = 1, 2, \ldots. \tag{4.5}$$

Note that the solution that corresponds to $n = 0$ is the trivial solution, $y = 0$.

The values of λ for which there is a nontrivial solution of this problem, namely $\lambda = n^2\pi^2/l^2$, are called the **eigenvalues** of the boundary value problem, whilst the corresponding solutions, $y = \sin\lambda^{1/2}x$, are the **eigenfunctions**. Note that there is an infinite sequence of eigenvalues, which are real and have magnitudes that tend to infinity as n tends to infinity. In terms of the physical problem, for a string of given line density ρ, length l and tension T, a nonequilibrium steady motion is only possible at a discrete set of angular velocities, the **eigenfrequencies**,

$$\omega = \frac{n\pi}{l}\sqrt{\frac{T}{\rho}}.$$

4.1 Inhomogeneous Linear Boundary Value Problems

Continuing with our example of a rotating string, let's consider what happens when there is a steady, imposed external force $TF(x)$ acting towards the x-axis. The linearized boundary value problem is then

$$\frac{d^2y}{dx^2} + \lambda y = F(x), \quad \text{subject to } y(0) = y(l) = 0. \tag{4.6}$$

We can solve this using the variation of parameters formula, (1.6), which gives

$$y(x) = A\cos\lambda^{1/2}x + B\sin\lambda^{1/2}x + \frac{1}{\lambda^{1/2}}\int_0^x F(s)\sin\lambda^{1/2}(x-s)\,ds.$$

To satisfy the boundary condition $y(0) = 0$ we need $A = 0$, whilst $y(l) = 0$ gives

$$B\sin\lambda^{1/2}l + \frac{1}{\lambda^{1/2}}\int_0^l F(s)\sin\lambda^{1/2}(x-s)\,ds = 0. \tag{4.7}$$

When λ is *not* an eigenvalue, $\sin\lambda^{1/2}l \neq 0$, and (4.7) has a *unique* solution

$$B = -\frac{1}{\lambda^{1/2}\sin\lambda^{1/2}l}\int_0^l F(s)\sin\lambda^{1/2}(x-s)\,ds.$$

However, when λ *is* an eigenvalue, we have $\sin\lambda^{1/2}l = 0$, so that there is a solution for arbitrary values of B, namely

$$y(x) = B\sin\lambda^{1/2}x + \frac{1}{\lambda^{1/2}}\int_0^x F(s)\sin\lambda^{1/2}(x-s)\,ds, \tag{4.8}$$

provided that

$$\int_0^l F(s)\sin\lambda^{1/2}(l-s)\,ds = 0. \tag{4.9}$$

If $F(s)$ does not satisfy this integral constraint there is *no* solution.

These cases, where there may be either no solution or many solutions depending on the form of $F(x)$, are in complete contrast to the solutions of an initial value problem, for which, as we shall see in Chapter 8, we can prove theorems that guarantee the existence and uniqueness of solutions, subject to a few mild conditions on the form of the ordinary differential equation. This situation also arises for any system that can be written in the form $A\mathbf{x} = \mathbf{b}$, where A is a linear operator. A famous result (see, for example, Courant and Hilbert, 1937) known as the **Fredholm alternative**, shows that *either* there is a unique solution of $A\mathbf{x} = \mathbf{b}$, *or* $A\mathbf{x} = \mathbf{0}$ has nontrivial solutions.

4.1 INHOMOGENEOUS LINEAR BOUNDARY VALUE PROBLEMS

In terms of the physics of the rotating string problem, if ω is not an eigenfrequency, for arbitrary forcing of the string there is a unique steady solution. If ω is an eigenfrequency and the forcing satisfies (4.9), there is a solution, (4.8), that is a linear combination of the eigensolution and the response to the forcing. The size of B depends upon how the steady state was reached, just as for the unforced problem. If ω is an eigenfrequency and the forcing does not satisfy (4.9) there is no steady solution. This reflects the fact that the forcing has a component that drives a response at the eigenfrequency. We say that there is a **resonance**. In practice, if the string were forced in this way, the amplitude of the motion would grow linearly, whilst varying spatially like $\sin \lambda^{1/2} x$, until the nonlinear terms in (4.1) were no longer negligible.

4.1.1 Solubility

As we have now seen, an inhomogeneous, linear boundary value problem may have no solutions. Let's examine this further for the general boundary value problem

$$(p(x)y'(x))' + q(x)y(x) = f(x) \quad \text{subject to} \quad y(a) = y(b) = 0. \qquad (4.10)$$

If $u(x)$ is a solution of the homogeneous problem, so that $(p(x)u'(x))' + q(x)u(x) = 0$ and $u(a) = u(b) = 0$, then multiplying (4.10) by $u(x)$ and integrating over the interval $[a, b]$ gives

$$\int_a^b u(x) \left\{ (p(x)y'(x))' + q(x)y(x) \right\} dx = \int_a^b u(x) f(x) \, dx. \qquad (4.11)$$

Now, using integration by parts,

$$\int_a^b u(x) \left(p(x)y'(x) \right)' dx = [u(x)p(x)y'(x)]_a^b - \int_a^b p(x)y'(x)u'(x) \, dx$$

$$= - \int_a^b p(x)y'(x)u'(x) \, dx,$$

using $u(a) = u(b) = 0$. Integrating by parts again gives

$$- \int_a^b p(x)y'(x)u'(x) \, dx$$

$$= - [p(x)u'(x)y(x)]_a^b + \int_a^b (p(x)u'(x))' y(x) \, dx = \int_a^b (p(x)u'(x))' y(x) \, dx,$$

using $y(a) = y(b) = 0$. Substituting this into (4.11) gives

$$\int_a^b y(x) \left\{ (p(x)u'(x))' + q(x)u(x) \right\} dx = \int_a^b u(x) f(x) \, dx,$$

and, since $u(x)$ is a solution of the homogeneous problem,

$$\int_a^b u(x) f(x) \, dx = 0. \qquad (4.12)$$

A necessary condition for there to be a solution of the inhomogeneous boundary value problem (4.10) is therefore (4.12), which we call a **solvability** or **solubility condition**. We say that the forcing term, $f(x)$, must be **orthogonal** to the solution of the homogeneous problem, a terminology that we will explore in more detail in Section 4.2.3. Note that variations in the form of the boundary conditions will give some variation in the form of the solvability condition.

4.1.2 The Green's Function

Let's now solve (4.10) using the variation of parameters formula, (1.6). This gives

$$y(x) = Au_1(x) + Bu_2(x) + \int_{s=a}^{x} \frac{f(s)}{W(s)} \{u_1(s)u_2(x) - u_1(x)u_2(s)\} \, ds, \quad (4.13)$$

where $u_1(x)$ and $u_2(x)$ are solutions of the homogeneous problem and $W(x) = u_1(x)u_2'(x) - u_1'(x)u_2(x)$ is the Wronskian of the homogeneous equation. The boundary conditions show that

$$Au_1(a) + Bu_2(a) = 0,$$

$$Au_1(b) + Bu_2(b) = \int_a^b \frac{f(s)}{W(s)} \{u_1(b)u_2(s) - u_1(s)u_2(b)\} \, ds.$$

Provided that $u_1(a)u_2(b) \neq u_1(b)u_2(a)$, we can solve these simultaneous equations and substitute back into (4.13) to obtain

$$y(x) = \int_a^b \frac{f(s)}{W(s)} \frac{\{u_1(b)u_2(s) - u_1(s)u_2(b)\}\{u_1(a)u_2(x) - u_1(x)u_2(a)\}}{u_1(a)u_2(b) - u_1(b)u_2(a)} \, ds$$

$$+ \int_a^x \frac{f(s)}{W(s)} \{u_1(s)u_2(x) - u_1(x)u_2(s)\} \, ds. \quad (4.14)$$

This form of solution, although correct, is not the most convenient one to use. To improve it, we note that the functions $v_1(x) = u_1(a)u_2(x) - u_1(x)u_2(a)$ and $v_2(x) = u_1(b)u_2(x) - u_1(x)u_2(b)$, which appear in (4.14) as a product, are linear combinations of solutions of the homogeneous problem, and are therefore themselves solutions of the homogeneous problem. They also satisfy $v_1(a) = v_2(b) = 0$. Because of the way that they appear in (4.14), it makes sense to look for a solution of the inhomogeneous boundary value problem in the form

$$y(x) = \int_a^b f(s) G(x, s) \, ds, \quad (4.15)$$

where

$$G(x, s) = \begin{cases} v_1(s)v_2(x) = G_<(x, s) & \text{for } a \leqslant s < x, \\ v_1(x)v_2(s) = G_>(x, s) & \text{for } x < s \leqslant b. \end{cases}$$

The function $G(x, s)$ is known as the **Green's function** for the boundary value problem. From the definition, it is clear that G is continuous at $s = x$, and that

4.1 INHOMOGENEOUS LINEAR BOUNDARY VALUE PROBLEMS

$y(a) = y(b) = 0$, since, for example, $G_>(a,s) = 0$ and $y(a) = \int_a^b G_>(a,s)f(s)\,ds = 0$. If we calculate the partial derivative with respect to x, we find that

$$G_x(x, s = x^-) - G_x(x, s = x^+) = v_1(x)v_2'(x) - v_1'(x)v_2(x) = W = \frac{C}{p(x)},$$

where, using Abel's formula, C is a constant.

For this to be useful, we must now show that $y(x)$, as defined by (4.15), actually satisfies the inhomogeneous differential equation, and also determine the value of C. To do this, we split the range of integration and write

$$y(x) = \int_a^x G_<(x,s)f(s)\,ds + \int_x^b G_>(x,s)f(s)\,ds.$$

If we now differentiate under the integral sign, we obtain

$$y'(x) = \int_a^x G_{<,x}(x,s)f(s)\,ds + G_<(x,x)f(x) + \int_x^b G_{>,x}(x,s)f(s)\,ds - G_>(x,x)f(x),$$

where $G_{<,x} = \partial G_</\partial x$. This simplifies, by virtue of the continuity of the Green's function at $x = s$, to give

$$y'(x) = \int_a^x G_{<x}(x,s)f(s)\,ds + \int_x^b G_{>x}(x,s)f(s)\,ds$$

$$= \int_a^x v_1(s)v_{2x}(x)f(s)\,ds + \int_x^b v_{1x}(x)v_2(s)f(s)\,ds,$$

and hence

$$(py')' = \int_a^x v_1(s)(pv_{2x})_x f(s)\,ds + p(x)v_1(x)v_{2x}(x)f(x)$$

$$+ \int_x^b (pv_{1x})_x v_2(s)f(s)\,ds - p(x)v_{1x}(x)v_2(x)f(x)$$

$$= \int_a^x v_1(s)(pv_{2x})_x f(s)\,ds + \int_x^b (pv_{1x})_x v_2(s)f(s)\,ds + Cf(x),$$

using the definition of C. If we substitute this into the differential equation (4.10), $(py')' + qy = f$, we obtain

$$\int_a^x v_1(s)(pv_{2x})_x f(s)\,ds + \int_x^b (pv_{1x})_x v_2(s)f(s)\,ds + Cf(x)$$

$$+ q(x)\left\{\int_a^x v_1(s)v_2(x)f(s)\,ds + \int_x^b v_1(x)v_2(s)f(s)\,ds\right\} = f(x).$$

Since v_1 and v_2 are solutions of the homogeneous problem, the integral terms vanish and, if we choose $C = 1$, our representation provides us with a solution of the differential equation.

As an example, consider the boundary value problem

$$y''(x) - y(x) = f(x) \text{ subject to } y(0) = y(1) = 0.$$

The solutions of the homogeneous problem are e^{-x} and e^x. Appropriate combinations of these that satisfy $v_1(0) = v_2(1) = 0$ are $v_1(x) = A \sinh x$ and $v_2(x) = B \sinh(1-x)$, which gives the Green's function

$$G(x, s) = \begin{cases} AB \sinh(1-x) \sinh s & \text{for } 0 \leqslant s < x, \\ AB \sinh(1-s) \sinh x & \text{for } x < s \leqslant 1, \end{cases}$$

which is continuous at $s = x$. In addition,

$$G_x(x, s = x^-) - G_x(x, s = x^+)$$

$$= -AB \cosh(1-x) \sinh x - AB \sinh(1-x) \cosh x = -AB \sinh 1.$$

Since $p(x) = 1$, we require $AB = -1/\sinh 1$, and the final Green's function is

$$G(x, s) = \begin{cases} -\dfrac{\sinh(1-x) \sinh s}{\sinh 1} & \text{for } 0 \leqslant s < x, \\ -\dfrac{\sinh(1-s) \sinh x}{\sinh 1} & \text{for } x < s \leqslant 1. \end{cases}$$

The solution of the inhomogeneous boundary value problem can therefore be written as

$$y(x) = -\int_0^x f(s) \frac{\sinh(1-x) \sinh s}{\sinh 1} \, ds - \int_x^1 f(s) \frac{\sinh(1-s) \sinh x}{\sinh 1} \, ds.$$

We will return to the subject of Green's functions in the next chapter.

4.2 The Solution of Boundary Value Problems by Eigenfunction Expansions

In the previous section, we developed the idea of a Green's function, with which we can solve inhomogeneous boundary value problems for linear, second order ordinary differential equations. We will now develop an alternative approach that draws heavily upon the ideas of linear algebra (see Appendix 1 for a reminder). Before we start, it is useful to be able to work with the simplest possible type of linear differential operator.

4.2.1 Self-Adjoint Operators

We define a linear differential operator, $L : C^2[a, b] \to C[a, b]$, as being in **self-adjoint form** if

$$L = \frac{d}{dx}\left(p(x)\frac{d}{dx}\right) + q(x), \tag{4.16}$$

where $p(x) \in C^1[a,b]$ and is strictly nonzero for all $x \in (a,b)$, and $q(x) \in C[a,b]$. The reasons for referring to such an operator as self-adjoint will become clear later in this chapter.

This definition encompasses a wide class of second order differential operators. For example, if

$$L^1 \equiv a_2(x)\frac{d^2}{dx^2} + a_1(x)\frac{d}{dx} + a_0(x) \tag{4.17}$$

is nonsingular on $[a,b]$, we can write it in self-adjoint form by defining (see Exercise 4.5)

$$p(x) = \exp\left(\int^x \frac{a_1(t)}{a_2(t)}\,dt\right), \quad q(x) = \frac{a_0(x)}{a_2(x)}\exp\left(\int^x \frac{a_1(t)}{a_2(t)}\,dt\right). \tag{4.18}$$

Note that $p(x) \neq 0$ for $x \in [a,b]$. By studying inhomogeneous boundary value problems of the form $Ly = f$, or

$$\frac{d}{dx}\left(p(x)\frac{dy}{dx}\right) + q(x)y = f(x), \tag{4.19}$$

we are therefore considering all second order, nonsingular, linear differential operators. For example, consider **Hermite's equation**,

$$\frac{d^2y}{dx^2} - 2x\frac{dy}{dx} + \lambda y = 0, \tag{4.20}$$

for $-\infty < x < \infty$. This is not in self-adjoint form, but, if we follow the above procedure, the self-adjoint form of the equation is

$$\frac{d}{dx}\left(e^{-x^2}\frac{dy}{dx}\right) + \lambda e^{-x^2} y = 0.$$

This can be simplified, and kept in self-adjoint form, by writing $u = e^{-x^2/2}y$, to obtain

$$\frac{d^2u}{dx^2} - (x^2 - 1)u = -\lambda u. \tag{4.21}$$

4.2.2 Boundary Conditions

To complete the definition of a boundary value problem associated with (4.19), we need to know the boundary conditions. In general these will be of the form

$$\alpha_1 y(a) + \alpha_2 y(b) + \alpha_3 y'(a) + \alpha_4 y'(b) = 0,$$
$$\beta_1 y(a) + \beta_2 y(b) + \beta_3 y'(a) + \beta_4 y'(b) = 0. \tag{4.22}$$

Since each of these is dependent on the values of y and y' at each end of $[a,b]$, we refer to these as **mixed** or **coupled** boundary conditions. It is unnecessarily complicated to work with the boundary conditions in this form, and we can start to simplify matters by deriving **Lagrange's identity**.

Lemma 4.1 (Lagrange's identity) *If L is the linear differential operator given by (4.16) on $[a,b]$, and if $y_1, y_2 \in C^2[a,b]$, then*

$$y_1(Ly_2) - y_2(Ly_1) = [p(y_1 y_2' - y_1' y_2)]'. \tag{4.23}$$

Proof From the definition of L,

$$y_1(Ly_2) - y_2(Ly_1) = y_1\left[(py_2')' + qy_2\right] - y_2\left[(py_1')' + qy_1\right]$$

$$= y_1(py_2')' - y_2(py_1')' = y_1(py_2'' + p'y_2') - y_2(py_1'' + p'y_1')$$

$$= p'(y_1 y_2' - y_1' y_2) + p(y_1 y_2'' - y_1'' y_2) = [p(y_1 y_2' - y_1' y_2)]'. \quad \square$$

Now recall that the space $C[a,b]$ is a real inner product space with a standard inner product defined by

$$\langle f, g \rangle = \int_a^b f(x)g(x)\,dx.$$

If we now integrate (4.23) over $[a,b]$ then

$$\langle y_1, Ly_2 \rangle - \langle Ly_1, y_2 \rangle = [p(y_1 y_2' - y_1' y_2)]_a^b. \tag{4.24}$$

This result can be used to motivate the following definitions. The **adjoint** operator to T, written $\bar{T}$, satisfies $\langle y_1, Ty_2 \rangle = \langle \bar{T}y_1, y_2 \rangle$ for all y_1 and y_2. For example, let's see if we can construct the adjoint to the operator

$$\mathcal{D} \equiv \frac{d^2}{dx^2} + \gamma \frac{d}{dx} + \delta,$$

with $\gamma, \delta \in \mathbb{R}$, on the interval $[0,1]$, when the functions on which $\mathcal{D}$ operates are zero at $x = 0$ and $x = 1$. After integrating by parts and applying these boundary conditions, we find that

$$\langle \phi_1, \mathcal{D}\phi_2 \rangle = \int_0^1 \phi_1\left(\phi_2'' + \gamma \phi_2' + \delta \phi_2\right) dx = \left[\phi_1 \phi_2'\right]_0^1 - \int_0^1 \phi_1' \phi_2'\,dx$$

$$+ \left[\gamma \phi_1 \phi_2\right]_0^1 - \int_0^1 \gamma \phi_1' \phi_2\,dx + \int_0^1 \delta \phi_1 \phi_2\,dx$$

$$= -\left[\phi_1' \phi_2\right]_0^1 + \int_0^1 \phi_1'' \phi_2\,dx - \int_0^1 \gamma \phi_1' \phi_2\,dx + \int_0^1 \delta \phi_1 \phi_2\,dx = \langle \bar{\mathcal{D}}\phi_1, \phi_2 \rangle,$$

where

$$\bar{\mathcal{D}} \equiv \frac{d^2}{dx^2} - \gamma \frac{d}{dx} + \delta.$$

4.2 EIGENFUNCTION EXPANSIONS

A linear operator is said to be **Hermitian**, or **self-adjoint**, if $\langle y_1, Ty_2 \rangle = \langle Ty_1, y_2 \rangle$ for all y_1 and y_2. It is clear from (4.24) that L is a Hermitian, or self-adjoint, operator if and only if

$$\left[p\left(y_1 y_2' - y_1' y_2\right)\right]_a^b = 0,$$

and hence

$$p(b)\{y_1(b)y_2'(b) - y_1'(b)y_2(b)\} - p(a)\{y_1(a)y_2'(a) - y_1'(a)y_2(a)\} = 0. \quad (4.25)$$

In other words, whether or not L is Hermitian depends only upon the boundary values of the functions in the space upon which it operates.

There are three different ways in which (4.25) can occur.

(i) $p(a) = p(b) = 0$. Note that this doesn't violate our definition of p as strictly nonzero on the open interval (a, b). This is the case of **singular boundary conditions**.

(ii) $p(a) = p(b) \neq 0$, $y_i(a) = y_i(b)$ and $y_i'(a) = y_i'(b)$. This is the case of **periodic boundary conditions**.

(iii) $\alpha_1 y_i(a) + \alpha_2 y_i'(a) = 0$ and $\beta_1 y_i(b) + \beta_2 y_i'(b) = 0$, with at least one of the α_i and one of the β_i nonzero. These conditions then have nontrivial solutions if and only if

$$y_1(a)y_2'(a) - y_1'(a)y_2(a) = 0, \quad y_1(b)y_2'(b) - y_1'(b)y_2(b) = 0,$$

and hence (4.25) is satisfied.

Conditions (iii), each of which involves y and y' at a single endpoint, are called **unmixed** or **separated**. We have therefore shown that our linear differential operator is Hermitian with respect to a pair of unmixed boundary conditions. The significance of this result becomes apparent when we examine the eigenvalues and eigenfunctions of Hermitian linear operators.

As an example of how such boundary conditions arise when we model physical systems, consider a string that is rotating (as in the example at the start of this chapter) or vibrating with its ends fixed. This leads to boundary conditions $y(0) = y(a) = 0$ – separated boundary conditions. In the study of the motion of electrons in a crystal lattice, the periodic conditions $p(0) = p(l)$, $y(0) = y(l)$ are frequently used to represent the repeating structure of the lattice.

4.2.3 Eigenvalues and Eigenfunctions of Hermitian Linear Operators

The **eigenvalues** and **eigenfunctions** of a Hermitian, linear operator L are the nontrivial solutions of $Ly = \lambda y$ subject to appropriate boundary conditions.

Theorem 4.1 *Eigenfunctions belonging to distinct eigenvalues of a Hermitian linear operator are orthogonal.*

Proof Let y_1 and y_2 be eigenfunctions that correspond to the distinct eigenvalues λ_1 and λ_2. Then

$$\langle Ly_1, y_2 \rangle = \langle \lambda_1 y_1, y_2 \rangle = \lambda_1 \langle y_1, y_2 \rangle,$$

and

$$\langle y_1, Ly_2 \rangle = \langle y_1, \lambda_2 y_2 \rangle = \lambda_2 \langle y_1, y_2 \rangle,$$

so that the Hermitian property $\langle Ly_1, y_2 \rangle = \langle y_1, Ly_2 \rangle$ gives

$$(\lambda_1 - \lambda_2) \langle y_1, y_2 \rangle = 0.$$

Since $\lambda_1 \neq \lambda_2$, $\langle y_1, y_2 \rangle = 0$, and y_1 and y_2 are orthogonal. $\square$

As we shall see in the next section, all of the eigenvalues of a Hermitian linear operator are real, a result that we will prove once we have defined the notion of a complex inner product.

If the space of functions $C^2[a, b]$ were of finite dimension, we would now argue that the orthogonal eigenfunctions generated by a Hermitian operator are linearly independent and can be used as a basis (or in the case of repeated eigenvalues, extended into a basis). Unfortunately, $C^2[a, b]$ is *not* finite dimensional, and we cannot use this argument. We will have to content ourselves with presenting a *credible* method for solving inhomogeneous boundary value problems based upon the ideas we have developed, and simply state a theorem that guarantees that the method will work in certain circumstances.

4.2.4 Eigenfunction Expansions

In order to solve the inhomogeneous boundary value problem given by (4.19) with $f \in C[a, b]$ and unmixed boundary conditions, we begin by finding the eigenvalues and eigenfunctions of L. We denote these eigenvalues by $\lambda_1, \lambda_2, \ldots, \lambda_n, \ldots$, and the eigenfunctions by $\phi_1(x), \phi_2(x), \ldots, \phi_n(x), \ldots$. Next, we expand $f(x)$ in terms of these eigenfunctions, as

$$f(x) = \sum_{n=1}^{\infty} c_n \phi_n(x). \tag{4.26}$$

By making use of the orthogonality of the eigenfunctions, after taking the inner product of (4.26) with ϕ_n, we find that the expansion coefficients are

$$c_n = \frac{\langle f, \phi_n \rangle}{\langle \phi_n, \phi_n \rangle}. \tag{4.27}$$

Next, we expand the solution of the boundary value problem in terms of the eigenfunctions, as

$$y(x) = \sum_{n=1}^{\infty} d_n \phi_n(x), \tag{4.28}$$

and substitute (4.27) and (4.28) into (4.19) to obtain

$$L\left[\sum_{n=1}^{\infty} d_n \phi_n(x)\right] = \sum_{n=1}^{\infty} c_n \phi_n(x).$$

From the linearity of L and the definition of ϕ_n this becomes

$$\sum_{n=1}^{\infty} d_n \lambda_n \phi_n(x) = \sum_{n=1}^{\infty} c_n \phi_n(x).$$

We have therefore constructed a solution of the boundary value problem with $d_n = c_n/\lambda_n$, *if* the series (4.28) converges and defines a function in $C^2[a, b]$. This process will work correctly and give a unique solution provided that none of the eigenvalues λ_n is zero. When $\lambda_m = 0$, there is no solution if $c_m \neq 0$ and an infinite number of solutions if $c_m = 0$, as we saw in Section 4.1.

Example

Consider the boundary value problem

$$-y'' = f(x) \text{ subject to } y(0) = y(\pi) = 0. \tag{4.29}$$

In this case, the eigenfunctions are solutions of

$$y'' + \lambda y = 0 \text{ subject to } y(0) = y(\pi) = 0,$$

which we already know to be $\lambda_n = n^2$, $\phi_n(x) = \sin nx$. We therefore write

$$f(x) = \sum_{n=1}^{\infty} c_n \sin nx,$$

and the solution of the inhomogeneous problem (4.29) is

$$y(x) = \sum_{n=1}^{\infty} \frac{c_n}{n^2} \sin nx.$$

In the case $f(x) = x$,

$$c_n = \frac{\int_0^\pi x \sin nx \, dx}{\int_0^\pi \sin^2 nx \, dx} = \frac{2(-1)^{n+1}}{n},$$

so that

$$y(x) = 2 \sum_{n=1}^{\infty} \frac{(-1)^{n+1}}{n^3} \sin nx.$$

We will discuss the convergence of this type of series, known as a Fourier series, in detail in Chapter 5.

This example is, of course, rather artificial, and we could have integrated (4.29) directly. There are, however, many boundary value problems for which this eigenfunction expansion method is the only way to proceed analytically, such as the example given in Section 3.9.1 on Bessel functions.

Example

Consider the inhomogeneous equation

$$(1-x^2)y'' - 2xy + 2y = f(x) \quad \text{on } -1 < x < 1, \tag{4.30}$$

with $f \in C[-1,1]$, subject to the condition that y should be bounded on $[-1,1]$. We begin by noting that there is a solubility condition associated with this problem. If $u(x)$ is a solution of the homogeneous problem, then, after multiplying through by u and integrating over $[-1,1]$, we find that

$$\left[u(1-x^2)y'\right]_{-1}^{1} - \left[u'(1-x^2)y\right]_{-1}^{1} = \int_{-1}^{1} u(x)f(x)\,dx.$$

If u and y are bounded on $[-1,1]$, the left hand side of this equation vanishes, so that $\int_{-1}^{1} u(x)f(x)\,dx = 0$. Since the Legendre polynomial, $u = P_1(x) = x$, is the bounded solution of the homogeneous problem, we have

$$\int_{-1}^{1} P_1(x)f(x)\,dx = 0.$$

Now, to solve the boundary value problem, we first construct the eigenfunction solutions by solving $Ly = \lambda y$, which is

$$(1-x^2)y'' - 2xy' + (2-\lambda)y = 0.$$

The choice $2 - \lambda = n(n+1)$, with n a positive integer, gives us Legendre's equation of integer order, which has bounded solutions $y_n(x) = P_n(x)$. These Legendre polynomials are orthogonal over $[-1,1]$ (as we shall show in Theorem 4.4), and form a basis for $C[-1,1]$. If we now write

$$f(x) = \sum_{m=0}^{\infty} A_m P_m(x),$$

where $A_1 = 0$ by the solubility condition, and then expand $y(x) = \sum_{m=0}^{\infty} B_m P_m(x)$, we find that

$$\{2 - m(m+1)\} B_m = A_m \quad \text{for } m \geq 0.$$

The required solution is therefore

$$y(x) = \frac{1}{2}A_0 + B_1 P_1(x) + \sum_{m=2}^{\infty} \frac{A_m}{2 - m(m+1)} P_m(x),$$

with B_1 an arbitrary constant.

Having seen that this method works, we can now state a theorem that gives the method a rigorous foundation.

Theorem 4.2 *If L is a nonsingular, linear differential operator defined on a closed interval $[a,b]$ and subject to unmixed boundary conditions at both endpoints, then*

(i) L has an infinite sequence of real eigenvalues $\lambda_0, \lambda_1, \ldots$, which can be ordered so that

$$|\lambda_0| < |\lambda_1| < \cdots < |\lambda_n| < \cdots$$

and

$$\lim_{n \to \infty} |\lambda_n| = \infty.$$

(ii) The eigenfunctions that correspond to these eigenvalues form a basis for $C[a, b]$, and the series expansion relative to this basis of a piecewise continuous function y with piecewise continuous derivative on $[a, b]$ converges uniformly to y on any subinterval of $[a, b]$ in which y is continuous.

We will not prove this result here.† Instead, we return to the equation, $Ly = \lambda y$, which defines the eigenfunctions and eigenvalues. For a self-adjoint, second order, linear differential operator, this is

$$\frac{d}{dx}\left(p(x)\frac{dy}{dx}\right) + q(x)y = \lambda y, \tag{4.31}$$

which, in its simplest form, is subject to the unmixed boundary conditions

$$\alpha_1 y(a) + \alpha_2 y'(a) = 0, \quad \beta_1 y(b) + \beta_2 y'(b) = 0, \tag{4.32}$$

with $\alpha_1^2 + \alpha_2^2 > 0$ and $\beta_1^2 + \beta_2^2 > 0$ to avoid a trivial condition. This is an example of a **Sturm–Liouville system**, and we will devote the rest of this chapter to a study of the properties of the solutions of such systems.

4.3 Sturm–Liouville Systems

In the first three chapters, we have studied linear second order differential equations. After examining some solution techniques that are applicable to such equations in general, we studied the particular cases of Legendre's equation and Bessel's equation, since they frequently arise in models of physical systems in spherical and cylindrical geometries. We saw that, in each case, we can construct a set of orthogonal solutions that can be used as the basis for a series expansion of the solution of the physical problem in question, namely the Fourier–Legendre and Fourier–Bessel series. In this chapter we will see that Legendre's and Bessel's equations are examples of **Sturm–Liouville equations**, and that we can deduce many properties of such equations independent of the functional form of the coefficients.

4.3.1 The Sturm–Liouville Equation

Sturm–Liouville equations are of the form

$$(p(x)y'(x))' + q(x)y(x) = -\lambda r(x)y(x), \tag{4.33}$$

† For a proof see Ince (1956).

which can be written more concisely as

$$Sy(x, \lambda) = -\lambda r(x) y(x, \lambda), \tag{4.34}$$

where the differential operator S is defined as

$$S\phi \equiv \frac{d}{dx}\left(p(x)\frac{d\phi}{dx}\right) + q(x)\phi. \tag{4.35}$$

This is a slightly more general equation than (4.31). In (4.33), the number λ is the eigenvalue, whose possible values, which may be complex, are critically dependent upon the given boundary conditions. It is often more important to know the properties of λ than it is to construct the actual solutions of (4.33).

We seek to solve the Sturm–Liouville equation, (4.33), on an open interval, (a, b), of the real line. We will also make some assumptions about the behaviour of the coefficients of (4.33) for $x \in (a, b)$, namely that

(i) $p(x)$, $q(x)$ and $r(x)$ are real-valued and continuous,
(ii) $p(x)$ is differentiable, $\tag{4.36}$
(iii) $p(x) > 0$ and $r(x) > 0$.

Some Examples of Sturm–Liouville Equations

Perhaps the simplest example of a Sturm–Liouville equation is **Fourier's equation**,

$$y''(x, \lambda) = -\lambda y(x, \lambda), \tag{4.37}$$

which has solutions $\cos(x\sqrt{\lambda})$ and $\sin(x\sqrt{\lambda})$. We discussed a physical problem that leads naturally to Fourier's equation at the start of this chapter, and we will meet another at the beginning of Chapter 5.

We can write Legendre's equation and Bessel's equation as Sturm–Liouville problems. Recall that Legendre's equation is

$$\frac{d^2 y}{dx^2} - \frac{2x}{1-x^2}\frac{dy}{dx} + \frac{\lambda}{1-x^2} y = 0,$$

and we are usually interested in solving this for $-1 < x < 1$. This can be written as

$$((1 - x^2) y')' = -\lambda y.$$

If $\lambda = n(n+1)$, we showed in Chapter 2 that this has solutions $P_n(x)$ and $Q_n(x)$. Similarly, Bessel's equation, which is usually solved for $0 < x < a$, is

$$x^2 y'' + xy' + (\lambda x^2 - \nu^2)\phi = 0.$$

This can be rearranged into the form

$$(xy')' - \frac{\nu^2}{x} y = -\lambda xy.$$

Again, from the results of Chapter 3, we know that this has solutions of the form $J_\nu(x\sqrt{\lambda})$ and $Y_\nu(x\sqrt{\lambda})$.

Although the Sturm–Liouville forms of these equations may look more cumbersome than the original forms, we will see that they are very convenient for the analysis that follows. This is because of the self-adjoint nature of the differential operator.

4.3.2 Boundary Conditions

We begin with a couple of definitions. The endpoint, $x = a$, of the interval (a, b) is a **regular endpoint** if a is finite and the conditions (4.36) hold on the closed interval $[a, c]$ for each $c \in (a, b)$. The endpoint $x = a$ is a **singular endpoint** if $a = -\infty$ or if a is finite but the conditions (4.36) do not hold on the closed interval $[a, c]$ for some $c \in (a, b)$. Similar definitions hold for the other endpoint, $x = b$. For example, Fourier's equation has regular endpoints if a and b are finite. Legendre's equation has regular endpoints if $-1 < a < b < 1$, but singular endpoints if $a = -1$ or $b = 1$, since $p(x) = 1 - x^2 = 0$ when $x = \pm 1$. Bessel's equation has regular endpoints for $0 < a < b < \infty$, but singular endpoints if $a = 0$ or $b = \infty$, since $q(x) = -\nu^2/x$ is unbounded at $x = 0$.

We can now define the types of boundary condition that can be applied to a Sturm–Liouville equation.

(i) On a finite interval, $[a, b]$, with regular endpoints, we prescribe unmixed, or separated, boundary conditions, of the form

$$\alpha_0 y(a, \lambda) + \alpha_1 y'(a, \lambda) = 0, \quad \beta_0 y(b, \lambda) + \beta_1 y'(b, \lambda) = 0. \tag{4.38}$$

These boundary conditions are said to be **real** if the constants α_0, α_1, β_0 and β_1 are real, with $\alpha_0^2 + \alpha_1^2 > 0$ and $\beta_0^2 + \beta_1^2 > 0$.

(ii) On an interval with one or two singular endpoints, the boundary conditions that arise in models of physical problems are usually boundedness conditions. In many problems, these are equivalent to **Friedrichs boundary conditions**, that for some $c \in (a, b)$ there exists $A \in \mathbb{R}^+$ such that

$$|y(x, \lambda)| \leqslant A \text{ for all } x \in (a, c],$$

and similarly if the other endpoint, $x = b$, is singular, there exists $B \in \mathbb{R}^+$ such that

$$|y(x, \lambda)| \leqslant B \text{ for all } x \in [c, b).$$

We can now define the **Sturm–Liouville boundary value problem** to be the Sturm–Liouville equation,

$$(p(x)y'(x))' + q(x)y(x) = -\lambda r(x) y(x) \text{ for } x \in (a, b),$$

where the coefficient functions satisfy the conditions (4.36), to be solved subject to a separated boundary condition at each regular endpoint and a Friedrichs boundary condition at each singular endpoint. Note that this boundary value problem is homogeneous and therefore always has the trivial solution, $y = 0$. A nontrivial solution, $y(x, \lambda) \not\equiv 0$, is an eigenfunction, and λ is the corresponding eigenvalue.

Some Examples of Sturm–Liouville Boundary Value Problems

Consider Fourier's equation,

$$y''(x, \lambda) = -\lambda y(x, \lambda) \text{ for } x \in (0, 1),$$

subject to the boundary conditions $y(0, \lambda) = y(1, \lambda) = 0$, which are appropriate since both endpoints are regular. The eigenfunctions of this system are $\sin \sqrt{\lambda_n} x$ for $n = 1, 2, \ldots$, with corresponding eigenvalues $\lambda = \lambda_n = n^2 \pi^2$.

Legendre's equation is

$$\{(1 - x^2) y'(x, \lambda)\}' = -\lambda y(x, \lambda) \text{ for } x \in (-1, 1).$$

Note that this is singular at both endpoints, since $p(\pm 1) = 0$. We therefore apply Friedrichs boundary conditions, for example with $c = 0$, in the form

$$|y(x, \lambda)| \leq A \text{ for } x \in (-1, 0], \quad |y(x, \lambda)| \leq B \text{ for } x \in [0, 1),$$

for some $A, B \in \mathbb{R}^+$. In Chapter 2 we used the method of Frobenius to construct the solutions of Legendre's equation, and we know that the only eigenfunctions bounded at both the endpoints are the Legendre polynomials, $P_n(x)$ for $n = 0, 1, 2, \ldots$, with corresponding eigenvalues $\lambda = \lambda_n = n(n + 1)$.

Let's now consider Bessel's equation with $\nu = 1$, over the interval $(0, 1)$,

$$(xy')' - \frac{y}{x} = -\lambda x y.$$

Because of the form of $q(x)$, $x = 0$ is a singular endpoint, whilst $x = 1$ is a regular endpoint. Suitable boundary conditions are therefore

$$|y(x, \lambda)| \leq A \text{ for } x \in \left(0, \tfrac{1}{2}\right], \quad y(1, \lambda) = 0,$$

for some $A \in \mathbb{R}^+$. In Chapter 3 we constructed the solutions of this equation using the method of Frobenius. The solution that is bounded at $x = 0$ is $J_1(x\sqrt{\lambda})$. The eigenvalues are solutions of

$$J_1(\sqrt{\lambda_n}) = 0,$$

which we write as $\lambda = \lambda_1^2, \lambda_2^2, \ldots$, where $J_1(\lambda_n) = 0$.

Finally, let's examine Bessel's equation with $\nu = 1$, but now for $x \in (0, \infty)$. Since both endpoints are now singular, appropriate boundary conditions are

$$|y(x, \lambda)| \leq A \text{ for } x \in \left(0, \tfrac{1}{2}\right], \quad |y(x, \lambda)| \leq B \text{ for } x \in \left[\tfrac{1}{2}, \infty\right),$$

for some $A, B \in \mathbb{R}^+$. The eigenfunctions are again $J_1(x\sqrt{\lambda})$, but now the eigenvalues lie on the half-line $[0, \infty)$. In other words, the eigenfunctions exist for all real, positive λ. The set of eigenvalues for a Sturm–Liouville system is often called the **spectrum**. In the first of the Bessel function examples above, we have a **discrete spectrum**, whereas for the second there is a **continuous spectrum**. We will focus our attention on problems that have a discrete spectrum only.

4.3.3 Properties of the Eigenvalues and Eigenfunctions

In order to further study the properties of the eigenfunctions and eigenvalues, we begin by defining the **inner product** of two complex-valued functions over an interval I to be

$$\langle \phi_1(x), \phi_2(x) \rangle = \int_I \phi_1^*(x) \phi_2(x) \, dx,$$

where a superscript asterisk denotes the complex conjugate. This means that the inner product has the properties

(i) $\langle \phi_1, \phi_2 \rangle = \langle \phi_2, \phi_1 \rangle^*$,
(ii) $\langle a_1 \phi_1, a_2 \phi_2 \rangle = a_1^* a_2 \langle \phi_1, \phi_2 \rangle$,
(iii) $\langle \phi_1, \phi_2 + \phi_3 \rangle = \langle \phi_1, \phi_2 \rangle + \langle \phi_1, \phi_3 \rangle$, $\langle \phi_1 + \phi_2, \phi_3 \rangle = \langle \phi_1, \phi_3 \rangle + \langle \phi_2, \phi_3 \rangle$,
(iv) $\langle \phi, \phi \rangle = \int_I |\phi|^2 \, dx \geqslant 0$, with equality if and only if $\phi(x) \equiv 0$ in I.

Note that this reduces to the definition of a real inner product if ϕ_1 and ϕ_2 are real. If $\langle \phi_1, \phi_2 \rangle = 0$ with $\phi_1 \not\equiv 0$ and $\phi_2 \not\equiv 0$, we say that ϕ_1 and ϕ_2 are **orthogonal**.

Let $y_1(x)$, $y_2(x) \in C^2[a,b]$ be twice-differentiable complex-valued functions. By integrating by parts, it is straightforward to show that (see Lemma 4.1)

$$\langle y_2, \mathcal{S} y_1 \rangle - \langle \mathcal{S} y_2, y_1 \rangle = \left[p(x) \left\{ y_1(x)(y_2^*(x))' - y_1'(x) y_2^*(x) \right\} \right]_\alpha^\beta, \tag{4.39}$$

which is known as **Green's formula**. The inner products are defined over a subinterval $[\alpha, \beta] \subset (a, b)$, so that we can take the limits $\alpha \to a^+$ and $\beta \to b^-$ when the endpoints are singular, and the Sturm–Liouville operator, $\mathcal{S}$, is given by (4.35). Now if $x = a$ is a regular endpoint and the functions y_1 and y_2 satisfy a separated boundary condition at a, then

$$p(a) \left\{ y_1(a)(y_2^*(a))' - y_1'(a) y_2^*(a) \right\} = 0. \tag{4.40}$$

If a is a finite singular endpoint and the functions y_1 and y_2 satisfy the Friedrichs boundary condition at a,

$$\lim_{x \to a^+} \left[p(x) \left\{ y_1(x)(y_2^*(x))' - y_1'(x) y_2^*(x) \right\} \right] = 0. \tag{4.41}$$

Similar results hold at $x = b$.

We can now derive several results concerning the eigenvalues and eigenfunctions of a Sturm–Liouville boundary value problem.

Theorem 4.3 *The eigenvalues of a Sturm–Liouville boundary value problem are real.*

Proof If we substitute $y_1(x) = y(x, \lambda)$ and $y_2(x) = y^*(x, \lambda)$ into Green's formula over the entire interval, $[a, b]$, we have

$$\langle y^*(x, \lambda), \mathcal{S} y(x, \lambda) \rangle - \langle \mathcal{S} y^*(x, \lambda), y(x, \lambda) \rangle$$

$$= \left[p(x) \left\{ y(x, \lambda)(y^*(x, \lambda))' - y'(x, \lambda) y^*(x, \lambda) \right\} \right]_a^b = 0,$$

making use of (4.40) and (4.41). Now, using the fact that the functions $y(x, \lambda)$ and $y^*(x, \lambda)$ are solutions of (4.33) and its complex conjugate, we find that

$$\int_a^b r(x) y(x, \lambda) y^*(x, \lambda)(\lambda - \lambda^*) \, dx = (\lambda - \lambda^*) \int_a^b r(x) |y(x, \lambda)|^2 \, dx = 0.$$

Since $r(x) > 0$ and $y(x, \lambda)$ is nontrivial, we must have $\lambda = \lambda^*$, and hence $\lambda \in \mathbb{R}$. $\square$

Theorem 4.4 *If $y(x, \lambda)$ and $y(x, \tilde{\lambda})$ are eigenfunctions of the Sturm–Liouville boundary value problem, with $\lambda \neq \tilde{\lambda}$, then these eigenfunctions are orthogonal over (a, b) with respect to the weighting function $r(x)$, so that*

$$\int_a^b r(x) y(x, \lambda) y(x, \tilde{\lambda}) \, dx = 0. \tag{4.42}$$

Proof Firstly, notice that the separated boundary condition, (4.38), at $x = a$ takes the form

$$\alpha_0 y_1(a) + \alpha_1 y_1'(a) = 0, \quad \alpha_0 y_2(a) + \alpha_1 y_2'(a) = 0. \tag{4.43}$$

Taking the complex conjugate of the second of these gives

$$\alpha_0 y_2^*(a) + \alpha_1 (y_2'(a))^* = 0, \tag{4.44}$$

since α_0 and α_1 are real. For the pair of equations $(4.43)_2$ and (4.44) to have a nontrivial solution, we need

$$y_1(a) (y_2'(a))^* - y_1'(a) y_2^*(a) = 0.$$

A similar result holds at the other endpoint, $x = b$. This clearly shows that

$$p(x) \left\{ y(x, \lambda) \left(y'(x, \tilde{\lambda}) \right)^* - y'(x, \lambda) \left(y(x, \tilde{\lambda}) \right)^* \right\} \to 0$$

as $x \to a$ and $x \to b$, so that, from Green's formula, (4.39),

$$\langle y(x, \tilde{\lambda}), \mathcal{S}y(x, \lambda) \rangle = \langle \mathcal{S}y(x, \tilde{\lambda}), y(x, \lambda) \rangle.$$

If we evaluate this formula, we find that

$$\int_a^b r(x) y(x, \lambda) y(x, \tilde{\lambda}) \, dx = 0,$$

so that the eigenfunctions associated with the distinct eigenvalues λ and $\tilde{\lambda}$ are orthogonal with respect to the weighting function $r(x)$. $\square$

Example

Consider Hermite's equation, (4.20). By using the method of Frobenius, we can show that there are polynomial solutions, $H_n(x)$, when $\lambda = 2n$ for $n = 0, 1, 2, \ldots$. For example, $H_0(x) = 1$, $H_1(x) = 2x$ and $H_2(x) = 4x^2 - 2$. The solutions of (4.21), the self-adjoint form of the equation, that are bounded at infinity for $\lambda = 2n$ then take the form $u_n = e^{-x^2/2}H_n(x)$, and, from Theorem 4.4, satisfy the orthogonality condition

$$\int_{-\infty}^{\infty} e^{-x^2} H_n(x) H_m(x)\, dx = 0 \text{ for } n \neq m.$$

4.3.4 Bessel's Inequality, Approximation in the Mean and Completeness

We can now define a sequence of **orthonormal eigenfunctions**

$$\phi_n(x) = \frac{\sqrt{r(x)}y(x, \lambda_n)}{\langle \sqrt{r(x)}y(x, \lambda_n), \sqrt{r(x)}y(x, \lambda_n)\rangle},$$

which satisfy

$$\langle \phi_n(x), \phi_m(x)\rangle = \delta_{nm}, \tag{4.45}$$

where δ_{nm} is the Kronecker delta. We will try to establish when we can write a piecewise continuous function $f(x)$ in the form

$$f(x) = \sum_{i=0}^{\infty} a_i \phi_i(x). \tag{4.46}$$

Taking the inner product of both sides of this series with $\phi_j(x)$ shows that

$$a_j = \langle f(x), \phi_j(x)\rangle, \tag{4.47}$$

using the orthonormality condition, (4.45). The quantities a_i are known as the **expansion coefficients**, or **generalized Fourier coefficients**. In order to motivate the infinite series expansion (4.46), we start by approximating $f(x)$ by a finite sum,

$$f_N(x) = \sum_{i=0}^{N} A_i \phi(x, \lambda_i),$$

for some finite N, where the A_i are to be determined so that this provides the most accurate approximation to $f(x)$. The error in this approximation is

$$R_N(x) = f(x) - \sum_{i=0}^{N} A_i \phi(x, \lambda_i).$$

We now try to minimize this error by minimizing its norm

$$\|R_N\|^2 = \langle R_N(x), R_N(x)\rangle = \int_a^b \left\{ f(x) - \sum_{i=0}^{N} A_i \phi_i(x) \right\}^2 dx,$$

which is the **mean square error** in the approximation. Now

$$\|R_N\|^2 = \left\langle f(x) - \sum_{i=0}^{N} A_i \phi_i(x), f(x) - \sum_{i=0}^{N} A_i \phi_i(x) \right\rangle$$

$$= \|f(x)\|^2 - \left\langle f(x), \sum_{i=0}^{N} A_i \phi_i(x) \right\rangle$$

$$- \left\langle \sum_{i=0}^{N} A_i \phi_i(x), f(x) \right\rangle + \left\langle \sum_{i=0}^{N} A_i \phi_i(x), \sum_{i=0}^{N} A_i \phi_i(x) \right\rangle.$$

We can now use the orthonormality of the eigenfunctions, (4.45), and the expression (4.47), which determines the coefficients a_i, to obtain

$$\|R_N(x)\|^2 = \|f(x)\|^2 - \sum_{i=0}^{N} A_i \langle f(x), \phi_i(x) \rangle$$

$$- \sum_{i=0}^{N} A_i^* \langle \phi_i(x), f(x) \rangle + \sum_{i=0}^{N} A_i^* A_i \langle \phi_i(x), \phi_i(x) \rangle$$

$$= \|f(x)\|^2 + \sum_{i=0}^{N} \{-A_i a_i - A_i^* a_i^* + A_i^* A_i\}$$

$$= \|f(x)\|^2 + \sum_{i=0}^{N} \{|A_i - a_i|^2 - |a_i|^2\}.$$

The error is therefore smallest when $A_i = a_i$ for $i = 0, 1, \ldots, N$, so the most accurate approximation is formed by simply truncating the series (4.46) after N terms. In addition, since the norm of $R_N(x)$ is positive,

$$\sum_{i=0}^{N} |a_i|^2 \leq \int_a^b |f(x)|^2 \, dx.$$

As the right hand side of this is independent of N, it follows that

$$\sum_{i=0}^{\infty} |a_i|^2 \leq \int_a^b |f(x)|^2 \, dx, \qquad (4.48)$$

which is **Bessel's inequality**. This shows that the sum of the squares of the expansion coefficients converges. Approximations by the method of least squares are often referred to as approximations **in the mean**, because of the way the error is minimized.

If, for a given orthonormal system, $\phi_1(x), \phi_2(x), \ldots$, any piecewise continuous function can be approximated in the mean to any desired degree of accuracy by choosing N large enough, then the orthonormal system is said to be **complete**. For complete orthonormal systems, $R_N(x) \to 0$ as $N \to \infty$, so that Bessel's inequality becomes an *equality*,

$$\sum_{i=0}^{\infty} |a_i|^2 = \int_a^b |f(x)|^2 \, dx, \qquad (4.49)$$

for every function $f(x)$.

The completeness of orthonormal systems, as expressed by

$$\lim_{N\to\infty} \int_a^b \left\{ f(x) - \sum_{i=0}^{N} a_i\phi_i(x) \right\}^2 dx = 0,$$

does not necessarily imply that $f(x) = \sum_{i=0}^{\infty} a_i\phi_i(x)$, in other words that $f(x)$ has an expansion in terms of the $\phi_i(x)$. If, however, the series $\sum_{i=0}^{\infty} a_i\phi_i(x)$ is uniformly convergent, then the limit and the integral can be interchanged, the expansion is valid, and we say that $\sum_{i=0}^{\infty} a_i\phi_i(x)$ **converges in the mean** to $f(x)$. The completeness of the system $\phi_1(x), \phi_2(x), \ldots$, should be seen as a necessary condition for the validity of the expansion, but, for an arbitrary function $f(x)$, the question of convergence requires a more detailed investigation.

The Legendre polynomials $P_0(x), P_1(x), \ldots$ on the interval $[-1, 1]$ and the Bessel functions $J_\nu(\lambda_1 x), J_\nu(\lambda_2 x), \ldots$ on the interval $[0, a]$ are both examples of complete orthogonal systems (they can easily be made orthonormal), and the expansions of Chapters 2 and 3 *are special cases of the more general results of this chapter.* For example, the Bessel functions $J_\nu(\sqrt{\lambda}x)$ satisfy the Sturm–Liouville equation, (4.33), with $p(x) = x$, $q(x) = -\nu^2/x$ and $r(x) = x$. They satisfy the orthogonality relation

$$\int_0^a x J_\nu(\sqrt{\mu}x) J_\nu(\sqrt{\lambda}x)\, dx = 0,$$

if λ and μ are distinct eigenvalues. Using the regular endpoint condition $J_\nu(\sqrt{\lambda}a) = 0$ and the singular endpoint condition at $x = 0$, the eigenvalues, that is the zeros of $J_\nu(x)$, can be written as $\sqrt{\lambda}a = \lambda_1 a, \lambda_2 a, \ldots$, so that $\sqrt{\lambda} = \lambda_i$ for $i = 1, 2, \ldots$, and we can write

$$f(x) = \sum_{i=1}^{\infty} a_i J_\nu(\lambda_i x),$$

with

$$a_i = \frac{2}{a^2 \{J_\nu'(\lambda_i a)\}^2} \int_0^a x J_\nu(\lambda_i x) f(x)\, dx,$$

consistent with (3.32).

4.3.5 Further Properties of Sturm–Liouville Systems

We conclude this section by investigating some of the qualitative properties of solutions of the Sturm–Liouville system (4.33). In particular, we will establish that the nth eigenfunction has n zeros in the open interval (a, b). We will take a geometrical point of view in order to establish this result, although we could have used an analytical framework. To achieve this, we introduce the **Prüfer substitution**,

$$p(x)y'(x) = R(x)\cos\theta(x), \quad y(x) = R(x)\sin\theta(x). \tag{4.50}$$

BOUNDARY VALUE PROBLEMS

The new dependent variables, R and θ, are then defined by

$$R^2 = y^2 + p^2(y')^2, \quad \theta = \tan^{-1}\left(\frac{y}{py'}\right) \tag{4.51}$$

and, by analogy with polar coordinates, we call R the **amplitude** and θ the **phase angle**. Nontrivial eigenfunction solutions of the Sturm–Liouville equation have $R > 0$, since $R = 0$ at any point $x = x_0$ would mean that $y(x_0) = y'(x_0) = 0$, and hence give the trivial solution for all x.

If we now write $\cot\theta(x) = py'/y$ and differentiate, we obtain

$$-\operatorname{cosec}^2\theta \frac{d\theta}{dx} = \frac{(py')'}{y} - \frac{p(y')^2}{y^2} = -\lambda r - q - \frac{1}{p}\cot^2\theta,$$

and hence

$$\frac{d\theta}{dx} = (q(x) + \lambda r(x))\sin^2\theta + \frac{1}{p(x)}\cos^2\theta. \tag{4.52}$$

If we now differentiate $(4.51)_1$, some simple manipulation gives

$$\frac{dR}{dx} = \frac{1}{2}\left\{\frac{1}{p(x)} - q(x) - \lambda r(x)\right\} R \sin 2\theta. \tag{4.53}$$

The Prüfer substitution has therefore changed our second order linear differential equation into a system of two first order nonlinear differential equations in R and θ over the interval $a < x < b$, (4.52) and (4.53). The equation for θ, (4.52), is however independent of R and is, as we shall see, relatively easy to analyze.

If we now consider the separated boundary conditions

$$\alpha_1 y(a) + \alpha_2 y'(a) = 0, \quad \beta_1 y(b) + \beta_2 y'(b) = 0,$$

we can define two phase angles, γ and δ, such that

$$\tan\gamma = \frac{y(a)}{p(a)y'(a)} = -\frac{\alpha_2}{p(a)\alpha_1} \quad \text{for } 0 \leqslant \gamma \leqslant \pi,$$

$$\tan\delta = \frac{y(b)}{p(b)y'(b)} = -\frac{\beta_2}{p(b)\beta_1} \quad \text{for } 0 \leqslant \delta \leqslant \pi,$$

and the eigenvalue problem that we have to solve is (4.52) subject to

$$\theta(a) = \gamma, \quad \theta(b) = \delta + n\pi \quad \text{for } n = 0, 1, 2, \ldots. \tag{4.54}$$

We need to add this multiple of $n\pi$ because of the periodicity of the tangent function.

We can infer the qualitative form of the solution of (4.52) by drawing its **direction field**. This is the set of small line segments of slope $d\theta/dx$ in the (θ, x) plane, as sketched in Figure 4.3. Note that $d\theta/dx = 1/p(x)$ at $\theta = n\pi$, which is independent of λ. In addition, $d\theta/dx = q(x) + \lambda p(x)$ at $\theta = \left(n + \frac{1}{2}\right)\pi$, which, for fixed x, increases with increasing λ. From Figure 4.4, which shows some typical solution curves for various values of λ, we can see that, for any initial condition $\theta(a) = \gamma$, $\theta(b)$ is an increasing function of λ. As λ increases from $-\infty$, there is a first value, $\lambda = \lambda_0$, for which $\theta(b) = \delta$. As λ increases further, there is a sequence of values for which $\theta = \delta + n\pi$ for $n = 1, 2, \ldots$. Each of these is associated with an eigenfunction, $y_n(x) = R_n(x) \sin \theta(x, \lambda_n)$. This has a zero when $\sin \theta = 0$ and, in the interval $\gamma \leqslant \theta \leqslant \delta + n\pi$, there are precisely n zeros at $\theta = \pi, 2\pi, \ldots, n\pi$.

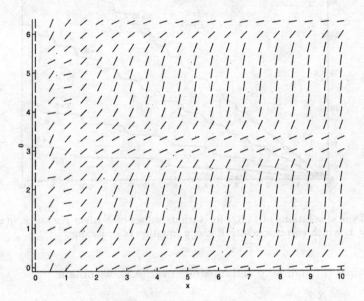

Fig. 4.3. The direction field (lines of slope $d\theta/dx$) for (4.52) when $p(x) = x$, $q(x) = -1/x$ and $r(x) = x$, which corresponds to Bessel's equation of order one.

Returning now to the example $y'' + \lambda y = 0$ subject to $y(0) = y(\pi) = 0$, we have eigenvalues $\lambda_n = n^2$ and eigenfunctions $\sin nx$ for $n = 1, 2, \ldots$. Because of the way we have labelled these, $\lambda_0 = 1$ and $y_0(x) = \sin x$ is the zeroth eigenfunction, which has no zeros in $0 < x < \pi$. We can also see that $\lambda_1 = 2^2$ and the first eigenfunction, $y_1(x) = \sin 2x$, has one zero in $0 < x < \pi$, at $x = \frac{\pi}{2}$, and so on. We can formalize this analysis as a theorem.

Theorem 4.5 *A regular Sturm–Liouville system has an infinite sequence of real eigenvalues, $\lambda_0 < \lambda_1 < \cdots < \lambda_n < \cdots$; with $\lambda_n \to \infty$ as $n \to \infty$. The corresponding eigenfunctions, $y_n(x)$, have n zeros in the interval $a < x < b$.*

We can prove another useful result if we add an additional constraint. If $q(x) < 0$, then all the eigenvalues are *positive*. To see why this should be so, consider the boundary value problem

$$\frac{d}{dx}\left(p(x)\frac{dy}{dx}\right) + (\lambda r(x) + q(x))\, y = 0 \text{ subject to } y(a) = y(b) = 0.$$

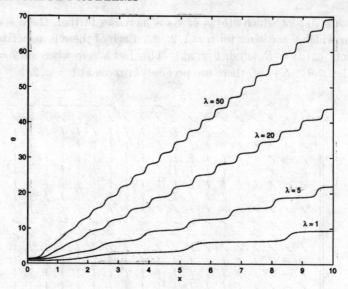

Fig. 4.4. Typical solutions of (4.52) for various values of λ when $p(x) = x$, $q(x) = -1/x$ and $r(x) = x$, which corresponds to Bessel's equation of order one.

If we multiply through by y and integrate over $[a, b]$, we obtain

$$\lambda \int_a^b r(x) y^2 \, dx = \int_a^b -q(x) y^2 \, dx - \int_a^b y \frac{d}{dx}\left(p(x) \frac{dy}{dx}\right) dx$$

$$= \int_a^b \left\{-q(x) y^2 + p(x) (y')^2\right\} dx,$$

using integration by parts. This shows that

$$\lambda = \int_a^b \left\{-q(x) y^2 + p(x) (y')^2\right\} dx \bigg/ \int_a^b r(x) y^2 \, dx \, ,$$

which is positive when p and r are positive and q is negative.

4.3.6 Two Examples from Quantum Mechanics

One of the areas of mathematical physics where Sturm–Liouville equations arise most often is **quantum mechanics**, the theory that governs the behaviour of matter on very small length scales. Three important postulates of quantum mechanics are (see Schiff, 1968):

(i) A system is completely specified by a **state function**, or **wave function**, $\psi(\mathbf{r}, t)$, with $\langle \psi, \psi \rangle = 1$. For example, if the system consists of a particle moving in an external potential, then $|\psi(\mathbf{r},t)|^2\, d^3\mathbf{r}$ is the probability of finding the particle in a small volume $d^3\mathbf{r}$ that surrounds the point $\mathbf{r}$, and hence we need

$$\iiint_{\mathbb{R}^3} |\psi(\mathbf{r})|^2\, d^3\mathbf{r} = \langle \psi, \psi \rangle = 1.$$

(ii) For every system there exists a certain Hermitian operator, H, called the **Hamiltonian operator**, such that

$$i\hbar \frac{\partial \psi}{\partial t} = H\psi,$$

where $2\pi\hbar \approx 6.62 \times 10^{-34}$ J s^{-1} is **Planck's constant**.

(iii) To each observable property of the system there corresponds a linear, Hermitian operator, A, and any measurement of the property gives one of the eigenvalues of A. For example, the operators that correspond to momentum and energy are $-i\hbar\nabla$ and $i\hbar\partial/\partial t$.

For a single particle of mass m moving in a potential field $V(\mathbf{r},t)$, the classical (as opposed to quantum mechanical) total energy is

$$E = p^2/2m + V(\mathbf{r},t), \tag{4.55}$$

where p is the momentum of the particle. This is just the sum of the kinetic and potential energies. To obtain the quantum mechanical analogue of this, we **quantize** the classical result (4.55) by substituting the appropriate operators for momentum and energy, and arrive at

$$i\hbar \frac{\partial \psi}{\partial t} = -\frac{\hbar^2}{2m}\nabla^2 \psi + V(\mathbf{r},t)\psi. \tag{4.56}$$

This is **Schrödinger's equation**, which governs the evolution of the wave function.

Let's look for a separable solution of (4.56) when V is independent of time. We write $\psi = u(\mathbf{r})T(t)$, and find that

$$i\hbar \frac{T'}{T} = -\frac{\hbar^2}{2mu}\nabla^2 u + V(\mathbf{r}) = E,$$

where E is the separation constant. Since $i\hbar T' = ET$,

$$i\hbar \frac{\partial \psi}{\partial t} = E\psi,$$

and hence E is the energy of the particle. The equation for u is then the **time-independent Schrödinger equation**,

$$-\frac{\hbar^2}{2m}\nabla^2 u + (V(\mathbf{r}) - E)\,u = 0. \tag{4.57}$$

We seek solutions of this equation subject to the conditions that u should be finite, and u and ∇u continuous throughout the domain of solution. After we impose appropriate boundary conditions, we will find that the energy of the system is not arbitrary, as classical physics would suggest. Instead, the energy must be one of the eigenvalues of the boundary value problem associated with (4.57), and hence the eigenvalues are of great interest. The energy is said to be **quantized**.

Example: A confined particle

Let's consider the one-dimensional problem of a particle of mass m confined in a region of zero potential by an infinite potential at $x = 0$ and $x = a$. What energies can the particle have?

Since the probability of finding the particle outside the region $0 < x < a$ is zero, we must have $\psi = 0$ there. By continuity, we therefore have $\psi = 0$ at $x = 0$ and $x = a$. We must therefore solve the eigenvalue problem

$$-\frac{\hbar^2}{2m}\frac{\partial^2 u}{\partial x^2} = Eu, \qquad (4.58)$$

subject to $u = 0$ at $x = 0$ and $x = a$. However, this is precisely the problem, given by (4.2), with which we began this chapter, with $\lambda = 2mE/\hbar^2$. We conclude that the allowed energies of the particle are $E = E_n = \hbar^2 \lambda_n/2m = \hbar^2 n^2 \pi^2/2ma^2$.

Example: The hydrogen atom

The hydrogen atom consists of an electron and a proton. Since the mass of the proton is much larger than that of the electron, let's assume that the proton is at rest. The steady state wave function, $\psi(\mathbf{r})$, then satisfies (4.57) which, after rescaling $\mathbf{r}$ and E to eliminate the constants, we write as

$$\nabla^2 \psi - 2V\psi = -2E\psi \quad \text{for } |\mathbf{r}| > 0. \qquad (4.59)$$

The potential $V(\mathbf{r}) = V(r) = -q^2/r$ is the **Coulomb potential** due to the electrical attraction between the proton and the electron, where $-q$ is the charge on the electron and q that on the proton.

We can now look for separable solutions in spherical polar coordinates (r, θ, ϕ) in the form $\psi = R(r)Y(s)\Theta(\phi)$, where $s = \cos\theta$. Substituting this into (4.59) gives us

$$r^2\frac{R''}{R} + 2r\frac{R'}{R} + 2r^2\left(E - V(r)\right) = -\frac{\{(1-s^2)Y'\}'}{Y} - \frac{1}{1-s^2}\frac{\Theta''}{\Theta} = \lambda,$$

for some separation constant λ. If we take $\Theta''/\Theta = -m^2$ with $m = 1, 2, \ldots$ for periodicity, we obtain $\Theta = A_m \cos m\phi + B_m \sin m\phi$. If $\lambda = n(n+1)$, with $n \geqslant m$, then the equation for Y is the associated Legendre equation, which we studied in Chapter 2. The bounded solution of this is $Y = C_n P_n^m(s)$, and we have a solution in the form

$$\psi(r, \theta, \phi) = R(r) P_n^m(s) \left(D_{n,m} \cos m\phi + E_{n,m} \sin m\phi\right),$$

where $D_{n,m}$ and $E_{n,m}$ are arbitrary constants and R satisfies the differential equation

$$r^2 R'' + 2rR' - n(n+1)R + 2r^2\left(E + \frac{q^2}{r}\right)R = 0.$$

We can simplify matters by defining $S = rR$, so that

$$S'' - \frac{n(n+1)}{r}S + 2\left(E + \frac{q^2}{r}\right)S = 0 \quad \text{for } r > 0, \qquad (4.60)$$

to be solved subject to the condition that $S/r \to 0$ as $r \to \infty$. This is an eigenvalue problem for E, the allowable energy levels of the electron in the hydrogen atom.

Some thought and experimentation leads one to try a solution in the form $S = r^{n+1}e^{-\mu r}$, which will satisfy (4.60) provided that $\mu = q^2/(n+1)$ and $E = -\frac{1}{2}\mu^2$. A reduction of order argument then shows that there are no other solutions with $E < 0$ that satisfy the condition at infinity. Each energy level, $E_N = -q^4/2N^2$ for $N = 1, 2, \ldots$, corresponds to a possible steady state for our model of the hydrogen atom, and is in agreement with the experimentally observed values. However, this is not the end of the matter, as it is possible to show that *every* positive value of E corresponds to a bounded, nontrivial solution of (4.60). In other words, there is both a continuous and a discrete part to the spectrum. These states with positive energy can be shown to be unstable, as the electron has too much energy.

Exercises

4.1 Use the eigenfunction expansion method to find a solution of the boundary value problem $y''(x) = -h(x)$ for $0 < x < 2\pi$, subject to the boundary conditions $y(0) = y(2\pi)$, $y'(0) = y'(2\pi)$, with $h \in C[0, 2\pi]$.

4.2 Find the Green's function for the boundary value problems

(a) $y''(x) = f(x)$ subject to $y(-1) = y(1) = 0$,
(b) $y''(x) + \omega^2 y(x) = f(x)$ subject to $y(0) = y(\pi/2) = 0$.

4.3 Comment on the difficulties that you face when trying to construct the Green's function for the boundary value problem

$$y''(x) + y(x) = f(x) \text{ subject to } y(a) = y'(b) = 0.$$

4.4 Show that when $y'' + \lambda y = 0$ for $0 < x < \pi$, the eigenvalues when (a) $y(0) = y'(\pi) = 0$, (b) $y'(0) = y(\pi) = 0$, (c) $y'(0) = y'(\pi) = 0$, are $\left(n + \frac{1}{2}\right)^2$, $\left(n + \frac{1}{2}\right)^2$ and n^2 respectively, where $n = 0, 1, 2, \ldots$. What are the corresponding eigenfunctions?

4.5 Show that the equation

$$\frac{d^2 y}{dx^2} + A(x)\frac{dy}{dx} + \{\lambda B(x) - C(x)\} y = 0$$

can be written in Sturm–Liouville form by defining $p(x) = \exp\left(\int A(x)\, dx\right)$. What are $q(x)$ and $r(x)$ in terms of A, B and C?

4.6 Write the generalized Legendre equation,

$$(1 - x^2)\frac{d^2 y}{dx^2} - 2x\frac{dy}{dx} + \left\{n(n+1) - \frac{m^2}{1-\mu^2}\right\} y = 0,$$

as a Sturm–Liouville equation.

4.7 Determine the eigenvalues, λ, of the fourth order equation $y^{(4)} + \lambda y = 0$ subject to $y(0) = y'(0) = y(\pi) = y'(\pi) = 0$ for $0 \leqslant x \leqslant \pi$.

122 BOUNDARY VALUE PROBLEMS

4.8 Consider the singular Sturm–Liouville system
$$\left(xe^{-x}y'\right)' + \lambda e^{-x}y = 0 \quad \text{for } x > 0,$$
with y bounded as $x \to 0$ and $e^{-x}y \to 0$ as $x \to \infty$. Show that when $\lambda = 0, 1, 2, \ldots$ there are polynomial eigenfunctions. These are known as the **Laguerre polynomials**.

4.9 Show that the boundary value problem
$$y''(x) + A(x)y'(x) + B(x)y(x) = C(x) \quad \text{for all } x \in (a,b),$$
subject to
$$\alpha_1 y(a) + \beta_1 y(b) + \tilde{\alpha}_1 y'(a) + \tilde{\beta}_1 y'(b) = \gamma_1$$
$$\alpha_2 y(a) + \beta_2 y(b) + \tilde{\alpha}_2 y'(a) + \tilde{\beta}_2 y'(b) = \gamma_2,$$
is self-adjoint provided that
$$\beta_1\tilde{\beta}_2 - \tilde{\beta}_1\beta_2 = (\alpha_1\tilde{\alpha}_2 - \tilde{\alpha}_1\alpha_2)\exp\left\{\int_a^b A(x)\,dx\right\}.$$

4.10 Show that
$$-(xy'(x))' = \lambda xy(x)$$
is self-adjoint on the interval $(0,1)$, with $x = 0$ a singular endpoint and $x = 1$ a regular endpoint with the condition $y(1) = 0$.

4.11 Find the eigenvalues of the system consisting of Fourier's equation and the conditions $y(0) - y'(0) = 0$ and $y(\pi) = 0$. Show that these eigenfunctions are orthogonal on the interval $(0, \pi)$.

4.12 Prove that $\sin mx$ has at least one zero between each pair of consecutive zeros of $\sin nx$, when $m > n$.

4.13 * Using the Sturm comparison theorem (see Exercise 1.14), show that every solution of Airy's equation, $y''(x) - xy(x) = 0$, vanishes infinitely often on the positive x-axis, and at most once on the negative x-axis.

CHAPTER FIVE

Fourier Series and the Fourier Transform

In order to motivate our discussion of Fourier series, we shall consider the solution of a diffusion problem. Let's suppose that we have a long, thin, cylindrical metal bar of length L whose curved sides and one end are insulated from its surroundings. Suppose also that, initially, the temperature of the bar is $T = T_0$, but that the uninsulated end is suddenly cooled to a temperature $T_1 < T_0$. How does the temperature in the metal bar vary for $t > 0$? Physical intuition suggests that heat will flow out of the end of the bar, and that, as time progresses, the temperature will approach T_1 throughout the bar.

In order to quantify this, we note that the temperature in the bar satisfies the one-dimensional diffusion equation

$$\frac{\partial T}{\partial t} = K \frac{\partial^2 T}{\partial x^2} \text{ for } 0 < x < L, \tag{5.1}$$

where x measures distance along the bar and K is its thermal diffusivity (see Section 2.6.1). The initial and boundary conditions are

$$T(x,0) = T_0, \quad T(0,t) = T_1, \quad \frac{\partial T}{\partial x}(L,t) = 0. \tag{5.2}$$

Before we solve this initial–boundary value problem, it is convenient to define dimensionless variables,

$$\hat{T} = \frac{T - T_1}{T_0 - T_1}, \quad \hat{x} = \frac{x}{L}, \quad \hat{t} = \frac{Kt}{L^2}.$$

In terms of these variables, (5.1) and (5.2) become

$$\frac{\partial \hat{T}}{\partial \hat{t}} = \frac{\partial^2 \hat{T}}{\partial \hat{x}^2} \text{ for } 0 < \hat{x} < 1, \tag{5.3}$$

$$\hat{T}(\hat{x},0) = 1, \quad \hat{T}(0,\hat{t}) = 0, \quad \frac{\partial \hat{T}}{\partial \hat{x}}(1,\hat{t}) = 0. \tag{5.4}$$

As you can see, by choosing appropriate dimensionless variables, we have managed to eliminate all of the physical constants from the problem.

The use of dimensionless variables has additional advantages, which makes it essential to use them in studying most mathematical models of real physical problems. Consider the length of the metal bar in the diffusion problem that we are studying. It is usual to measure lengths in terms of metres, and, at first sight, it seems reasonable to say that a metal bar with length 100 m is long, whilst a bar of length 10^{-2} m = 1 cm is short. In effect, we choose a bar of length 1 m as a reference relative to which we measure the length of our actual bar. Is this a reasonable thing to do? In fact, the only length that is defined in the problem is the length of the bar itself. This is the most sensible length with which to make the problem dimensionless, and leads to a bar that lies between $\hat{x} = 0$ and $\hat{x} = 1$. For any given problem, it is essential to choose a length scale that is relevant to the physics of the problem itself, not some basically arbitrary length, such as 1 m. For example, problems in celestial mechanics are often best made dimensionless using the mean distance from the Earth to the Sun, whilst problems in molecular dynamics may be made dimensionless using the mean atomic separation in a hydrogen atom. These length scales are enormously different from 1 m. The same argument applies to all of the dimensional, physical quantities that are used in a mathematical model. By dividing each variable by a suitable constant with the same dimensions, we can state that a variable is small or large in a meaningful way. For example, using the dimensionless time, $\hat{t} = Kt/L^2$, which we defined above, the solution when $\hat{t} \ll 1$, the small time solution, really does represent the behaviour of the temperature in the metal bar when diffusion has had little effect on the initial state, *over the length scale, L, that characterizes the full length of the bar*. The other advantage of using dimensionless variables is that any physical constants that remain explicitly in the dimensionless problem appear in dimensionless groups, which can themselves be said to be large or small in a meaningful way. No dimensionless groups appear in the simple diffusion problem given by (5.3) and (5.4), and we will defer any further discussion of them until Chapter 11.

We can now continue with our study of the diffusion of heat in a metal bar by seeking a separable solution, $\hat{T}(\hat{x}, \hat{t}) = X(\hat{x})F(\hat{t})$. On substituting this into (5.3) we obtain

$$\frac{X''}{X} = \frac{\dot{F}}{F} = -k^2, \quad \text{the separation constant.}$$

The equation $\dot{F} = -k^2 F$ has solution $F = ae^{-k^2\hat{t}}$, whilst $X'' + k^2 X = 0$, Fourier's equation, (4.37), which we studied in Chapter 4, has solution

$$X = A \sin k\hat{x} + B \cos k\hat{x}.$$

The condition $\hat{T}(0, \hat{t}) = 0$ means that $X(0) = 0$, and hence that $B = 0$. Similarly, the condition that there should be no flux of heat through the bar at $\hat{x} = 1$ leads to $kA \cos k = 0$, and hence shows that

$$k = \frac{\pi}{2} + n\pi \quad \text{for } n = 0, 1, 2, \ldots.$$

There is therefore a countably-infinite sequence of solutions,

$$\hat{T}_n = A_n \exp\left\{-\pi^2\left(n + \frac{1}{2}\right)^2 \hat{t}\right\} \sin\left\{\pi\left(n + \frac{1}{2}\right)\hat{x}\right\}.$$

Since this is a linear problem, the general solution is

$$\hat{T}(\hat{x},\hat{t}) = \sum_{n=0}^{\infty} A_n \exp\left\{-\pi^2 \left(n+\frac{1}{2}\right)^2 \hat{t}\right\} \sin\left\{\pi \left(n+\frac{1}{2}\right)\hat{x}\right\}. \quad (5.5)$$

We are now left with the task of determining the constants A_n.

The only information we have not used is the initial condition, $\hat{T}(\hat{x},0) = 1$, which shows that

$$\sum_{n=0}^{\infty} A_n \sin\left\{\pi \left(n+\frac{1}{2}\right)\hat{x}\right\} = 1. \quad (5.6)$$

We now multiply this expression through by $\sin\left\{\pi \left(m+\frac{1}{2}\right)\hat{x}\right\}$ and integrate over the length of the bar. After noting that

$$\int_0^1 \sin\left\{\pi \left(n+\frac{1}{2}\right)\hat{x}\right\} \sin\left\{\pi \left(m+\frac{1}{2}\right)\hat{x}\right\} d\hat{x}$$

$$= \int_0^1 \frac{1}{2}\left[\cos\left\{\pi (m-n)\hat{x}\right\} - \cos\left\{\pi (m+n+1)\hat{x}\right\}\right] d\hat{x} = 0 \text{ for } m \neq n, \quad (5.7)$$

and

$$\int_0^1 \sin^2\left\{\pi \left(n+\frac{1}{2}\right)\hat{x}\right\} d\hat{x} = \int_0^1 \frac{1}{2}\left[1 - \cos\left\{\pi (2m+1)\hat{x}\right\}\right] d\hat{x} = \frac{1}{2} \text{ for } m = n, \quad (5.8)$$

we conclude that

$$A_n = 2 \int_0^1 \sin\left\{\pi \left(n+\frac{1}{2}\right)\hat{x}\right\} d\hat{x} = \frac{4}{(2n+1)\pi},$$

and hence that

$$\hat{T}(\hat{x},\hat{t}) = \sum_{n=0}^{\infty} \frac{4}{(2n+1)\pi} \exp\left\{-\pi^2 \left(n+\frac{1}{2}\right)^2 \hat{t}\right\} \sin\left\{\pi \left(n+\frac{1}{2}\right)\hat{x}\right\}. \quad (5.9)$$

This is a **Fourier series** solution, and is shown in Figure 5.1 at various times. We produced Figure 5.1 by plotting the MATLAB function

```
function heat = heat(x,t)
acc = 10^-8; n=ceil(sqrt(-log(acc)/t)-0.5);
N = 0:n; a = 4*exp(-pi^2*(N+0.5).^2*t)/pi./(2*N+1);
for k = 0:n
    X(:,k+1) = sin(pi*(k+0.5)*x(:));
end
heat = X*a';
```

This adds enough terms of the series that $\exp\{-\pi^2(n+\frac{1}{2})^2 t\}$ is less than some small number ($acc = 10^{-8}$) in the final term. Note that the function ceil rounds its argument upwards to the nearest integer (the **ceiling** function).

It is clear that $\hat{T} \to 0$, and hence $T \to T_1$, as $\hat{t} \to \infty$, as expected, and that the

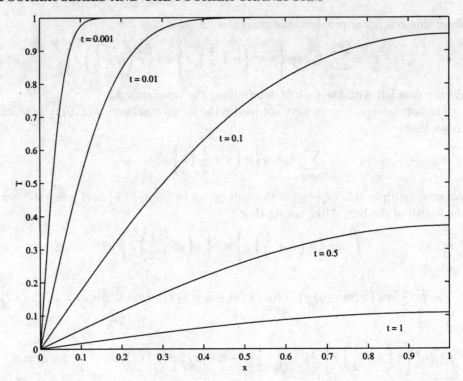

Fig. 5.1. The solution of the initial–boundary value problem, (5.3) and (5.4), given by (5.9), at various times.

heat flows out of the cool end of the bar, $\hat{x} = 0$. It is precisely problems of this sort, which were first solved by Fourier himself in the 1820s, that led to the development of the theory of Fourier series. Note also that (5.9) with $\hat{t} = 0$ and $\hat{x} = 1/2$ and 1 leads to the interesting results

$$1 - \frac{1}{3} + \frac{1}{5} - \frac{1}{7} + \frac{1}{9} - \frac{1}{11} + \cdots = \frac{\pi}{4}, \quad 1 + \frac{1}{3} - \frac{1}{5} - \frac{1}{7} + \frac{1}{9} + \frac{1}{11} - \cdots = \frac{\pi}{2\sqrt{2}}.$$

That the basis functions are orthogonal, as given by (5.7) and (5.8), should come as no surprise, since Fourier's equation is a Sturm–Liouville equation, and the Fourier series is just another example of a series expansion in terms of an orthogonal basis, similar to the Fourier–Legendre and Fourier–Bessel series. We have developed the Fourier series solution of this problem without too much attention to mathematical rigour. However, a number of questions arise. Under what conditions does a series of the form

$$\sum_{n=0}^{\infty} A_n \sin\left\{\pi \left(n + \frac{1}{2}\right) \hat{x}\right\}$$

converge, and is this convergence uniform? Does every function have a convergent Fourier series representation? These questions can also be asked of expansions in terms of other sequences of orthogonal functions, such as the Fourier–Legendre and Fourier–Bessel series. The technical details are, however, rather more straightforward for the Fourier series.

5.1 General Fourier Series

The most general form for a Fourier series representation of a function, $f(t)$, with period T is

$$f(t) = \frac{1}{2}A_0 + \sum_{n=1}^{\infty} \left\{ A_n \cos\left(\frac{2\pi nt}{T}\right) + B_n \sin\left(\frac{2\pi nt}{T}\right) \right\}, \quad (5.10)$$

and, using the method described above, the **Fourier coefficients**, A_n and B_n, are given by

$$A_n = \frac{2}{T} \int_{-\frac{1}{2}T}^{\frac{1}{2}T} \cos\left(\frac{2\pi nt}{T}\right) f(t)\, dt, \quad B_n = \frac{2}{T} \int_{-\frac{1}{2}T}^{\frac{1}{2}T} \sin\left(\frac{2\pi nt}{T}\right) f(t)\, dt. \quad (5.11)$$

Note that if f is an odd function of t, $A_n = 0$, since $\cos(2\pi nt/T)$ is an even function of t. This means that the resulting Fourier series is a sum of just the odd functions $\sin(2\pi nt/T)$, a **Fourier sine series**. Similarly, if f is an even function of t, the resulting expansion is a **Fourier cosine series**. Equations (5.10) and (5.11) can also be written in a more compact, complex form as

$$f(t) = \sum_{n=-\infty}^{\infty} C_n e^{2\pi i nt/T}, \quad (5.12)$$

with complex Fourier coefficients

$$C_n = \frac{1}{2}(A_n - iB_n) = \frac{1}{T} \int_{-\frac{1}{2}T}^{\frac{1}{2}T} e^{-2\pi i nt/T} f(t)\, dt. \quad (5.13)$$

As we have seen, Fourier series can arise as solutions of differential equations, but they are also useful for representing periodic functions in general. For example, consider the function of period 2π, defined by

$$f(t) = t \text{ for } -\pi < t < \pi, \quad f(t) = f(t \pm 2n\pi) \text{ for } n = 1, 2, \ldots,$$

which is piecewise continuous. Using (5.10) and (5.11), we conclude that

$$f(t) = \sum_{n=1}^{\infty} \frac{2}{n}(-1)^{n+1} \sin nt = 2\left(\sin t - \frac{1}{2}\sin 2t + \frac{1}{3}\sin 3t - \cdots\right). \quad (5.14)$$

The partial sums of this series,

$$f_N(t) = \sum_{n=1}^{N} \frac{2}{n}(-1)^{n+1} \sin nt,$$

128   FOURIER SERIES AND THE FOURIER TRANSFORM

are shown in Figure 5.2 for various N. An inspection of this figure suggests that the series does indeed converge, but that the convergence is not uniform. At the points of discontinuity, $t = \pm\pi$, (remember that $f(t)$ has period 2π, and therefore $f \to \pi$ as $t \to \pi^-$, but $f \to -\pi$ as $t \to \pi^+$) the series (5.14) gives the value zero, the mean of the two limits as t approaches π from above and below. In the neighbourhood of $t = \pm\pi$, the difference between the partial sums, f_N, and $f(t)$ appears not to become smaller as N increases, but the size of the region where this occurs decreases – a sure sign of nonuniform convergence. This nonuniform, oscillatory behaviour close to discontinuities is known as **Gibbs' phenomenon**, which we also met briefly in Chapter 3.

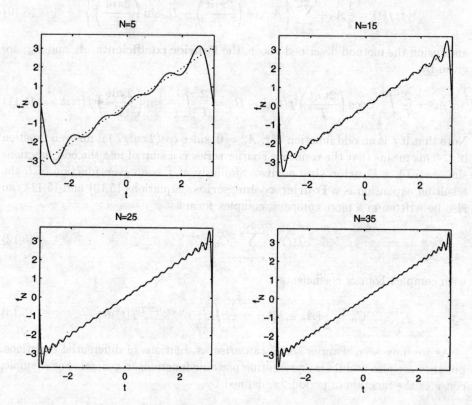

Fig. 5.2. The partial sums of the Fourier series for t, (5.14), for $N = 5$, 15, 25 and 35. The function being approximated is shown as a dotted line.

Before we proceed, we need to construct the **Dirichlet kernel** and prove the **Riemann–Lebesgue lemma**, both of which are crucial in what follows.

Lemma 5.1 (The Dirichlet kernel) *Let f be a bounded function of period T*

with $f \in PC[a,b]$†, and let

$$S_n(f,t) = \sum_{m=-n}^{n} C_m e^{2\pi imt/T} \qquad (5.15)$$

be the nth partial sum of the Fourier series of $f(t)$, with C_m given by (5.13). If we define the **Dirichlet kernel**, $D_n(t)$, to be a function of period T given for $-\frac{1}{2}T < t < \frac{1}{2}T$ by

$$D_n(t) = \begin{cases} \dfrac{\sin\{(2n+1)\frac{\pi t}{T}\}}{\sin(\frac{\pi t}{T})} & \text{when } t \neq 0, \\ 2n+1 & \text{when } t = 0, \end{cases} \qquad (5.16)$$

which is illustrated in Figure 5.3 for $T = 2\pi$, then

$$S_n(f,t) = \frac{1}{T} \int_{-\frac{1}{2}T}^{\frac{1}{2}T} f(\tau) D_n(t-\tau) \, d\tau. \qquad (5.17)$$

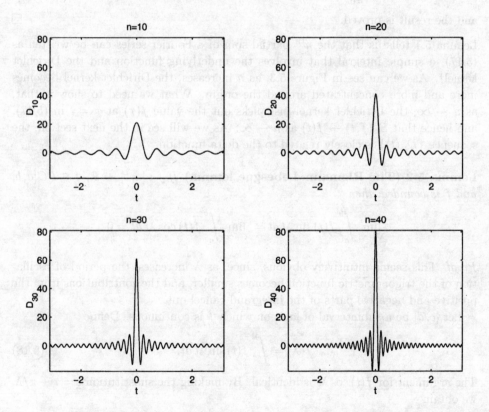

Fig. 5.3. The Dirichlet kernel, D_n for $n = 10, 20, 30$ and 40 and $T = 2\pi$.

† See Appendix 2.

Proof From (5.13) and (5.15),

$$S_n(f,t) = \sum_{m=-n}^{n} \left(\frac{1}{T}\int_{-\frac{1}{2}T}^{\frac{1}{2}T} f(\tau)e^{-2\pi in\tau/T}\,d\tau\right)e^{2\pi int/T}$$

$$= \frac{1}{T}\int_{-\frac{1}{2}T}^{\frac{1}{2}T} f(\tau)\left(\sum_{m=-n}^{n} e^{2\pi in(t-\tau)/T}\right)d\tau = \frac{1}{T}\int_{-\frac{1}{2}T}^{\frac{1}{2}T} f(\tau)D_n(t-\tau)\,d\tau,$$

where

$$D_n(t) = \sum_{m=-n}^{n} e^{2\pi int/T}.$$

Clearly $D_n(0) = 2n+1$. When $t \neq 0$, putting $m = r - n$ gives the simple geometric progression

$$D_n(t) = e^{-2\pi int/T}\sum_{r=0}^{2n} e^{2\pi irt/T} = e^{-2\pi int/T}\frac{e^{4\pi i(n+1/2)t/T}-1}{e^{2\pi it/T}-1} = \frac{\sin\left\{(2n+1)\frac{\pi t}{T}\right\}}{\sin\left(\frac{\pi t}{T}\right)}$$

and the result is proved. $\square$

Lemma 5.1 tells us that the nth partial sum of a Fourier series can be written as (5.17), a simple integral that involves the underlying function and the Dirichlet kernel†. As we can see in Figure 5.3, as n increases, the Dirichlet kernel becomes more and more concentrated around the origin. What we need to show is that, as $n \to \infty$, the Dirichlet kernel just picks out the value $f(\tau)$ at $\tau = t$ in (5.17), and hence that $S_n(f,t) \to f(t)$ as $n \to \infty$. As we will see in the next section, the sequence $\{D_n(t)\}$ is closely related to the delta function.

Lemma 5.2 (The Riemann–Lebesgue lemma) *If $f : [a,b] \to \mathbb{R}$, $f \in PC[a,b]$ and f is bounded, then*

$$\lim_{\lambda \to \infty} \int_a^b f(t) \sin \lambda t\, dt = \lim_{\lambda \to \infty} \int_a^b f(t) \cos \lambda t\, dt = 0.$$

Proof This seems intuitively obvious, since, as λ increases, the period of oscillation of the trigonometric function becomes smaller, and the contributions from the positive and negative parts of the integrand cancel out.

Let $[c,d]$ be a subinterval of $[a,b]$ on which f is continuous. Define

$$I(\lambda) = \int_c^d f(t)\sin \lambda t\, dt. \tag{5.18}$$

The argument for $f(t)\cos \lambda t$ is identical. By making the substitution $t = \tau + \pi/\lambda$, we obtain

$$I(\lambda) = -\int_{c-\pi/\lambda}^{d-\pi/\lambda} f\left(\tau + \frac{\pi}{\lambda}\right)\sin \lambda \tau\, d\tau. \tag{5.19}$$

† As we shall see, this is actually a convolution integral.

Adding (5.18) and (5.19) gives

$$2I(\lambda) = -\int_{c-\pi/\lambda}^{c} f\left(t+\frac{\pi}{\lambda}\right) \sin \lambda t \, dt + \int_{d-\pi/\lambda}^{d} f(t) \sin \lambda t \, dt$$

$$+ \int_{c}^{d-\pi/\lambda} \left\{ f(t) - f\left(t+\frac{\pi}{\lambda}\right) \right\} \sin \lambda t \, dt.$$

Let K be the maximum value of $|f|$ on $[c,d]$, which we know exists by Theorem A2.1. If we also assume that λ is large enough that $\pi/\lambda \leqslant d-c$, then, remembering that $|\sin \lambda t| \leqslant 1$,

$$|I(\lambda)| \leqslant \frac{K\pi}{\lambda} + \frac{1}{2} \int_{c}^{d-\pi/\lambda} \left| f(t) - f\left(t+\frac{\pi}{\lambda}\right) \right| dt. \tag{5.20}$$

Now, since f is continuous on the *closed* interval $[c,d]$, it is also uniformly continuous there and, given any constant ϵ, we can find a constant λ_0 such that

$$\left| f(t) - f\left(t+\frac{\pi}{\lambda}\right) \right| < \frac{\epsilon}{d-c-\pi/\lambda} \quad \forall \lambda > \lambda_0.$$

Since we can also choose λ_0 so that $K\pi/\lambda < \epsilon/2 \; \forall \; \lambda > \lambda_0$, (5.20) shows that $|I(\lambda)| < \epsilon$, and hence that $I(\lambda) \to 0$ as $\lambda \to \infty$. Applying this result to all of the subintervals of $[a,b]$ on which f is continuous completes the proof. $\square$

Theorem 5.1 (The Fourier theorem) *If f and f' are bounded functions of period T with f, $f' \in PC[a,b]$, then the right hand side of (5.12), with C_n given by (5.13), converges pointwise to*

$$\frac{1}{2}\left\{\lim_{\tau \to t^-} f(\tau) + \lim_{\tau \to t^+} f(\tau)\right\} \quad \text{for } -\tfrac{1}{2}T < t < \tfrac{1}{2}T,$$

$$\frac{1}{2}\left\{\lim_{\tau \to -\frac{1}{2}T^+} f(\tau) + \lim_{\tau \to \frac{1}{2}T^-} f(\tau)\right\} \quad \text{for } t = -\tfrac{1}{2}T \text{ or } \tfrac{1}{2}T.$$

Note that, at points where $f(t)$ is continuous, (5.10) converges pointwise to $f(t)$.

Proof At any point $t = t_0 \in (-\tfrac{1}{2}T, \tfrac{1}{2}T)$,

$$f \to f_- \text{ as } t \to t_0^-, \quad f \to f_+ \text{ as } t \to t_0^+,$$

since f is piecewise continuous, with $f_- = f_+$ if f is continuous at $t = t_0$. Similarly, since f' is piecewise continuous, f has well-defined left and right derivatives

$$f'_-(t_0) = \lim_{h \to 0} \frac{f_- - f(t_0 - h)}{h}, \quad f'_+(t_0) = \lim_{h \to 0} \frac{f(t_0 + h) - f_+}{h},$$

with $f'_-(t_0) = f'_+(t_0)$ if f' is continuous at $t = t_0$. By the mean value theorem (Theorem A2.2), for h small enough that f is continuous when $t_0 - h \leqslant t < t_0$,

$$f_- - f(t_0 - h) = f'(c)h \quad \text{for some } t_0 - h < c < t_0.$$

Since f' is bounded, there exists some M such that
$$|f_- - f(t_0 - h)| \leq \frac{1}{2}Mh.$$
After using a similar argument for $t > t_0$, we arrive at
$$|f_- - f(t_0 - h)| + |f(t_0 + h) - f_+| \leq Mh \tag{5.21}$$
for all $h > 0$ such that f is continuous when $t_0 - h < t_0 < t_0 + h$.

Now, in (5.17) we make the change of variable $\tau = t + \tau'$ and, using the facts that $D_n(t) = D_n(-t)$, and f and D_n have period T, we find that
$$S_n(f,t) = \frac{1}{T}\int_{-\frac{1}{2}T}^{\frac{1}{2}T} f(t+\tau')D(\tau')\,d\tau'$$
$$= \frac{1}{T}\int_{-\frac{1}{2}T}^{\frac{1}{2}T} \frac{1}{2}\{f(t+\tau') + f(t-\tau')\}D(\tau')\,d\tau',$$
and hence that
$$S_n(f,t) - \frac{1}{2}(f_- + f_+) = \frac{1}{T}\int_{-\frac{1}{2}T}^{\frac{1}{2}T} g(t,\tau')\sin\left\{(2n+1)\frac{\pi\tau'}{T}\right\}d\tau', \tag{5.22}$$
where
$$g(t,\tau') = \frac{f(t+\tau') - f_+ + f(t-\tau') - f_-}{2\sin(\pi\tau'/T)}.$$

Considered as a function of τ', $g(t,\tau')$ is bounded and piecewise continuous, except possibly at $\tau' = 0$. However, for sufficiently small τ', (5.21) shows that
$$|g(t,\tau')| \leq \frac{M|\tau'|}{2|\sin(\pi\tau'/T)|},$$
which is bounded as $\tau' \to 0$. The function g therefore satisfies the conditions of the Riemann–Lebesgue lemma, and we conclude from (5.22) that $S_n \to \frac{1}{2}(f_- + f_+)$ as $n \to \infty$ and the result is proved. By exploiting the periodicity of f, the same proof works for $t = -\frac{1}{2}T$ or $\frac{1}{2}T$ with minor modifications. $\square$

Theorem 5.2 (Uniform convergence of Fourier series) *For a function $f(t)$ that satisfies the conditions of Theorem 5.1, in any closed subinterval of $[-\frac{1}{2}T, \frac{1}{2}T]$, the right hand side of (5.12), with C_n given by (5.13), converges uniformly to $f(t)$ if and only if $f(t)$ is continuous there.*

We will not give a proof, but note that this is consistent with our earlier discussion of Gibbs' phenomenon.

Finally, returning to the question of whether all bounded, piecewise continuous, periodic functions have a convergent Fourier series expansion, we note that this difficult question was not finally resolved until 1964, when Carleson showed that such functions exist whose Fourier series expansions diverge at a finite or countably-infinite set of points (for example, the rational numbers). Of course, as we have seen in Theorem 5.1, these functions cannot possess a bounded piecewise continuous derivative, even though they are continuous, and are clearly rather peculiar (see, for example, Figure A2.1). However, such functions do arise in the theory of Brownian motion and stochastic processes, which are widely applicable to real world problems, for example in models of financial derivatives. For further details, see Körner (1988), and references therein.

5.2 The Fourier Transform

In order to motivate the definition of the Fourier transform, consider (5.12) and (5.13), which define the complex form of the Fourier series expansion of a periodic function, $f(t)$. What happens as the period, T, tends to infinity? Combining (5.12) and (5.13) gives

$$f(t) = \frac{1}{T} \sum_{n=-\infty}^{\infty} e^{2\pi i n t/T} \int_{-\frac{1}{2}T}^{\frac{1}{2}T} e^{-2\pi i n t'/T} f(t') \, dt'.$$

If we now let $k_n = 2\pi n/T$ and $\Delta k_n = k_n - k_{n-1} = 2\pi/T$, we have

$$f(t) = \frac{1}{2\pi} \sum_{n=-\infty}^{\infty} \Delta k_n e^{i k_n t} \int_{-\frac{1}{2}T}^{\frac{1}{2}T} e^{-i k_n t'} f(t') \, dt'.$$

If we now momentarily abandon mathematical rigour, as $T \to \infty$, k_n becomes a continuous variable and the summation becomes an integral, so that we obtain

$$f(t) = \frac{1}{2\pi} \int_{-\infty}^{\infty} e^{ikt} \int_{-\infty}^{\infty} e^{-ikt'} f(t') \, dt' \, dk, \tag{5.23}$$

which is known as the **Fourier integral**. We will prove this result rigorously later. If we now define the **Fourier transform** of $f(t)$ as

$$\mathcal{F}[f(t)] = \tilde{f}(k) = \int_{-\infty}^{\infty} e^{ikt} f(t) \, dt, \tag{5.24}$$

we immediately have an **inversion formula**,

$$f(t) = \frac{1}{2\pi} \int_{-\infty}^{\infty} e^{-ikt} \tilde{f}(k) \, dk. \tag{5.25}$$

Note that some texts alter the definition of the Fourier transform (and its inverse) to take account of the factor of $1/2\pi$ in a different way, for example with the transform and its inverse each multiplied by a factor of $1/\sqrt{2\pi}$. This is a minor, but irritating, detail, which does not affect the basic ideas.

The Fourier transform maps a function of t to a function of k. In the same way as the Fourier series expansion of a periodic function decomposes the function into its constituent harmonic parts, the Fourier transform produces a function of a continuous variable whose value indicates the frequency content of the original function. This has led to the widespread use of the Fourier transform to analyze the form of time-varying signals, for example in electrical engineering and seismology. We will be more concerned here with the use of the Fourier transform to solve partial differential equations.

Before we can proceed, we need to give (5.24) and (5.25) a firmer mathematical basis. What restrictions must we place on the function $f(t)$ for its Fourier transform to exist? This is a difficult question, and our approach is to treat $f(t)$ as a **generalized function**. The advantage of this is that *every* generalized function has a Fourier transform and an inverse Fourier transform, and that the ordinary functions in whose Fourier transforms we are interested form a subset of the generalized functions. We will not go into great detail, for which the reader is referred to Lighthill (1958), the classic, and very accessible, introduction to the subject.

5.2.1 Generalized Functions

We begin with some definitions. A **good** function, $g(x)$, is a function in $C^\infty(\mathbb{R})$ that decays rapidly enough that $g(x)$ and all of its derivatives tend to zero faster than $|x|^{-N}$ as $x \to \pm\infty$ for all $N > 0$. For example, e^{-x^2} and $\mathrm{sech}^2 x$ are good functions.

A sequence of good functions, $\{f_n(x)\}$, is said to be **regular** if, for any good function $F(x)$,

$$\lim_{n\to\infty} \int_{-\infty}^{\infty} f_n(x) F(x) \, dx \qquad (5.26)$$

exists. For example $f_n(x) = G(x)/n$ is a regular sequence for any good function $G(x)$, with

$$\lim_{n\to\infty} \int_{-\infty}^{\infty} f_n(x) F(x) \, dx = \lim_{n\to\infty} \frac{1}{n} \int_{-\infty}^{\infty} G(x) F(x) \, dx = 0.$$

Two regular sequences of good functions are **equivalent** if, for any good function $F(x)$, the limit (5.26) exists and is the same for each sequence. For example, the regular sequences $\{G(x)/n\}$ and $\{G(x)/n^2\}$ are equivalent.

A **generalized function**, $f(x)$, is a regular sequence of good functions, and two generalized functions are equal if their defining sequences are equivalent. Generalized functions are therefore only defined in terms of their action on integrals of good functions, with

$$\int_{-\infty}^{\infty} f(x) F(x) \, dx = \lim_{n\to\infty} \int_{-\infty}^{\infty} f_n(x) F(x) \, dx,$$

for any good function $F(x)$.

If $f(x)$ is an ordinary function such that $(1+x^2)^{-N} f(x)$ is integrable from $-\infty$ to ∞ for some N, then the generalized function $f(x)$ equivalent to the ordinary function is defined as any sequence of good functions $\{f_n(x)\}$ such that, for any good function $F(x)$,

$$\lim_{n\to\infty} \int_{-\infty}^{\infty} f_n(x) F(x) \, dx = \int_{-\infty}^{\infty} f(x) F(x) \, dx.$$

For example, the generalized function equivalent to zero can be represented by either of the sequences $\{G(x)/n\}$ and $\{G(x)/n^2\}$.

The zero generalized function is very simple. Let's consider some more useful generalized functions.

— The **unit function**, $I(x)$, is defined so that for any good function $F(x)$,

$$\int_{-\infty}^{\infty} I(x)F(x)\,dx = \int_{-\infty}^{\infty} F(x)\,dx.$$

A useful sequence of good functions that defines the unit function is $\{e^{-x^2/4n}\}$. The unit function is the generalized function equivalent to the ordinary function $f(x) = 1$.

— The **Heaviside function**, $H(x)$, is defined so that for any good function $F(x)$,

$$\int_{-\infty}^{\infty} H(x)F(x)\,dx = \int_{0}^{\infty} F(x)\,dx.$$

The generalized function $H(x)$ is equivalent to the ordinary step function†

$$H(x) = \begin{cases} 0 & \text{for } x < 0, \\ 1 & \text{for } x > 0. \end{cases}$$

— The **sign function**, $\text{sgn}(x)$, is defined so that for any good function $F(x)$,

$$\int_{-\infty}^{\infty} \text{sgn}(x)F(x)\,dx = \int_{0}^{\infty} F(x)\,dx - \int_{-\infty}^{0} F(x)\,dx.$$

Then $\text{sgn}(x)$ can be identified with the ordinary function

$$\text{sgn}(x) = \begin{cases} -1 & \text{for } x < 0, \\ 1 & \text{for } x > 0. \end{cases}$$

In fact, $\text{sgn}(x) = 2H(x) - I(x)$, since we can note that

$$\int_{-\infty}^{\infty} \{2H(x) - I(x)\} g(x)\,dx = 2\int_{-\infty}^{\infty} H(x)g(x)\,dx - \int_{-\infty}^{\infty} I(x)g(x)\,dx$$

$$= 2\int_{0}^{\infty} g(x)\,dx - \int_{-\infty}^{\infty} g(x)\,dx = \int_{0}^{\infty} g(x)\,dx - \int_{-\infty}^{0} g(x)\,dx,$$

using the definition of the Heaviside and unit functions. This is just the definition of the function $\text{sgn}(x)$.

— The **Dirac delta function**, $\delta(x)$, is defined so that for any good function $F(x)$,

$$\int_{-\infty}^{\infty} \delta(x)F(x)\,dx = F(0).$$

No ordinary function can be equivalent to the delta function. We can see from

---

† Since generalized functions are only defined through their action on integrals of good functions, the value of H at $x = 0$ does not have any significance in this context.

(5.17) and Theorem 5.1 that a sequence based on the Dirichlet kernel, $f_n(x) = D_n(x)/2\pi = \sin\left\{(2n+1)\frac{\pi t}{T}\right\}/2\pi \sin\left(\frac{\pi t}{T}\right)$, satisfies

$$F(0) = \lim_{n\to\infty} \int_{-\frac{1}{2}T}^{\frac{1}{2}T} f_n(x) F(x)\, dx.$$

As we can see in Figure 5.3, the function becomes more and more concentrated about the origin as n increases, effectively plucking the value of $F(0)$ out of the integral. However, this only works on a finite domain. On an infinite domain, the sequence

$$f_n(x) = \left(\frac{n}{\pi}\right)^{1/2} e^{-nx^2}, \qquad (5.27)$$

which is illustrated in Figure 5.4 for various n, is a useful way of defining $\delta(x)$.

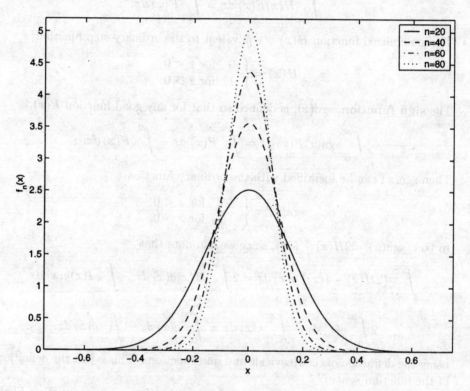

Fig. 5.4. The sequence (5.27), which is equivalent to $\delta(x)$.

5.2.2 Derivatives of Generalized Functions

Derivatives of generalized functions are defined by the derivative of any of the equivalent sequences of good functions. Since we can integrate by parts using any member of the sequence, we can also, formally, do so for the equivalent generalized function, treating it as if it were zero at infinity. For example

$$\int_{-\infty}^{\infty} \delta'(x)F(x)\,dx = -\int_{-\infty}^{\infty} \delta(x)F'(x)\,dx = -F'(0).$$

This does not allow us to represent $\delta'(x)$ in terms of other generalized functions, but does show us how it acts in an integral, which is all we need to know.

We can also show that $H'(x) = \delta(x)$, since

$$\int_{-\infty}^{\infty} H'(x)F(x)\,dx = -\int_{-\infty}^{\infty} H(x)F'(x)\,dx = -\int_{0}^{\infty} F'(x)\,dx = -[F(x)]_{0}^{\infty} = F(0).$$

Another useful result is

$$f(x)\delta(x) = f(0)\delta(x).$$

The proof is straightforward, since

$$\int_{-\infty}^{\infty} f(x)\delta(x)F(x)\,dx = f(0)F(0),$$

and

$$\int_{-\infty}^{\infty} f(0)\delta(x)F(x)\,dx = \int_{-\infty}^{\infty} \delta(x)[f(0)F(x)]dx = f(0)F(0).$$

We can define the modulus function in terms of the function $\text{sgn}(x)$ through $|x| = x\,\text{sgn}(x)$. Now consider the derivative of the modulus function. Using the product rule, which is valid because it works for the equivalent sequence of good functions,

$$\frac{d}{dx}|x| = \frac{d}{dx}\{x\,\text{sgn}(x)\} = x\frac{d}{dx}\{\text{sgn}(x)\} + \text{sgn}(x)\frac{d}{dx}(x).$$

We can now use the fact that $\text{sgn}(x) = 2H(x) - I(x)$ to show that

$$\frac{d}{dx}|x| = x\frac{d}{dx}\{2H(x) - I(x)\} + \text{sgn}(x) = 2x\delta(x) + \text{sgn}(x) = \text{sgn}(x),$$

since $x\delta(x) = 0$.

5.2.3 Fourier Transforms of Generalized Functions

As we have seen, the Fourier transform of a function $f(x)$ is defined as

$$\mathcal{F}[f(x)] = \int_{-\infty}^{\infty} e^{ikx}f(x)\,dx.$$

For example, consider the Fourier transform of the function $e^{-|x|}$, which, using the definition, is

$$\mathcal{F}\left[e^{-|x|}\right] = \int_{-\infty}^{0} e^{x}e^{ikx}\,dx + \int_{0}^{\infty} e^{-x}e^{ikx}\,dx$$

$$= \frac{1}{1+ik} - \frac{1}{-1+ik} = \frac{2}{1+k^2}. \qquad (5.28)$$

We would also like to know the Fourier transform of a constant, c. However, it is not clear whether

$$\mathcal{F}[c] = c \int_{-\infty}^{\infty} e^{ikx} \, dx$$

is a well-defined integral. Instead we note that, treated as a generalized function, $c = cI(x)$, and we can deal with the Fourier transform of an equivalent sequence instead, for example

$$\mathcal{F}[ce^{-x^2/4n}] = c \int_{-\infty}^{\infty} e^{ikx - x^2/4n} \, dx = ce^{-nk^2} \int_{-\infty}^{\infty} e^{-\left(x/2\sqrt{n} - ik\sqrt{n}\right)^2} \, dx.$$

By writing $z = x - 2ikn$ and deforming the contour of integration in the complex plane (see Exercise 5.10 for an alternative method), we find that

$$\mathcal{F}[ce^{-x^2/4n}] = ce^{-nk^2} \int_{-\infty}^{\infty} e^{-z^2/4n} \, dz = 2\pi c \sqrt{\frac{n}{\pi}} e^{-nk^2},$$

using (3.5). Since $\{e^{-x^2/4n}\}$ is a sequence equivalent to the unit function, and $\{\sqrt{\frac{n}{\pi}} e^{-nk^2}\}$ is a sequence equivalent to the delta function, we conclude that

$$\mathcal{F}[c] = 2\pi c \delta(k).$$

Another useful result is that, if $\mathcal{F}[f(x)] = \tilde{f}(k)$, $\mathcal{F}[f(ax)] = \tilde{f}(k/a)/a$ for $a > 0$. To prove this, note that from the definition

$$\mathcal{F}[f(ax)] = \int_{-\infty}^{\infty} e^{ikx} f(ax) \, dx.$$

Making the change of variable $y = ax$ gives

$$\mathcal{F}[f(ax)] = \frac{1}{a} \int_{-\infty}^{\infty} e^{iky/a} f(y) \, dy,$$

which is equal to $\tilde{f}(k/a)/a$ as required. As an example of how this result can be used, we know that $\mathcal{F}[e^{-|x|}] = 2/(1+k^2)$, so

$$\mathcal{F}[e^{-a|x|}] = \frac{1}{a} \frac{2}{1 + \left(\frac{k}{a}\right)^2} = \frac{2a}{a^2 + k^2}.$$

Finally, the Fourier transformation is clearly a linear transformation, so that

$$\mathcal{F}[\alpha f + \beta g] = \alpha \mathcal{F}[f] + \beta \mathcal{F}[g]. \tag{5.29}$$

This means that linear combinations of functions can be transformed separately.

5.2.4 The Inverse Fourier Transform

If we can show that (5.25) holds for all good functions, it follows that it holds for all generalized functions. We begin with a useful lemma.

Lemma 5.3 *The Fourier transform of a good function is a good function.*

5.2 THE FOURIER TRANSFORM

Proof If $f(x)$ is a good function, its Fourier transform clearly exists and is given by

$$\mathcal{F}[f] = \tilde{f}(k) = \int_{-\infty}^{\infty} e^{ikx} f(x)\, dx.$$

If we differentiate p times and integrate by parts N times we find that

$$\left|\tilde{f}^{(p)}(k)\right| \leq \left|\frac{(-1)^N}{(ik)^N} \int_{-\infty}^{\infty} e^{ikx} \frac{d^N}{dx^N}\{(ix)^p f(x)\}\, dx\right|$$

$$\leq \frac{1}{|k|^N} \int_{-\infty}^{\infty} \left|\frac{d^N}{dx^N}\{x^p f(x)\}\right| dx.$$

All derivatives of $\tilde{f}$ therefore decay at least as fast as $|k|^{-N}$ as $|k| \to \infty$ for any $N > 0$, and hence $\tilde{f}$ is a good function. $\square$

Theorem 5.3 (The Fourier inversion theorem) *If $f(x)$ is a good function with Fourier transform*

$$\tilde{f}(k) = \int_{-\infty}^{\infty} e^{ikx} f(x)\, dx,$$

then the inverse Fourier transform is given by

$$f(x) = \frac{1}{2\pi} \int_{-\infty}^{\infty} e^{-ikx} \tilde{f}(k)\, dk.$$

Proof Firstly, we note that for $\epsilon > 0$,

$$\mathcal{F}\left[e^{-\epsilon x^2} \tilde{f}(-x)\right] = \int_{-\infty}^{\infty} e^{ikx - \epsilon x^2} \left\{\int_{-\infty}^{\infty} e^{-ixt} f(t)\, dt\right\} dx.$$

Since f is a good function, we can exchange the order of integration and arrive at

$$\mathcal{F}\left[e^{-\epsilon x^2} \tilde{f}(-x)\right] = \int_{-\infty}^{\infty} f(t) \int_{-\infty}^{\infty} e^{i(k-t)x - \epsilon x^2}\, dx\, dt$$

$$= \int_{-\infty}^{\infty} f(t) e^{-(k-t)^2/4\epsilon} \int_{-\infty}^{\infty} \exp\left\{-\left(\epsilon^{1/2} x - i\frac{(k-t)}{2\epsilon}\right)^2\right\} dx\, dt.$$

Now, by making the change of variable $x = \bar{x} + i(k-t)/2\epsilon^{3/2}$, checking that this change of contour is possible in the complex x-plane, we find that

$$\int_{-\infty}^{\infty} \exp\left\{-\left(\epsilon^{1/2} x - i\frac{(k-t)}{2\epsilon}\right)^2\right\} dx = \int_{-\infty}^{\infty} e^{-\epsilon \bar{x}^2}\, d\bar{x} = \sqrt{\frac{\pi}{\epsilon}}.$$

This means that

$$\mathcal{F}\left[e^{-\epsilon x^2} \tilde{f}(-x)\right] = \sqrt{\frac{\pi}{\epsilon}} \int_{-\infty}^{\infty} e^{-(k-t)^2/4\epsilon} f(t)\, dt.$$

In addition,

$$\sqrt{\frac{\pi}{\epsilon}} \int_{-\infty}^{\infty} e^{-(k-t)^2/4\epsilon} \, dt = \sqrt{\frac{\pi}{\epsilon}} \int_{-\infty}^{\infty} e^{-t^2/4\epsilon} \, dt = 2\pi,$$

so we can write

$$\frac{1}{2\pi} \mathcal{F}\left[e^{-\epsilon x^2} \tilde{f}(-x)\right] - f(k) = \frac{1}{2\pi} \sqrt{\frac{\pi}{\epsilon}} \int_{-\infty}^{\infty} e^{-(k-t)^2/4\epsilon} \{f(t) - f(k)\} \, dt.$$

Since f is a good function,

$$\left|\frac{f(t) - f(k)}{t - k}\right| \leq \max_{x \in \mathbb{R}} |f'(x)|,$$

and hence

$$\left|\frac{1}{2\pi} \mathcal{F}\left[e^{-\epsilon x^2} \tilde{f}(-x)\right] - f(k)\right| \leq \frac{1}{2\pi} \sqrt{\frac{\pi}{\epsilon}} \max_{x \in \mathbb{R}} |f'(x)| \int_{-\infty}^{\infty} e^{-(k-t)^2/4\epsilon} |t - k| \, dt$$

$$= \frac{1}{2\pi} \sqrt{\frac{\pi}{\epsilon}} \max_{x \in \mathbb{R}} |f'(x)| \, 4\epsilon \int_{-\infty}^{\infty} e^{-X^2} |X| \, dX \to 0 \text{ as } \epsilon \to 0.$$

We conclude that

$$f(k) = \frac{1}{2\pi} \mathcal{F}\left[\tilde{f}(-x)\right] = \frac{1}{2\pi} \int_{-\infty}^{\infty} e^{ikx} \int_{-\infty}^{\infty} e^{-ixt} f(t) \, dt.$$

This is precisely the Fourier integral, (5.23), and hence the result is proved. □

5.2.5 Transforms of Derivatives and Convolutions

Fourier transforms are an appropriate tool for solving differential equations on the unbounded domain $-\infty < x < \infty$. In order to proceed, we need to be able to find the Fourier transform of a derivative. For good functions, this can be done using integration by parts. We find that

$$\mathcal{F}[f'(x)] = \int_{-\infty}^{\infty} f'(x) e^{ikx} \, dx = -ik \int_{-\infty}^{\infty} f(x) e^{ikx} \, dx = -ik\mathcal{F}[f],$$

since the good function f must tend to zero as $x \to \pm\infty$. Since generalized functions are defined in terms of sequences of good functions, this result also holds for generalized functions. Similarly, the second derivative is

$$\mathcal{F}[f''(x)] = -k^2 \mathcal{F}[f].$$

We can also define the **convolution** of two functions f and g as

$$f * g = \int_{-\infty}^{\infty} f(y) g(x - y) \, dy,$$

which is a function of x only. Note that $f * g = g * f$. As we shall see below, the solutions of differential equations can often be written as a convolution because of the key property

$$\mathcal{F}[f * g] = \mathcal{F}[f] \mathcal{F}[g].$$

5.3 GREEN'S FUNCTIONS REVISITED

To derive this, we write down the definition of the Fourier transform of $f * g$,

$$\mathcal{F}[f*g] = \int_{-\infty}^{\infty} e^{ikx} \left\{ \int_{-\infty}^{\infty} f(y) g(x-y)\, dy \right\} dx.$$

Note that the factor e^{ikx} is independent of y and so can be taken inside the inner integral to give

$$\mathcal{F}[f*g] = \int_{x=-\infty}^{\infty} \int_{y=-\infty}^{\infty} e^{ikx} f(y) g(x-y)\, dy\, dx.$$

Since the limits of integration are independent of one another, we can exchange the order of integration so that

$$\mathcal{F}[f*g] = \int_{y=-\infty}^{\infty} \int_{x=-\infty}^{\infty} e^{ikx} f(y) g(x-y)\, dx\, dy.$$

Now $f(y)$ is independent of x, and can be extracted from the inner integral to give

$$\mathcal{F}[f*g] = \int_{y=-\infty}^{\infty} f(y) \left\{ \int_{x=-\infty}^{\infty} e^{ikx} g(x-y)\, dx \right\} dy.$$

By making the transformation $z = x - y$ (so that $dz = dx$) in the inner integral, we have

$$\mathcal{F}[f*g] = \int_{y=-\infty}^{\infty} f(y) \left\{ \int_{z=-\infty}^{\infty} e^{ik(z+y)} g(z)\, dz \right\} dy,$$

and now extracting the factor e^{iky}, since this is independent of z, allows us to write

$$\mathcal{F}[f*g] = \int_{y=-\infty}^{\infty} f(y) e^{iky}\, dy \int_{z=-\infty}^{\infty} e^{ikz} g(z)\, dz = \mathcal{F}[f]\mathcal{F}[g].$$

5.3 Green's Functions Revisited

Let's, for the moment, forget our previous definition of a Green's function (Section 4.1.2), and define it instead, for a linear operator L with domain $-\infty < x < \infty$, as the solution of the differential equation $LG = \delta(x)$ subject to $G \to 0$ as $|x| \to \infty$. If we assume that G is a good function, we can use the Fourier transform to find G. For example, consider the operator

$$L \equiv \frac{d^2}{dx^2} - 1,$$

so that $LG = \delta$ is

$$\frac{d^2 G}{dx^2} - G = \delta(x).$$

We will solve this equation subject to $G \to 0$ as $|x| \to \infty$. Taking the Fourier transform of both sides of this equation and exploiting the linearity of the transform, (5.29), gives

$$\mathcal{F}\left[\frac{d^2 G}{dx^2}\right] - \mathcal{F}[G] = \mathcal{F}[\delta(x)].$$

Firstly we need to determine the Fourier transform of the delta function. From the definition
$$\mathcal{F}[\delta(x)] = \int_{-\infty}^{\infty} e^{ikx} \delta(x)\,dx = \left(e^{ikx}\right)\big|_{x=0} = 1.$$

Therefore
$$-k^2 \mathcal{F}[G] - \mathcal{F}[G] = 1,$$

using the fact that $\mathcal{F}[G''] = -k^2 \mathcal{F}[G]$. After rearrangement this becomes
$$\mathcal{F}[G] = -\frac{1}{k^2+1},$$

which we know from (5.28) means that
$$G(x) = -\frac{1}{2} e^{-|x|}.$$

Why should we want to construct such a Green's function? As before, the answer is, to be able to solve the inhomogeneous differential equation. Suppose that we need to solve $L\phi = P$. The solution is $\phi = G * P$. To see this, note that

$$L(G * P) = L \int_{-\infty}^{\infty} G(x-y) P(y)\,dy = \int_{-\infty}^{\infty} LG(x-y) P(y)\,dy$$

$$= \int_{-\infty}^{\infty} \delta(x-y) P(y)\,dy = P(x).$$

Once we know the Green's function we can therefore write down the solution of $L\phi = P$ as $\phi = G * P$.

There are both differences and similarities between this definition of the Green's function, which is sometimes referred to as the **free space Green's function**, and the definition that we gave in Section 4.1.2. Firstly, the free space Green's function depends *only* on x, whilst the other definition depends upon both x and s. We can see some similarities by considering the self-adjoint problem

$$\frac{d}{dx}\left(p(x) \frac{dG}{dx}\right) + q(x) G = \delta(x).$$

On any interval that does not contain the point $x = 0$, G is clearly the solution of the homogeneous problem, as before, and is continuous there. If we integrate between $x = -\epsilon$ and $x = \epsilon$, we get

$$\left[p(x) \frac{dG}{dx}\right]_{-\epsilon}^{\epsilon} + \int_{-\epsilon}^{\epsilon} q(x) G(x)\,dx = \int_{-\epsilon}^{\epsilon} \delta(x)\,dx = 1.$$

If $p(x)$ is continuous at $x = 0$, when we take the limit $\epsilon \to 0$ this reduces to

$$\left[\frac{dG}{dx}\right]_{x=0^-}^{x=0^+} = \frac{1}{p(0)},$$

which, apart from a sign difference, is the result that we obtained in Section 4.1.2.

5.4 SOLUTION OF LAPLACE'S EQUATION USING FOURIER TRANSFORMS

The end result from constructing the Green's function is the same as in Section 4.1.2. We can express the solution of the inhomogeneous boundary value problem as an integral that involves the Green's function and the inhomogeneous term. We will return to consider the Green's function for partial differential equations in Section 5.5.

5.4 Solution of Laplace's Equation Using Fourier Transforms

Let's consider the problem of solving Laplace's equation, $\nabla^2 \phi = 0$, for $y > 0$ subject to $\phi = f(x)$ on $y = 0$ and ϕ bounded as $y \to \infty$. Boundary value problems for partial differential equations where the value of the dependent variable is prescribed on the boundary are often referred to as **Dirichlet problems**. We can solve this Dirichlet problem using a Fourier transform with respect to x. We define

$$\tilde{\phi}(k, y) = \int_{-\infty}^{\infty} e^{ikx} \phi(x, y)\, dx = \mathcal{F}[\phi].$$

We begin by taking the Fourier transform of Laplace's equation, noting that

$$\mathcal{F}\left[\frac{\partial^2 \phi}{\partial y^2}\right] = \frac{\partial^2}{\partial y^2} \mathcal{F}[\phi] = \frac{\partial^2 \tilde{\phi}}{\partial y^2},$$

which is easily verified from the definition of the transform. This gives us

$$\mathcal{F}\left[\frac{\partial^2 \phi}{\partial y^2} + \frac{\partial^2 \phi}{\partial x^2}\right] = \frac{\partial^2 \tilde{\phi}}{\partial y^2} - k^2 \tilde{\phi} = 0.$$

This has solution $\tilde{\phi} = A(k)e^{|k|y} + B(k)e^{-|k|y}$. However, we require that ϕ is bounded as $y \to \infty$, which gives $A(k) = 0$, and hence

$$\tilde{\phi}(k, y) = B(k)e^{-|k|y}.$$

It now remains to satisfy the condition at $y = 0$, namely $\phi(x, 0) = f(x)$. We take the Fourier transform of this condition to give $\tilde{\phi}(k, 0) = \tilde{f}(k)$, where $\tilde{f}(k) = \mathcal{F}[f]$. By putting $y = 0$ we therefore find that $B(k) = \tilde{f}(k)$, so that

$$\tilde{\phi}(k, y) = \tilde{f}(k)e^{-|k|y}.$$

We can invert this Fourier transform of the solution using the convolution theorem. Since

$$\mathcal{F}^{-1}[\tilde{f}\,\tilde{g}] = \mathcal{F}^{-1}[\mathcal{F}[f]\,\mathcal{F}[g]] = f * g,$$

the solution is just the convolution of the boundary condition, $f(x)$, with the inverse transform of $\tilde{g}(k) = e^{-|k|y}$. To find $g(x) = \mathcal{F}^{-1}[\tilde{g}(k)] = \mathcal{F}^{-1}[e^{-|k|y}]$, we note from (5.28) that $\mathcal{F}[e^{-|x|}] = 2/(1 + k^2)$, and exploit the fact that $\mathcal{F}[f(ax)] = \tilde{f}(k/a)/a$, so that $\mathcal{F}[e^{-a|x|}] = 2a/(a^2 + k^2)$, and hence $e^{-a|x|} = \mathcal{F}^{-1}[2a/(a^2 + k^2)]$. Using the formula for the inverse transform gives

$$e^{-a|x|} = \frac{1}{2\pi} \int_{-\infty}^{\infty} e^{-ikx} \frac{2a}{a^2 + k^2}\, dk.$$

We can exploit the similarity between the Fourier transform and its inverse by making the transformation $k \mapsto -x$, $x \mapsto k$, which leads to

$$e^{-a|k|} = \frac{1}{2\pi}\int_{-\infty}^{\infty} e^{ikx}\frac{2a}{a^2+x^2}\,dx = \frac{1}{2\pi}\mathcal{F}\left[\frac{2a}{a^2+x^2}\right],$$

and hence

$$\mathcal{F}^{-1}\left[e^{-|k|y}\right] = \frac{y}{\pi(y^2+x^2)}.$$

In the definition of the convolution we use the variable ξ as the dummy variable rather than y to avoid confusion with the spatial coordinate in the problem, so that

$$f*g = \int_{-\infty}^{\infty} f(\xi)g(x-\xi)\,d\xi.$$

The solution ϕ can now be written as the convolution integral

$$\phi(x,y) = \int_{-\infty}^{\infty} f(\xi)\frac{y}{\pi\{y^2+(x-\xi)^2\}}\,d\xi = \frac{y}{\pi}\int_{-\infty}^{\infty}\frac{f(\xi)}{y^2+(x-\xi)^2}\,d\xi. \qquad (5.30)$$

As an example, consider the two-dimensional, inviscid, irrotational flow in the upper half plane (see Section 2.6.2), driven by the Dirichlet condition $\phi(x,0) = f(x)$, where

$$f(x) = \begin{cases} 1 & \text{for } -1 < x < 1, \\ 0 & \text{elsewhere.} \end{cases}$$

Using the formula we have derived,

$$\phi = \frac{y}{\pi}\int_{-1}^{1}\frac{d\xi}{y^2+(x-\xi)^2} = \frac{y}{\pi}\left[\frac{1}{y}\tan^{-1}\left(\frac{\xi-x}{y}\right)\right]_{-1}^{1}$$

$$= \frac{1}{\pi}\left\{\tan^{-1}\left(\frac{1-x}{y}\right) + \tan^{-1}\left(\frac{1+x}{y}\right)\right\}. \qquad (5.31)$$

Figure 5.5 shows some contours of equal ϕ.†

Finally, let's consider the **Neumann problem** for Laplace's equation in the upper half plane, $y > 0$. This is the same as the Dirichlet problem except that the boundary condition is in terms of a derivative, with $\partial\phi/\partial y = f(x)$ at $y = 0$. As before we find that $\tilde{\phi}(k,y) = B(k)e^{-|k|y}$, and the condition at $y = 0$ tells us that

$$B(k) = -\frac{\tilde{f}(k)}{|k|},$$

and hence

$$\tilde{\phi}(k,y) = -\tilde{f}(k)\frac{e^{-|k|y}}{|k|}.$$

In order to invert this Fourier transform we recall that

$$\int_{-\infty}^{\infty} e^{ikx}\frac{2y}{y^2+x^2}\,dx = 2\pi e^{-|k|y},$$

† See Section 2.6.2 for details of how to create this type of contour plot in MATLAB.

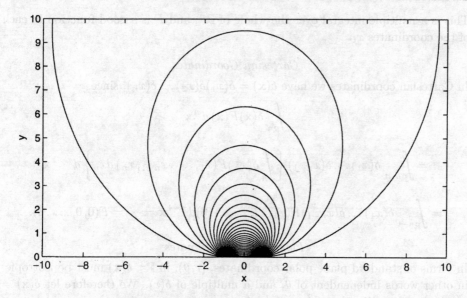

Fig. 5.5. Contours of constant potential function, ϕ, given by (5.31).

and integrate both sides with respect to y to obtain

$$\int_{x=-\infty}^{\infty} e^{ikx} \log(y^2 + x^2)\, dx = -2\pi \frac{e^{-|k|y}}{|k|}.$$

In other words

$$-\frac{e^{-|k|y}}{|k|} = \frac{1}{2\pi} \mathcal{F}\left[\log\left(y^2 + x^2\right)\right],$$

and hence the solution is

$$\phi(x,y) = \frac{1}{2\pi} \int_{-\infty}^{\infty} f(\xi) \log\{y^2 + (x-\xi)^2\}\, d\xi.$$

5.5 Generalization to Higher Dimensions

The theory of Fourier transforms and generalized functions can be extended to higher dimensions. This allows us to use the techniques that we have developed above to solve other partial differential equations.

5.5.1 The Delta Function in Higher Dimensions

If we let $\mathbf{x} = (x_1, x_2, \ldots, x_n)$ be a vector in $\mathbb{R}^n$, we can define the delta function in $\mathbb{R}^n$, $\delta(\mathbf{x})$, through the integral

$$\int_{\mathbb{R}^n} \delta(\mathbf{x}) F(\mathbf{x})\, d^n\mathbf{x} = F(\mathbf{0}).$$

Cartesian Coordinates

In Cartesian coordinates we have $\delta(\mathbf{x}) = \delta(x_1)\delta(x_2)\ldots\delta(x_n)$, since

$$\int_{\mathbb{R}^n} \delta(\mathbf{x}) F(\mathbf{x}) d^n\mathbf{x}$$

$$= \int_{\mathbb{R}^{n-1}} \delta(x_1)\ldots\delta(x_{n-1}) \left\{ \int_{\mathbb{R}} \delta(x_n) F(x_1,\ldots,x_{n-1},x_n) dx_n \right\} d^{n-1}\mathbf{x}$$

$$= \int_{\mathbb{R}^{n-1}} \delta(x_1)\ldots\delta(x_{n-1}) F(x_1,\ldots,x_{n-1},0) d^{n-1}\mathbf{x} = \cdots = F(0,0,\ldots,0).$$

Plane Polar Coordinates

In terms of standard plane polar coordinates, (r,θ), in $\mathbb{R}^2$, $\delta(\mathbf{x})$ must be isotropic, in other words independent of θ, and a multiple of $\delta(r)$. We therefore let $\delta(\mathbf{x}) = a(r)\delta(r)$ and, noting that $d^2\mathbf{x} = r\, dr\, d\theta$,

$$\int_{\theta=0}^{2\pi} \int_{r=0}^{\infty} \delta(\mathbf{x}) d^2\mathbf{x} = 2\pi \int_0^{\infty} a(r)\delta(r) r\, dr.$$

By symmetry,

$$\int_0^{\infty} \delta(r) dr = \frac{1}{2},$$

so we can take $a(r) = 1/\pi r$, and hence

$$\delta(\mathbf{x}) = \frac{1}{\pi r} \delta(r).$$

Spherical Polar Coordinates

In $\mathbb{R}^3$, the isotropic volume element can be written as $d^3\mathbf{x} = 4\pi r^2\, dr$, the volume of a thin, spherical shell, where r is now $|\mathbf{x}|$, the distance from the origin. Again, the delta function must be a multiple of $\delta(r)$, and the same argument gives

$$\delta(\mathbf{x}) = \frac{1}{2\pi r^2} \delta(r).$$

5.5.2 Fourier Transforms in Higher Dimensions

If $f(\mathbf{x}) = f(x_1, x_2, \ldots, x_n)$ and $\mathbf{k} = (k_1, k_2, \ldots, k_n)$, we can define the Fourier transform of f as

$$\tilde{f}(\mathbf{k}) = \int_{\mathbb{R}^n} f(\mathbf{x}) e^{i\mathbf{k}\cdot\mathbf{x}} d^n\mathbf{x}. \qquad (5.32)$$

For example, in $\mathbb{R}^3$

$$\tilde{f}(k_1,k_2,k_3) = \int_{x_3=-\infty}^{\infty} \int_{x_2=-\infty}^{\infty} \int_{x_1=-\infty}^{\infty} f(x_1,x_2,x_3) e^{i(k_1 x_1 + k_2 x_2 + k_3 x_3)} dx_1\, dx_2\, dx_3.$$

5.5 GENERALIZATION TO HIGHER DIMENSIONS

We can proceed as we did for one-dimensional functions and show, although we will not provide the details, that:

(i) If $f(\mathbf{x})$ is a generalized function, then its Fourier transform exists as a generalized function.
(ii) The inversion formula is

$$f(\mathbf{x}) = \frac{1}{(2\pi)^n} \int_{\mathbb{R}^n} \tilde{f}(\mathbf{k}) e^{-i\mathbf{k}\cdot\mathbf{x}} d^n\mathbf{k}. \tag{5.33}$$

The inversion of Fourier transforms in higher dimensions is considerably more difficult than in one dimension, as we shall see.

Example: Laplace's equation

Let's try to construct the free space Green's function for Laplace's equation in $\mathbb{R}^3$. We introduced the idea of a Green's function for a linear operator in Section 5.3. In this case, we seek a solution of

$$\nabla^2 G = \delta(\mathbf{x}) \quad \text{subject to } G \to 0 \text{ as } |\mathbf{x}| \to \infty. \tag{5.34}$$

The Fourier transform of G is

$$\tilde{G}(k_1, k_2, k_3) = \int_{-\infty}^{\infty} \int_{-\infty}^{\infty} \int_{-\infty}^{\infty} G(x_1, x_2, x_3) e^{i(k_1 x_1 + k_2 x_2 + k_3 x_3)} \, dx_1 \, dx_2 \, dx_3,$$

so that

$$F\left[\frac{\partial G}{\partial x_1}\right] = \int_{-\infty}^{\infty}\int_{-\infty}^{\infty}\int_{-\infty}^{\infty} \frac{\partial G}{\partial x_1}(x_1, x_2, x_3) e^{i(k_1 x_1 + k_2 x_2 + k_3 x_3)} \, dx_1 \, dx_2 \, dx_3$$

$$= \int_{-\infty}^{\infty}\int_{-\infty}^{\infty} \left\{ \left[e^{i(k_1 x_1 + k_2 x_2 + k_3 x_3)} G\right]_{-\infty}^{\infty} \right.$$

$$\left. - \int_{-\infty}^{\infty} G i k_1 e^{i(k_1 x_1 + k_2 x_2 + k_3 x_3)} \, dx_1 \right\} dx_2 \, dx_3,$$

after integrating by parts. Since G vanishes at infinity, we conclude that

$$\left[\frac{\partial G}{\partial x_1}\right] = -ik_1 \int_{-\infty}^{\infty}\int_{-\infty}^{\infty}\int_{-\infty}^{\infty} G e^{i(k_1 x_1 + k_2 x_2 + k_3 x_3)} \, dx_1 \, dx_2 \, dx_3 = -ik_1 \tilde{G}.$$

This result is, of course, analogous to the one-dimensional result that we derived in Section 5.2.5.

If we now take the Fourier transform of (5.34), we find that

$$\{(-ik_1)^2 + (-ik_2)^2 + (-ik_3)^2\}\tilde{G} = \int_{\mathbb{R}^3} \delta(\mathbf{x}) e^{i\mathbf{k}\cdot\mathbf{x}} d^3\mathbf{x} = 1,$$

and hence $\tilde{G} = -1/|\mathbf{k}|^2$. The inversion formula, (5.33), then shows that

$$G(\mathbf{x}) = -\frac{1}{8\pi^3} \int_{\mathbb{R}^3} \frac{e^{-i\mathbf{k}\cdot\mathbf{x}}}{|\mathbf{k}|^2} d^3\mathbf{k}. \tag{5.35}$$

In order to evaluate this integral, we need to introduce spherical polar coordinates

(k, θ, ϕ), with the line $\theta = 0$ in the **x**-direction. Then $\mathbf{k} \cdot \mathbf{x} = |\mathbf{k}| |\mathbf{x}| \cos \theta = kr \cos \theta$ and (5.35) becomes

$$G(\mathbf{x}) = -\frac{1}{8\pi^3} \int_{k=0}^{\infty} \int_{\theta=0}^{\pi} \int_{\phi=0}^{2\pi} \frac{e^{-ikr\cos\theta}}{k^2} k^2 \sin\theta \, dk \, d\theta \, d\phi$$

$$= -\frac{1}{4\pi^2} \int_{k=0}^{\infty} \left[\frac{e^{-ikr\cos\theta}}{ikr} \right]_{\theta=0}^{\pi} dk = -\frac{1}{2\pi^2} \int_{k=0}^{\infty} \frac{\sin kr}{kr} dk$$

$$= -\frac{1}{2\pi^2 r} \int_{z=0}^{\infty} \frac{\sin z}{z} dz = -\frac{1}{4\pi r},$$

using the standard result, $\int_0^\infty (\sin z/z) \, dz = \pi/2$ (see, for example, Ablowitz and Fokas, 1997).

Now that we know the Green's function, we are able to solve the inhomogeneous problem

$$\nabla^2 \phi = Q(\mathbf{x}) \text{ subject to } \phi \to 0 \text{ as } |\mathbf{x}| \to \infty, \qquad (5.36)$$

in terms of a convolution integral, by a direct generalization of the results presented in Section 5.2.5. We find that

$$\phi = -\frac{1}{4\pi} \int_{\mathbb{R}^3} \frac{Q(\mathbf{y})}{|\mathbf{x} - \mathbf{y}|} d^3\mathbf{y}.$$

This result is fundamental in the theory of electrostatics, where Q is the distribution of charge density, and ϕ the corresponding electrical potential.

Example: The wave equation

Let's try to solve the three-dimensional wave equation†,

$$\nabla^2 u = \frac{1}{c^2} \frac{\partial^2 u}{\partial t^2} \text{ for } t > 0 \text{ and } \mathbf{x} \in \mathbb{R}^3, \qquad (5.37)$$

subject to the initial conditions

$$u(x, y, z, 0) = 0, \quad \frac{\partial u}{\partial t}(x, y, z, 0) = f(x, y, z). \qquad (5.38)$$

The Fourier transform of (5.37) is

$$\frac{\partial^2 \tilde{u}}{\partial t^2} = -c^2 k^2 \tilde{u}, \qquad (5.39)$$

where $\tilde{u}$ is the Fourier transform of u and $k^2 = k_1^2 + k_2^2 + k_3^2$. The initial conditions, (5.38), then become

$$\tilde{u}(\mathbf{k}, 0) = 0, \quad \frac{\partial \tilde{u}}{\partial t}(\mathbf{k}, 0) = \tilde{f}(\mathbf{k}).$$

The general solution of (5.39) is

$$\tilde{u} = A(\mathbf{k})e^{ickt} + B(\mathbf{k})e^{-ickt},$$

† See Section 3.9.1 for a derivation of the wave equation in two dimensions, and Billingham and King (2001) for a derivation of (5.37) for sound waves.

and when we enforce the initial conditions, this becomes

$$\tilde{u} = \frac{\tilde{f}(\mathbf{k})}{2ick}\left(e^{ickt} - e^{-ickt}\right).$$

The inversion formula, (5.33), then shows that

$$u(\mathbf{x},t) = \frac{1}{8\pi^3}\int_{\mathbb{R}^3}\frac{\tilde{f}(\mathbf{k})}{2ick}\left\{\exp\left(ickt - i\mathbf{k}\cdot\mathbf{x}\right) - \exp\left(-ickt - i\mathbf{k}\cdot\mathbf{x}\right)\right\}d^3\mathbf{k}. \quad (5.40)$$

For any given value of $\mathbf{k}$, the terms $\exp(\pm ickt - i\mathbf{k}\cdot\mathbf{x})$ represent **plane travelling wave solutions** of the three-dimensional wave equation, since they remain constant on the planes $\mathbf{k}\cdot\mathbf{x} = \pm ckt$, which move perpendicular to themselves, in the direction $\pm\mathbf{k}$, at speed c. Since the integral (5.40) is a weighted sum of these plane waves at different wavenumbers $\mathbf{k}$, we can see that the Fourier transform has revealed that the solution of (5.37) subject to (5.38) can be written as a continuous spectrum of plane wave solutions travelling in *all* directions. It is somewhat easier to interpret solutions like (5.40) in the large time limit, $t \gg 1$, with $|\mathbf{x}|/t$ fixed. We will do this in Section 11.2.2 using the method of stationary phase.

Exercises

5.1 Determine the Fourier series expansion of $f(x)$ for $-\pi < x < \pi$, where (a) $f(x) = e^x$, (b) $f(x) = |\sin x|$, (c) $f(x) = x^2$.

5.2 Determine the Fourier series expansion of the function

$$f(x) = \begin{cases} 0 & \text{for } -\pi < x \leqslant 0, \\ x & \text{for } 0 < x < \pi. \end{cases}$$

Using this series, show that

$$1 + \frac{1}{3^2} + \frac{1}{5^2} + \cdots = \frac{\pi^2}{8}.$$

5.3 Show that

(a) $\delta(\alpha x) = \frac{1}{|\alpha|}\delta(x)$,

(b) $\delta(x^2 - a^2) = \frac{1}{2a}\left\{\delta(x+a) + \delta(x-a)\right\}.$

5.4 Express $\delta(ax+b)$ in the form $\mu\delta(x+\sigma)$ for appropriately chosen constants μ and σ.

5.5 What is the general solution of the equation $(x-a)f(x) = b$, if f is a generalized function?

5.6 If $g(x) = 0$ at the points x_i for $i = 1, 2, \ldots, n$, find an expression for $\delta(g(x))$ in terms of the sequence of delta functions $\{\delta(x-x_i)\}$.

5.7 Show that $\frac{d}{dx}\text{sgn}(x) = 2\delta(x)$. What is $\frac{d}{dx}e^{-|x|}$?

5.8 Calculate the Fourier transforms of

(a)
$$f(x) = \begin{cases} 0 & \text{for } x < -1 \text{ and } x > 1, \\ -1 & \text{for } -1 < x < 0, \\ 1 & \text{for } 0 < x < 1, \end{cases}$$

(b)
$$f(x) = \begin{cases} e^x & \text{for } x < 0, \\ 0 & \text{for } x > 0, \end{cases}$$

(c)
$$f(x) = \begin{cases} xe^{-x} & \text{for } x > 0, \\ 0 & \text{for } x < 0. \end{cases}$$

5.9 Show that the Fourier transform has the 'shifting property',
$$\mathcal{F}[f(x-a)] = e^{ika}\mathcal{F}[f(x)].$$

5.10 If
$$J(k) = \int_{-\infty}^{\infty} e^{-(x-ik/2)^2} dx,$$
show by differentiation under the integral sign that $dJ/dk = 0$. Verify that $J(0) = \sqrt{\pi}$ and hence evaluate $\mathcal{F}\left[e^{-x^2}\right]$.

5.11 Use Fourier transforms to show that the solution of the initial boundary value problem,
$$\frac{\partial u}{\partial t} = k\frac{\partial^2 u}{\partial x^2} - \lambda u, \quad \text{for } -\infty < x < \infty, t > 0,$$
with λ and k real constants, subject to
$$u(x,0) = f(x), \quad u \to 0 \text{ as } |x| \to \infty,$$
can be written in convolution form as
$$u(x,t) = \frac{e^{-\lambda t}}{\sqrt{4\pi kt}} \int_{-\infty}^{\infty} \exp\left\{-\frac{(x-y)^2}{4kt}\right\} f(y)\, dy.$$

5.12 For $n \in \mathbb{N}$ and $k \in \mathbb{Z}$, evaluate the integral
$$\int_{-\pi}^{\pi} e^{kx} e^{inx}\, dx,$$
and hence show that
$$\int_{-\pi}^{\pi} e^{kx} \cos nx\, dx = \frac{(-1)^{n+1} 2n \sinh k\pi}{k^2 + n^2}.$$
Obtain the Fourier series of the function (of period 2π) defined by
$$f(x) = \cosh kx \quad \text{for } -\pi < x < \pi.$$

Find the value of
$$\sum_{n=1}^{\infty} \frac{1}{k^2+n^2}$$
as a function of k and use (4.49) to evaluate
$$\sum_{n=1}^{\infty} \frac{1}{(k^2+n^2)^2}.$$

5.13   The **forced wave equation** is
$$\frac{\partial^2 \phi}{\partial x^2} - \frac{1}{c_0^2}\frac{\partial^2 \phi}{\partial t^2} = Q(x,t),$$
where Q is the forcing function, which satisfies $Q \to 0$ as $|x| \to \infty$.

(a) Show that if $Q = q(x)e^{-i\omega t}$ then a separable solution exists with $\phi = e^{-i\omega t} f(x)$ and $f'' + k_0^2 f = q(x)$, where $k_0 = \omega/c_0$.

(b) Find the Green's function solution of $G'' + k_0^2 G = \delta(x)$ that behaves like $e^{\pm ik_0 x}$ as $x \to \pm\infty$. What is the physical significance of these boundary conditions?

(c) Show that
$$\phi = e^{-i\omega t} G * q$$
$$= \frac{e^{-i\omega t}}{2ik_0}\left\{\int_{-\infty}^{x} q(y)e^{ik_0(x-y)}\,dy + \int_{x}^{\infty} q(y)e^{-ik_0(x-y)}\,dy\right\}.$$

(d) Show that, as $x \to \infty$,
$$\phi \sim \frac{e^{-i(\omega t - k_0 x)}}{2ik_0}\int_{-\infty}^{\infty} q(y)e^{-ik_0 y}\,dy.$$

What is the physical interpretation of this result?

5.14   The free space Green's function for the **modified Helmholtz equation** in three dimensions satisfies
$$(\nabla^2 - m^2)\,G = \delta(x_1, x_2, x_3),$$
with $G \to 0$ as $|x| \to \infty$. Use Fourier transforms to show that
$$G(\mathbf{x}) = -\frac{1}{8\pi^3}\int_{\mathbb{R}^3} \frac{e^{-i\mathbf{k}\cdot\mathbf{x}}}{k^2+m^2}\,d^3\mathbf{k}.$$
Use contour integration to show that this can be simplified to
$$G(\mathbf{x}) = -\frac{1}{4\pi r}e^{-mr},$$
where $r = |\mathbf{x}|$. Verify by direct substitution that this function satisfies the modified Helmholtz equation.

CHAPTER SIX

Laplace Transforms

Integral transforms of the form

$$T[f(x)] = \int_I K(x,k)f(x)\,dx,$$

where I is some interval on the real line, $K(x,k)$ is the **kernel** of the transform and k is the **transform variable**, are useful tools for solving linear differential equations. Other types of equation, such as linear integral equations, can also be solved using integral transforms. In Chapter 5 we met the Fourier transform, for which the kernel is e^{ikx} and $I = \mathbb{R}$. The Fourier transform allows us to solve linear boundary value problems whose domain of solution is the whole real line. In this chapter we will study the **Laplace transform**, for which the usual notation for the original variable is t and for the transform variable is s, the kernel is e^{-st} and $I = \mathbb{R}^+ = [0,\infty)$. This transform allows us to solve linear initial value problems, with t representing time. As we shall see, it is closely related to the Fourier transform.

6.1 Definition and Examples

The Laplace transform of $f(t)$ is

$$\mathcal{L}[f(t)] = F(s) = \int_0^\infty e^{-st} f(t)\,dt. \qquad (6.1)$$

We will consider for what values of s the integral is convergent later in the chapter, and begin with some examples.

Example 1

Consider $f(t) = e^{kt}$, where k is a constant. Substituting this into (6.1), we have

$$\mathcal{L}[e^{kt}] = \int_0^\infty e^{-st} e^{kt}\,dt = \int_0^\infty e^{-(s-k)t}\,dt = \left[\frac{e^{-(s-k)t}}{-(s-k)}\right]_0^\infty.$$

Now, in order that $e^{-(s-k)t} \to 0$ as $t \to \infty$, and hence that the integral converges, we need $s > k$. If s is a complex-valued variable, which is how we will need to treat s when we use the inversion formula, (6.5), we need $\mathrm{Re}(s) > \mathrm{Re}(k)$. This shows that

$$\mathcal{L}[e^{kt}] = \frac{1}{s-k} \quad \text{for } \mathrm{Re}(s) > \mathrm{Re}(k).$$

Example 2

Consider $f(t) = \cos \omega t$, where ω is a constant. Since $\cos \omega t = \frac{1}{2}(e^{i\omega t} + e^{-i\omega t})$,

$$\mathcal{L}[\cos \omega t] = \frac{1}{2} \int_0^\infty \left(e^{(i\omega - s)t} + e^{-(i\omega + s)t} \right) ds = \frac{1}{2} \left(\frac{1}{s - i\omega} + \frac{1}{s + i\omega} \right),$$

provided $\text{Re}(s) > 0$. We can combine the two fractions to show that

$$\mathcal{L}[\cos \omega t] = \frac{s}{s^2 + \omega^2} \quad \text{for} \quad \text{Re}(s) > 0.$$

It is easy to show that $\mathcal{L}[\sin \omega t] = \omega/(s^2 + \omega^2)$ using the same technique. In the same vein,

$$\mathcal{L}[\cosh(at)] = \frac{s}{s^2 - a^2}, \quad \mathcal{L}[\sinh(at)] = \frac{a}{s^2 - a^2}.$$

Example 3

Consider $f(t) = t^n$ for $n \in \mathbb{R}$. By definition,

$$\mathcal{L}[t^n] = \int_0^\infty e^{-st} t^n \, dt.$$

If we let $x = st$ so that $dx = s \, dt$, we obtain

$$\mathcal{L}[t^n] = \int_0^\infty e^{-x} \left(\frac{x}{s} \right)^n \frac{1}{s} \, dx = \frac{1}{s^{n+1}} \int_0^\infty e^{-x} x^n \, dx = \frac{\Gamma(n+1)}{s^{n+1}}.$$

If n is an integer, this gives $\mathcal{L}[t^n] = n!/s^{n+1}$, a result that we could have obtained directly using integration by parts.

In order to use the Laplace transform, we will need to know how to invert it, so that we can determine $f(t)$ from a given function $F(s)$. For the Fourier transform, this inversion process involves the integral formula (5.25). The inverse of a Laplace transform is rather similar, and involves an integral in the complex s-plane. We will return to give an outline derivation of the inversion formula, (6.5), in Section 6.4. For the moment, we will deal with the problem of inverting Laplace transforms by trying to recognize the function $f(t)$ from the form of $F(s)$. As we shall see, there are a lot of techniques available that allow us to invert the Laplace transform in an elementary manner.

6.1.1 The Existence of Laplace Transforms

So far, we have rather glossed over the question of when the Laplace transform of a particular function actually exists. We have, however, discussed in passing that the real part of s needs to be greater than some constant value in some cases. In fact the definition of the Laplace transform is an improper integral and should be written as

$$\mathcal{L}[f(t)] = \lim_{N \to \infty} \int_0^N e^{-st} f(t) \, dt.$$

154 LAPLACE TRANSFORMS

We will now fix matters by defining a class of functions for which the Laplace transform does exist. We say that a function $f(t)$ is of **exponential order** on $0 \leqslant t < \infty$ if there exist constants A and b such that $|f(t)| < Ae^{bt}$ for $t \in [0, \infty)$.

Theorem 6.1 (Lerch's theorem) *A piecewise continuous function of exponential order on $[0, \infty)$ has a Laplace transform.*

The proof of this is rather technical, and the interested reader is referred to Kreider, Kuller, Ostberg and Perkins (1966).

Theorem 6.2 *If f is of exponential order on $[0, \infty)$, then $\mathcal{L}[f] \to 0$ as $|s| \to \infty$.*

Proof Consider the definition of $\mathcal{L}[f(t)]$ and its modulus

$$|\mathcal{L}[f(t)]| = \left|\int_0^\infty e^{-st} f(t)\, dt\right| \leqslant \int_0^\infty \left|e^{-st} f(t)\right| dt = \int_0^\infty e^{-st} |f(t)|\, dt,$$

using the triangle inequality. If f is of exponential order,

$$|\mathcal{L}[f(t)]| < \int_0^\infty e^{-st} A e^{bt}\, dt = \int_0^\infty A e^{(b-s)t}\, dt$$

$$= \left[\frac{A}{b-s} e^{(b-s)t}\right]_0^\infty = \lim_{Y \to \infty} \left[\frac{A}{b-s} e^{(b-s)Y} - \frac{A}{b-s}\right].$$

Provided $s > b$, we therefore have

$$|\mathcal{L}[f(t)]| < \frac{A}{s-b},$$

and hence $|\mathcal{L}[f(t)]| \to 0$ as $|s| \to \infty$. □

Conversely, if $\lim_{s \to \infty} F(s) \neq 0$, then $F(s)$ is not the Laplace transform of a function of exponential order. For example, $s^2/(s^2+1) \to 1$ as $s \to \infty$, and is not therefore the Laplace transform of a function of exponential order.

In contrast, if f is not of exponential order and grows too quickly as $t \to \infty$, the integral will not converge. For example, consider the Laplace transform of the function e^{t^2},

$$\mathcal{L}\left[e^{t^2}\right] = \lim_{N \to \infty} \int_0^N e^{t^2} e^{-st}\, dt.$$

It should be clear that e^{t^2} grows faster than e^{st} for all s, so that the integrand diverges, and hence that the Laplace transform of e^{t^2} does not exist.

6.2 Properties of the Laplace Transform

Theorem 6.3 (Linearity) *The Laplace transform and inverse Laplace transform are linear operators.*

6.2 PROPERTIES OF THE LAPLACE TRANSFORM

Proof By definition,

$$\mathcal{L}[\alpha f(t) + \beta g(t)] = \int_0^\infty e^{-st}(\alpha f(t) + \beta g(t))\, dt$$

$$= \alpha \int_0^\infty e^{-st} f(t)\, dt + \beta \int_0^\infty e^{-st} g(t)\, dt = \alpha \mathcal{L}[f(t)] + \beta \mathcal{L}[g(t)],$$

so the Laplace transform is a linear operator. Taking the inverse Laplace transform of both sides gives

$$\alpha f(t) + \beta g(t) = \mathcal{L}^{-1}\left[\alpha \mathcal{L}[f(t)] + \beta \mathcal{L}[g(t)]\right].$$

Since we can write $f(t)$ as $\mathcal{L}^{-1}[\mathcal{L}[f(t)]]$ and similarly for $g(t)$, this gives

$$\alpha \mathcal{L}^{-1}[\mathcal{L}[f(t)]] + \beta \mathcal{L}^{-1}[\mathcal{L}[g(t)]] = \mathcal{L}^{-1}\left[\alpha \mathcal{L}[f(t)] + \beta \mathcal{L}[g(t)]\right].$$

If we now define $F(s) = \mathcal{L}[f]$ and $G(s) = \mathcal{L}[g]$, we obtain

$$\alpha \mathcal{L}^{-1}[F] + \beta \mathcal{L}^{-1}[G] = \mathcal{L}^{-1}\left[\alpha F + \beta G\right],$$

so the inverse Laplace transform is also a linear operator. □

This is extremely useful, since we can calculate, for example $\mathcal{L}[2t^2 - t + 1]$ and $\mathcal{L}[e^{3t} + \cos 2t]$ easily in terms of the Laplace transforms of their constituent parts.

As we mentioned earlier, the Laplace inversion formula involves complex integration, which we would prefer to avoid when possible. Often we can recognize the constituents of an expression whose inverse Laplace transform we seek. For example, consider $\mathcal{L}^{-1}[1/(s^2 - 5s + 6)]$. We can proceed by splitting this rational function into its partial fractions representation and exploit the linearity of the inverse Laplace transform. We have

$$\mathcal{L}^{-1}\left[\frac{1}{s^2 - 5s + 6}\right] = \mathcal{L}^{-1}\left[\frac{1}{(s-2)(s-3)}\right]$$

$$= \mathcal{L}^{-1}\left[-\frac{1}{s-2} + \frac{1}{s-3}\right] = -\mathcal{L}^{-1}\left[\frac{1}{s-2}\right] + \mathcal{L}^{-1}\left[\frac{1}{s+3}\right] = -e^{2t} + e^{3t}.$$

We pause here to note that the inversion of Laplace transforms using standard forms is only possible because the operation is a bijection, that is, it is one-to-one and onto. For every function $f(t)$ the Laplace transform $\mathcal{L}[f(t)]$ is uniquely defined and vice versa. This is a direct consequence of Lerch's theorem, 6.1.

Theorem 6.4 (First shifting theorem) *If $\mathcal{L}[f(t)] = F(s)$ for $\mathrm{Re}(s) > b$, then $\mathcal{L}[e^{at} f(t)] = F(s-a)$ for $\mathrm{Re}(s) > a + b$.*

Proof By definition,

$$\mathcal{L}[e^{at} f(t)] = \int_0^\infty e^{-st} e^{at} f(t)\, dt = \int_0^\infty e^{-(s-a)t} f(t)\, dt = F(s-a),$$

provided that $\mathrm{Re}(s - a) > b$. □

Example 1

Consider the function $f(t) = e^{3t}\cos 4t$. We recall from the previous section that

$$\mathcal{L}[\cos 4t] = \frac{s}{s^2+4^2} \quad \text{for } \operatorname{Re}(s) > 0.$$

Using Theorem 6.4,

$$\mathcal{L}\left[e^{3t}\cos 4t\right] = \frac{s-3}{(s-3)^2+4^2} \quad \text{for } \operatorname{Re}(s) > 3.$$

Example 2

Consider the function $F(s) = s/(s^2+s+1)$. What is its inverse Laplace transform? We begin by completing the square of the denominator, which gives

$$F(s) = \frac{s}{\left(s+\frac{1}{2}\right)^2 + \frac{3}{4}} = \frac{s+\frac{1}{2}}{\left(s+\frac{1}{2}\right)^2 + \frac{3}{4}} - \frac{\frac{1}{2}}{\left(s+\frac{1}{2}\right)^2 + \frac{3}{4}},$$

and hence

$$\mathcal{L}^{-1}\left[\frac{s}{s^2+s+1}\right] = \mathcal{L}^{-1}\left[\frac{s+\frac{1}{2}}{\left(s+\frac{1}{2}\right)^2+\frac{3}{4}}\right] - \mathcal{L}^{-1}\left[\frac{1}{\sqrt{3}}\frac{\frac{\sqrt{3}}{2}}{\left(s+\frac{1}{2}\right)^2+\frac{3}{4}}\right].$$

Using the first shifting theorem then gives us

$$\mathcal{L}^{-1}\left[\frac{s}{s^2+s+1}\right] = e^{-t/2}\cos\frac{\sqrt{3}t}{2} - \frac{1}{\sqrt{3}}e^{-t/2}\sin\frac{\sqrt{3}t}{2}.$$

Theorem 6.5 (Second shifting theorem) *If the Laplace transform of $f(t)$ is $F(s)$, then the Laplace transform of the function $g(t) = H(t-a)f(t-a)$ is $e^{-sa}F(s)$, where H is the Heaviside step function.*

Proof By definition,

$$\mathcal{L}[g(t)] = \int_0^\infty g(t)e^{-st}\,dt = \int_a^\infty f(t-a)e^{-st}\,dt.$$

By writing $\tau = t - a$, this becomes

$$\mathcal{L}[g(t)] = \int_0^\infty f(\tau)e^{-s\tau}e^{-sa}\,d\tau = e^{-sa}F(s),$$

since the definition of the Laplace transform, (6.1), can be written in terms of any dummy variable of integration. □

For example, to determine the inverse transform of the function e^{-3s}/s^3, we firstly note that $\mathcal{L}[t^2] = 2/s^3$, using Example 3 of Section 6.1. The second shifting theorem then shows immediately that

$$\mathcal{L}^{-1}\left[\frac{e^{-3s}}{s^3}\right] = \frac{1}{2}H(t-3)(t-3)^2.$$

6.3 The Solution of Ordinary Differential Equations Using Laplace Transforms

In order to be able to take the Laplace transform of a differential equation, we will need to be able to calculate the Laplace transform of the derivative of a function. By definition,

$$\mathcal{L}[f'] = \int_0^\infty e^{-st} f'(t)\, dt.$$

After integrating by parts, we find that

$$\mathcal{L}[f'] = \left[e^{-st} f(t)\right]_0^\infty - \int_0^\infty -s e^{-st} f(t)\, dt.$$

At this stage, we will assume that the values of s are restricted so that $e^{-st} f(t) \to 0$ as $t \to \infty$. This means that

$$\mathcal{L}[f'] = s\mathcal{L}[f] - f(0). \tag{6.2}$$

A useful corollary of (6.2) is that, if

$$g(t) = \int_0^t f(\tau)\, d\tau,$$

so that, except where $f(t)$ is discontinuous, $g'(t) = f(t)$, we have $\mathcal{L}[f] = s\mathcal{L}[g] - g(0)$. Since $g(0) = 0$ by definition,

$$\mathcal{L}[g] = \mathcal{L}\left[\int_0^t f(\tau)\, d\tau\right] = \frac{1}{s}\mathcal{L}[f],$$

and hence

$$\mathcal{L}^{-1}\left[\frac{1}{s} F(s)\right] = \int_0^t f(\tau)\, d\tau.$$

This can be useful for inverting Laplace transforms, for example,

$$F(s) = \frac{1}{s(s^2 + \omega^2)}.$$

We know that $\mathcal{L}[\sin \omega t]/\omega = 1/(s^2 + \omega^2)$ so that

$$f(t) = \mathcal{L}^{-1}\left[\frac{1}{s}\frac{1}{s^2 + \omega^2}\right] = \mathcal{L}^{-1}\left[\frac{1}{s}\mathcal{L}\left[\frac{1}{\omega}\sin \omega t\right]\right]$$

$$= \int_0^t \frac{1}{\omega} \sin \omega \tau \, d\tau = \frac{1}{\omega^2}(1 - \cos \omega t).$$

Let's now try to solve the simple differential equation

$$\frac{dy}{dt} - 2y = 0,$$

subject to the initial condition $y(0) = 1$. Of course, it is trivial to solve this separable equation, but it is a useful illustrative example. We begin by taking the Laplace transform of the differential equation, which gives

$$\mathcal{L}\left[\frac{dy}{dt}\right] - 2\mathcal{L}[y] = sY(s) - y(0) - 2Y(s) = 0,$$

where $Y(s) = \mathcal{L}[y(t)]$. Using the initial condition and manipulating the equation gives

$$Y(s) = \frac{1}{s-2},$$

which is easily inverted to give $y(t) = e^{2t}$.

Many of the differential equations that we will try to solve are of second order, so we need to determine $\mathcal{L}[f'']$. If we introduce the function $g(t) = f'(t)$, (6.2) shows that

$$\mathcal{L}[g'] = sG(s) - g(0),$$

where $G(s) = \mathcal{L}[g] = \mathcal{L}[f'] = sF(s) - f(0)$. We conclude that

$$\mathcal{L}[f''] = \mathcal{L}[g'] = s^2 F(s) - sf(0) - f'(0). \tag{6.3}$$

We can obtain the same result by integrating the definition of the Laplace transform of f'' twice by parts. It is also straightforward to show by induction that

$$\mathcal{L}[f^{(n)}] = s^n F(s) - s^{n-1} f(0) - s^{n-2} f'(0) - \cdots - f^{(n-1)}(0).$$

For example, we can now solve the differential equation

$$\frac{d^2 y}{dt^2} - 5\frac{dy}{dt} + 6y = 0,$$

subject to the initial conditions $y(0) = 0$ and $y'(0) = 1$ using Laplace transforms. We find that

$$s^2 Y(s) - sy(0) - y'(0) - 5(sY(s) - y(0)) + 6Y = 0.$$

Using the initial conditions then shows that

$$Y(s) = \frac{1}{s^2 - 5s + 6}.$$

In order to invert the Laplace transform we split the fraction into its constituent partial fractions,

$$Y(s) = \frac{1}{s-3} - \frac{1}{s-2},$$

which immediately shows that

$$y(t) = e^{3t} - e^{2t}.$$

Let's now consider the solution of the coupled system of equations

$$\frac{dy_1}{dt} + y_1 = y_2, \quad \frac{dy_2}{dt} - y_2 = y_1,$$

subject to the initial conditions that $y_1(0) = y_2(0) = 1$. Although we could

6.3 THE SOLUTION OF ORDINARY DIFFERENTIAL EQUATIONS

combine these two first order equations to obtain a second order equation, we will solve them directly using Laplace transforms. The transform of the equations is

$$sY_1 - 1 + Y_1 = Y_2, \quad sY_2 - 1 - Y_2 = Y_1,$$

where $\mathcal{L}[y_j(t)] = Y_j(s)$. The solution of these algebraic equations is

$$Y_1(s) = \frac{s}{s^2 - 2}, \quad Y_2(s) = \frac{s+2}{s^2 - 2},$$

which we can easily invert by inspection to give

$$y_1(t) = \cosh \sqrt{2}t, \quad y_2(t) = \cosh \sqrt{2}t + \sqrt{2} \sinh \sqrt{2}t.$$

Note that the reduction of a system of differential equations to a system of algebraic equations is the key benefit of these transform methods.

Of course, for simple scalar equations, it is easier to use the standard solution techniques described in Appendix 5. The real power of the Laplace transform method lies in the solution of problems for which the inhomogeneity is not of a simple form, for example

$$\frac{d^2y}{dt^2} + \frac{dy}{dt} = \delta(t-1).$$

In order to take the Laplace transform of this equation, we need to know the Laplace transform of the delta function. We can calculate this directly, as

$$\mathcal{L}[\delta(t-a)] = \int_0^\infty \delta(t-a)e^{-st}\,dt = e^{-sa},$$

so that the differential equation becomes

$$s^2 Y(s) - sy(0) - y'(0) + sY(s) - y(0) = e^{-s},$$

and hence

$$Y(s) = \frac{e^{-s} + (s+1)y(0) + y'(0)}{s(s+1)}.$$

Judicious use of the two shifting theorems now allows us to invert this Laplace transform. Note that

$$\mathcal{L}^{-1}\left[\frac{1}{s}\right] = 1,$$

and, using the first shifting theorem,

$$\mathcal{L}^{-1}\left[\frac{1}{s+1}\right] = e^{-t}.$$

This means that

$$\mathcal{L}^{-1}\left[\frac{1}{s(s+1)}\right] = \mathcal{L}^{-1}\left[\frac{1}{s} - \frac{1}{s+1}\right] = 1 - e^{-t},$$

and, using the second shifting theorem,

$$\mathcal{L}^{-1}\left[\frac{e^{-s}}{s(s+1)}\right] = H(t-1)(1 - e^{-(t-1)}).$$

LAPLACE TRANSFORMS

Combining all of these results, and using the linearity of the Laplace transform, shows that

$$y(t) = y(0) + y'(0)(1 - e^{-t}) + H(t-1)(1 - e^{-(t-1)}).$$

6.3.1 The Convolution Theorem

The Laplace transform of the convolution of two functions plays an important role in the solution of inhomogeneous differential equations, just as it did for Fourier transforms. However, when using Laplace transforms, we define the convolution of two functions as

$$f * g = \int_0^t f(\tau) g(t - \tau) \, d\tau.$$

With this definition, if $f(t)$ and $g(t)$ have Laplace transforms, then

$$\mathcal{L}[f * g] = \mathcal{L}[f(t)] \mathcal{L}[g(t)].$$

We can show this by taking the Laplace transform of the convolution integral, which is itself a function of t, to obtain

$$\mathcal{L}[f * g] = \int_0^\infty e^{-st} \int_0^t f(\tau) g(t - \tau) \, d\tau \, dt.$$

Since e^{-st} is independent of τ it can be moved inside the inner integral so that we have

$$\mathcal{L}[f * g] = \int_0^\infty \int_0^t e^{-st} f(\tau) g(t - \tau) \, d\tau \, dt.$$

We now note that the domain of integration is a triangular region delimited by the lines $\tau = 0$ and $\tau = t$, as shown in Figure 6.1. If we switch the order of integration, this becomes

$$\mathcal{L}[f * g] = \int_{\tau=0}^{\tau=\infty} \int_{t=\tau}^{\infty} e^{-st} f(\tau) g(t - \tau) \, dt \, d\tau.$$

If we now introduce the variable $z = t - \tau$, so that $dz = dt$, we can transform the inner integral into

$$\mathcal{L}[f * g] = \int_0^\infty f(\tau) \int_0^\infty e^{-s(z+\tau)} g(z) \, dz \, d\tau,$$

and hence

$$\mathcal{L}[f * g] = \int_{\tau=0}^{\infty} f(\tau) e^{-s\tau} \, d\tau \int_{z=0}^{\infty} e^{-sz} g(z) \, dz = \mathcal{L}[f] \, \mathcal{L}[g].$$

This result is most useful in the form

$$\mathcal{L}^{-1}[F(s) G(s)] = \mathcal{L}^{-1}[F(s)] * \mathcal{L}^{-1}[G(s)].$$

6.3 THE SOLUTION OF ORDINARY DIFFERENTIAL EQUATIONS

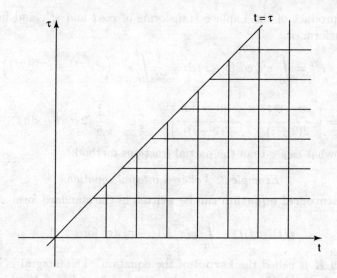

Fig. 6.1. The domain of integration of the Laplace transform of a convolution integral.

Example 1

Consider the Laplace transform $F(s) = 1/(s-2)(s-3)$. This rational function has two easily recognizable constituents, namely $1/(s-2) = \mathcal{L}[e^{2t}]$ and $1/(s-3) = \mathcal{L}[e^{3t}]$, and hence

$$\mathcal{L}^{-1}\left[\frac{1}{(s-2)(s-3)}\right] = \mathcal{L}^{-1}\left[\frac{1}{s-2}\right] * \mathcal{L}^{-1}\left[\frac{1}{s-3}\right] = e^{2t} * e^{3t}.$$

We now need to calculate the convolution integral, which is

$$e^{2t} * e^{3t} = \int_{\tau=0}^{t} e^{2\tau} e^{3(t-\tau)} \, d\tau = e^{3t} \int_{\tau=0}^{t} e^{-\tau} \, d\tau$$

$$= e^{3t} \left[-e^{-\tau}\right]_0^t = e^{3t}\left(-e^{-t} + 1\right) = -e^{2t} + e^{3t}.$$

This is, of course, the same result as we obtained using partial fractions.

Example 2

Consider the Laplace transform

$$F(s) = \frac{s}{(s^2+1)(s-2)}.$$

162 LAPLACE TRANSFORMS

This is the product of the Laplace transforms of $\cos t$ and e^{2t}, and hence is the Laplace transform of

$$\cos t * e^{2t} = \int_0^t e^{2\tau} \cos(t-\tau)\, d\tau = \frac{1}{2}\int_0^t e^{2\tau}\left(e^{i(t-\tau)} + e^{-i(t-\tau)}\right) d\tau$$

$$= \left[\frac{e^{i(t-\tau)+2\tau}}{2(2+i)} + \frac{e^{-i(t-\tau)+2\tau}}{2(2-i)}\right]_0^t = \frac{2}{5}e^{2t} - \frac{1}{5}(2\cos t - \sin t).$$

This is somewhat easier than the partial fractions method.

Example 3: Volterra integral equations

A **Volterra integral equation** can be written in the standard form

$$y(t) = f(t) + \int_0^t y(\tau) K(t-\tau)\, d\tau \quad \text{for } t > 0. \tag{6.4}$$

The function K is called the **kernel** of the equation. The integral is in the form of a convolution and, if we treat t as time, is an integral over the history of the solution. The integral equation (6.4) can therefore be written as

$$y = f + y * K.$$

If we now take a Laplace transform of this equation, we obtain

$$Y = F + \mathcal{L}[y * K] = F + Y\mathcal{L}[K],$$

and hence

$$Y = \frac{F}{1 - \mathcal{L}[K]},$$

where $F(s) = \mathcal{L}[f]$ and $Y = \mathcal{L}[y]$. For example, to solve

$$y(t) = 1 + \int_0^t (t-\tau) y(\tau)\, d\tau,$$

for which $f(t) = 1$ and $K(t) = t$, we note that $F = 1/s$ and $\mathcal{L}[K] = 1/s^2$, and hence

$$Y(s) = \frac{s}{s^2 - 1}.$$

This gives us the solution, $y(t) = \cosh t$.

6.4 The Inversion Formula for Laplace Transforms

We have now seen that many Laplace transforms can be inverted from a knowledge of the transforms of a few common functions along with the linearity of the Laplace transform and the first and second shifting theorems. However, these techniques are often inadequate to invert Laplace transforms that arise as solutions of more complicated problems, in particular of partial differential equations. We will need an inversion formula. We can derive this in an informal way using the Fourier integral, (5.23).

6.4 THE INVERSION FORMULA FOR LAPLACE TRANSFORMS

Let $g(t)$ be a function of exponential order, in particular with γ the smallest real number such that $e^{-\gamma t}g(t)$ is bounded as $t \to \infty$. As we have seen, $g(t)$ therefore has a Laplace transform $G(s)$, which exists for $\mathrm{Re}(s) > \gamma$. We now define $h(t) = e^{-\gamma t}g(t)H(t)$. Since $h(t)$ is bounded as $t \to \infty$, the Fourier integral (5.23) exists, and shows that

$$h(t) = \frac{1}{2\pi}\int_{-\infty}^{\infty} e^{-ikt}\int_{-\infty}^{\infty} e^{ikT}h(T)\,dT\,dk,$$

and hence that

$$e^{-\gamma t}g(t) = \frac{1}{2\pi}\int_{-\infty}^{\infty} e^{-ikt}\int_{0}^{\infty} e^{-(\gamma-ik)T}g(T)\,dT\,dk.$$

If we now make the change of variable $s = \gamma - ik$, so that $ds = -i\,dk$, we find that†

$$g(t) = \frac{1}{2\pi i}\int_{\gamma-i\infty}^{\gamma+i\infty} e^{st}\int_{0}^{\infty} e^{-sT}g(T)\,dT\,ds.$$

Finally, from the definition of the Laplace transform, we arrive at the inversion formula, sometimes called the **Bromwich inversion integral**,

$$g(t) = \frac{1}{2\pi i}\int_{\gamma-i\infty}^{\gamma+i\infty} e^{st}G(s)\,ds. \tag{6.5}$$

Note that the contour of integration is a vertical line in the complex s-plane. Since $G(s)$ is only guaranteed to exist for $\mathrm{Re}(s) > \gamma$, this contour lies to the right of any singularities of $G(s)$. It is often possible to simplify (6.5) by closing the contour of integration using a large semicircle in the left half plane. As we shall see in the following examples, if the contour can be closed in this way, the result will depend crucially on the residues at any poles of $e^{st}G(s)$.

Example 1

We start with the simple case $G(s) = \beta/(s - \alpha)$. Consider the integral

$$I(t) = \frac{1}{2\pi i}\int_{C} e^{st}\frac{\beta}{s-\alpha}\,ds, \tag{6.6}$$

where the closed contour C is shown in Figure 6.2. By the residue theorem, $I(t)$ is equal to the sum of the residues at any poles of $e^{st}G(s)$ enclosed by C. Let's assume that the contour encloses the simple pole of $G(s)$ at $s = \alpha$, and hence that the straight boundary of C lies to the right of the pole at $s = \alpha$. The residue theorem then shows that $I(t) = \beta e^{\alpha t}$. As $b \to \infty$, the semicircular part of the contour becomes large and, since $|G(s)|$ is algebraically small when $|s| \gg 1$, we conclude from Jordan's lemma (see Section A6.4) that the integral along the semicircle tends to zero. On the straight part of the contour C, as $b \to \infty$ we recover the inversion integral (6.5), so that $I(t) = g(t)$, and hence the inverse Laplace transform is $g(t) = \beta e^{\alpha t}$, as we would expect.

---

† This is the point at which the derivation is informal, since we have not shown that this change in the contour of integration is possible.

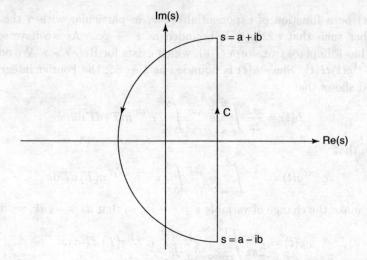

Fig. 6.2. The contour C used to evaluate the inverse Laplace transform of $\beta/(s-\alpha)$.

Example 2

Let's try to determine the inverse Laplace transform of $G(s) = 1/(s^2+1)$. In this case $G(s)$ has simple poles at $s = \pm i$, since

$$G(s) = \frac{1}{2i}\left(\frac{1}{s-i} - \frac{1}{s+i}\right).$$

Choosing the contour C as we did in Example 1, we again find that $g(t)$ is the sum of the residues at the two simple poles of $e^{st}G(s)$, and hence that, as expected,

$$g(t) = \frac{1}{2i}\left(e^{it} - e^{-it}\right) = \sin t.$$

Example 3

We now consider a Laplace transform that has a nonsimple pole, $G(s) = 1/(s-\alpha)^3$. The simplest way to calculate the residue of $e^{st}G(s)$ at $s = \alpha$ is to note that

$$\frac{e^{st}}{(s-\alpha)^3} = \frac{e^{\alpha t}}{(s-\alpha)^3}e^{(s-\alpha)t} = \frac{e^{\alpha t}}{(s-\alpha)^3}\left(1 + (s-\alpha)t + \frac{1}{2}(s-\alpha)^2 t^2 + \cdots\right),$$

and hence that $g(t) = \frac{1}{2}t^2 e^{\alpha t}$. We can check this result by noting that $\mathcal{L}[t^2] = \Gamma(3)/s^3 = 2/s^3$ and using the first shifting theorem.

Example 4

Consider the inverse Laplace transform of $G(s) = e^{-\alpha s^{1/2}}/s$. Since $G(s)$ contains a fractional power, $s^{1/2}$, the point $s = 0$ is a branch point, and the definition of

6.4 THE INVERSION FORMULA FOR LAPLACE TRANSFORMS

$G(s)$ is incomplete. We need to introduce a branch cut in order to make $G(s)$ single-valued. It is convenient to place the branch cut along the negative real axis, so that, if $s = |s|e^{i\theta}$ with $-\pi < \theta < \pi$, $s^{1/2} = +\sqrt{|s|}e^{i\theta/2}$ and the real part of $s^{1/2}$ is positive. We cannot now integrate $e^{st}G(s)$ along any contour that crosses the negative real axis, such as C. Because of this, we use C_B, which avoids the branch cut, as shown in Figure 6.3. This contour also includes a small circle around the origin of radius ϵ, since $G(s)$ has a simple pole there, and is often referred to as a **keyhole contour**.

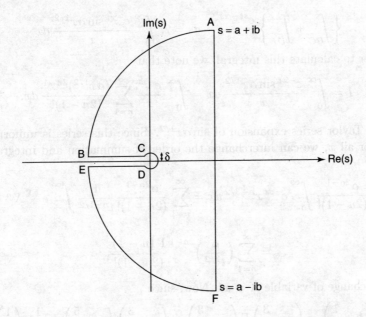

Fig. 6.3. The keyhole inversion contour, C_B, used for inverting Laplace transforms with a branch cut along the negative real axis.

Since $e^{st}G(s)$ is analytic within C_B, the integral around C_B is zero, by Cauchy's theorem (see Appendix 6). On the circular arcs AB and EF, $G(s) \to 0$ exponentially fast as $b \to \infty$, since $s^{1/2}$ has positive real part, and hence the contributions to the integral from these arcs tend to zero. As before, the integral along the straight contour AF tends to $g(t)$ as $b \to \infty$, by the inversion formula (6.5). We conclude that

$$g(t) = -\lim_{\substack{b \to \infty \\ \epsilon \to 0}} \frac{1}{2\pi i} \left\{ \int_{BC} + \int_{CD} + \int_{DE} \right\} \frac{e^{st - \alpha s^{1/2}}}{s} \, ds. \tag{6.7}$$

Let's consider the contributions from the lines BC and DE. We can parameterize these lines as $s = xe^{\pm i\pi}$ respectively. In this way, we ensure that we use the correct value of $s^{1/2}$ on either side of the branch cut. Along BC, $s^{1/2} = x^{1/2}e^{i\pi/2} = ix^{1/2}$,

166   LAPLACE TRANSFORMS

and hence, in the limit as $\epsilon \to 0$ and $b \to \infty$

$$\int_{BC} \frac{e^{st-\alpha s^{1/2}}}{s} ds = \int_{\infty}^{0} \frac{e^{-xt-\alpha i x^{1/2}}}{x} dx.$$

Similarly

$$\int_{DE} \frac{e^{st-\alpha s^{1/2}}}{s} ds = \int_{0}^{\infty} \frac{e^{-xt+\alpha i x^{1/2}}}{x} dx,$$

and hence

$$\left\{\int_{BC} + \int_{DE}\right\} \frac{e^{st-\alpha s^{1/2}}}{s} ds = 2i \int_{0}^{\infty} \frac{e^{-xt}\sin \alpha x^{1/2}}{x} dx. \qquad (6.8)$$

In order to calculate this integral, we note that

$$I = \int_{0}^{\infty} \frac{e^{-xt}\sin \alpha x^{1/2}}{x} dx = \int_{0}^{\infty} \frac{e^{-xt}}{x} \sum_{n=1}^{\infty} \frac{(\alpha x^{1/2})^{2n-1}}{(2n-1)!} dx,$$

using the Taylor series expansion of $\sin \alpha x^{1/2}$. Since this series is uniformly convergent for all x, we can interchange the order of summation and integration, so that

$$I = \sum_{n=1}^{\infty} \frac{\alpha^{2n-1}}{(2n-1)!} \int_{0}^{\infty} e^{-xt} x^{n-3/2} dx = \sum_{n=1}^{\infty} \frac{\alpha^{2n-1}}{(2n-1)!} \frac{1}{t^{n-1/2}} \int_{0}^{\infty} e^{-X} X^{n-3/2} dX$$

$$= \sum_{n=1}^{\infty} \left(\frac{\alpha}{t^{1/2}}\right)^{2n-1} \frac{\Gamma\left(n-\frac{1}{2}\right)}{(2n-1)!},$$

using the change of variable $X = xt$. Now, since

$$\Gamma\left(n-\frac{1}{2}\right) = \left(n-\frac{3}{2}\right)\Gamma\left(n-\frac{3}{2}\right) = \left(n-\frac{3}{2}\right)\left(n-\frac{5}{2}\right)\cdots\frac{1}{2}\Gamma\left(\frac{1}{2}\right)$$

$$= \frac{1}{2^{n-1}}(2n-3)(2n-5)\cdots 3 \cdot 1 \cdot \sqrt{\pi},$$

we find that

$$\frac{\Gamma\left(n-\frac{1}{2}\right)}{(2n-1)!} = \frac{\sqrt{\pi}}{2^{n-1}} \frac{(2n-3)(2n-5)\cdots 3 \cdot 1}{(2n-1)(2n-2)\cdots 3 \cdot 2 \cdot 1}$$

$$= \frac{\sqrt{\pi}}{2^{n-1}(2n-1)} \frac{1}{(2n-2)(2n-4)\cdots 4 \cdot 2} = \frac{\sqrt{\pi}}{2^{2(n-1)}(2n-1)(n-1)!},$$

and hence

$$I = 2 \sum_{n=1}^{\infty} \left(\frac{\alpha}{2t^{1/2}}\right)^{2n-1} \frac{\sqrt{\pi}}{(2n-1)(n-1)!} = 2\sqrt{\pi} \int_{0}^{\alpha/2t^{1/2}} \sum_{n=1}^{\infty} \frac{(s^2)^{n-1}}{(n-1)!} ds$$

$$= 2\sqrt{\pi} \int_{0}^{\alpha/2t^{1/2}} e^{-s^2} ds = \pi \operatorname{erf}\left(\frac{\alpha}{2t^{1/2}}\right),$$

6.4 THE INVERSION FORMULA FOR LAPLACE TRANSFORMS

where erf(x) is the **error function**, defined by

$$\operatorname{erf}(x) = \frac{2}{\sqrt{\pi}} \int_0^x e^{-q^2} \, dq.$$

We can parameterize the small circle CD using $s = \epsilon e^{i\theta}$, where θ runs from π to $-\pi$. On this curve we have $s^{1/2} = \epsilon^{1/2} e^{i\theta/2}$, so that

$$\int_{CD} \frac{e^{st - \alpha s^{1/2}}}{s} \, ds = \int_{\pi}^{-\pi} \frac{e^{\epsilon t e^{i\theta} - \alpha \epsilon^{1/2} e^{i\theta/2}}}{\epsilon e^{i\theta}} \, i\epsilon e^{i\theta} \, d\theta = \int_{\pi}^{-\pi} e^{\epsilon t e^{i\theta} - \alpha \epsilon^{1/2} e^{i\theta/2}} \, i \, d\theta.$$

As $\epsilon \to 0$ we therefore have

$$\int_{CD} \frac{e^{st - \alpha s^{1/2}}}{s} \, ds = \int_{\pi}^{-\pi} i \, d\theta = -2\pi i. \tag{6.9}$$

If we now use (6.8) and (6.9) in (6.7), we conclude that

$$\mathcal{L}^{-1}\left[\frac{e^{-\alpha s^{1/2}}}{s}\right] = 1 - \operatorname{erf}\left(\frac{\alpha}{2t^{1/2}}\right) = \operatorname{erfc}\left(\frac{\alpha}{2t^{1/2}}\right),$$

where erfc(x) is the **complementary error function**, defined by

$$\operatorname{erfc}(x) = \frac{2}{\sqrt{\pi}} \int_x^{\infty} e^{-q^2} \, dq.$$

Laplace transforms of the error function and complementary error function arise very frequently when solving diffusion problems, as we shall see in the following example.

Example 5: Flow due to an impulsively-started flat plate

Let's consider the two-dimensional flow of a semi-infinite expanse of viscous fluid caused by the sudden motion of a flat plate in its own plane. We will use Cartesian coordinates with the x-axis lying in the plane of the plate and the y-axis pointing into the semi-infinite body of fluid. This is a uni-directional flow with velocity $u(x, y, t)$ in the x-direction only and associated scalar pressure field $p(x, y, t)$. The continuity equation, $u_x + v_y = 0$, which we derived in Chapter 2, shows that the streamwise velocity, u, is solely a function of y and t. We now consider the streamwise momentum within a small element of fluid, as shown in Figure 6.4. Balancing forces in the x-direction, and taking the limit $\delta x, \delta y, \delta t \to 0$, we find that

$$\rho \frac{Du}{Dt} = \rho \left(\frac{\partial u}{\partial t} + u \frac{\partial u}{\partial x} \right) = -\frac{\partial p}{\partial x} + \frac{\partial \tau}{\partial y},$$

where ρ is the density and τ is the shear stress. Here we have used the convective derivative, which we derived in Chapter 2. For a one-dimensional flow, it is found experimentally that $\tau = \mu \partial u / \partial y$, where μ is the **dynamic viscosity** (see Acheson, 1990). To be consistent with $u = u(y, t)$, we now insist that $\partial/\partial x \equiv 0$. This reduces the x-momentum equation to $u_t = \nu u_{yy}$, where we have introduced the quantity $\nu = \mu/\rho$, the **kinematic viscosity**. Flows with high values of ν are extremely

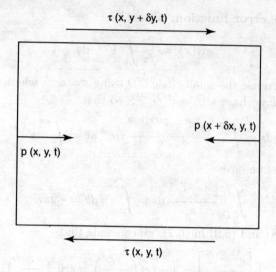

Fig. 6.4. The x-momentum balance on a small element of fluid.

viscous, for example tar, lava and mucus, whilst those with low viscosity include air, water and inert gases. For example, the kinematic viscosity of lava is around 10 m^2 s^{-1}, whilst the values for air and water at room temperature and pressure, 10^{-6} m^2 s^{-1} and 1.5×10^{-5} m^2 s^{-1}, respectively, are very similar.

Our final initial–boundary value problem is

$$\frac{\partial u}{\partial t} = \nu \frac{\partial^2 u}{\partial y^2}, \tag{6.10}$$

to be solved subject to

$$u = 0 \quad \text{when } t = 0 \text{ for } y > 0, \tag{6.11}$$

$$u = \mathcal{U}, \quad \text{at } y = 0 \text{ for } t > 0, \tag{6.12}$$

$$u \to 0, \quad \text{as } y \to \infty \text{ for } t > 0. \tag{6.13}$$

This is known as **Rayleigh's problem**. Equation (6.10) states that the only process involved in this flow is the diffusion of x-momentum into the bulk of the fluid. Of course, this initial–boundary value problem can also be thought of as modelling other diffusive systems, for example, a semi-infinite bar of metal, insulated along its sides, suddenly heated up at one end.

In order to solve this initial–boundary value problem, we will take a Laplace transform with respect to time, so that

$$\mathcal{L}[u(y,t)] = U(y,s) = \int_0^\infty e^{-st} u(y,t) \, dt.$$

6.4 THE INVERSION FORMULA FOR LAPLACE TRANSFORMS

Taking two derivatives of this definition with respect to y shows that

$$\mathcal{L}\left[\frac{\partial^2 u}{\partial y^2}\right] = \frac{\partial^2 U}{\partial y^2}.$$

In general, y-derivatives are not affected by a Laplace transform with respect to t. After using (6.2) to determine the Laplace transform of $\partial u/\partial t$, (6.10) becomes

$$sU = \nu \frac{\partial^2 U}{\partial y^2}. \tag{6.14}$$

The transform variable only appears as a parameter in this equation, which therefore has the solution

$$U(y,s) = A(s)e^{s^{1/2}y/\nu^{1/2}} + B(s)e^{-s^{1/2}y/\nu^{1/2}}.$$

Since $u \to 0$ as $y \to \infty$, we must also have $U \to 0$ as $y \to \infty$, so that $A(s) = 0$. We now need to transform the boundary condition (6.12), which gives

$$U(0,s) = \int_0^\infty u(0,t)e^{-st}\,dt = \int_0^\infty \mathcal{U}e^{-st}\,dt = \frac{\mathcal{U}}{s},$$

and hence $B(s) = \mathcal{U}/s$. We conclude that

$$U(y,s) = \frac{\mathcal{U}}{s}e^{-s^{1/2}y/\nu^{1/2}}.$$

From the result of the previous example, we find that

$$u(y,t) = \mathcal{U}\,\mathrm{erfc}\left(\frac{y}{2\sqrt{\nu t}}\right).$$

Some typical velocity profiles are shown in Figure 6.5. These are easy to plot, since the error and complementary error functions are available as `erf` and `erfc` in MATLAB.

We can also consider what happens when the velocity of the plate is a function of time. All we need to change is the boundary condition (6.12), which becomes

$$u = \mathcal{U}f(t), \quad \text{at } y = 0 \text{ for } t > 0. \tag{6.15}$$

The Laplace transform of this condition is $U(0,s) = \mathcal{U}F(s)$, where $F(s)$ is the Laplace transform of $f(t)$. Now $B(s) = F(s)$, and hence

$$U(y,s) = \mathcal{U}F(s)e^{-s^{1/2}y/\nu^{1/2}} = \mathcal{U}sF(s)\frac{e^{-s^{1/2}y/\nu^{1/2}}}{s}. \tag{6.16}$$

Using (6.2), we can see that $sF(s) = \mathcal{L}[f'(t)] + f(0)$, and hence

$$U(y,s) = \mathcal{U}\left\{\mathcal{L}[f'(t)] + f(0)\right\}\mathcal{L}\left[\mathrm{erfc}\left(\frac{y}{2\sqrt{\nu t}}\right)\right]. \tag{6.17}$$

We can now invert this Laplace transform using the convolution theorem, to give

$$u(y,t) = \mathcal{U}\left[\int_{\tau=0}^{t}\left\{f'(t-\tau) + f(0)\right\}\mathrm{erfc}\left(\frac{y}{2\sqrt{\nu \tau}}\right)d\tau\right]. \tag{6.18}$$

170   LAPLACE TRANSFORMS

Exact solutions of the equations that govern the flow of a viscous fluid, the **Navier–Stokes equations**, are very useful for verifying that numerical solution methods are working correctly. Their stability can also be studied more easily than for flows where no analytical solution is available.

By evaluating (6.18) numerically, we can calculate the flow profiles for any $f(t)$. A MATLAB function that evaluates $u(y,t)$, for any specified $f(t)$, is

```
function rayleigh(yout,tout)

for t = tout
    u = [];
    for y = yout
        u = [u quadl(@integrand,0,t,10^-4,0,t,y)];
    end
    plot(u,yout), xlabel('u'), ylabel('y')
    title(strcat('t = ',num2str(t))), Xlim([0 1])
    pause(0.5)
end

function integrand = integrand(tau,t,y)
nu = 1;
df = f(t-tau);
integrand = df.*erfc(y/2./(eps+sqrt(nu*tau)));

function f = f(t)
df = 2*cos(2*t); f0 = 0; f = df + f0;
```

This uses the MATLAB function `quadl` to evaluate the integral correct to four decimal places, and then plots the solution at the points given in `yout` at the times given in `tout`. In the example shown, $f(t) = \sin 2t$, which corresponds to an oscillating plate, started from rest. Note that we add the inbuilt small quantity `eps` to the denominator of the argument of the error function to avoid MATLAB producing lots of irritating division by zero warnings. In fact, MATLAB is able to evaluate $\mathtt{erfc(y/0)} = \mathtt{erfc(Inf)} = 0$ for y positive.

As a final example, let's consider what we would do if we did not have access to a computer, but wanted to know what happens when the plate oscillates, with $f(t) = \sin \omega t$. Since $F(s) = \omega/(s^2 + \omega^2)$, (6.16), along with the inversion formula (6.5), gives

$$u(y,t) = \frac{1}{2\pi i} \int_{\gamma-i\infty}^{\gamma+i\infty} \frac{\omega e^{st-y\sqrt{s/\nu}}}{s^2 + \omega^2} \, ds, \qquad (6.19)$$

where $\gamma > 0$, since the integrand has poles at $s = \pm i\omega$. Although (6.18) is the most convenient form to use for numerical integration, (6.19) gives us the most helpful way of approaching this specific problem. We proceed as we did in Example 4, evaluating the integral on the contour C_B shown in Figure 6.3. The analysis

6.4 THE INVERSION FORMULA FOR LAPLACE TRANSFORMS

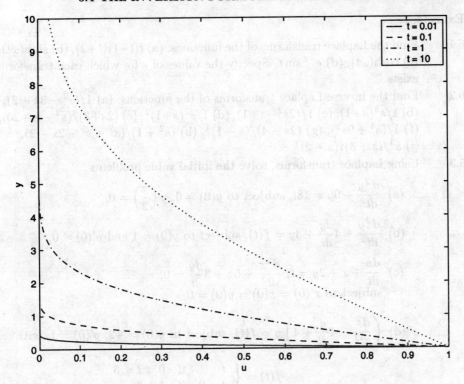

Fig. 6.5. The flow due to an impulsively-started flat plate when $\mathcal{U} = 1$ and $\nu = 1$.

proceeds exactly as it did in Example 4, except that now the integral around C_B is equal to the sum of the residues of the integrand at the poles at $s = \pm i\omega$. In addition, the integral around the small circle CD tends to zero as $\epsilon \to 0$. After taking all of the contributions into account (see Exercise 6.8), we find that

$$u(y,t) = \frac{\omega}{\pi} \int_0^\infty \frac{e^{-\sigma t} \sin\left(\sqrt{\frac{\sigma}{\nu}}y\right)}{\sigma^2 + \omega^2} \, d\sigma + e^{-\sqrt{\frac{\omega}{2\nu}}y} \sin\left\{\omega t - \sqrt{\frac{\omega}{2\nu}}y\right\}. \qquad (6.20)$$

The first term, which comes from the branch cut integrals, tends to zero as $t \to \infty$, and represents the initial transient that arises because the plate starts at rest. The second term, which comes from the residues at the poles, represents the oscillatory motion of the fluid, whose phase changes with y and whose amplitude decays exponentially fast with y. This second term therefore gives the large time behaviour of the fluid. You can see what this solution looks like by running the MATLAB function that we gave above.

Exercises

6.1 Find the Laplace transforms of the functions, (a) $t(t+1)(t+2)$, (b) $\sinh(\omega t)$, (c) $\cosh(\omega t)$, (d) $e^{-t}\sin t$. Specify the values of s for which each transform exists.

6.2 Find the inverse Laplace transforms of the functions, (a) $1/(s^2 - 3s + 2)$, (b) $1/s^3(s+1)$, (c) $1/(2s^2 - s + 1)$, (d) $1/s^2(s+1)^2$, (e) $(2s+3)/(s^2 - 4s + 20)$, (f) $1/(s^4 + 9s^2)$, (g) $(2s - 4)/(s - 1)^4$, (h) $(s^2 + 1)/(s^3 - s^2 + 2s - 2)$, (i) $s^3/(s+3)^2(s+2)^2$.

6.3 Using Laplace transforms, solve the initial value problems

(a) $\dfrac{d^2 y}{dt^2} + 9y = 18t$, subject to $y(0) = 0$, $y\left(\dfrac{\pi}{2}\right) = 0$,

(b) $\dfrac{d^2 y}{dt^2} - 4\dfrac{dy}{dt} + 3y = f(t)$, subject to $y(0) = 1$ and $y'(0) = 0$,

(c) $\dfrac{dx}{dt} + x + 2y = t$, $\dfrac{d^2 x}{dt^2} + 5x + 3\dfrac{dy}{dt} = 0$,
subject to $x'(0) = x(0) = y(0) = 0$,

(d) $\left(\dfrac{d^2}{dt^2} - 4\dfrac{d}{dt} + 4\right) y = f(t)$, subject to $y(0) = -2$, $y'(0) = 1$, with

$$f(t) = \begin{cases} t & \text{if } 0 \leq t \leq 3, \\ t+2 & \text{if } t \geq 3. \end{cases}$$

(e) $\dfrac{dx}{dt} + x + \dfrac{dy}{dt} = 0$, $\dfrac{dx}{dt} - x + 2\dfrac{dy}{dt} = e^{-t}$, subject to $x(0) = y(0) = 1$.

(f) $\dfrac{d^2 x}{dt^2} = -2x + y$, $\dfrac{d^2 y}{dt^2} = x - 2y$, subject to the initial conditions $x(0) = y(0) = 1$ and $x'(0) = y'(0) = 0$.

6.4 Using Laplace transforms, solve the integral and integro-differential equations

(a) $y(t) = t + \dfrac{1}{6}\displaystyle\int_0^t y(\tau)(t - \tau)^3 \, d\tau$,

(b) $\displaystyle\int_0^t y(\tau)\cos(t - \tau) \, d\tau = \dfrac{dy}{dt}$, subject to $y(0) = 1$.

6.5 Show that $\mathcal{L}[tf(t)] = -dF/ds$, where $F(s) = \mathcal{L}[f(t)]$. Hence solve the initial value problem

$$\dfrac{d^2 x}{dt^2} + 2t\dfrac{dx}{dt} - 4x = 1, \quad \text{subject to } x(0) = x'(0) = 0.$$

6.6 Determine the inverse Laplace transform of

$$F(s) = \dfrac{s}{(s^2 + 1)(s - 2)}$$

using (a) the inversion formula (6.5), and (b) the residue theorem.

6.7 Find the inverse Laplace transform of

$$F(s) = \frac{1}{(s-1)(s^2+1)},$$

by (a) expressing $F(s)$ as partial fractions and inverting the constituent parts and (b) using the convolution theorem.

6.8 Evaluate the integral (6.19), and hence verify that (6.20) holds.

6.9 Consider the solution of the diffusion equation

$$\frac{\partial \phi}{\partial t} = D \frac{\partial^2 \phi}{\partial y^2}$$

for $t > 0$ and $y \in [0, L]$, subject to the initial conditions $\phi(y, 0) = 0$ and the boundary conditions $\phi(0, t) = \phi(L, t) = 1$. Show that the Laplace transform of the solution is

$$\Phi(y, s) = \frac{1}{s(1 + e^{-\kappa L})} \left(e^{-\kappa y} + e^{-\kappa(L-y)} \right),$$

where $\kappa = \sqrt{s/D}$. By expanding the denominator, show that the solution can be written as

$$\phi(y, t) = \sum_{j=0}^{\infty} (-1)^j \left[\mathrm{erfc}\left(\frac{y + jL}{2\sqrt{Dt}} \right) - \mathrm{erfc}\left(\frac{(j+1)L - y}{2\sqrt{Dt}} \right) \right].$$

6.10 (a) Show that a small displacement, $y(x, t)$, of a uniform string with constant tension T and line density ρ subject to a uniform gravitational acceleration, g, downwards satisfies

$$\frac{\partial^2 y}{\partial t^2} = c^2 \frac{\partial^2 y}{\partial x^2} - g,$$

where $c = \sqrt{T/\rho}$ (see Section 3.9.1).

(b) Such a string is semi-infinite in extent, and has $y = y_t = 0$ for $x \geq 0$ when $t = 0$. Use Laplace transforms to determine the solution when the string is fixed at $x = 0$, satisfies $y_x \to 0$ as $x \to \infty$ and is allowed to fall under gravity. Sketch the solution, and explain how and why the qualitative form of the solution depends upon the sign of $x - ct$.

6.11 **Project** The voltage, $v(t)$, in an RLC circuit with implied current $i(t)$ is given by the solution of the differential equation

$$C \frac{d^2 v}{dt^2} + \frac{1}{R} \frac{dv}{dt} + \frac{v}{L} = \frac{di}{dt},$$

where C is the capacitance, R the resistance and L the inductance.

(a) Solve this equation using Laplace transforms when $i(t) = H(t-1) - H(t)$.

(b) Write a MATLAB script that inverts the Laplace transform of this equation directly, so that it works when $i(t)$ is supplied by an input routine input.m.

(c) Compare your results for various values of the parameters R, L and C.

(d) Extend this further to consider what happens when the implied current is periodic.

CHAPTER SEVEN

Classification, Properties and Complex Variable Methods for Second Order Partial Differential Equations

In this chapter, we will consider a class of partial differential equations, of which Laplace's equation, the diffusion equation and the wave equation are the canonical examples. We will discuss what sort of boundary conditions are appropriate, and why solutions of these equations have their distinctive properties. We will also demonstrate that complex variable methods are powerful tools for solving boundary value problems for Laplace's equation, with particular application to certain problems in fluid mechanics.

7.1 Classification and Properties of Linear, Second Order Partial Differential Equations in Two Independent Variables

Consider a second order linear partial differential equation in two independent variables, which we can write as

$$a(x,y)\frac{\partial^2 \phi}{\partial x^2} + 2b(x,y)\frac{\partial^2 \phi}{\partial x \partial y} + c(x,y)\frac{\partial^2 \phi}{\partial y^2}$$
$$+ d_1(x,y)\frac{\partial \phi}{\partial x} + d_2(x,y)\frac{\partial \phi}{\partial y} + d_3(x,y)\phi \doteq f(x,y). \quad (7.1)$$

Equations of this type arise frequently in mathematical modelling, as we have already seen. We will show that the first three terms of (7.1) allow us to classify the equation into one of three distinct types: **elliptic**, for example Laplace's equation, (2.13), **parabolic**, for example the diffusion equation, (2.12), or **hyperbolic**, for example the wave equation, (3.39). Each of these types of equation has distinctive properties. These mathematical properties are related to the physical properties of the system that the equation models.

7.1.1 Classification

We would like to know about properties of (7.1) that are unchanged by an invertible change of coordinates, since these must be of fundamental significance, and not just a result of our choice of coordinate system. We can write this change of coordinates as†

$$(x,y) \mapsto (\xi(x,y), \eta(x,y)), \quad \text{with } \frac{\partial(\xi,\eta)}{\partial(x,y)} \neq 0. \tag{7.2}$$

In particular, if (7.1) is a model for a physical system, a change of coordinates should not affect its qualitative behaviour. Writing $\phi(x,y) \equiv \psi(\xi,\eta)$ and using subscripts to denote partial derivatives, we find that

$$\phi_x = \xi_x \psi_\xi + \eta_x \psi_\eta, \quad \phi_{xx} = \xi_x^2 \psi_{\xi\xi} + 2\xi_x \eta_x \psi_{\xi\eta} + \eta_x^2 \psi_{\eta\eta} + \xi_{xx} \psi_\xi + \eta_{xx} \psi_\eta,$$

and similarly for the other derivatives. Substituting these into (7.1) gives us

$$A\psi_{\xi\xi} + 2B\psi_{\xi\eta} + C\psi_{\eta\eta} + b_1(\xi,\eta)\psi_\xi + b_2(\xi,\eta)\psi_\eta + b_3(\xi,\eta)\psi = g(\xi,\eta), \tag{7.3}$$

where

$$A = a\xi_x^2 + 2b\xi_x\xi_y + c\xi_y^2,$$

$$B = a\xi_x\eta_x + b(\eta_x\xi_y + \xi_x\eta_y) + c\xi_y\eta_y, \tag{7.4}$$

$$C = a\eta_x^2 + 2b\eta_x\eta_y + c\eta_y^2.$$

We do not need to consider the other coefficient functions here. We can express (7.4) in a concise matrix form as

$$\begin{pmatrix} A & B \\ B & C \end{pmatrix} = \begin{pmatrix} \xi_x & \eta_x \\ \xi_y & \eta_y \end{pmatrix} \begin{pmatrix} a & b \\ b & c \end{pmatrix} \begin{pmatrix} \xi_x & \xi_y \\ \eta_x & \eta_y \end{pmatrix}, \tag{7.5}$$

which shows that

$$\det \begin{pmatrix} A & B \\ B & C \end{pmatrix} = \det \begin{pmatrix} a & b \\ b & c \end{pmatrix} \left(\frac{\partial(\xi,\eta)}{\partial(x,y)} \right)^2. \tag{7.6}$$

This shows that the sign of $ac - b^2$ is independent of the choice of coordinate system, which allows us to classify the equation.

— An **elliptic** equation has $ac > b^2$, for example, Laplace's equation

$$\frac{\partial^2 \phi}{\partial x^2} + \frac{\partial^2 \phi}{\partial y^2} = 0.$$

— A **parabolic** equation has $ac = b^2$, for example, the diffusion equation

$$K\frac{\partial^2 \phi}{\partial x^2} - \frac{\partial \phi}{\partial y} = 0.$$

- † Note that

$$\frac{\partial(\xi,\eta)}{\partial(x,y)} = \det \begin{pmatrix} \xi_x & \xi_y \\ \eta_x & \eta_y \end{pmatrix}$$

is the **Jacobian** of the transformation. The Jacobian is the factor by which the transformation changes infinitesimal volume elements.

7.1 CLASSIFICATION OF PARTIAL DIFFERENTIAL EQUATIONS

— A **hyperbolic** equation has $ac < b^2$, for example, the wave equation

$$\frac{1}{c^2}\frac{\partial^2 \phi}{\partial x^2} - \frac{\partial^2 \phi}{\partial y^2} = 0.$$

Note that, although these three examples are of the given type throughout the (x,y)-plane, equations of mixed type are possible. For example, **Tricomi's equation**, $\phi_{xx} = x\phi_{yy}$, is elliptic for $x < 0$ and hyperbolic for $x > 0$.

7.1.2 Canonical Forms

Any equation of the form given by (7.1) can be written in **canonical form** by choosing the **canonical coordinate system**, in terms of which the second derivatives appear in the simplest possible way.

Hyperbolic Equations: $ac < b^2$

In this case, we can factorize A and C to give

$$A = a\xi_x^2 + 2b\xi_x\xi_y + c\xi_y^2 = (p_1\xi_x + q_1\xi_y)(p_2\xi_x + q_2\xi_y),$$

$$C = a\eta_x^2 + 2b\eta_x\eta_y + c\eta_y^2 = (p_1\eta_x + q_1\eta_y)(p_2\eta_x + q_2\eta_y),$$

with the two factors not multiples of each other. We can then choose ξ and η so that

$$p_1\xi_x + q_1\xi_y = p_2\eta_x + q_2\eta_y = 0,$$

and hence $A = C = 0$. This means that

ξ is constant on curves with $\dfrac{dy}{dx} = \dfrac{q_1}{p_1}$, η is constant on curves with $\dfrac{dy}{dx} = \dfrac{q_2}{p_2}$.

We can therefore write $p_1 dy - q_1 dx = p_2 dy - q_2 dx = 0$, and hence

$$(p_1 dy - q_1 dx)(p_2 dy - q_2 dx) = 0,$$

which gives

$$a\,dy^2 - 2b\,dx\,dy + c\,dx^2 = 0. \tag{7.7}$$

As we shall see, this is the easiest equation to use to determine (ξ, η). We call (ξ, η) the **characteristic coordinate system**, in terms of which (7.1) takes its canonical form

$$\psi_{\xi\eta} + b_1(\xi,\eta)\psi_\xi + b_2(\xi,\eta)\psi_\eta + b_3(\xi,\eta)\psi = g(\xi,\eta). \tag{7.8}$$

The curves where ξ is constant and the curves where η is constant are called the **characteristic curves**, or simply **characteristics**. As we shall see, it is the existence, or nonexistence, of characteristic curves for the three types of equation that determines the distinctive properties of their solutions. We discussed the reduction of the wave equation to this canonical form in Section 3.9.1.

As a less trivial example, consider the hyperbolic equation

$$\phi_{xx} - \mathrm{sech}^4 x\,\phi_{yy} = 0. \tag{7.9}$$

Equation (7.7) shows that the characteristics are given by

$$dy^2 - \mathrm{sech}^4 x\, dx^2 = (dy + \mathrm{sech}^2 x\, dx)(dy - \mathrm{sech}^2 x\, dx) = 0,$$

and hence

$$\frac{dy}{dx} = \pm \mathrm{sech}^2 x.$$

The characteristics are therefore $y \pm \tanh x = $ constant, and the characteristic coordinates are $\xi = y + \tanh x$, $\eta = y - \tanh x$. On writing (7.9) in terms of these variables, with $\phi(x,y) = \psi(\xi,\eta)$, we find that its canonical form is

$$\psi_{\xi\eta} = \frac{(\eta - \xi)(\psi_\xi - \psi_\eta)}{4 - (\xi - \eta)^2}, \tag{7.10}$$

in the domain $(\eta - \xi)^2 < 4$.

Parabolic Equations: $ac = b^2$

In this case,

$$A = a\xi_x^2 + 2b\xi_x\xi_y + c\xi_y^2 = (p\xi_x + q\xi_y)^2,$$

$$C = a\eta_x^2 + 2b\eta_x\eta_y + c\eta_y^2 = (p\eta_x + q\eta_y)^2,$$

so we can only construct one set of characteristic curves. We therefore take ξ to be constant on the curves $p\,dy - q\,dx = 0$. This gives us $A = 0$ and, since $AC = B^2$, $B = 0$. For any set of curves where η is constant that is never parallel to the characteristics, C does not vanish, and the canonical form is

$$\psi_{\eta\eta} + b_1(\xi,\eta)\psi_\xi + b_2(\xi,\eta)\psi_\eta + b_3(\xi,\eta)\psi = g(\xi,\eta). \tag{7.11}$$

We can now see that the diffusion equation is in canonical form.

As a further example, consider the parabolic equation

$$\psi_{xx} + 2\mathrm{cosec}\, y\, \phi_{xy} + \mathrm{cosec}^2 y\, \phi_{yy} = 0. \tag{7.12}$$

The characteristic curves satisfy

$$dy^2 - 2\mathrm{cosec}\, y\, dx\, dy + \mathrm{cosec}^2 y\, dx^2 = (dy - \mathrm{cosec}\, y\, dx)^2 = 0,$$

and hence

$$\frac{dy}{dx} = \mathrm{cosec}\, y.$$

The characteristic curves are therefore given by $x + \cos y = $ constant, and we can take $\xi = x + \cos y$ as the characteristic coordinate. A suitable choice for the other coordinate is $\eta = y$. On writing (7.12) in terms of these variables, with $\phi(x,y) = \psi(\xi,\eta)$, we find that its canonical form is

$$\psi_{\eta\eta} = \sin^2 \eta \cos \eta\, \psi_\xi, \tag{7.13}$$

in the whole (ξ,η)-plane.

7.1 CLASSIFICATION OF PARTIAL DIFFERENTIAL EQUATIONS

Elliptic Equations: $ac > b^2$

In this case, we can make neither A nor C zero, since no real characteristic curves exist. Instead, we can simplify by making $A = C$ and $B = 0$, so that the second derivatives form the Laplacian, $\nabla^2 \psi$, and the canonical form is

$$\psi_{\xi\xi} + \psi_{\eta\eta} + b_1(\xi,\eta)\psi_\xi + b_2(\xi,\eta)\psi_\eta + b_3(\xi,\eta)\psi = g(\xi,\eta). \qquad (7.14)$$

Clearly, Laplace's equation is in canonical form.

In order to proceed, we must solve

$$A - C = a\left(\xi_x^2 - \eta_x^2\right) + 2b\left(\xi_x\xi_y - \eta_x\eta_y\right) + c\left(\xi_y^2 - \eta_y^2\right) = 0,$$

$$B = a\xi_x\eta_x + b\left(\eta_x\xi_y + \xi_x\eta_y\right) + c\xi_y\eta_y = 0.$$

We can do this by defining $\chi = \xi + i\eta$, and noting that these two equations form the real and imaginary parts of

$$a\chi_x^2 + 2b\chi_x\chi_y + c\chi_y^2 = 0,$$

and hence that

$$\frac{\chi_x}{\chi_y} = \frac{-b \pm i\sqrt{ac - b^2}}{a}. \qquad (7.15)$$

Now χ is constant on curves given by $\chi_y\, dy + \chi_x\, dx = 0$, and hence, from (7.15), on

$$\frac{dy}{dx} = \frac{b \mp i\sqrt{ac - b^2}}{a}. \qquad (7.16)$$

By solving (7.16) we can deduce ξ and η. For example, consider the elliptic equation

$$\phi_{xx} + \operatorname{sech}^4 x \, \phi_{yy} = 0. \qquad (7.17)$$

In this case, $\chi = \xi + i\eta$ is constant on the curves given by

$$\frac{dy}{dx} = \pm i \operatorname{sech}^2 x,$$

and hence $y \mp i\tanh x =$ constant. We can therefore take $\chi = y + i\tanh x$, and hence $\xi = y$, $\eta = \tanh x$. On writing (7.17) in terms of these variables, with $\phi(x,y) = \psi(\xi,\eta)$, we find that its canonical form is

$$\psi_{\xi\xi} + \psi_{\eta\eta} = \frac{2\eta}{1 - \eta^2}\psi_\eta, \qquad (7.18)$$

in the domain $|\eta| < 1$.

We can now describe some of the properties of the three different types of equation. For more detailed information, the reader is referred to Kevorkian (1990) and Carrier and Pearson (1988).

7.1.3 Properties of Hyperbolic Equations

Hyperbolic equations are distinguished by the existence of two sets of characteristics. This allows us to establish two key properties. Firstly, characteristics are carriers of information. For the wave equation, we saw in Section 3.9.1 that solutions propagate at speeds $\pm c$, which corresponds to propagation on the characteristic curves, $x \pm ct = \text{constant}$. Indeed, the use of the independent variable t, time, instead of y suggests that the equation has an evolutionary nature. More specifically, consider the **Cauchy problem** for a hyperbolic equation of the form (7.1). In a Cauchy problem, we specify a curve C in the (x,y)-plane upon which we know the **Cauchy data**; ϕ and the derivative of ϕ normal to C, $\partial \phi/\partial n$. The initial value problem for the wave equation that we studied in Section 3.9.1 is a Cauchy problem. Does this problem have a solution in general? Let's assume that ϕ and $\partial\phi/\partial n$ on C can be expanded as power series, and that the functions that appear as coefficients in (7.1) can also be expanded as power series in the neighbourhood of C. We can then write

$$\phi(\xi,\eta) = \phi(\xi,0) + \eta \frac{\partial \phi}{\partial \eta}(\xi,0) + \frac{1}{2!}\eta^2 \frac{\partial^2 \phi}{\partial \eta^2}(\xi,0) + \cdots,$$

where the orthogonal coordinate system (ξ,η) is set up so that $\eta = 0$ is the curve C, which is then parameterized by ξ. As we have seen, we can write (7.1) in terms of this new coordinate system as (7.3). We know $\phi(\xi,0)$ and the normal derivative, $\partial \phi/\partial \eta(\xi,0)$, and, *provided that the coefficient of $\partial^2 \phi/\partial \eta^2$ does not vanish on C*, we can deduce $\partial^2 \phi/\partial \eta^2(\xi,0)$ from (7.3). When does the coefficient of $\partial^2\phi/\partial\eta^2$ vanish on C? Precisely when C is a characteristic curve. Provided that C is *not* a characteristic curve, we can also deduce higher derivatives from derivatives of (7.3), and hence construct a power series solution, valid in the neighbourhood of C. The formal statement of this informally presented procedure is the **Cauchy–Kowalewski theorem**, a local existence theorem (see Garabedian, 1964, for a formal statement and proof). The effects of the initial data propagate into the (x,y)-plane on the characteristics, so it is inconsistent to specify initial data upon a characteristic curve. In addition, the solution at any point (x,y) is only dependent on the initial conditions that lie between the two characteristics through (x,y), as shown in Figure 7.1. For the wave equation, this is immediately obvious from d'Alembert's solution, (3.43).

Secondly, discontinuities in the second derivative of ϕ can propagate on characteristic curves. To see this, consider a curve C in the (x,y)-plane, not necessarily a characteristic, given by $\xi(x,y) = \xi_0$. Suppose that $\phi_{\xi\xi}$ is not continuous on C, but that ϕ satisfies the hyperbolic equation on either side of C. Can we choose ξ in such a way that the equation is satisfied, even though $\phi_{\xi\xi}(\xi_0^+,\eta) \neq \phi_{\xi\xi}(\xi_0^-,\eta)$? If we evaluate the equation on either side of the curve and subtract, we find that

$$A(\xi_0,\eta)\left\{\phi_{\xi\xi}(\xi_0^+,\eta) - \phi_{\xi\xi}(\xi_0^-,\eta)\right\} = 0.$$

We can therefore satisfy the equation if $A(\xi_0,\eta) = 0$, and hence C is a characteristic curve. We conclude that discontinuities in the second derivative can propagate on characteristic curves. In general, ϕ and its first derivatives are continuous, but for

7.1 CLASSIFICATION OF PARTIAL DIFFERENTIAL EQUATIONS

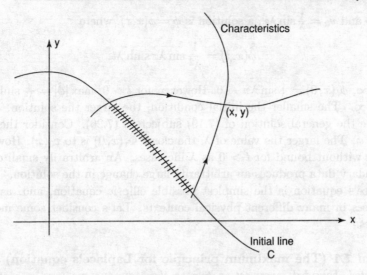

Fig. 7.1. The domain of dependence of the solution of a hyperbolic equation. The solution at (x, y) depends only upon the initial data on the marked part of the initial line, C.

the wave equation, in which only second derivatives appear, discontinuities in ϕ and its derivatives can also propagate on characteristics, as shown in Figure 3.10.

7.1.4 Properties of Elliptic Equations

For elliptic equations, A and C are never zero, and there are no characteristics. Solutions are therefore infinitely-differentiable. Moreover, there is no timelike variable; for example, Laplace's equation is written in terms of the spatial variables, x and y. For physical problems that can be modelled using elliptic equations, boundary value problems, rather than initial value problems, usually arise naturally; for example the steady state diffusion problem discussed in Section 2.6.1. In fact, the solution of the Cauchy problem for elliptic equations does not depend continuously on the initial conditions, and is therefore not a sensible representation of any real, physical problem.† We can easily demonstrate this for Laplace's equation.

Consider the Cauchy, or initial value, problem

$$\phi_{xx} + \phi_{tt} = 0 \quad \text{for } t \geqslant 0, \ -\infty < x < \infty, \tag{7.19}$$

subject to

$$\phi(x, 0) = \phi_0(x), \quad \phi_t(x, 0) = v_0(x). \tag{7.20}$$

† We say that the problem is **ill-posed**.

If $\phi_0 = 0$ and $v_0 = \frac{1}{\lambda}\sin\lambda x$, a solution is $\phi = \hat{\phi}(x,t)$, where

$$\hat{\phi}(x,t) = \frac{1}{\lambda^2}\sin\lambda x \sinh\lambda t.$$

As $\lambda \to \infty$, $\hat{\phi}_t(x,0) = \frac{1}{\lambda}\sin\lambda x \to 0$. However, for $t > 0$, $\max|\hat{\phi}| = \frac{1}{\lambda^2}\sinh\lambda t \to \infty$ as $\lambda \to \infty$. The smaller the initial condition, the larger the solution. Now, let $\Phi(x,t)$ be the general solution of (7.19) subject to (7.20). Consider the function $\hat{\Phi} = \Phi + \hat{\phi}$. The larger the value of λ, the closer $\hat{\Phi}_t(x,0)$ is to $v_0(x)$. However, $|\hat{\Phi}|$ increases without bound for $t > 0$ as λ increases. An arbitrarily small change in the boundary data produces an arbitrarily large change in the solution.

Laplace's equation is the simplest possible elliptic equation, and, as we have seen, arises in many different physical contexts. Let's consider some more of its properties.

Theorem 7.1 (The maximum principle for Laplace's equation) *Let D be a connected, bounded, open set in two or three dimensions, and ϕ a solution of Laplace's equation in D. Then ϕ attains its maximum and minimum values on ∂D, the boundary of D, and nowhere in the interior of D, unless ϕ is a constant.*

Proof The idea of the proof is straightforward. For example in two dimensions, at a local maximum in the interior of D, $\phi_x = \phi_y = 0$, $\phi_{xx} \leq 0$ and $\phi_{yy} \leq 0$. At a maximum with $\phi_{xx} < 0$ or $\phi_{yy} < 0$, we have $\nabla^2\phi = \phi_{xx} + \phi_{yy} < 0$, and ϕ cannot be a solution of Laplace's equation. However, it is possible to have $\phi_{xx} = \phi_{yy} = 0$ at an interior local maximum, for example, if the Taylor series expansion close to $(x,y) = (x_0,y_0)$ is $\phi = (x-x_0)^4 + (y-y_0)^4 + \cdots$. Although this is clearly not a solution of Laplace's equation, we do need to do a little work in order to exclude this possibility in general.

Let $\psi = \phi + \epsilon|\mathbf{x}|^2$, with $\epsilon > 0$ and $|\mathbf{x}|^2 = x^2 + y^2$ in two dimensions and $|\mathbf{x}|^2 = x^2 + y^2 + z^2$ in three dimensions. We then have

$$\nabla^2\psi = \nabla^2\phi + \epsilon\nabla^2|\mathbf{x}|^2 = k\epsilon > 0,$$

where $k = 4$ in two dimensions and $k = 6$ in three dimensions. Since $\nabla^2\psi \leq 0$ at an interior local maximum, we conclude that ψ has no local maximum in the interior of D, and hence that ψ must attain its maximum value on the boundary of D, say at $\mathbf{x} = \mathbf{x}_0$. But, by definition

$$\phi(\mathbf{x}) \leq \psi(\mathbf{x}) \leq \psi(\mathbf{x}_0) = \phi(\mathbf{x}_0) + \epsilon|\mathbf{x}_0|^2 \leq \max_{\partial D}\phi + \epsilon l^2,$$

where l is the greatest distance from ∂D to the origin. Since this is true for all $\epsilon > 0$, we can make ϵ arbitrarily small, and conclude that

$$\phi(\mathbf{x}) \leq \max_{\partial D}\phi \text{ for all } \mathbf{x} \in D.$$

Similarly, $-\phi$, which also satisfies Laplace's equation, attains its maximum value on ∂D, and hence ϕ attains its minimum value on ∂D. $\square$

7.1 CLASSIFICATION OF PARTIAL DIFFERENTIAL EQUATIONS

We can use the maximum principle to show that the Dirichlet problem for Laplace's equation has a unique solution.

Theorem 7.2 *The Dirichlet problem*

$$\nabla^2 \phi = 0 \quad \text{for } \mathbf{x} \in D, \tag{7.21}$$

with

$$\phi(\mathbf{x}) = f(\mathbf{x}) \quad \text{on } \partial D, \tag{7.22}$$

has a unique solution.

Proof Let ϕ_1 and ϕ_2 be solutions of (7.21) subject to (7.22). Then $\psi = \phi_1 - \phi_2$ satisfies

$$\nabla^2 \psi = 0 \quad \text{for } \mathbf{x} \in D, \tag{7.23}$$

with

$$\psi(\mathbf{x}) = 0 \quad \text{on } \partial D. \tag{7.24}$$

By Theorem 7.1, ψ attains its maximum and minimum values of ∂D, so that $0 \leqslant \psi(\mathbf{x}) \leqslant 0$ for $\mathbf{x} \in D$, and hence $\psi \equiv 0$. This means that $\phi_1 = \phi_2$, and hence that there is unique solution. □

The maximum principle can also be used to study other elliptic equations. For example, consider the boundary value problem

$$\nabla^2 \phi = -1 \quad \text{for } \mathbf{x} \in X = \left\{ (x, y) \,\bigg|\, \frac{|x|}{a} + \frac{|y|}{b} < 1 \right\}, \tag{7.25}$$

with $a > b > 0$, subject to

$$\phi = 0 \quad \text{on } \partial X. \tag{7.26}$$

How big is $\phi(0,0)$? We can obtain some bounds on this quantity by transforming (7.25) into Laplace's equation. We define $\psi = (x^2 + y^2)/4$, so that $\nabla^2 \psi = 1$, and let $\hat{\phi} = \phi + \psi$, so that $\nabla^2 \hat{\phi} = 0$. We can also see that $\hat{\phi} = \psi$ on ∂X, and that $\hat{\phi}(0,0) = \phi(0,0)$. Theorem 7.1 then shows that

$$\min_{\partial X} \psi \leqslant \phi(0,0) \leqslant \max_{\partial X} \psi.$$

We can therefore bound $\phi(0,0)$ using the maximum and minimum values of $\psi = (x^2 + y^2)/4$ on the boundary of X. The maximum value is clearly $a^2/4$. Since X is symmetric about both coordinate axes, we can find the minimum value of ψ by determining the value of the radius r_0 for which the circle $x^2 + y^2 = r_0^2$ just touches the straight line $x/a + y/b = 1$, which forms the boundary of X in the quadrant $x > 0$, $y > 0$. A little algebra shows that $r_0^2 = a^2 b^2/(a^2 + b^2)$, and hence that

$$\frac{a^2 b^2}{4(a^2 + b^2)} \leqslant \phi(0,0) \leqslant \frac{a^2}{4}.$$

Finally, the other type of boundary value problem that often arises for Laplace's equation is the Neumann problem (see Section 5.4), where the normal derivative of ϕ is specified on the boundary of the solution domain. Such problems can only be solved if the boundary data satisfies a solubility condition.

Theorem 7.3 *A necessary condition for the existence of a solution of the Neumann problem*

$$\nabla^2 \phi = 0 \quad \text{for } \mathbf{x} \in D \subset \mathbb{R}^3, \tag{7.27}$$

with

$$\frac{\partial \phi}{\partial n}(\mathbf{x}) = g(\mathbf{x}) \quad \text{on } \partial D, \tag{7.28}$$

is

$$\int_{\partial D} g(\mathbf{x}) \, d^2 \mathbf{x} = 0. \tag{7.29}$$

Proof If we integrate (7.27) over D and use the divergence theorem and (7.28), we find that

$$\int_D \nabla^2 \phi \, d^3 \mathbf{x} = \int_{\partial D} \mathbf{n} \cdot \nabla \phi \, d^2 \mathbf{x} = \int_{\partial D} g(\mathbf{x}) \, d^2 \mathbf{x} = 0.$$

□

This solubility condition has a simple interpretation in terms of inviscid, incompressible, irrotational fluid flow, which we introduced in Section 2.6.2. In this case, the Neumann problem specifies a steady flow in D with the normal velocity of the fluid on ∂D given by $g(\mathbf{x})$. Since the fluid is incompressible, such a flow can only exist if the total flux into D through ∂D is zero, as expressed by (7.29).

7.1.5 Properties of Parabolic Equations

Parabolic equations have just one set of characteristics. For example, for the diffusion equation, $K\phi_{xx} = \phi_t$ with $K > 0$, t is constant on the characteristic curves. As we have seen, any localized disturbance is therefore felt everywhere in $-\infty < x < \infty$ instantaneously, as we can see from Example 5 of Section 6.4. In addition, solutions are infinitely-differentiable with respect to x. We can also prove a maximum principle for the diffusion equation.

Theorem 7.4 (The maximum principle for the diffusion equation) *Let ϕ be a solution of the diffusion equation. Consider the domain $0 \leqslant x \leqslant L$, $0 \leqslant t \leqslant T$. Then ϕ attains its maximum value on $x = 0$, $0 \leqslant t \leqslant T$, or $x = L$, $0 \leqslant t \leqslant T$, or $t = 0$, $0 \leqslant x \leqslant L$.*

Proof This is very similar to the proof of the maximum principle for Laplace's equation, and we leave it as Exercise 7.8(a). □

7.1 CLASSIFICATION OF PARTIAL DIFFERENTIAL EQUATIONS

We can use this maximum principle to show that the solution of the initial–boundary value problem given by

$$\frac{\partial \phi}{\partial t} = K \frac{\partial^2 \phi}{\partial x^2} \quad \text{for } t > 0,\ 0 < x < L, \tag{7.30}$$

subject to

$$\phi(x,0) = \phi_0(x) \quad \text{for } 0 \leqslant x \leqslant L, \tag{7.31}$$

and

$$\phi(0,t) = \phi_1(t),\ \phi(L,t) = \phi_2(t) \quad \text{for } t > 0, \tag{7.32}$$

with ϕ_0, ϕ_1 and ϕ_2 prescribed functions, is unique. We leave the details as Exercise 7.8(b).

We have now seen that parabolic equations have a single set of characteristics, and that for a canonical example, the diffusion equation, there is a maximum principle, so that parabolic equations share some of the features of both hyperbolic and elliptic equations. We end this section by stating a useful theorem that holds for reaction–diffusion equations, which are parabolic. We will meet reaction–diffusion equations frequently in Part 2.

Theorem 7.5 (A comparison theorem for reaction–diffusion equations)
Consider the reaction–diffusion equation

$$\phi_t = K \nabla^2 \phi + f(\phi, \mathbf{x}, t) \quad \text{for } \mathbf{x} \in D,\ t > 0, \tag{7.33}$$

with $K > 0$ and f a smooth function. If $\bar{\phi}(\mathbf{x}, t)$ is a bounded function that satisfies

$$\bar{\phi}_t \geqslant K \nabla^2 \bar{\phi} + f(\bar{\phi}, \mathbf{x}, t) \quad \text{for } \mathbf{x} \in D,\ t > 0,$$

we say that $\bar{\phi}$ is a supersolution of (7.33). Similarly, if $\underline{\phi}(\mathbf{x}, t)$ is a bounded function that satisfies

$$\underline{\phi}_t \leqslant K \nabla^2 \underline{\phi} + f(\underline{\phi}, \mathbf{x}, t) \quad \text{for } \mathbf{x} \in D,\ t > 0,$$

we say that $\underline{\phi}$ is a subsolution of (7.33). If there also exist constants α and β with $\alpha^2 + \beta^2 \neq 0$, such that

$$\alpha \bar{\phi} - \beta \frac{\partial \bar{\phi}}{\partial n} \geqslant \alpha \underline{\phi} - \beta \frac{\partial \underline{\phi}}{\partial n} \quad \text{for } \mathbf{x} \in \partial D,\ t > 0,$$

and

$$\bar{\phi}(\mathbf{x}, 0) \geqslant \underline{\phi}(\mathbf{x}, 0) \quad \text{for } \mathbf{x} \in D,$$

then

$$\bar{\phi}(\mathbf{x}, t) \geqslant \underline{\phi}(\mathbf{x}, t) \quad \text{for } \mathbf{x} \in D,\ t > 0.$$

We will not prove this result here (see Grindrod, 1991, for further details).

To see how this theorem can be used, consider the case $f = \phi(1 - \phi)$, with $\partial \phi / \partial n = 0$ on ∂D. This problem arises in a model for the propagation of chemical waves (see Billingham and King, 2001). If $0 \leqslant \phi(\mathbf{x}, 0) \leqslant 1$, then by taking firstly $\bar{\phi} = 1$, $\underline{\phi} = \phi$, and secondly $\bar{\phi} = \phi$, $\underline{\phi} = 0$, we find that $0 \leqslant \phi(\mathbf{x}, t) \leqslant 1$ for $t \geqslant 0$.

In other words, the comparison theorem allows us to determine upper and lower bounds on the solution (see Exercise 7.9 for another example).

7.2 Complex Variable Methods for Solving Laplace's Equation

In Section 2.6.2 we described how steady, inviscid, incompressible, irrotational fluid flow, commonly referred to as **ideal** fluid flow, past a rigid body can be modelled as a boundary value problem for Laplace's equation, $\nabla^2 \phi = 0$, in the fluid, with $\partial \phi / \partial n = 0$ on the boundary of the body, where ϕ is the velocity potential and $\mathbf{u} = \nabla \phi$ the velocity field. Appropriate conditions at infinity also need to be prescribed, for example, a uniform stream. Once ϕ is known, we can determine the pressure using Bernoulli's equation, (2.18). In general, solutions of this boundary value problem are hard to find, except for the simplest body shapes. However, two-dimensional flow is an exception, since powerful complex variable methods can, without too much effort, give simple descriptions of the flow past many simply-connected body shapes, such as aircraft wing sections, channels with junctions and flows with free surfaces (see, for example, Milne-Thompson, 1960).

7.2.1 The Complex Potential

Recall from Section 2.6.2 that the velocity in a two-dimensional ideal fluid flow, $\mathbf{u} = (u, v) = (\phi_x, \phi_y)$, satisfies the continuity equation, $u_x + v_y = 0$. We can therefore introduce a **stream function**, $\psi(x, y)$, defined by $u = \psi_y$, $v = -\psi_x$, which, for sufficiently smooth functions, satisfies the continuity equation identically. Elementary calculus shows that the stream function is constant on any streamline in the flow, and the change in ψ between any two streamlines is equal to the flux of fluid between them. If we now look at the definitions we have for the components of velocity, for compatibility we need $\phi_x = \psi_y$ and $\phi_y = -\psi_x$. These are the Cauchy–Riemann equations for the complex function $w = \phi + i\psi$ and, with our smoothness assumption, imply that $w(z)$ is an *analytic* function of the complex variable $z = x + iy$ in the domain occupied by the fluid.† The quantity $w(z)$ is called the **complex potential** for the flow and, for simple flows, can be found easily. We can also see that

$$\frac{dw}{dz} = \phi_x + i\psi_x = u - iv = qe^{-i\theta},$$

where q is the magnitude of the velocity and θ the angle that it makes with the x-axis, so that, once we know w, we can easily compute the components of the velocity, and vice versa.

Let's consider some examples.

(i) A uniform stream flowing at an angle α to the horizontal has $u = U \cos \alpha$, $v = U \sin \alpha$, so that

$$\frac{dw}{dz} = u - iv = Ue^{-i\alpha},$$

† See Appendix 6 for a reminder of some basic ideas in the theory of complex variables.

and hence $w = Ue^{-i\alpha}z$.

(ii) A **point vortex** is a system of concentric circular streamlines centred on $z = z_0$, with complex potential

$$w = \frac{i\kappa}{2\pi} \log(z - z_0).$$

If we now define the **circulation** to be $\int_C \mathbf{u} \cdot \mathbf{dr}$, where C is any closed contour that encloses $z = z_0$, the point vortex has circulation κ. By taking real and imaginary parts, we can see that $\phi = -\kappa\theta/2\pi$, $\psi = \kappa \log r/2\pi$, where r is the polar coordinate centred on $z = z_0$. The streamlines are therefore given by $\psi = $ constant, and hence $r = $ constant, a family of concentric circles, as expected.

(iii) A **point source** of fluid at $z = z_0$ ejects m units of fluid per unit area per unit time. Its complex potential is

$$w = \frac{m}{2\pi} \log(z - z_0),$$

for a source at $z = z_0$. The streamlines are straight lines passing through $z = z_0$ and, since

$$\frac{dw}{dz} = qe^{-i\theta} = \frac{m}{2\pi} \frac{1}{z - z_0},$$

we have that $q = m/2\pi r$, where r is the polar coordinate centred at $z = z_0$, and that the flow is purely radial. It is simple to show that the flow through any closed curve, C, that encloses z_0 is

$$\int_C \mathbf{u} \cdot \mathbf{n}\, dl = m,$$

as expected.

A word of warning is required here. The first of these complex potentials is analytic over the whole z-plane. The second and third fail to be analytic at $z = z_0$. In reality, we need to invoke some extra physics close to this point. For example, any real source of fluid will be of finite size, and, close to $z = z_0$, we need to take this into account. For a point vortex, close to the singularity the effect of viscosity becomes important, and we need to include it. In each case, this can be done by defining an inner asymptotic region, using the methods that we will discuss in Chapter 12. Alternatively, if there are no sources or vortices in the flow domain that we are considering, but we wish to use sources and vortices to model the flow, the points where the complex potential is not analytic must be excluded from the flow domain. We will see an example of this in the next section.

7.2.2 Simple Flows Around Blunt Bodies

Since Laplace's equation is linear, complex potentials can be added together to produce more complicated fluid flow fields. For example, consider a uniform stream of magnitude U flowing parallel to the x-axis and a source of strength $2\pi m$ situated on the x-axis at $z = a$. The complex potential is $w = Uz + m\log(z-a)$, from which we can deduce that

$$\psi = Uy + m\tan^{-1}\left(\frac{y}{x-a}\right). \tag{7.34}$$

This is an odd function of y, so that the velocity field associated with it is an even function of y, as we would expect from the symmetry of the flow. The x-axis ($y = 0$) is a streamline with $\psi = 0$, whilst for $x > 0$ there is another streamline with $\psi = 0$, given by

$$y = (a-x)\tan\left(\frac{Uy}{m}\right), \tag{7.35}$$

which is a blunt-nosed curve. The streamlines are shown in Figure 7.2. As there is no flow across a streamline, we can replace it with the surface of a rigid body without affecting the flow pattern. The complex potential (7.34) therefore represents the flow past a blunt body of the form given by (7.35). At the point $z = a - m/U$, $dw/dz = 0$, and hence both components of the velocity vanish there. This is called a **stagnation point** of the flow. Note that the singularity due to the point source does not lie within the flow domain, and can be disregarded.

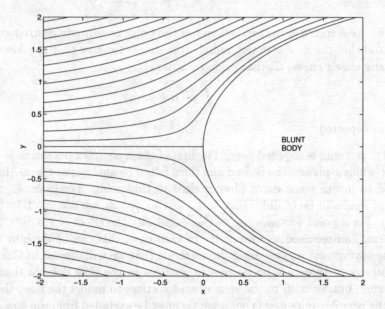

Fig. 7.2. The flow past a blunt body given by the stream function (7.34) with $U = m = a = 1$. There is a stagnation point at the origin.

Adding potentials is one way to construct ideal fluid flows around rigid bodies. However, since it is an inverse method (for a given flow field, we can introduce a rigid body bounded by any streamline), it has its drawbacks. We would prefer a

7.2 COMPLEX VARIABLE METHODS FOR SOLVING LAPLACE'S EQUATION

direct method, whereby we can calculate a stream function for a given rigid body. Before we describe such a method, we need the **circle theorem**.

Theorem 7.6 (The circle theorem) *If $w = f(z)$ is a complex potential for a flow with no singularities in the domain $|z| \leq a$, then $w = f(z) + f^*(a^2/z)$ is a complex potential for the same flow obstructed by a circle of radius a centred at the origin.*

Proof Firstly, since $f(z)$ is analytic in $|z| < a$, $f^*(a^2/z)$ is analytic in $|z| > a$, as a^2/z is just an inversion of the point z. Secondly, on the boundary of the circle, $z = ae^{i\theta}$ with $0 \leq \theta \leq 2\pi$, so that $f(z) + f^*(a^2/z) = f(ae^{i\theta}) + f^*(ae^{-i\theta})$, which, as it is the sum of complex conjugate functions, is real. Therefore, $\psi = 0$ on $z = ae^{i\theta}$, and the boundary of the circle is a streamline. Finally, since $w \sim f(z) + f^*(0)$ as $|z| \to \infty$, the flow in the far field is given by $f(z)$. □

We can illustrate this theorem by finding the complex potential for a uniform stream of strength U flowing past a circle of radius a. The complex potential for the stream is $w = Ue^{-i\alpha}z$, which has no singularities in $|z| \leq a$. By the circle theorem, the complex potential for the flow when the circular boundary is introduced is

$$w = Ue^{-i\alpha}z + Ue^{i\alpha}\frac{a^2}{z}.$$

When $\alpha = 0$, the flow is symmetric about the x-axis, with

$$\frac{dw}{dz} = U\left(1 - \frac{a^2}{z^2}\right),$$

and $dw/dz \sim U$ as $|z| \to \infty$, as expected. By taking real and imaginary parts, we can recover the velocity field, potential and stream function, which we discussed in Section 2.6.2. Before leaving this problem, we note that the complex potential is *not* unique, since we can add a point vortex, centred upon $z = 0$, and still have the circle as a streamline, without introducing a singularity in the flow domain. When $\alpha = 0$, this gives the potential

$$w = U\left(z + \frac{a^2}{z}\right) + \frac{i\kappa}{2\pi}\log z. \tag{7.36}$$

The introduction of this nonzero circulation about the circle has important consequences, as it breaks the symmetry (see Figure 7.3), and, as we shall now show, causes a nonzero lift force on the body.

In order to calculate the force exerted by an ideal fluid flow upon a rigid body, consider an infinitesimal element of arc on the surface of the body, ds. Let the tangent to this element of arc make an angle θ with the x-axis. The force on this infinitesimal arc is due to the pressure, and given by

$$-p\sin\theta \, ds + ip\cos\theta \, ds = ipe^{i\theta}ds.$$

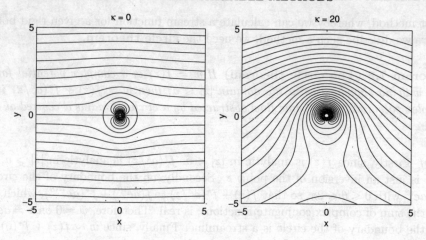

Fig. 7.3. Flow past a circle, with no circulation and with $\kappa = 20$.

Since $dz = dx + i\,dy = \cos\theta\,ds + i\sin\theta\,ds = e^{i\theta}\,ds$, the total force on the body is

$$F_x + iF_y = \int_{\text{body}} ipe^{i\theta}\,ds = \int_{\text{body}} ip\,dz.$$

Now, from Bernoulli's equation, (2.18), $p = p_0 - \frac{1}{2}\rho q^2$ and $qe^{-i\theta} = dw/dz$, where θ is the angle of the flow, since this is equal to the tangential angle at the surface of the rigid body, so that

$$F_x + iF_y = \int_{\text{body}} i\left\{ p_0 - \frac{1}{2}\rho\left|\frac{dw}{dz}\right|^2 \right\} dz.$$

As p_0 is a constant,

$$F_x + iF_y = -\frac{1}{2}\rho i \int_{\text{body}} \left|\frac{dw}{dz}\right|^2 dz.$$

To get a more usable result, it is convenient to manipulate this. Firstly, taking the complex conjugate,

$$F_x - iF_y = \frac{1}{2}\rho i \int_{\text{body}} \left|\frac{dw}{dz}\right|^2 dz^*.$$

Now note that

$$\left|\frac{dw}{dz}\right|^2 = \frac{dw}{dz}\left(\frac{dw}{dz}\right)^* = \frac{dw}{dz}\frac{dw^*}{dz^*},$$

so that

$$\left|\frac{dw}{dz}\right|^2 dz^* = \frac{dw}{dz}\,dw^*.$$

7.2 COMPLEX VARIABLE METHODS FOR SOLVING LAPLACE'S EQUATION

Since the boundary of the body is a streamline, $dw^* = dw$, so that $dw^* = (dw/dz)dz$, and hence

$$F_x - iF_y = \frac{1}{2}\rho i \int_{\text{body}} \left(\frac{dw}{dz}\right)^2 dz. \tag{7.37}$$

In this form, the force can usually be evaluated using the residue theorem. For the example of flow around a circle with circulation κ, we have

$$F_x - iF_y = \frac{1}{2}\rho i \int_{\text{body}} \left\{ U\left(1 - \frac{a^2}{z^2}\right) + \frac{i\kappa}{2\pi z} \right\}^2 dz.$$

After evaluating the residue, which is just due to the simple pole, we find that $F_x - iF_y = -i\rho U\kappa$, so that there is a vertical force of magnitude $\rho U\kappa$, which tries to lift the body in the direction of increasing y. This lift force arises from the pressure distribution around the circle, with pressures at the lower surface of the body higher than those on the upper surface. As we shall see in Example 4 in the next section, it is also the circulation around an aerofoil that provides the lift.

7.2.3 Conformal Transformations

Suppose we have a flow in the z-plane past a body whose shape cannot immediately be seen to be the streamline of a simple flow. If we can transform this problem into a new plane, the ζ-plane, where the shape of the body is simpler, such as a half-plane or a circle, we may be able to solve for the flow in the ζ-plane, and then invert the transformation to get the flow in the original, z-plane. Specifically, if we seek $w = w(z)$, the complex potential in the z-plane, and we have a transformation, $\zeta = f(z)$, which maps the surface of the rigid body, and the flow outside it, onto a flow outside a half-plane or circle in the ζ-plane, then, by defining $W(\zeta) = w(z)$, we have a correspondence between flow and geometry in both planes. Any streamline in the z-plane transforms to a streamline in the ζ-plane because of this correspondence in complex potentials. The complex velocity is

$$\frac{dw}{dz} = \frac{dw}{d\zeta}\frac{d\zeta}{dz} = \frac{dw}{d\zeta}f'(z).$$

If $|f'(z)|$ is bounded and nonzero, except perhaps for isolated points on the boundary of the domain to be transformed, we say that the transformation between the z- and ζ-planes is **conformal**, and a unique inverse transformation can be defined. Conformal mappings are so called because they also preserve the angle between line segments except at any isolated points where $f'(z)$ is zero or infinity (see Exercise 7.10). A consequence of this is that if we want to map a domain whose boundary has corners to a domain with a smooth boundary, we must have $f'(z)$ equal to zero or infinity at these corner points. Such points can cause difficulties. For example, if $|f'(z)| = \infty$, we could induce an infinite velocity in the z-plane, which is unphysical.

Although beautifully simple in principle, there is a practical difficulty with this method: how can we construct the transformation, $f(z)$? Fortunately, dictionaries of common conformal transformations are available (see, for example, Milne-Thompson, 1952). Let's try to make things clearer with some examples.

Example 1: Mapping a three-quarter-plane onto a half-plane

Consider the three-quarter-plane that lies in the z-plane shown in Figure 7.4(a). How can we map this onto the half-plane in the ζ-plane shown in Figure 7.4(b)? Let's try the transformation $\zeta = Az^n$, so that $\arg \zeta = \arg A + n \arg z$. On BC, $\arg z = -\pi/2$, and we require B′C′, the image of BC, to have $\arg \zeta = 0$, which shows that $\arg A = n\pi/2$. On AB, $\arg z = \pi$, and we require A′B′, the image of AB, to have $\arg \zeta = \pi$. This means that $\pi = \arg A + n\pi$ and hence $n = 2/3$, $\arg A = \pi/3$. This means that the family of transformations $\zeta = |A|e^{i\pi/3}z^{2/3}$ will map the three-quarter-plane to the half-plane. If we further require that the image of $z = -1$ should be $\zeta = -1$, then $|A| = 1$, and $\zeta = e^{i\pi/3}z^{2/3}$. Note that this transformation is not conformal at $z = 0$, since the angle of the boundary changes there.

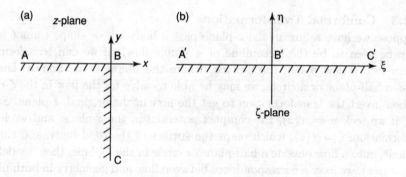

Fig. 7.4. Mapping a three-quarter-plane to a half-plane.

Example 2: Mapping a strip onto a half-plane

Consider the strip of width 2 that lies in the z-plane shown in Figure 7.5(a). How can we map this onto the half-plane in the ζ-plane shown in Figure 7.5(b)? Let's try the transformation $z = K \log \zeta + L$. On A′B′, $\arg \zeta = 0$, so if $\zeta = \xi + i\eta$, $z = K \log |\xi| + L$. If K is real, the choice $L = -i$ makes z range from $-i - \infty$ to $-i+\infty$, as required. On C′D′, $\arg \zeta = \pi$, so $z = K \log |\xi| + i\pi K - i$. If we now choose $i\pi K = 2i$, then z ranges from $i - \infty$ to $i + \infty$, as required. The transformation is therefore

$$z = \frac{2}{\pi} \log \zeta - i, \quad \zeta = ie^{\pi z/2},$$

which is conformal in the finite z-plane.

7.2 COMPLEX VARIABLE METHODS FOR SOLVING LAPLACE'S EQUATION

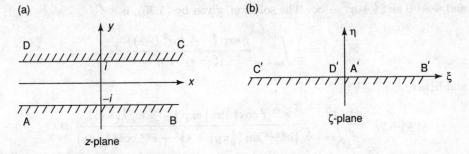

Fig. 7.5. Mapping a strip to a half-plane.

We now pause to note that, in both of these examples, we can use the conformal mapping to solve the Dirichlet problem for Laplace's equation, in the three-quarter-plane and strip respectively, by mapping to the half-plane. We can solve Laplace's equation in the half-plane using the formula (5.30), which we derived using Fourier transforms. We can then easily write down the result in the z-plane using the inverse transformation. To justify this, note that if $\phi(x,y) = \Phi(x(\xi,\eta), y(\xi,\eta))$, we can differentiate to show that

$$\frac{\partial^2 \phi}{\partial x^2} + \frac{\partial^2 \phi}{\partial y^2} = \left|\frac{d\zeta}{dz}\right|^2 \left(\frac{\partial^2 \Phi}{\partial \xi^2} + \frac{\partial^2 \Phi}{\partial \eta^2}\right), \qquad (7.38)$$

so that, provided $\zeta = f(z)$ is conformal, a function that is harmonic in the (ξ, η)-plane is harmonic in the (x,y)-plane, and vice versa (see Exercise 7.11).

We can illustrate this by solving a Dirichlet problem for Laplace's equation in a strip, namely

$$\nabla^2 \phi = 0 \quad \text{for } -\infty < x < \infty \text{ and } -1 < y < 1, \qquad (7.39)$$

subject to

$$\phi(x,-1) = 0, \quad \phi(x,1) = e^{-x^2} \quad \text{for } -\infty < x < \infty. \qquad (7.40)$$

We know that the transformation that maps this strip in the $z = x + iy$-plane to the upper-half-plane in the $\zeta = \xi + i\eta$-plane is $\zeta = ie^{\pi z/2}$, and hence

$$\xi = -e^{\pi x/2} \sin\left(\frac{1}{2}\pi y\right), \quad \eta = e^{\pi x/2} \cos\left(\frac{1}{2}\pi y\right).$$

In the ζ-plane, (7.39) and (7.40) become

$$\frac{\partial^2 \Phi}{\partial \xi^2} + \frac{\partial^2 \Phi}{\partial \eta^2} = 0 \quad \text{for } -\infty < \xi < \infty \text{ and } 0 < \eta < \infty,$$

subject to

$$\phi(x,0) = \begin{cases} 0 & \text{for } 0 < \xi < \infty, \\ \exp\left\{-\frac{4}{\pi^2} \log^2(-\xi)\right\} & \text{for } -\infty < \xi < 0, \end{cases}$$

and $\Phi \to 0$ as $\xi^2 + \eta^2 \to \infty$. The solution, given by (5.30), is

$$\Phi(\xi, \eta) = \frac{1}{\pi} \int_{s=-\infty}^{0} \frac{\eta \exp\left\{-\frac{4}{\pi^2} \log^2(-s)\right\}}{(\xi - s)^2 + \eta^2} \, ds,$$

and hence

$$\phi(x, y) = \frac{1}{\pi} \int_{s=-\infty}^{0} \frac{e^{\pi x/2} \cos\left(\frac{1}{2}\pi y\right) \exp\left\{-\frac{4}{\pi^2} \log^2(-s)\right\}}{\left\{e^{\pi x/2} \sin\left(\frac{1}{2}\pi y\right) + s\right\}^2 + e^{\pi x} \cos^2\left(\frac{1}{2}\pi y\right)} \, ds.$$

Example 3: Flow past a flat plate

Consider a plate of length $2a$ positioned perpendicular to a uniform flow with speed U, as shown in Figure 7.6(a) in the z-plane. We want to map the exterior of the plate to the exterior of a circle in the ζ-plane, as shown in Figure 7.6(b). This can be done using

$$z = \frac{1}{2}i\left(\zeta + \frac{a^2}{\zeta}\right). \tag{7.41}$$

To show this, we write the surface of the circle as $\zeta = ae^{i\theta}$, so that $z = ia\cos\theta$. The points labelled in Figure 7.6(a) in the z-plane then map to the corresponding points labelled in Figure 7.6(b). Following the sequence A'B'C'D'E', the fluid is on the right hand side in the ζ-plane, and consequently is on the right hand side in the z-plane, so that the outside of the circle maps to the outside of the plate. As $|\zeta| \to \infty$, $z \sim \frac{1}{2}i\zeta$, so that

$$\frac{dw}{dz} \sim -2i\frac{dw}{d\zeta}.$$

Since we require that $dw/dz \sim U$, we must have $dw/d\zeta \sim Ui/2$, so that the stream at infinity in the ζ-plane is half the strength of the one in the z-plane, and has been rotated through $\pi/2$ radians.

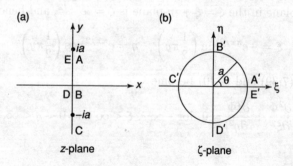

Fig. 7.6. Mapping a flat plate to a unit circle.

7.2 COMPLEX VARIABLE METHODS FOR SOLVING LAPLACE'S EQUATION

We can now solve the flow problem in the ζ-plane using the circle theorem, to give

$$w(\zeta) = \frac{1}{2}Ue^{i\pi/2}\zeta + \frac{1}{2}Ue^{-i\pi/2}\frac{a^2}{\zeta}. \tag{7.42}$$

Next, we need to write this potential in terms of z. From (7.41),

$$\zeta^2 - \frac{2z}{i}\zeta + a^2 = 0,$$

and hence

$$\zeta = -i\left(\sqrt{z^2 + a^2} + z\right),$$

choosing this root so that $|\zeta| \to \infty$ as $|z| \to \infty$. From the potential (7.42), we find that

$$w = \frac{1}{2}U\left(\sqrt{z^2 + a^2} + z + \frac{a^2}{\sqrt{z^2 + a^2} + z}\right). \tag{7.43}$$

Although this result is correct, it does give rise to infinite velocities at $z = \pm ia$, the ends of the plate, which are not physical, and arise because of the sharpness of the boundary at these points. In reality, viscosity will become important close to these points, and lead to finite velocities.

Example 4: Flow past an aerofoil

As we have seen, the transformation

$$z = \frac{1}{2}\left(\zeta + \frac{1}{\zeta}\right) \tag{7.44}$$

maps the exterior of a unit circle in the ζ-plane to the exterior of a flat plate in the z-plane. In contrast, the exterior of circles of radius $a > 1$ whose centres are *not* at the origin, but which still pass through the point $\zeta = 1$ and enclose the point $\zeta = -1$, are mapped to shapes, known as **Joukowski profiles**, that look rather like aerofoils. These shapes have two distinguishing characteristics: a blunt nose, or leading edge, and a sharp tail, or trailing edge, at $z = \zeta = 1$, as shown in Figure 7.7.

Let's now consider the flow of a stream of strength U, at an angle α to the horizontal, around a Joukowski profile. Since $dz/d\zeta = 0$ at $\zeta = 1$, the transformation is not conformal at the trailing edge, and the velocity will be infinite there, unless we can make the flow in the transform plane have a stagnation point there. This can be achieved by including an appropriate circulation around the aerofoil. We must choose the strength of the circulation to make the velocity finite at the trailing edge. This is called the **Kutta condition** on the flow.

If the centre of the circle in the ζ-plane is at $\zeta = \zeta_c$, we must begin by making another conformal transformation, $\bar{\zeta} = (\zeta - \zeta_c)/a$, so that we can use the circle theorem. The flow in the $\bar{\zeta}$-plane is a uniform stream and a point vortex, so that

$$\bar{w}(\bar{\zeta}) = Ve^{-i\beta}\bar{\zeta} + Ve^{i\beta}\frac{1}{\bar{\zeta}} + \frac{i\kappa}{2\pi}\log\bar{\zeta}.$$

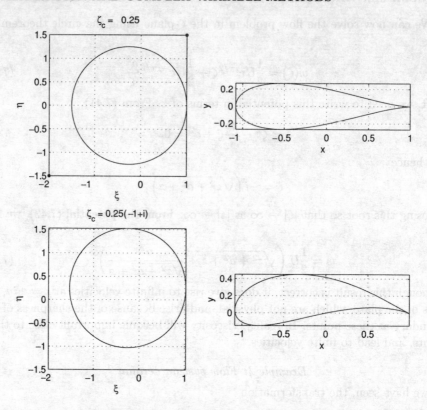

Fig. 7.7. Two examples of Joukowski profiles in the z-plane, and the corresponding circles in the ζ-plane, where $\zeta = \xi + i\eta$ and $z = x + iy$.

To determine V, we note that

$$z = \frac{1}{2}\left(a\bar{\zeta} + \zeta_c + \frac{1}{a\bar{\zeta} + \zeta_c}\right),$$

and hence that $z \sim a\bar{\zeta}/2$ as $|\bar{\zeta}| \to \infty$, so that we need $Ue^{-i\alpha} = 2Ve^{-i\beta}/a$, and hence $V = aU/2$ and $\beta = \alpha$. The complex potential is therefore

$$\bar{w}(\bar{\zeta}) = \frac{1}{2}aU\left(e^{-i\alpha}\bar{\zeta} + e^{i\alpha}\frac{1}{\bar{\zeta}}\right) + \frac{i\kappa}{2\pi}\log\bar{\zeta}, \tag{7.45}$$

and hence

$$\frac{d\bar{w}}{d\bar{\zeta}} = \frac{1}{2}aU\left(e^{-i\alpha} - e^{i\alpha}\frac{1}{\bar{\zeta}^2}\right) + \frac{i\kappa}{2\pi\bar{\zeta}}. \tag{7.46}$$

We can therefore make the trailing edge, $\bar{\zeta} = (1 - \zeta_c)/a \equiv \bar{\zeta}_s$, a stagnation point

by choosing
$$\kappa = i\pi a U \left(e^{-i\alpha} \bar{\zeta}_s - e^{i\alpha} \frac{1}{\bar{\zeta}_s} \right). \tag{7.47}$$

This expression is real, because $|\bar{\zeta}_s| = 1$ and hence the quantities $e^{-i\alpha}\bar{\zeta}_s$ and $e^{i\alpha}/\bar{\zeta}_s$ are complex conjugate.

In order to calculate the force on the Joukowski profile, we need to evaluate

$$F_x - iF_y = \frac{1}{2}\rho i \int_{\text{body}} \left(\frac{dw}{dz}\right)^2 dz = \frac{1}{2}\rho i \int_{|\bar{\zeta}|=1} \left(\frac{d\bar{w}}{d\bar{\zeta}}\right)^2 \frac{d\bar{\zeta}}{dz} d\bar{\zeta}$$

$$= \frac{1}{2}\rho i a \int_{|\bar{\zeta}|=1} \frac{\left\{\frac{1}{2}U\left(e^{-i\alpha} - e^{i\alpha}/\bar{\zeta}^2\right) + i\kappa/2\pi a\bar{\zeta}\right\}^2}{\frac{1}{2}\left\{1 - 1/(a\bar{\zeta} + \zeta_c)^2\right\}} d\bar{\zeta}.$$

This integral can be evaluated using residue calculus. Since we know that the integrand is analytic for $|\bar{\zeta}| > 1$, we can evaluate the residue at the point at infinity by making the transformation $\bar{\zeta} = 1/s$ and evaluating the residue at $s = 0$. We find, after some algebra, that

$$F_x - iF_y = -i\rho\kappa U e^{-i\alpha},$$

or equivalently,

$$F_x + iF_y = \rho\kappa U e^{i(\pi/2 + \alpha)}.$$

From this, we can see that the lift force on the aerofoil is directed perpendicular to the incoming stream. The streamlines for the flow around the aerofoils illustrated in Figure 7.7 are shown in Figure 7.8 with $\alpha = \pi/4$.

Exercises

7.1 Find and sketch the regions in the (x, y)-plane where the equation
$$(1 + x)\phi_{xx} + 2xy\phi_{xy} + y^2\phi_{yy} = 0$$
is elliptic, parabolic and hyperbolic.

7.2 Determine the type of each of the equations
   (a) $\phi_{xx} - 5\phi_{xy} + 5\phi_{yy} = 0$,
   (b) $\phi_{xx} - 2\phi_{xy} + 3\phi_{yy} + 24\phi_y + 5\phi = 0$,
   (c) $\phi_{xx} + 6y\phi_{xy} + 9y^2\phi_{yy} + 4\phi = 0$,

and reduce them to canonical form.

7.3 Determine the characteristics of Tricomi's equation, $\phi_{xx} = x\phi_{yy}$, in the hyperbolic region, $x > 0$. Transform the equation to canonical form in (a) the hyperbolic region and (b) the elliptic region.

7.4 Show that any solution, $\phi(x,t)$, of the one-dimensional wave equation, (3.39), satisfies the difference equation $y_1 - y_2 + y_3 - y_4 = 0$, where y_1, y_2, y_3 and y_4 are the values of the solution at successive corners of any quadrilateral whose edges are characteristics.

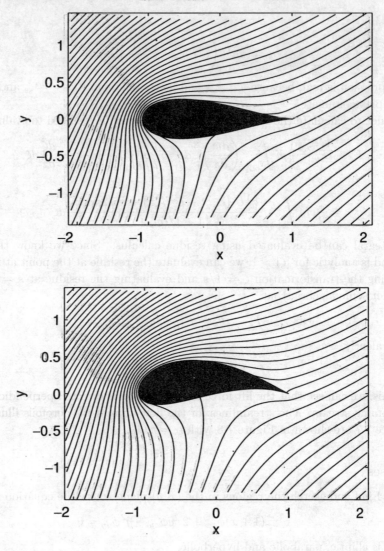

Fig. 7.8. The streamlines for flow past the two examples of Joukowski profiles shown in Figure 7.7, with $\alpha = \pi/4$.

7.5 Consider the initial–boundary value problem

$$c^{-2}\phi_{tt} + 2k\phi_t - \phi_{xx} = 0 \text{ for } t > 0,\, x > 0,$$

subject to

$$\phi(x,0) = \phi_t(x,0) = 0 \text{ for } x \geqslant 0,$$

$$\phi(0,t) = \frac{1}{2}t^2 \text{ for } t \geq 0.$$

By differentiating the canonical form of the equation, obtain an equation for the jump in a second derivative of ϕ across the characteristic curve $x = ct$. By solving this equation, show that the jump in ϕ_{tt} across $x = ct$ is e^{-kcx}.

7.6 Consider the boundary value problem given by (7.25) and (7.26). By using suitable functions $\psi = Ax^2 + By^2$ that satisfy $\nabla^2 \psi = 1$ with $A > 0$ and $B > 0$, show that

$$\frac{a^2 b^2}{2(a+b)^2} \leq \phi(0,0) \leq \frac{a^2 b^2}{2(a^2+b^2)}.$$

7.7 The function $\phi(r,\theta)$ satisfies Laplace's equation in region $a < r < b$, where r and θ are polar coordinates and a and b are positive constants. Using the maximum principle and the fact that $\nabla^2(\log r) = 0$, show that

$$m(R) \log\left(\frac{b}{a}\right) \leq m(b) \log\left(\frac{R}{a}\right) - m(a) \log\left(\frac{R}{b}\right),$$

where $m(R)$ is the maximum value of ϕ on the circle $r = R$, with $a < R < b$.

7.8 (a) Prove Theorem 7.4. (b) Hence show that the solution of the initial–boundary value problem given by (7.30) to (7.32) is unique.

7.9 Consider the initial–boundary value problem

$$\phi_t = K\nabla^2 \phi - \phi^3 \text{ for } \mathbf{x} \in D, \, t > 0,$$

with $k > 0$, subject to

$$\phi(\mathbf{x},0) = \phi_0(\mathbf{x}) \text{ for } \mathbf{x} \in D,$$

and

$$\frac{\partial \phi}{\partial n} = 0 \text{ for } \mathbf{x} \in \partial D.$$

Use Theorem 7.5 to show that $\phi \to 0$ as $t \to \infty$, uniformly in D.

7.10 Show that the angle between two line segments that intersect at $z = z_0$ in the complex z-plane is unchanged by a conformal mapping $\zeta = f(z)$, provided that $f'(z_0)$ is bounded and nonzero.

7.11 Show that, if $\phi(x,y) = \Phi(x(\xi,\eta), y(\xi,\eta))$, $\zeta = \xi + i\eta$ and $z = x + iy$, then

$$\frac{\partial^2 \phi}{\partial x^2} + \frac{\partial^2 \phi}{\partial y^2} = \left|\frac{d\zeta}{dz}\right|^2 \left(\frac{\partial^2 \Phi}{\partial \xi^2} + \frac{\partial^2 \Phi}{\partial \eta^2}\right).$$

7.12 Show that the mapping $\zeta = \sin(\pi z / 2k)$ maps a semi-infinite strip in the z-plane to the upper half ζ-plane.

7.13 Show that the image of a circle of radius r in the z-plane under the transformation $\zeta = (z + 1/z)/2$ is an ellipse.

CLASSIFICATION AND COMPLEX VARIABLE METHODS

7.14 Find the function $T(x,y)$ that is harmonic in the lens-shaped region defined by the intersection of the circles $|z - i| = 1$ and $|z - 1| = 1$ and takes the value 0 on the upper circular arc and 1 on the lower circular arc.

Hint: Consider the effect of the transformation $\zeta = z/(z - 1 - i)$ on this region.

7.15 Find the lift on an aerofoil section that consists of a unit circle centred on the origin, joined at $z = 1$ to a horizontal flat plate of unit length, when a unit stream is incident at an angle of $45°$. *Hint:* Firstly apply a Joukowski transformation to map the section to a straight line, and then use an inverse Joukowski transformation to map this straight line to a circle.

Part Two
Nonlinear Equations and Advanced Techniques

Part Two

CHAPTER EIGHT

Existence, Uniqueness, Continuity and Comparison of Solutions of Ordinary Differential Equations

Up to this point in the book we have implicitly assumed that each of the differential equations that we have been studying has a solution of the appropriate form. This is not always the case. For example, the equation

$$\frac{dy}{dt} = \sqrt{y^2 - 1}$$

has no real solution for $|y| < 1$. When we ask whether an equation actually has a solution, we are considering the question of **existence** of solutions. A related issue is that of **uniqueness** of solutions. If we have a differential equation for which solutions exist, does the prescription of initial conditions specify a unique solution? Not necessarily. Consider the equation

$$\frac{dy}{dt} = 3y^{2/3} \text{ subject to } y(0) = 0 \text{ for } y > 0.$$

If we integrate this separable equation we obtain $y = (t-c)^3$. The initial condition gives $c = 0$ and the solution $y = t^3$. However, it is fairly obvious that $y = 0$ is another solution that satisfies the initial condition. In fact, this equation has an infinite number of solutions that satisfy the initial condition, namely

$$y = \begin{cases} 0 & \text{for } 0 \leqslant t \leqslant c, \\ (t-c)^3 & \text{for } t > c, \end{cases}$$

for any $c \geqslant 0$.

Another question that we should consider arises from the process of mathematical modelling. If a differential equation provides a model of a real physical system, how sensitive are the solutions of the differential equation to changes in the initial conditions? If we performed two experiments with nearly identical initial conditions (exact repetition of an experiment being impossible in practice), should we expect nearly identical outcomes?

In this chapter we will prove some rigorous results related to the issues we have discussed above.

8.1 Local Existence of Solutions

We will start with the simplest type of scalar differential equation by proving a **local existence theorem** for

$$\frac{dy}{dt} = f(y,t) \text{ subject to } y(t_0) = y_0 \text{ in the domain } |t - t_0| < \alpha. \qquad (8.1)$$

Here, $\alpha > 0$ defines the size of the region where we will be able to show that a solution exists. We begin by defining a closed rectangle,

$$R = \{(y,t) \mid |y - y_0| \leqslant b, \ |t - t_0| \leqslant a\},$$

centred upon the initial point, (y_0, t_0), within which we will make certain assumptions about the behaviour of f. If we integrate (8.1) with respect to t and apply the initial condition, we obtain

$$y(t) = y_0 + \int_{t_0}^{t} f(y(s), s) \, ds. \qquad (8.2)$$

This is an integral equation for the unknown solutions, $y = y(t)$, which is equivalent to the original differential equation. Our strategy now is to use this to produce a set of **successive approximations** to the solution that we seek. We will do this using the initial condition as the starting point.

We define

$$y_0(t) = y_0,$$
$$y_1(t) = y_0 + \int_{t_0}^{t} f(y_0, s) \, ds,$$
$$y_2(t) = y_0 + \int_{t_0}^{t} f(y_1(s), s) \, ds, \qquad (8.3)$$
$$\vdots$$
$$y_{k+1}(t) = y_0 + \int_{t_0}^{t} f(y_k(s), s) \, ds.$$

As an example of how this works, consider the simple differential equation $y' = y$ subject to $y(0) = 2$. In this case,

$$y_0(t) = 2,$$
$$y_1(t) = 2 + \int_0^t 2 \, ds = 2(1 + t),$$
$$y_2(t) = 2 + \int_0^t 2(1 + s) \, ds = 2\left(1 + t + \frac{1}{2}t^2\right),$$
$$\vdots$$
$$y_{k+1}(t) = 2 \sum_{n=0}^{k+1} \frac{1}{n!} t^n.$$

As $k \to \infty$, $\{y_k(t)\} \to 2e^t$, the correct solution. In general, if the sequence of functions (8.3) converges uniformly to a limit, $\{y_k(t)\} \to y_\infty(t)$, then $y_\infty(t)$ is a continuous solution of the integral equation. For continuous $f(y,t)$, we can then

8.1 LOCAL EXISTENCE OF SOLUTIONS

differentiate under the integral sign, to show that y is a solution of the original differential equation, (8.1). We will shortly prove that, in order to ensure that the sequence (8.3) converges to a continuous function, it is sufficient that $f(y,t)$ and $\frac{\partial f}{\partial y}(y,t)$ are continuous in R. In other words, we need $f, \partial f/\partial y \in C^0(R)$.†

As a preliminary to the main proof, recall Theorem A2.1, which says that a function continuous on a bounded region is itself bounded, so that there exist strictly positive, real constants M and K such that $|f(y,t)| < M$ and $\left|\frac{\partial f}{\partial y}(y,t)\right| < K$ for all $(y,t) \in R$. If (y_1,t) and (y_2,t) are two points in R, Theorem A2.2, the mean value theorem, states that

$$f(y_2,t) - f(y_1,t) = \frac{\partial f}{\partial y}(c,t)(y_2 - y_1) \quad \text{for some } c \text{ with } y_1 < c < y_2.$$

Since $(c,t) \in R$, we have $\left|\frac{\partial f}{\partial y}(c,t)\right| < K$, and hence

$$|f(y_2,t) - f(y_1,t)| < K|y_2 - y_1| \quad \forall (y_2,t),(y_1,t) \in R. \tag{8.4}$$

Functions that satisfy the inequality (8.4) are said to satisfy a **Lipschitz condition** in R. It is possible for a function to satisfy a Lipschitz condition in a region R without having a continuous partial derivative everywhere in R. For example, $f(y,t) = t|y|$ satisfies $|f(y_2,t) - f(y_1,t)| < |y_2 - y_1|$ in the unit square centred on the origin, so that it is Lipschitz with $K = 1$. However, $\partial f/\partial y$ is not continuous on the line $y = 0$. Our assumption about the continuity of the first partial derivative automatically leads to functions that satisfy a Lipschitz condition, which is the key to proving the main result of this section.

As we stated earlier, we are going to use the successive approximations (8.3) to establish the existence of a solution. Prior to using this, we must show that the elements of this sequence are well-defined. Specifically, if $y_{k+1}(t)$ is to be defined on some interval I, we must establish that the point $(y_k(s), s)$ remains in the rectangle $R = \{(y,t) \mid |y - y_0| \leqslant b, |t - t_0| \leqslant a\}$ for all $s \in I$.

Lemma 8.1 *If $\alpha = \min(a, b/M)$, then the successive approximations,*

$$y_0(t) = y_0, \quad y_{k+1}(t) = y_0 + \int_{t_0}^{t} f(y_k(s), s)\, ds$$

are well-defined in the interval $I = \{t \mid |t - t_0| < \alpha\}$, and on this interval $|y_k(t) - y_0| < M|t - t_0| \leqslant b$, where $|f| < M$.

Proof We proceed by induction. It is clear that $y_0(t)$ is defined on I, as it is constant. We assume that

$$y_n(t) = y_0 + \int_{t_0}^{t} f(y_{n-1}(s), s)\, ds$$

---

† If these conditions hold in a larger domain, D, we can use the local result repeatedly until the solution moves out of D.

is well-defined on I, so that the point $(y_n(t), t)$ remains in R for $t \in I$. By definition,
$$y_{n+1}(t) = y_0 + \int_{t_0}^t f(y_n(s), s)\, ds,$$
so we have $y_{n+1}(t)$ defined in I. Now
$$|y_{n+1}(t) - y_0| = \left|\int_{t_0}^t f(y_n(s), s)\, ds\right| \leqslant \int_{t_0}^t |f(y_n(s), s)|\, ds \leqslant M|t - t_0| \leqslant M\alpha \leqslant b,$$
which is what we claimed. □

To see rather less formally why we chose $\alpha = \min(a, b/M)$, note that the condition $|f(y, t)| < M$ implies that the solution of the differential equation, $y = y(t)$, cannot cross lines of slope M or $-M$ through the initial point (y_0, t_0), as shown in Figure 8.1. The relationship $|y_k(t) - y_0| < M|t - t_0|$, which we established in Lemma 8.1, means that the successive approximations cannot cross these lines either. These lines intersect the boundary of the rectangle R, and the length of the interval I depends upon whether they meet the horizontal sides ($\alpha = b/M$) or the vertical sides ($\alpha = a$), as shown in Figure 8.1.

We can now proceed to establish the main result of this section.

Theorem 8.1 (Local existence) *If f and $\partial f/\partial y$ are in $C^0(R)$, then the successive approximations $y_k(t)$, defined by (8.3), converge on I to a solution of the differential equation $y' = f(y, t)$ that satisfies the initial condition $y(t_0) = y_0$.*

Proof We begin with the identity
$$y_j(t) = y_0(t) + \{y_1(t) - y_0(t)\} + \{y_2(t) - y_1(t)\} + \cdots + \{y_j(t) - y_{j-1}(t)\}$$
$$= y_0(t) + \sum_{n=0}^{j-1} \{y_{n+1}(t) - y_n(t)\}. \quad (8.5)$$

In order to use this, we need to estimate the value of $y_{n+1}(t) - y_n(t)$. Using the definition (8.3), we have, for $n \geqslant 1$ and $|t - t_0| < \alpha$,
$$|y_{n+1}(t) - y_n(t)| = \left|\int_{t_0}^t \{f(y_n(s), s) - f(y_{n-1}(s), s)\}\, ds\right|$$
$$\leqslant \int_{t_0}^t |f(y_n(s), s) - f(y_{n-1}(s), s)|\, ds \leqslant K \int_{t_0}^t |y_n(s) - y_{n-1}(s)|\, ds.$$
For $n = 0$ we have
$$|y_1(t) - y_0(t)| = \left|\int_{t_0}^t f(y_0(s), s)\, ds\right| \leqslant M|t - t_0|,$$
by using the continuity bound on f. If we now repeatedly use these inequalities, it is straightforward to show that
$$|y_{n+1}(t) - y_n(t)| \leqslant \frac{MK^n|t - t_0|^{n+1}}{(n+1)!} \quad \text{for } n \geqslant 1,\ |t - t_0| \leqslant \alpha,$$

8.1 LOCAL EXISTENCE OF SOLUTIONS

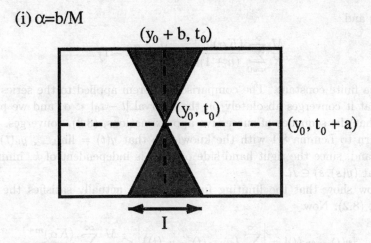

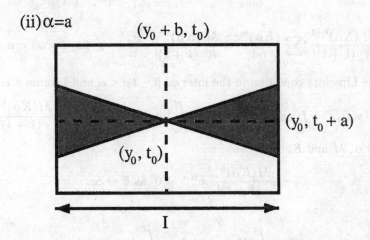

Fig. 8.1. The two cases that determine the length of the interval, α.

and hence that

$$|y_{n+1}(t) - y_n(t)| \leq \frac{M(K\alpha)^{n+1}}{K(n+1)!}. \tag{8.6}$$

If we denote by $y(t)$ the limit of the sequence $\{y_k(t)\}$, we now have, from (8.5),

$$y(t) = y_0(t) + \sum_{n=0}^{\infty} \{y_{n+1}(t) - y_n(t)\}. \tag{8.7}$$

Each term in this infinite series is dominated by $M(K\alpha)^{n+1}/K(n+1)!$, a positive constant; and

$$\frac{M}{K} \sum_{n=0}^{\infty} \frac{(K\alpha)^{n+1}}{(n+1)!} = \frac{M}{K} \left(e^{K\alpha} - 1\right), \tag{8.8}$$

which is a finite constant. The comparison theorem applied to the series in (8.7) shows that it converges absolutely on the interval $|t - t_0| < \alpha$, and we have now proved that the sequence of successive approximations, (8.3), converges. We can now return to Lemma 8.1 with the knowledge that $y(t) = \lim_{k \to \infty} y_k(t)$ is well-defined, and, since the right hand side of (8.8) is independent of k, immediately claim that $(y(s), s) \in R$.

We now show that the limiting function, $y(t)$, actually satisfies the integral equation, (8.2). Now

$$|y(t) - y_k(t)| = \left|\sum_{n=k}^{\infty} \{y_{n+1}(t) - y_n(t)\}\right| \leq \frac{M}{K} \sum_{m=k}^{\infty} \frac{(K\alpha)^{m+1}}{(m+1)!}$$

$$\leq \frac{M}{K} \frac{(K\alpha)^{k+1}}{(k+1)!} \sum_{m=0}^{\infty} \frac{(K\alpha)^m}{m!} \leq \frac{M}{K} \frac{(K\alpha)^{k+1}}{(k+1)!} e^{K\alpha} = \epsilon_k \text{ for } |t - t_0| < \alpha. \tag{8.9}$$

Using the Lipschitz condition in the interval $|t - t_0| < \alpha$ and Lemma 8.1,

$$\left|\int_{t_0}^{t} \{f(y(s), s) - f(y_k(s), \dot s)\} \, ds\right| \leq K \int_{t_0}^{t} |y(s) - y_k(s)| \, ds \leq \alpha \frac{M(K\alpha)^{k+1}}{(k+1)!} e^{K\alpha}.$$

For fixed α, M and K,

$$\alpha \frac{M(K\alpha)^{k+1}}{(k+1)!} e^{K\alpha} \to 0 \text{ as } k \to \infty,$$

so that

$$\lim_{k \to \infty} \int_{t_0}^{t} f(y_k(s), s) \, ds = \int_{t_0}^{t} f(y(s), s) \, ds,$$

and consequently $y(t)$ satisfies (8.2). □

To prove that $y(t)$ is continuous on $|t - t_0| < \alpha$, we can use some of the auxiliary results that we derived above. Consider

$$y(t+h) - y(t) = y(t+h) - y_k(t+h) + y_k(t+h) - y_k(t) + y_k(t) - y(t),$$

so that, using the triangle inequality,

$$|y(t+h) - y(t)| \leq |y(t+h) - y_k(t+h)| + |y_k(t+h) - y_k(t)| + |y_k(t) - y(t)|.$$

By choosing k sufficiently large, using the estimate (8.9), we have

$$|y_k(t) - y(t)| \leq \epsilon_k,$$

so that

$$|y(t+h) - y(t)| \leq 2\epsilon_k + |y_k(t+h) - y_k(t)| \leq \epsilon,$$

for sufficiently small h, using the continuity of $y_k(t)$. Hence, $y(t)$ is continuous, as we claimed at the beginning of this section, and the equivalence of the integral and differential equations can be established by differentiation under the integral sign.

Remarks

(i) For a particular differential equation, Theorem 8.1 is easy to use. For example, if $y' = y + t$, both $f(y,t) = y + t$ and $\partial f/\partial y = 1$ are continuous for $-\infty < t < \infty$, $-\infty < y < \infty$, so the successive approximations are guaranteed to converge in this domain. If $y' = 3y^{2/3}$, $f(y,t) = 3y^{2/3}$, which is continuous in $-\infty < y < \infty$, $-\infty < t < \infty$. However, its partial derivative, $\partial f/\partial y = 2y^{-1/3}$, is discontinuous at $y = 0$. Consequently, the successive approximations are not guaranteed to converge. This should be no surprise, since, as we noted in the introduction to this chapter, the solution of this equation is not unique for some initial data.

(ii) Consider the equation $y' = y^2$ subject to $y(0) = 1$. Since this equation satisfies the conditions of Theorem 8.1, the successive approximations will converge, and therefore a solution exists in some rectangle containing the initial point, namely

$$R = \{(y,t) \mid |t| \leqslant a, |y - 1| \leqslant b\}.$$

Let's now try to determine the values of the constants a and b. Since $M = \max_R y^2 = (1 + b)^2$, $\alpha = \min(a, b/(1 + b)^2)$. From simple calculus,

$$\max_{b > 0} \frac{b}{(1 + b)^2} = \frac{1}{4},$$

so that, independent of the particular choice of a, $\alpha \leqslant \frac{1}{4}$. Theorem 8.1 therefore shows that a solution exists for $|t| \leqslant \frac{1}{4}$. In fact, the solution, $y = 1/(1-t)$, exists for $-\infty < t < 1$. It is rather more difficult to determine regions of **global existence** like this, than it is to establish **local existence**, the point of Theorem 8.1, and we will not attempt it here.

(iii) We can also consider the local existence of solutions of systems of first order equations in the form

$$\mathbf{y}' = \mathbf{f}(\mathbf{y}, t), \tag{8.10}$$

where

$$\mathbf{y} = \begin{pmatrix} y_1 \\ y_2 \\ \vdots \\ y_n \end{pmatrix}, \quad \mathbf{f} = \begin{pmatrix} f_1(y_1, y_2, \ldots, y_n, t) \\ f_2(y_1, y_2, \ldots, y_n, t) \\ \vdots \\ f_n(y_1, y_2, \ldots, y_n, t) \end{pmatrix}$$

are vectors with n components and t is a real scalar. The vector form of the Lipschitz condition can be established by requiring $\mathbf{f}$ and $\partial \mathbf{f}/\partial y_i$ to be continuous with respect to all of their components and t, so that

$$\left\|\frac{\partial \mathbf{f}}{\partial y_i}\right\| \leqslant K$$

for all $(\mathbf{y}, t) \in D \subset \mathbb{R}^{n+1}$, where $\|.\|$ is a suitable vector norm. It then follows that for any points $(\mathbf{y}, t)$ and $(\mathbf{z}, t)$ in D, $\|\mathbf{f}(\mathbf{y}, t) - \mathbf{f}(\mathbf{z}, t)\| \leqslant K\|\mathbf{y} - \mathbf{z}\|$. The proof then follows that of Theorem 8.1 step by step and line by line, with modulus signs replaced by vector norms.

(iv) For a selection of results similar to, but in some ways better than, Theorem 8.1, see Coddington and Levinson (1955).

8.2 Uniqueness of Solutions

We will now show that, *under the same conditions as Theorem 8.1*, the solution of the initial value problem (8.1) is **unique**. In order to prove this, we need a result called **Gronwall's inequality**.

Lemma 8.2 (Gronwall's inequality) *If $f(t)$ and $g(t)$ are non-negative functions on the interval $\alpha \leqslant t \leqslant \beta$, L is a non-negative constant and*

$$f(t) \leqslant L + \int_\alpha^t f(s)g(s)\,ds \quad \text{for } t \in [\alpha, \beta],$$

then

$$f(t) \leqslant L \exp\left\{\int_\alpha^t g(s)\,ds\right\} \quad \text{for } t \in [\alpha, \beta].$$

Proof Define

$$h(t) = L + \int_\alpha^t f(s)g(s)\,ds,$$

so that $h(\alpha) = L$. By hypothesis, $f(t) \leqslant h(t)$, and by the fundamental theorem of calculus, since $g(t) \geqslant 0$, we have

$$h'(t) = f(t)g(t) \leqslant h(t)g(t) \quad \text{for } t \in [\alpha, \beta].$$

Using the idea of an integrating factor, this can be rewritten as

$$\frac{d}{dt}\left[h(t)\exp\left\{-\int_\alpha^t g(s)\,ds\right\}\right] \leqslant 0.$$

Integrating from α to t gives

$$h(t)\exp\left\{-\int_\alpha^t g(s)\,ds\right\} - L \leqslant 0,$$

and, since $f(t) \leqslant h(t)$,

$$f(t) \leqslant L \exp\left\{\int_\alpha^t g(s)\,ds\right\}.$$

□

8.3 DEPENDENCE OF THE SOLUTION ON THE INITIAL CONDITIONS

Theorem 8.2 (Uniqueness) *If f, $\frac{\partial f}{\partial y} \in C^0(R)$, then the solution of the initial value problem $y' = f(y,t)$ subject to $y(t_0) = y_0$ is unique on $|t - t_0| < \alpha$.*

Proof We proceed by contradiction. Suppose that there exist *two* solutions, $y = y_1(t)$ and $y = y_2(t)$, with $y_1(t_0) = y_2(t_0) = y_0$. These functions satisfy the integral equations

$$y_1(t) = y_0 + \int_{t_0}^{t} f(y_1(s), s)\, ds, \quad y_2(t) = y_0 + \int_{t_0}^{t} f(y_2(s), s)\, ds, \qquad (8.11)$$

and hence the points $(y_i(s), s)$ lie in the region R. Taking the modulus of the difference of these leads to

$$|y_1(t) - y_2(t)| = \left| \int_{t_0}^{t} \{f(y_1(s), s) - f(y_2(s), s)\}\, ds \right|$$

$$\leqslant \int_{t_0}^{t} |\{f(y_1(s), s) - f(y_2(s), s)\}|\, ds \leqslant \int_{t_0}^{t} K |y_1(s) - y_2(s)|\, ds \quad \text{for } |t - t_0| \leqslant \alpha,$$

where $K > 0$ is the Lipschitz constant for the region R. We can now apply Gronwall's inequality to the non-negative function $|y_1(t) - y_2(t)|$ with $L = 0$ and $g(s) = K > 0$ to conclude that $|y_1(t) - y_2(t)| \leqslant 0$. However, since the modulus of a function cannot be negative, $|y_1(t) - y_2(t)| = 0$, and hence $y_1(t) \equiv y_2(t)$. □

Just as we saw earlier when discussing the existence of solutions, this uniqueness result can be extended to systems of ordinary differential equations. Under the conditions given in Remark (iii) in the previous section, the solution of (8.10) exists and is unique.

8.3 Dependence of the Solution on the Initial Conditions

A solution of $y' = f(y,t)$ that passes through the initial point (y_0, t_0) depends continuously on the three variables t_0, y_0 and t. For example, $y' = 3y^{2/3}$ subject to $y(t_0) = y_0$ has solution $y = (t - t_0 + y_0^{1/3})^3$, which depends continuously on t_0, y_0 and t. Earlier, we hypothesized that solutions of identical differential equations with initial conditions that are close to each other should remain close, at least for values of t close to the initial point, $t = t_0$. We can now prove a theorem about this.

Theorem 8.3 *Let $y = y_0(t)$ be the solution of $y' = f(y,t)$ that passes through the initial point (t_0, y_0) and $y = y_1(t)$ the solution that passes through (t_1, y_1). If f, $\partial f/\partial y \in C^0(R)$ and both $y_0(t)$ and $y_1(t)$ exist on some interval $\alpha < t < \beta$ with t_0, $t_1 \in (\alpha, \beta)$, then $\forall \epsilon > 0\ \exists \delta > 0$ such that if $|t_0 - t_1| < \delta$ and $|y_0 - y_1| < \delta$ then $|y_0(t) - y_1(t)| < \epsilon\ \forall t \in (\alpha, \beta)$.*

EXISTENCE, UNIQUENESS, CONTINUITY AND COMPARISON

Proof Since $y_1(t)$ and $y_2(t)$ satisfy the integral equations (8.11), we can take their difference and, by splitting up the range of integration, write†

$$y_0(t) - y_1(t) = y_0 - y_1 + \int_{t_0}^{t_1} f(y_0(s), s)\, ds + \int_{t_1}^{t} \{f(y_0(s), s) - f(y_1(s), s)\}\, ds.$$

Taking the modulus gives

$$|y_0(t) - y_1(t)| \leq |y_0 - y_1| + \left|\int_{t_0}^{t_1} f(y_0(s), s)\, ds\right| + \left|\int_{t_1}^{t} \{f(y_0(s), s) - f(y_1(s), s)\}\, ds\right|$$

$$\leq |y_0 - y_1| + M(t_1 - t_0) + \int_{t_1}^{t} K|y_0(s) - y_1(s)|\, ds,$$

by using the upper bound on f and the Lipschitz condition on R. If $|t_1 - t_0| < \delta$ and $|y_1 - y_0| < \delta$, this reduces to

$$|y_0(t) - y_1(t)| \leq (M+1)\delta + \int_{t_1}^{t} K|y_0(s) - y_1(s)|\, ds.$$

We can now apply Gronwall's inequality to obtain

$$|y_0(t) - y_1(t)| \leq (M+1)\delta \exp\left\{\int_{t_1}^{t} K\, ds\right\} = (M+1)\delta \exp\{K(t - t_1)\},$$

and, since $|t - t_1| < \beta - \alpha$, we have

$$|y_0(t) - y_1(t)| \leq (M+1)\delta \exp\{K(\beta - \alpha)\}.$$

If we now choose $\delta < \epsilon \exp\{-K(\beta - \alpha)\}/(M+1)$, we have $|y_0(t) - y_1(t)| < \epsilon$, and the proof is complete. □

8.4 Comparison Theorems

Since a large class of differential equations cannot be solved explicitly, it is useful to have a method of placing bounds on the solution of a given equation by comparing it with solutions of a related equation that is simpler to solve. For example, $y' = e^{-ty}$ subject to $y(0) = 1$ is a rather nasty nonlinear differential equation, and the properties of its solution are not immediately obvious. However, since $0 \leq e^{-ty} \leq 1$ for $0 \leq t < \infty$ and $0 \leq y < \infty$, it would seem plausible that $0 \leq y' \leq 1$. Integration of this traps the solution of the initial value problem in the range $1 \leq y \leq 1 + t$. In order to prove this result rigorously, we need some preliminary results and ideas. Note that, to simplify things, we will prove all of the results in this section for one side of a point only. Firstly, we need some definitions.

A function $F(y, t)$ satisfies a **one-sided Lipschitz condition** in a domain D if, for some constant K,

$$y_2 > y_1 \Rightarrow F(y_2, t) - F(y_1, t) \leq K(y_2 - y_1) \quad \text{for } (y_2, t), (y_1, t) \in D.$$

† We have assumed that $t_1 > t_0$, but the argument needs only a slight modification if this is not the case.

8.4 COMPARISON THEOREMS

If $\sigma(t)$ is differentiable for $t > a$, then $\sigma'(t) \leq K$ is called a **differential inequality**. Note that if $\sigma'(t) \leq 0$ for $t \geq a$, then $\sigma(t) \leq \sigma(a)$.

Lemma 8.3 *If $\sigma(t)$ is a differentiable function that satisfies the differential inequality $\sigma'(t) \leq K\sigma(t)$ for $t \in [a, b]$ and some constant K, then $\sigma(t) \leq \sigma(a)e^{K(t-a)}$ for $t \in [a, b]$.*

Proof We write the inequality as $\sigma'(t) - K\sigma(t) \leq 0$ and multiply through by $e^{-Kt} > 0$ to obtain

$$\frac{d}{dt}\left(e^{-Kt}\sigma(t)\right) \leq 0.$$

This leads to $e^{-Kt}\sigma(t) \leq e^{-Ka}\sigma(a)$, and hence the result. Note that this is really just a special case of Gronwall's inequality. $\square$

Lemma 8.4 *If $g(t)$ is a solution of $y' = F(y,t)$ for $t \geq a$ and $f(t)$ is a function that satisfies the differential inequality $f'(t) \leq F(f(t),t)$ for $t \geq a$ with $f(a) = g(a)$, then, provided that F satisfies a one-sided Lipschitz condition for $t \geq a$, $f(t) \leq g(t)$ for all $t \geq a$.*

Proof We will proceed by contradiction. Suppose that $f(t_1) > g(t_1)$ for some $t_1 > a$. Define t_0 to be the largest t in the interval $a \leq t \leq t_1$ such that $f(t) \leq g(t)$, and hence $f(t_0) = g(t_0)$. Define $\sigma(t) = f(t) - g(t)$, so that $\sigma(t) \geq 0$ for $t_0 \leq t \leq t_1$ and $\sigma(t_0) = 0$. The one-sided Lipschitz condition shows that

$$\sigma'(t) = f'(t) - g'(t) \leq F(f(t),t) - F(g(t),t) \leq K\{f(t) - g(t)\} = K\sigma(t),$$

and hence $\sigma'(t) \leq K\sigma(t)$ for $t > t_0$. By Lemma 8.3, $\sigma(t) \leq \sigma(t_0)e^{K(t-t_0)} \leq 0$. However, since $\sigma(t)$ is non-negative for $t \geq t_0$, $\sigma(t) \equiv 0$. This contradicts the hypothesis that $\sigma(t_1) = f(t_1) - g(t_1) > 0$, and hence the result is proved. $\square$

Theorem 8.4 *Let $f(t)$ and $g(t)$ be solutions of the differential equations $y' = F(y,t)$ and $z' = G(z,t)$ respectively, on the strip $a \leq t \leq b$, with $f(a) = g(a)$. If $F(y,t) \leq G(y,t)$ and F or G is one-sided Lipschitz on this strip, then $f(t) \leq g(t)$ for $a \leq t \leq b$*

Proof Since $y' = F(y,t) \leq G(y,t)$, g satisfies a differential inequality of the form described in Lemma 8.4, and f satisfies the differential equation. This gives us the result immediately. $\square$

Note that Theorem 8.4 and the preceding two lemmas can be made two-sided in the neighbourhood of the initial point with minor modifications.

Comparison theorems, such as Theorem 8.4, are of considerable practical use. Consider as an example the initial value problem

$$y' = t^2 + y^2 \quad \text{subject to } y(0) = 1. \tag{8.12}$$

It is known, either from analysis of the solution or by numerical integration, that the solution of this equation 'blows up' to infinity at some finite value of $t = t_\infty$. We can estimate the position of this blowup point by comparing the solution of (8.12) with the solutions of similar, but simpler, differential equations. Since $t^2 + y^2 \geq y^2$ for $0 \leq t < \infty$, the solution of (8.12) is bounded below by the solution of $y' = y^2$ subject to $y(0) = 1$, which is $y = 1/(1-t)$. Since this blows up when $t = 1$, we must have $t_\infty \leq 1$. Also, since $t^2 + y^2 \leq 1 + y^2$ for $0 \leq t \leq 1$, an upper bound is provided by the solution of $y' = 1 + y^2$ subject to $y(0) = 1$, namely $y = \tan(t + \frac{\pi}{4})$. This blows up when $t = \pi/4$. By sandwiching the solution of (8.12) between two simpler solutions, we are able to conclude that the blowup time satisfies $\frac{\pi}{4} \leq t_\infty \leq 1$.

Comparison theorems are also available for second order differential equations. These take a particularly simple form for the equation

$$y''(x) + g(x)y(x) = 0, \tag{8.13}$$

from which the first derivative of y is absent. In fact, we can transform any linear, second order, ordinary differential equation into this form, as we shall see in Section 12.2.7. We will compare the solution of (8.13) with that of

$$z''(x) + h(x)z(x) = 0, \tag{8.14}$$

and prove the following theorem.

Theorem 8.5 *If $g(x) < h(x)$ for $x \geq x_0$, $y(x)$ is the solution of (8.13) with $y(x_0) = y_0 \neq 0$, $y'(x_0) = y_1$, these conditions being such that $y(x) > 0$ for $x_0 \leq x \leq x_1$, and $z(x)$ is the solution of (8.14) with $z(x_0) = y_0$ and $z'(x_0) = y_1$, then $y(x) > z(x)$ for $x_0 < x \leq x_1$, provided that $z(x) > 0$ on this interval.*

Proof From (8.13) and (8.14),

$$y''z - yz'' = (h - g)yz.$$

Integrating this equation from x_0 to $x < x_1$, we obtain

$$y'(x)z(x) - y(x)z'(x) = \int_{x_0}^{x} (h(t) - g(t))y(t)z(t)\, dt.$$

By hypothesis, the right hand side of this is positive. Also, by direct differentiation,

$$\frac{d}{dx}\left(\frac{y(x)}{z(x)}\right) = \frac{y'(x)z(x) - y(x)z'(x)}{z^2(x)} > 0,$$

so that y/z is an increasing function of x. Since $y(x_0)/z(x_0) = 1$, we have $y(x) > z(x)$ for $x_0 < x \leq x_1$. □

As an example of the use of Theorem 8.5, consider Airy's equation,

$$\ddot{y}(t) - ty(t) = 0$$

(see Sections 3.8 and 11.2), in $-1 < t < 0$ with $y(0) = 1$ and $\dot{y}(0) = 0$. By making the transformation $t \mapsto -x$, we arrive at $y''(x) + xy(x) = 0$ in $0 < x < 1$, with $y(0) = 1$ and $y'(0) = 0$. The solution of this is positive for $0 \leq x < x_1$. If we consider $z''(x) + z(x) = 0$ with $z(0) = 1$ and $z'(0) = 0$, then clearly $z = $

$\cos x$, which is positive for $0 \leqslant x < \pi/2$, and hence for $0 \leqslant x \leqslant 1$. Since the equations and boundary conditions that govern $y(x)$ and $z(x)$ satisfy the conditions of Theorem 8.5, we conclude that $y(x) > \cos x$ for $0 < x < 1$. In fact, we can see from the exact solution,

$$y(x) = \frac{\text{Bi}'(0)\text{Ai}(-x) - \text{Ai}'(0)\text{Bi}(-x)}{\text{Bi}'(0)\text{Ai}(0) - \text{Ai}'(0)\text{Bi}(0)},$$

shown in Figure 8.2, that $x_1 \approx 1.986$.

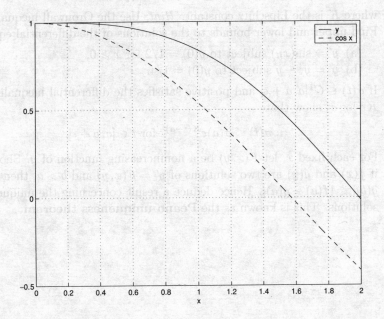

Fig. 8.2. The exact solution of $y'' + xy = 0$ subject to $y(0) = 1$ and $y'(0) = 0$.

Exercises

8.1 Determine the integral equations equivalent to the initial value problems

(a) $y' = t^2 + y^4$ subject to $y(0) = 1$,

(b) $y' = y + t$ subject to $y(0) = 0$,

and determine the first two successive approximations to the solution.

8.2 Show that the functions

(a) $f(y,t) = te^{-y^2}$ for $|t| \leqslant 1$, $|y| < \infty$,

(b) $f(y,t) = t^2 + y^2$ for $|t| \leqslant 2$, $|y| \leqslant 3$,

are Lipschitz in the regions indicated, and find the Lipschitz constant, K, in each case.

8.3 How could successive approximations to the solution of $y' = 3y^{2/3}$ fail to converge to a solution?

8.4 Let $f(y,t)$ and $g(y,t)$ be continuous and satisfy a Lipschitz condition with respect to y in a region D. Suppose $|f(y,t) - g(y,t)| < \epsilon$ in D for some $\epsilon > 0$. If $y_1(t)$ is a solution of $y' = f(y,t)$ and $y_2(t)$ is a solution of $y' = g(y,t)$, such that $|y_2(t_0) - y_1(t_0)| < \delta$ for some t_0 and $\delta > 0$, show that, for all t for which $y_1(t)$ and $y_2(t)$ both exist,

$$|y_2(t) - y_1(t)| \leq \delta \exp(K|t - t_0|) + \frac{\epsilon}{K}\{\exp(K|t - t_0|) - 1\},$$

where K is the Lipschitz constant. *Hint:* Use the Gronwall inequality.

8.5 Find upper and lower bounds to the solutions of the differential equations

(a) $y' = \sin(xy)$ subject to $y(0) = 1/2$ for $x \geq 0$,
(b) $y' = y^3 - y$ subject to $y(0) = 1/4$.

8.6 If $\sigma(t) \in C^1[a, a+\epsilon]$ and positive satisfies the differential inequality $\sigma' \leq K\sigma \log \sigma$, show that

$$\sigma(t) \leq \sigma(a)e^{K(t-a)} \quad \text{for } t \in [a, a+\epsilon].$$

8.7 For each fixed x, let $F(x, y)$ be a nonincreasing function of y. Show that, if $f(x)$ and $g(x)$ are two solutions of $y' = F(x, y)$ and $b > a$, then $|f(b) - g(b)| \leq |f(a) - g(a)|$. Hence deduce a result concerning the uniqueness of solutions. This is known as the **Peano uniqueness theorem**.

CHAPTER NINE

Nonlinear Ordinary Differential Equations: Phase Plane Methods

9.1 Introduction: The Simple Pendulum

Ordinary differential equations can be used to model many different types of physical system. We now know a lot about second order linear ordinary differential equations. For example, simple harmonic motion,

$$\frac{d^2\theta}{dt^2} + \omega^2\theta = 0, \qquad (9.1)$$

describes many physical systems that oscillate with small amplitude θ. The general solution is $\theta = A\sin\omega t + B\cos\omega t$, where A and B are constants that can be fixed from the initial values of θ and $d\theta/dt$. The solution is an oscillatory function of t. Note that we can also write this as $\theta = Ce^{i\omega t} + De^{-i\omega t}$, where C and D are complex constants. In the real world, the physics of a problem is rarely as simple as this. Let's consider the frictionless simple pendulum, shown in Figure 9.1. A mass, m, is attached to a light, rigid rod of length l, which can rotate without friction about the point O.

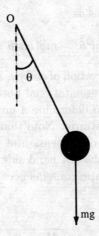

Fig. 9.1. A simple pendulum.

Using Newton's second law on the force perpendicular to the rod gives

$$-mg\sin\theta = ml\frac{d^2\theta}{dt^2},$$

and hence

$$\frac{d^2\theta}{dt^2} + \omega^2 \sin\theta = 0, \qquad (9.2)$$

where $\omega^2 = g/l$ and t is time. For oscillations of small amplitude, $\theta \ll 1$, so that $\sin\theta \sim \theta$, and we obtain simple harmonic motion, (9.1). If θ is not small, we must study the full equation of motion, (9.2), which is **nonlinear**. Do we expect the solutions to be qualitatively different to those of simple harmonic motion? If we push a pendulum hard enough, we should be able to make it swing round and round its point of support, with θ increasing continuously with t (remember there is no friction), so we would hope that (9.2) has solutions of this type.

In general, nonlinear ordinary differential equations cannot be solved analytically, but for equations like (9.2), where the first derivative, $d\theta/dt$ does not appear explicitly, an analytical solution is available. Using the notation $\dot\theta = d\theta/dt$, the trick is to treat $\dot\theta$ as a function of θ instead of t. Note that

$$\frac{d^2\theta}{dt^2} = \frac{d}{dt}\left(\frac{d\theta}{dt}\right) = \frac{d\dot\theta}{dt} = \frac{d\theta}{dt}\frac{d\dot\theta}{d\theta} = \dot\theta\frac{d\dot\theta}{d\theta} = \frac{d}{d\theta}\left(\frac{1}{2}\dot\theta^2\right).$$

This allows us to write (9.2) as

$$\frac{d}{d\theta}\left(\frac{1}{2}\dot\theta^2\right) = -\omega^2 \sin\theta,$$

which we can integrate once to give

$$\frac{1}{2}\dot\theta^2 = \omega^2 \cos\theta + \text{constant}.$$

Using $\omega^2 = g/l$, we can write this as

$$\frac{1}{2}ml^2\dot\theta^2 - mgl\cos\theta = E. \qquad (9.3)$$

This is just a statement of conservation of energy, E, with the first term representing kinetic energy, and the second, gravitational potential energy. Systems like (9.2), which can be integrated once to determine a **conserved quantity**, here energy, E, are called **conservative systems**. Note that if we try to account for a small amount of friction at the point of suspension of the pendulum, we need to add a term proportional to $d\theta/dt$ to the left hand side of (9.2). The system is then no longer conservative, with dramatic consequences for the motion of the pendulum (see Exercise 9.6).

From (9.3) we can see that

$$\frac{d\theta}{dt} = \pm\sqrt{\frac{2E}{ml^2} + \frac{2g}{l}\cos\theta}.$$

9.1 INTRODUCTION: THE SIMPLE PENDULUM

We can integrate this to arrive at the implicit solution

$$t = \pm \int_{\theta_0}^{\theta} \frac{d\theta'}{\sqrt{\frac{2E}{ml^2} + \frac{2g}{l} \cos \theta'}}, \qquad (9.4)$$

where θ_0 is the angle of the pendulum when $t = 0$. Note that the two constants of integration are E and θ_0, the initial energy and angle of the pendulum. Equation (9.4) is a simple representation of the solution, which, if necessary, we can write in terms of Jacobian elliptic functions (see Section 9.4), so now everything is clear ... , except of course that it isn't! Presumably equation (9.4) gives oscillatory solutions for small initial angles and kinetic energies, and solutions with θ increasing with t for large enough initial energies, but this doesn't really leap off the page at you. From (9.4) we have a *quantitative* expression for the solution, but we are really more interested in the *qualitative* nature of the solution.

Let's go back to (9.3) and write

$$\dot{\theta}^2 = \frac{2E}{ml^2} + \frac{2g}{l} \cos \theta.$$

Graphs of $\dot{\theta}^2$ as a function of θ are shown in Figure 9.2(a) for different values of E.

— For $E > mgl$ the curves lie completely above the θ-axis.
— For $-mgl < E < mgl$ the curves intersect the θ-axis.
— For $E < -mgl$ the curves lie completely below the θ-axis (remember, $-1 \leqslant \cos \theta \leqslant 1$).

We can now determine $\dot{\theta}$ as a function of θ by taking the square root of the curves in Figure 9.2(a) to obtain the curves in Figure 9.2(b), remembering to take both the positive and negative square root.

— For $E > mgl$, the curves lie either fully above or fully below the θ-axis.
— For $-mgl < E < mgl$, only finite portions of the graph of $\dot{\theta}^2$ lie above the θ-axis, so the square root gives finite, closed curves.
— For $E < -mgl$, there is no real solution. This corresponds to the fact that the pendulum always has a gravitational potential energy of at least $-mgl$, so we must have $E \geqslant -mgl$.

As we shall see later, the solution with $E = mgl$ is an important one. The graph of $\dot{\theta}^2$ just touches the θ-axis at $\theta = \pm(2n-1)\pi$, for $n = 1, 2, \ldots$, and taking the square root gives the curves that pass through these points shown in Figure 9.2(b).

How do θ and $\dot{\theta}$ vary along these solution curves as t increases? If $\dot{\theta}$ is positive, θ increases with t, and vice versa (remember, $\dot{\theta} = d\theta/dt$ is, by definition, the rate at which θ changes with t). This allows us to add arrows to Figure 9.2(b), indicating in which direction the solution changes with time. We have now constructed our first **phase portrait** for a nonlinear ordinary differential equation. The $(\theta, \dot{\theta})$-plane is called the **phase plane**. Each of the solution curves represents a possible solution of (9.2), and is known as an **integral path** or **trajectory**. If we know the initial conditions, $\theta = \theta_0$, $\dot{\theta} = \dot{\theta}_0 = \sqrt{\frac{2E}{ml^2} + \frac{2g}{l} \cos \theta_0}$, the integral path that passes through the point $(\theta_0, \dot{\theta}_0)$ when $t = 0$ represents the solution. Finally, note that, since $\theta = \pi$ is equivalent to $\theta = -\pi$, we only need to consider the phase portrait for $-\pi \leqslant \theta \leqslant \pi$. Alternatively, we can cut the phase plane along the lines $\theta = -\pi$ and $\theta = \pi$ and join them up, so that the integral paths lie on the surface of a cylinder.

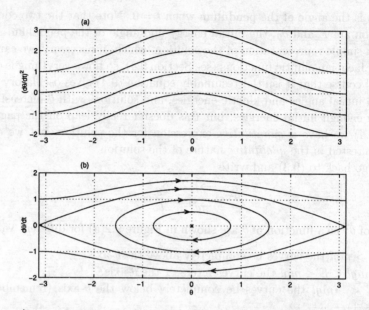

Fig. 9.2. (a) $\dot{\theta}^2$ as a function of θ for different values of E. (b) $\dot{\theta}$ as a function of θ – the phase portrait for the simple pendulum. The cross at $(0, 0)$ indicates an equilibrium solution. The arrows indicate the path followed by the solution as t increases. Note that the phase portrait for $|\theta| \geqslant \pi$ is the periodic extension of the phase portrait for $|\theta| \leqslant \pi$.

Having gone to the trouble of constructing a phase portrait for (9.2), does it tell us what the pendulum does in a form more digestible than that given by (9.4)? Let's consider the three qualitatively different types of integral path shown in Figure 9.2(b).

(i) **Equilibrium solutions** The points $\dot{\theta} = 0$, $\theta = 0$ or $\pm \pi$, represent the two equilibrium solutions of (9.2). The point $(0, 0)$ is the equilibrium with the pendulum vertically downward, $(\pi, 0)$ the equilibrium with the pendulum vertically upward. Points close to $(0, 0)$ lie on small closed trajectories close to $(0, 0)$. This indicates that $(0, 0)$ is a **stable equilibrium point**, since a small change in the state of the system away from equilibrium leads to solutions that remain close to equilibrium. If you cough on a pendulum hanging downwards, you will only excite a small oscillation. In contrast, points close to $(\pi, 0)$ lie on trajectories that take the solution far away from $(\pi, 0)$, and we say that this is an **unstable equilibrium point**. If you cough on a pendulum balanced precariously above its point of support, it will fall. Of course, in practice it is impossible to balance a pendulum in

9.1 INTRODUCTION: THE SIMPLE PENDULUM

this way, precisely because the equilibrium is unstable. We will refine our definitions of stability and instability in Chapter 13.

(ii) **Periodic solutions** Integral paths with $-mgl < E < mgl$ are closed, and represent periodic solutions. They are often referred to as **limit cycles** or **periodic orbits**. The pendulum swings back and forth without reaching the upward vertical. These orbits are stable, since nearby orbits remain nearby. Note that the frequency of the oscillation depends upon its amplitude, a situation that is typical for nonlinear oscillators. For the simple pendulum, the amplitude of the motion is $\theta_{max} = \cos^{-1}(E/mgl)$, and (9.4) shows that the period of the motion, T, is given by

$$T = 2 \int_{-\theta_{max}}^{\theta_{max}} \frac{d\theta}{\sqrt{\frac{2E}{ml^2} + \frac{2g}{l}\cos\theta}}.$$

Small closed trajectories in the neighbourhood of the equilibrium point at $(0,0)$ are described by simple harmonic motion, (9.1). Note that, in contrast to the full nonlinear system, the frequency of simple harmonic motion is independent of its amplitude. The idea of **linearizing** a nonlinear system of ordinary differential equations close to an equilibrium point in order to determine what the phase portrait looks like there, is one that we will return to later.

In terms of the phase portrait on the cylindrical surface, trajectories with $E > mgl$ are also stable periodic solutions, looping round and round the cylinder. The pendulum has enough kinetic energy to swing round and round its point of support.

(iii) **Heteroclinic solutions** The two integral paths that connect $(-\pi, 0)$ and $(\pi, 0)$ have $E = mgl$. The path with $\dot\theta \geqslant 0$ has $\theta \to \pm\pi$ as $t \to \pm\infty$ (we will prove this later). This solution represents a motion where the pendulum swings around towards the upward vertical, and is just caught between falling back and swinging over. This is known as a **heteroclinic path**, since it connects different equilibrium points.[†] Heteroclinic paths are important, because they represent the boundaries between qualitatively different types of behaviour. They are also unstable, since nearby orbits behave qualitatively differently. Here, the heteroclinic orbits separate motions where the pendulum swings back and forth from motions where it swings round and round (different types of periodic solution). In terms of the phase portrait on a cylindrical surface, we can consider these paths to be **homoclinic paths**, since they connect an equilibrium point to itself.[‡]

We have now seen that, if we can determine the qualitative nature of the phase portrait of a second order nonlinear ordinary differential equation and sketch it, we can extract a lot of information about the qualitative behaviour of its solutions. This information provides rather more insight than the analytical solution, (9.4),

[†] Greek hetero = different
[‡] Greek homo = same

into the behaviour of the physical system that the equation models. For nonconservative equations, we cannot integrate the equation directly to get at the equation of the integral paths, and we have to be rather more cunning in order to sketch the phase portrait. We will develop methods to tackle second order, nonconservative, ordinary differential equations in Section 9.3. Before that, we will consider the simpler case of first order nonlinear ordinary differential equations.

9.2 First Order Autonomous Nonlinear Ordinary Differential Equations

An **autonomous** ordinary differential equation is one in which the independent variable does not appear explicitly, for example (9.2). The equations $\ddot{\theta} + t^2\theta = 0$ and $\dot{x} = t - x^2$ are **nonautonomous**. Note, however, that an nth order nonautonomous ordinary differential equation can always be written as a $(n+1)^{\text{th}}$ order system of autonomous ordinary differential equations. For example $\dot{x} = t - x^2$ is equivalent to $\dot{x} = y - x^2$, $\dot{y} = 1$ with $y = 0$ when $t = 0$.

In this section we focus on the qualitative behaviour of solutions of first order, autonomous, ordinary differential equations, which can always be written as

$$\frac{dx}{dt} = \dot{x} = X(x), \qquad (9.5)$$

with $X(x)$ a given function of x. Of course, such an equation is separable, with the solution subject to $x(0) = x_0$ given by

$$t = \int_{x_0}^{x} \frac{dx'}{X(x')}.$$

As we found in the previous section, solving the equation analytically is not necessarily the easiest way to determine the qualitative behaviour of the system (see Exercise 9.2).

9.2.1 The Phase Line

Consider the graph of the function $X(x)$. An example is shown in Figure 9.3(a). If $X(x_1) = 0$, then $\dot{x} = 0$, and hence $x = x_1$ is an equilibrium solution of (9.5). For the example shown in Figure 9.3 there are three equilibrium points, at $x = x_1$, x_2 and x_3. We can also see that $\dot{x} = X(x) < 0$, and hence x is a decreasing function of t for $x < x_1$ and $x_2 < x < x_3$. Similarly, $\dot{x} = X(x) > 0$, and hence x increases as t increases, for $x_1 < x < x_2$ and $x > x_3$. By analogy with the phase plane, where we constructed the phase portrait for a second order system in the previous section, we can draw a **phase line** for this first order equation, as shown in Figure 9.3(b). The arrows indicate whether x increases or decreases with t. Clearly, the different types of behaviour that are possible are rather limited by the constraint that the trajectories lie in a single dimension. Both for this example and in general, trajectories either enter an equilibrium point or head off to infinity.

In particular, periodic solutions are not possible.† Solutions that begin close to $x = x_1$ or $x = x_3$ move away from the equilibrium point, so that $x = x_1$ and $x = x_3$ are unstable. In contrast, $x = x_2$ is a stable equilibrium point, since solutions that start close to it approach it.

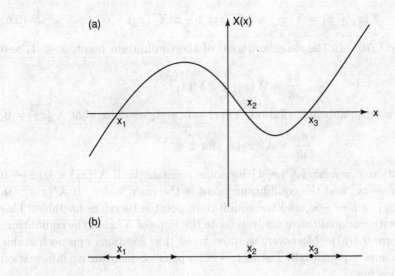

Fig. 9.3. (a) An example of a graph of $\dot{x} = X(x)$. (b) The equivalent phase line.

From Figure 9.3(b) we can see that

$$x \to -\infty \text{ as } t \to \infty \quad \text{when } x_0 < x_1,$$
$$x \to x_2 \text{ as } t \to \infty \quad \text{when } x_1 < x_0 < x_3,$$
$$x \to \infty \text{ as } t \to \infty \quad \text{when } x_0 > x_3.$$

The set $D_{-\infty} = \{x \mid x < x_1\}$ is called the **domain of attraction** or **basin of attraction** of minus infinity. Similarly, $D_2 = \{x \mid x_1 < x < x_3\}$ is the domain of attraction of x_2 and $D_\infty = \{x \mid x > x_3\}$ that of plus infinity. All of this information has come from a qualitative analysis of the graph of $\dot{x} = X(x)$.

9.2.2 Local Analysis at an Equilibrium Point

If $x = x_1$ is an equilibrium solution of (9.5), what happens *quantitatively* close to x_1? Firstly, let's move the equilibrium point to the origin by defining $\bar{x} = x - x_1$,

---

† If there is some geometrical periodicity in the problem so that, for example, by analogy with the simple pendulum, $-\pi \leqslant x \leqslant \pi$ and $x = -\pi$ is equivalent to $x = \pi$, we can construct a phase loop, around which the solution can orbit indefinitely.

so that
$$\frac{d\bar{x}}{dt} = X(x_1 + \bar{x}).$$

We can expand $X(x_1 + \bar{x})$ as a Taylor series,
$$X(x_1 + \bar{x}) = X(x_1) + \bar{x}X'(x_1) + \frac{1}{2}\bar{x}^2 X''(x_1) + \cdots, \tag{9.6}$$

where $X' = dX/dx$. In the neighbourhood of the equilibrium point, $\bar{x} \ll 1$, and hence
$$\frac{d\bar{x}}{dt} \approx X(x_1) + \bar{x}X'(x_1).$$

Since $x = x_1$ is an equilibrium solution, $X(x_1) = 0$, and, assuming that $X'(x_1) \neq 0$,
$$\frac{d\bar{x}}{dt} \approx X'(x_1)\bar{x} \quad \text{for } \bar{x} \ll 1.$$

This has solution $\bar{x} = k \exp\{X'(x_1)t\}$ for some constant k. If $X'(x_1) < 0$, $\bar{x} \to 0$ ($x \to x_1$) as $t \to \infty$, and the equilibrium point is therefore stable. If $X'(x_1) > 0$, $\bar{x} \to 0$ ($x \to x_1$) as $t \to -\infty$, and the equilibrium point is therefore unstable. This is consistent with our qualitative analysis (note the slope of $X(x)$ at the equilibrium points in Figure 9.3(a)). Moreover, we now know that solutions approach stable equilibrium points exponentially fast as $t \to \infty$, a piece of quantitative information well worth knowing.

If $X'(x_1) \neq 0$, we say that $x = x_1$ is a **hyperbolic equilibrium point**, and this analysis determines how solutions behave in its neighbourhood. In particular, solutions that do not start at a hyperbolic equilibrium point cannot reach it in a finite time. If $X'(x_1) = 0$, $x = x_1$ is a **nonhyperbolic equilibrium point**, and we need to retain more terms in the Taylor expansion of X, (9.6), in order to sort out what happens close to the equilibrium point. We will consider this further in Chapter 13.

9.3 Second Order Autonomous Nonlinear Ordinary Differential Equations

Any second order, autonomous, ordinary differential equation can be written as a system of two first order equations, in the form
$$\dot{x} = X(x,y), \quad \dot{y} = Y(x,y). \tag{9.7}$$

For example, consider (9.2), which governs the motion of a simple pendulum. Let's define $x = \theta$ and $y = \dot{x}$. Now, $\dot{y} = \ddot{x} = \ddot{\theta} = -(g/l)\sin\theta$, and hence
$$\dot{x} = y, \quad \dot{y} = -\frac{g}{l}\sin x.$$

From now on, we will assume that the right hand sides in (9.7) are continuously differentiable†, and hence that a solution exists and is unique in the sense of Theorem 8.2, for all of the systems that we study.

† Continuously differentiable functions are continuous and have derivatives with respect to the independent variables that are continuous.

9.3.1 The Phase Plane

In Section 9.1, we saw how the solutions of the equations of motion of a simple pendulum can be represented in the phase plane. Let's now consider the phase plane for a general second order system.

An integral path is a solution $(x(t), y(t))$ of (9.7), plotted in the (x, y)-plane, or phase plane. The slope of an integral path is

$$\frac{dy}{dx} = \frac{dy/dt}{dx/dt} = \frac{\dot{y}}{\dot{x}} = \frac{Y(x,y)}{X(x,y)}.$$

This slope is uniquely defined at all points $x = x_0$ and $y = y_0$ at which $X(x_0, y_0)$ and $Y(x_0, y_0)$ are not both zero. These are known as **ordinary points**. Since the solution through such a point exists and is unique, we deduce that integral paths cannot cross at ordinary points. This is possibly the most useful piece of information that you need to bear in mind when sketching phase portraits. *Integral paths cannot cross at ordinary points!*

Points (x_0, y_0) that are not ordinary have $X(x_0, y_0) = Y(x_0, y_0) = 0$, and are therefore equilibrium points. In other words, $\dot{x} = \dot{y} = 0$ when $x = x_0$, $y = y_0$, and this point represents an equilibrium solution of (9.7). At an equilibrium point, the slope, dy/dx, of the integral paths is not well-defined, and we deduce that *integral paths can meet only at equilibrium points, and only as $t \to \pm\infty$*. This will become clearer in the next section.

9.3.2 Equilibrium Points

In order to determine what the phase portrait looks like close to an equilibrium point, we proceed as we did for first order systems. We begin by shifting the origin to an equilibrium point at (x_0, y_0), using

$$x = x_0 + \bar{x}, \quad y = y_0 + \bar{y}.$$

Now we Taylor expand the functions X and Y, remembering that $X(x_0, y_0) = Y(x_0, y_0) = 0$, to obtain

$$X(x_0 + \bar{x}, y_0 + \bar{y}) = \bar{x}\frac{\partial X}{\partial x}(x_0, y_0) + \bar{y}\frac{\partial X}{\partial y}(x_0, y_0) + \cdots,$$

$$Y(x_0 + \bar{x}, y_0 + \bar{y}) = \bar{x}\frac{\partial Y}{\partial x}(x_0, y_0) + \bar{y}\frac{\partial Y}{\partial y}(x_0, y_0) + \cdots.$$

In the neighbourhood of the equilibrium point, $\bar{x} \ll 1$ and $\bar{y} \ll 1$, and hence

$$\frac{d\bar{x}}{dt} \approx \bar{x}X_x(x_0, y_0) + \bar{y}X_y(x_0, y_0),$$

$$\frac{d\bar{y}}{dt} \approx \bar{x}Y_x(x_0, y_0) + \bar{y}Y_y(x_0, y_0), \tag{9.8}$$

where we have used the notation $X_x = \partial X/\partial x$. The most convenient way of writing this is in matrix form, as

$$\frac{d\mathbf{u}}{dt} = J(x_0, y_0)\mathbf{u}, \tag{9.9}$$

where

$$\mathbf{u} = \begin{pmatrix} \bar{x} \\ \bar{y} \end{pmatrix}, \quad J(x_0, y_0) = \begin{pmatrix} X_x(x_0, y_0) & X_y(x_0, y_0) \\ Y_x(x_0, y_0) & Y_y(x_0, y_0) \end{pmatrix}.$$

We call $J(x_0, y_0)$ the **Jacobian matrix** at the equilibrium point (x_0, y_0). Equation (9.9) represents a pair of linear, first order ordinary differential equations with constant coefficients, which suggests that we can look for solutions of the form $\mathbf{u} = \mathbf{u}_0 e^{\lambda t}$, where $\mathbf{u}_0$ and λ are constants to be determined. Substituting this form of solution into (9.9) gives

$$\lambda \mathbf{u}_0 = J \mathbf{u}_0. \tag{9.10}$$

This is an eigenvalue problem. There are two eigenvalues, $\lambda = \lambda_1$ and $\lambda = \lambda_2$, possibly equal, possibly complex, and corresponding unit eigenvectors $\mathbf{u}_0 = \mathbf{u}_1$ and $\mathbf{u}_0 = \mathbf{u}_2$. These provide us with two possible solutions of (9.9), $\mathbf{u} = \mathbf{u}_1 e^{\lambda_1 t}$ and $\mathbf{u} = \mathbf{u}_2 e^{\lambda_2 t}$, so that the general solution is the linear combination

$$\mathbf{u} = A_1 \mathbf{u}_1 e^{\lambda_1 t} + A_2 \mathbf{u}_2 e^{\lambda_2 t}, \tag{9.11}$$

for any constants A_1 and A_2.

When the eigenvalues λ_1 and λ_2 are real, distinct and nonzero, so are the eigenvectors $\mathbf{u}_1$ and $\mathbf{u}_2$, and (9.11) suggests that the form of the solution is simpler if we make a linear transformation so that $\mathbf{u}_1$ and $\mathbf{u}_2$ lie along the coordinate axes. Such a transformation is given by $\mathbf{u} = P\mathbf{v}$, where $\mathbf{v} = (\hat{x}, \hat{y})$ and $P = (\mathbf{u}_1, \mathbf{u}_2)$ is a matrix with columns given by $\mathbf{u}_1$ and $\mathbf{u}_2$. Substituting this into (9.9) gives

$$P\dot{\mathbf{v}} = JP\mathbf{v},$$

$$\dot{\mathbf{v}} = P^{-1}JP\mathbf{v} = \Lambda \mathbf{v},$$

where

$$\Lambda = \begin{pmatrix} \lambda_1 & 0 \\ 0 & \lambda_2 \end{pmatrix}.$$

Therefore $\dot{\hat{x}} = \lambda_1 \hat{x}$, $\dot{\hat{y}} = \lambda_2 \hat{y}$, so that

$$\hat{x} = k_1 e^{\lambda_1 t}, \quad \hat{y} = k_2 e^{\lambda_2 t}, \tag{9.12}$$

with k_1 and k_2 constants, and hence

$$\hat{y} = k_3 \hat{x}^{\lambda_2/\lambda_1}, \tag{9.13}$$

where $k_3 = k_1^{\lambda_2/\lambda_1}/k_2$. This is the equation of the integral paths in the transformed coordinates. Note that the transformed coordinate axes are integral paths. There are now three main cases to consider.

(i) **Distinct, real, positive eigenvalues** $(\lambda_1, \lambda_2 > 0)$ The solution (9.12) shows that $\hat{x}, \hat{y} \to 0$ as $t \to -\infty$, exponentially fast, so the equilibrium point is unstable. The equation of the integral paths, (9.13), then shows that they all meet at the equilibrium point at the origin. The local phase portrait is sketched in Figure 9.4, in both the transformed and untransformed coordinate systems. This type of equilibrium point is called an **unstable node**.

Remember, this is the phase portrait close to the equilibrium point. As the integral paths head away from the equilibrium point, the linearization that we have performed in order to obtain (9.9) becomes inappropriate and we must consider how this local phase portrait fits into the full, global picture. Figure 9.4(a) illustrates the situation when $\lambda_2 > \lambda_1$. This means that $\hat{y}$ grows more quickly than $\hat{x}$ as t increases, and thereby causes the integral paths to bend as they do. Figure 9.4(b) shows that when $\lambda_2 > \lambda_1$, the solution grows more rapidly in the $\mathbf{u}_2$-direction than the $\mathbf{u}_1$-direction.

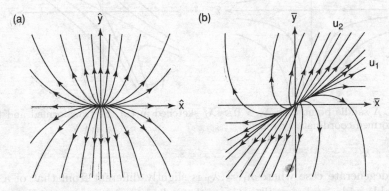

Fig. 9.4. An unstable node with $\lambda_2 > \lambda_1$ sketched in (a) the transformed and (b) the untransformed coordinate systems.

(ii) **Distinct, real, negative eigenvalues** ($\lambda_1, \lambda_2 < 0$) In this case, all we have to do is consider the situation with the sense of t reversed, and we recover the previous case. The situation is as shown in Figure 9.4, but with the arrows reversed. This type of equilibrium point is called a **stable node**.

(iii) **Real eigenvalues of opposite sign** ($\lambda_1 \lambda_2 < 0$) In this case, the coordinate axes are the only integral paths in the $(\hat{x}, \hat{y})$-plane that enter the equilibrium point. On the other integral paths, given by (9.13), $\hat{x} \to \pm\infty$ as $\hat{y} \to 0$, and vice versa, as shown in Figure 9.5(a). When $\lambda_2 > 0 > \lambda_1$, $\hat{x} \to 0$ and $\hat{y} \to \pm\infty$ as $t \to \infty$. The integral paths in the directions of the eigenvectors $\mathbf{u}_1$ and $\mathbf{u}_2$, shown in Figure 9.5(b), are therefore the only ones that enter the equilibrium point, and are called the **stable and unstable separatrices**†. This type of equilibrium point is called a **saddle point**. The separatrices of saddle points usually represent the boundaries between different types of behaviour in a phase portrait. See if you can spot the saddle points in Figure 9.2(b). Remember, although the separatrices are straight trajectories

---

† singular form, separatrix.

in the neighbourhood of the saddle point, they will start to bend as they head away from it and become governed by the full nonlinear equations.

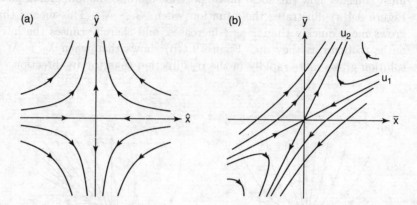

Fig. 9.5. A saddle point with $\lambda_2 > 0 > \lambda_1$ sketched in (a) the transformed and (b) the untransformed coordinate systems.

The degenerate case where $\lambda_1 = \lambda_2$ is slightly different from that of a stable or unstable node, and we will not consider it here, but refer you to Exercise 9.4. However, it should be clear that when $\lambda_1 = \lambda_2 > 0$ the equilibrium point is unstable, whilst for $\lambda_1 = \lambda_2 < 0$ it is stable.

Let's now consider what the phase portrait looks like when the eigenvalues λ_1 and λ_2 are complex. Since the eigenvalues are the solutions of a quadratic equation with real coefficients, they must be complex conjugate, so we can write $\lambda = \alpha \pm i\beta$, with α and β real. The general solution, (9.11), then becomes

$$\mathbf{u} = e^{\alpha t}\left(A_1 \mathbf{u}_1 e^{i\beta t} + A_2 \mathbf{u}_2 e^{-i\beta t}\right). \qquad (9.14)$$

There are two cases to consider.

(i) **Eigenvalues with strictly positive real part** ($\alpha > 0$) The term $e^{\alpha t}$ in (9.14) means that the solution grows exponentially with t, so that the equilibrium point is unstable. The remaining term in (9.14) is oscillatory, and we conclude that the integral paths spiral away from the equilibrium point, as shown in Figure 9.6. This type of equilibrium point is called an **unstable spiral** or **unstable focus**. To determine whether the sense of rotation is clockwise or anticlockwise, it is easiest just to consider the sign of $d\bar{y}/dt$ on the positive $\bar{x}$-axis. When $\bar{y} = 0$, (9.8) shows that $d\bar{y}/dt = X_x(x_0, y_0)\bar{x}$, and hence the spiral is anticlockwise if $X_x(x_0, y_0) > 0$, clockwise if $X_x(x_0, y_0) < 0$.†

† If $X_x(x_0, y_0) = 0$, try looking on the positive $\bar{y}$-axis.

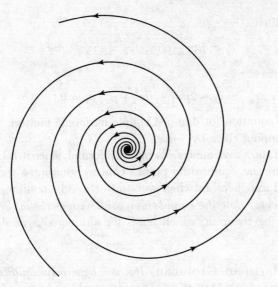

Fig. 9.6. An unstable, anticlockwise spiral.

(ii) **Eigenvalues with strictly negative real part** ($\alpha < 0$) This is just the same as the previous case, but with the sense of t reversed. This is a **stable spiral** or **stable focus**, and has the phase portrait shown in Figure 9.6, but with the arrows reversed.

Note that, for complex conjugate eigenvalues, we can transform the system into a more convenient form by defining the transformation matrix, P, slightly differently. If we choose either of the eigenvectors, for example $\mathbf{u}_1$, and write $\lambda_1 = \mu + i\omega$, we can define

$$P = (\text{Im}(\mathbf{u}_1), \text{Re}(\mathbf{u}_1)),$$

and $\mathbf{u} = P\mathbf{v}$. As before, $\dot{\mathbf{v}} = P^{-1}JP\mathbf{v}$, but now

$$JP = J\left(\text{Im}(\mathbf{u}_1), \text{Re}(\mathbf{u}_1)\right) = (\text{Im}(J\mathbf{u}_1), \text{Re}(J\mathbf{u}_1)) = (\text{Im}(\lambda_1 \mathbf{u}_1), \text{Re}(\lambda_1 \mathbf{u}_1))$$

$$= (\mu \text{Im}(\mathbf{u}_1) + \omega \text{Re}(\mathbf{u}_1), \mu \text{Re}(\mathbf{u}_1) - \omega \text{Im}(\mathbf{u}_1)) = P \begin{pmatrix} \mu & -\omega \\ \omega & \mu \end{pmatrix},$$

and hence

$$P^{-1}JP = \begin{pmatrix} \mu & -\omega \\ \omega & \mu \end{pmatrix}.$$

If $\mathbf{v} = (v_1, v_2)^T$, then
$$\dot{v}_1 = \mu v_1 - \omega v_2, \quad \dot{v}_2 = \omega v_1 + \mu v_2,$$
and, eliminating v_2,
$$\ddot{v}_1 - 2\mu \dot{v}_1 + (\omega^2 + \mu^2) v_1 = 0.$$

This is the usual equation for damped simple harmonic motion, with natural frequency ω and damping given by $-\mu$.

Note that all of the above analysis is rather informal, since it is based on a simple linearization about the equilibrium point. Can we guarantee that the behaviour that we have deduced persists when we study the full, nonlinear system, (9.7)? Yes, we can, *provided that the equilibrium point is hyperbolic*. We can formalize this in the following theorem, which holds for autonomous nonlinear systems of arbitrary order.

Theorem 9.1 (Hartman–Grobman) *If $\bar{\mathbf{x}}$ is a hyperbolic equilibrium point of the nth order system $\dot{\mathbf{x}} = \mathbf{f}(\mathbf{x})$, then there is homeomorphism (a mapping that is one-to-one, onto and continuous and has a continuous inverse) from $\mathbb{R}^n$ to $\mathbb{R}^n$ defined in a neighbourhood of $\bar{\mathbf{x}}$ that maps trajectories of the nonlinear system to trajectories of the local linearized system.*

We will not give the proof, but refer the interested reader to the original papers by Hartman (1960) and Grobman (1959).

Each of the five cases we have discussed above is indeed an example of a hyperbolic equilibrium point. If at least one eigenvalue has zero real part, the equilibrium point is said to be nonhyperbolic. As we shall see in Chapter 13, the behaviour in the neighbourhood of a nonhyperbolic equilibrium point is determined by higher order, nonlinear, terms in the Taylor expansions of X and Y.

An important example where it may not be necessary to consider higher order terms in the Taylor expansions is when the eigenvectors are purely imaginary, with $\lambda = \pm i\beta$. In this case, (9.14) with $\alpha = 0$ shows that the solution is purely oscillatory. The integral paths are closed and consist of a set of nested limit cycles around the equilibrium point. This type of equilibrium point is called a **centre**. However, the effect of higher order, nonlinear, terms in the Taylor expansions may be to make the local solutions spiral into or out of the equilibrium point. We say that the equilibrium point is a **linear centre**, but a **nonlinear spiral**. On the other hand, there are many physical systems where a linear centre persists, even in the presence of the higher order, nonlinear terms, and is called a **nonlinear centre**.

As an example, consider the simple pendulum, with
$$\dot{x} = y, \quad \dot{y} = -\frac{g}{l} \sin x.$$

This has
$$J = \begin{pmatrix} 0 & 1 \\ -\frac{g}{l} \cos x & 0 \end{pmatrix} \quad \text{and} \quad J(0,0) = \begin{pmatrix} 0 & 1 \\ -\frac{g}{l} & 0 \end{pmatrix}.$$

The eigenvalues of $J(0,0)$ are $\lambda = \pm i\sqrt{g/l}$. As we can see in Figure 9.2, the phase portrait close to the origin is indeed a nonlinear centre. What is it about this system that allows the centre to persist in the full nonlinear analysis? Well, the obvious answer is that there is a conserved quantity, the energy, which parameterizes the set of limit cycles. However, another way of looking at this is to ask whether the system has any symmetries. Here we know that if we map $x \mapsto -x$ and $t \mapsto -t$, reflecting the phase portrait in the y-axis and reversing time, the equations are unchanged, and hence the phase portrait should not change. Under this transformation, a stable spiral would become an unstable spiral, and vice versa, because of the reversal of time. Therefore, since the phase portrait should not change, the origin cannot be a spiral and must be a nonlinear centre.

9.3.3 An Example from Mechanics

Consider a rigid block of mass M attached to a horizontal spring. The block lies flat on a horizontal conveyor belt that moves at speed U and tries to carry the block away with it, as shown in Figure 9.7. From Newton's second law of motion and Hooke's law,

$$M\ddot{x} = F(\dot{x}) - k(x - x_e),$$

where x is the length of the spring, x_e is the equilibrium length of the spring, k is the spring constant, $F(\dot{x})$ is the frictional force exerted on the block by the conveyor belt, and a dot denotes d/dt. We model the frictional force as

$$F(\dot{x}) = \begin{cases} F_0 & \text{for } \dot{x} < U, \\ -F_0 & \text{for } \dot{x} > U, \end{cases}$$

with F_0 a constant force. When $\dot{x} = U$, the block moves at the same speed as the conveyor belt, and this occurs when $k|x - x_e| < F_0$. In other words, the force exerted by the spring must exceed the frictional force for the block to move. This immediately gives us a solution, $\dot{x} = U$ for $x_e - F_0/k \leq x \leq x_e + F_0/k$.

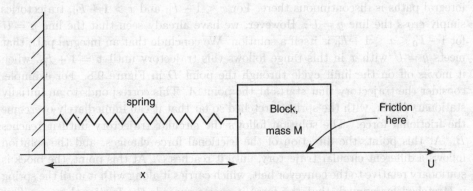

Fig. 9.7. A spring-mounted, rigid block on a conveyor belt.

Our model involves five physical parameters, M, k, x_e, F_0 and U. If we now define dimensionless variables $\bar{x} = x/x_e$ and $\bar{t} = t/\sqrt{M/k}$, we obtain

$$\ddot{\bar{x}} = \bar{F} - \bar{x} + 1, \tag{9.15}$$

where

$$\bar{F}(\dot{\bar{x}}) = \begin{cases} \bar{F}_0 & \text{for } \dot{\bar{x}} < \bar{U}, \\ -\bar{F}_0 & \text{for } \dot{\bar{x}} > \bar{U}. \end{cases} \tag{9.16}$$

We also have the possible solution $\dot{\bar{x}} = \bar{U}$ for $1 - \bar{F}_0 \leqslant \bar{x} \leqslant 1 + \bar{F}_0$. There are now just two dimensionless parameters,

$$\bar{F}_0 = \frac{F_0}{kx_e}, \quad \bar{U} = \frac{U}{x_e}\sqrt{\frac{M}{k}}.$$

We can now write (9.15) as the system

$$\dot{x} = y, \quad \dot{y} = F(y) - x + 1. \tag{9.17}$$

We have left out the overbars for notational convenience. This system has a single equilibrium point at $x = 1 + F_0$, $y = 0$. Since $y = 0 < \bar{U}$, the system is linear in the neighbourhood of this point, with

$$\begin{pmatrix} \dot{x} \\ \dot{y} \end{pmatrix} = \begin{pmatrix} 0 & 1 \\ -1 & 0 \end{pmatrix} \begin{pmatrix} x \\ y \end{pmatrix} + \begin{pmatrix} 0 \\ F_0 + 1 \end{pmatrix}.$$

The Jacobian matrix has eigenvalues $\pm i$, so the equilibrium point is a linear centre. In fact, since $\dot{x}(x - 1 - F) + y\dot{y} = 0$, we can integrate to obtain

$$\begin{cases} \{x - (1 + F_0)\}^2 + y^2 = \text{constant} & \text{for } y < U, \\ \{x - (1 - F_0)\}^2 + y^2 = \text{constant} & \text{for } y > U. \end{cases} \tag{9.18}$$

The solutions for $y \neq U$ are therefore concentric circles, and we conclude that the equilibrium point remains a centre when we take into account the effect of the nonlinear terms. The phase portrait is sketched in Figure 9.8.

We now need to take some care with integral paths that meet the line $y = U$. Since the right hand side of (9.17) is discontinuous at $y = U$, the slope of the integral paths is discontinuous there. For $x < 1 - F_0$ and $x > 1 + F_0$, trajectories simply cross the line $y = U$. However, we have already seen that the line $y = U$ for $1 - F_0 \leqslant x \leqslant 1 + F_0$ is itself a solution. We conclude that an integral path that meets $y = U$ with x in this range follows this trajectory until $x = 1 + F_0$, when it moves off on the limit cycle through the point D in Figure 9.8. For example, consider the trajectory that starts at the point A. This corresponds to an initially stationary block, with the spring stretched so far that it can immediately overcome the frictional force. The solution follows the circular trajectory until it reaches B. At this point, the direction of the frictional force changes, and the solution follows a different circular trajectory, until it reaches C. At this point, the block is stationary relative to the conveyor belt, which carries it along with it until the spring is stretched far enough that the force it exerts exceeds the frictional force. This occurs at the point D. Thereafter, the solution remains on the periodic solution through D. On this periodic solution, the speed of the block is always less than that of the conveyor belt, so the frictional force remains constant, and the block undergoes simple harmonic motion.

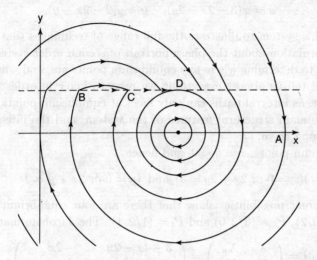

Fig. 9.8. The phase portrait for a spring-mounted, rigid block on a conveyor belt.

9.3.4 Example: Population Dynamics

Consider two species of animals that live on an island with populations $X(t)$ and $Y(t)$. If P is a typical size of a population, we can define $x(t) = X(t)/P$ and $y(t) = Y(t)/P$ as dimensionless measures of the population of each species, and regard x and y as continuous functions of time, t. A model for the way in which the two species interact with each other and their environment is

$$\dot{x} = x\left(A + a_1 x + b_1 y\right), \quad \dot{y} = y\left(B + b_2 x + a_2 y\right), \qquad (9.19)$$

where A, B, a_1, b_1, a_2 and b_2 are constants. We can interpret each of these equations as

Rate of change of population = Present population × (Birth rate − Death rate).

Let's now consider what each of the terms that model the difference between the birth and death rates represents. If $A > 0$, the population of species x grows when $x \ll 1$ and $y \ll 1$, so we can say that x does not rely on eating species y to survive. In contrast, if $A < 0$, the population of x dies out, and therefore x must need to eat species y to survive. The term $a_1 x$ represents the effect of overcrowding and competition for resources within species x, so we require that $a_1 < 0$. The term $b_1 y$ represents the interaction between the two species. If species x eats species y, $b_1 > 0$, so that the more of species y that is available, the faster the population of x grows. If species y competes with x for the available resources, $b_1 < 0$.

We will consider two species that do not eat each other, but compete for resources, for example sheep and goats (see Exercise 9.10 for an example of a predator–prey system). Specifically, we will study the typical system

$$\dot{x} = x(3 - 2x - 2y), \quad \dot{y} = y(2 - 2x - y). \tag{9.20}$$

We will use this system to illustrate the full range of techniques that are available to obtain information about the phase portrait of second order systems. The first thing to do is to determine where any equilibrium points are, and what their types are. This will tell us almost everything we need to know in order to sketch the phase portrait, as integral paths can only meet at equilibrium points. Equilibrium points are the main structural features of the system, and the integral paths are organized around them.

At equilibrium points, $\dot{x} = \dot{y} = 0$, and hence

$$(x = 0 \text{ or } 2x + 2y = 3) \text{ and } (y = 0 \text{ or } 2x + y = 2).$$

The four different possibilities show that there are four equilibrium points, $P_1 = (0,0)$, $P_2 = (0,2)$, $P_3 = (3/2, 0)$ and $P_4 = (1/2, 1)$. The Jacobian matrix is

$$J = \begin{pmatrix} X_x & X_y \\ Y_x & Y_y \end{pmatrix} = \begin{pmatrix} 3 - 4x - 2y & -2x \\ -2y & 2 - 2x - 2y \end{pmatrix},$$

and hence

$$J(0,0) = \begin{pmatrix} 3 & 0 \\ 0 & 2 \end{pmatrix}, \quad J(0,2) = \begin{pmatrix} -1 & 0 \\ -4 & -2 \end{pmatrix},$$

$$J(3/2, 0) = \begin{pmatrix} -3 & -3 \\ 0 & -1 \end{pmatrix}, \quad J(1/2, 1) = \begin{pmatrix} -1 & -1 \\ -2 & -1 \end{pmatrix}.$$

Three of these matrices have at least one off-diagonal element equal to zero, so their eigenvalues can be read directly from the diagonal elements. P_1 has eigenvalues 3 and 2 and is therefore an unstable node. P_2 has eigenvalues -1 and -2 and is therefore a stable node. P_3 has eigenvalues -3 and -1 and is therefore also a stable node. We could work out the direction of the eigenvectors for these three equilibrium points, but they are not really important for sketching the phase portrait. The final equilibrium point, P_4, has eigenvalues $\lambda = \lambda_\pm = -1 \pm \sqrt{2}$. Since $\sqrt{2} > 1$, P_4 has one positive and one negative eigenvalue, and is therefore a saddle point. For saddle points it is usually important to determine the directions of the eigenvectors, since these determine the directions in which the separatrices leave the equilibrium point. These separatrices are the only points that meet P_4, and we will see that determining their global behaviour is the key to sketching the full phase portrait. The unit eigenvectors of P_4 are $\mathbf{u}_\pm = (\sqrt{1/3}, \mp\sqrt{2/3})^T$.

Next, we can consider the **nullclines**. These are the lines on which either $\dot{x} = 0$ or $\dot{y} = 0$. The vertical nullclines, where $\dot{x} = 0$ (only y is varying so the integral paths are vertical on the nullcline), are given by

$$x = 0 \text{ or } 2x + 2y = 3.$$

9.3 SECOND ORDER EQUATIONS

The horizontal nullclines, where $\dot{y} = 0$, are given by

$$y = 0 \quad \text{or} \quad 2x + y = 2.$$

In general, the nullclines are *not* integral paths. However, in this case, since $\dot{x} = 0$ when $x = 0$ and $\dot{y} = 0$ when $y = 0$, the coordinate axes are themselves integral paths. This is clear from the biology of the problem, since, if species x is absent initially, there is no way for it to appear spontaneously, and similarly for species y.

All of the information we have amassed so far is shown in Figure 9.9(a). We will confine our attention to the positive quadrant, because we are not interested in negative populations. Moreover, the fact that the coordinate axes are integral paths prevents any trajectory from leaving the positive quadrant. Note that the equilibrium points lie at the points where the nullclines intersect. The directions of the arrows are determined by considering the signs of $\dot{x}$ and $\dot{y}$ at any point. For example, on the x-axis, $\dot{x} = x(3 - 2x)$, so that $\dot{x} > 0$, which means that x is increasing with t, for $0 < x < 3/2$, and $\dot{x} < 0$, which means that x is decreasing with t, for $x > 3/2$. From all of this local information, and the fact that integral paths can only cross at equilibrium points, we can sketch a plausible and consistent phase portrait, as shown in Figure 9.9(b). Apart from the stable separatrices of the saddle point P_4, labelled S_1 and S_2, all trajectories asymptote to either P_2 or P_3 as $t \to \infty$. These separatrices therefore represent the boundary between two very different types of behaviour. Depending upon the initial conditions, either species x or species y dies out completely. Although neither species eats the other, by competing for the same resources there can only be one eventual winner in the competition to stay alive. Although this is obviously a very simple model, the displacement of the native red squirrel by the grey squirrel in Britain, and the extinction of Neanderthal man after the arrival of modern man in Europe, are examples of situations where two species competed for the same resources. It is not necessary that one of the species kills the other directly. Simply a sufficient advantage in numbers or enough extra efficiency in exploiting natural resources could have been enough for modern man to consign Neanderthal man to oblivion.

Returning to our simple model, there are some questions that we really ought to answer before we can be confident that Figure 9.9(b) is an accurate sketch of the phase portrait of (9.20). Firstly, can we be sure that there are no limit cycle solutions? An associated question is, if we think that a system does possess a limit cycle, can we prove it? Secondly, since the position of the stable separatrices of P_4 is so important, can we prove that we have sketched them correctly? More specifically, how do we know that S_1 originates at P_1 and that S_2 originates at infinity? Finally, what does the phase portrait look like far from the origin? We will consider some mathematical tools for answering these questions in the next four sections. We can, however, also test whether our phase portrait is correct by solving equations (9.19) numerically in MATLAB. We must first define the MATLAB function

PHASE PLANE METHODS

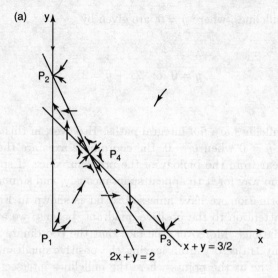

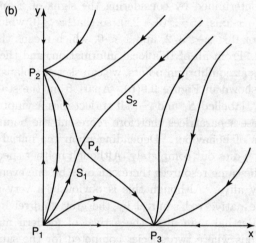

Fig. 9.9. (a) Local information at the equilibrium points and the nullclines, and (b) the full phase portrait for the population dynamical system (9.20).

```
function dy = population(t,y)
dy(1) = y(1)*(3-2*y(1)-2*y(2));
dy(2) = y(2)*(2-2*y(1)- y(2));
dy=dy';
```

which returns a column vector with elements given by the right hand sides of (9.19). Then [t y] = ode45(@population, [0 10], [1 2]) integrates the equations numerically, in this case for $0 \leqslant t \leqslant 10$, with initial conditions $x = $ y(1) $= 1$, $y = $ y(2) $= 2$. The results can then be displayed with plot(t,y). Figure 9.10 shows the effect of holding the initial value of x fixed, and varying the initial value of y. For small enough initial values of y, $y \to 0$ as $t \to \infty$. However, when the initial value of y is sufficiently large, the type of behaviour changes, and $x \to 0$ as $t \to \infty$, consistent with the phase portrait sketched in Figure 9.9(b).

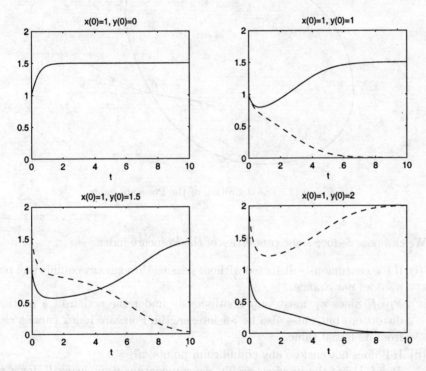

Fig. 9.10. The solution of (9.19) for various initial conditions. The dashed line is y, the solid line is x.

9.3.5 The Poincaré Index

The Poincaré index of an equilibrium point provides a very simple way of determining whether a given equilibrium point or collection of equilibrium points can be surrounded by one or more limit cycles.

Consider the second order system of nonlinear, autonomous ordinary differential equations given by (9.7). Let Γ be any smooth, closed, nonself-intersecting curve in the (x, y)-phase plane, *not necessarily an integral path*, that does not pass through any equilibrium points, as shown in Figure 9.11. At any point P lying on Γ, there

238 PHASE PLANE METHODS

is a unique integral path through P, with a well-defined direction that we can characterize using the angle, $\psi(P)$, that it makes with the horizontal ($\tan\psi(P) = Y/X$). The angle ψ varies continuously as P moves around Γ. If ψ changes by an amount $2n_\Gamma\pi$ as P makes one complete, anticlockwise circuit around Γ, we call n_Γ the **Poincaré index**, which must be an integer.

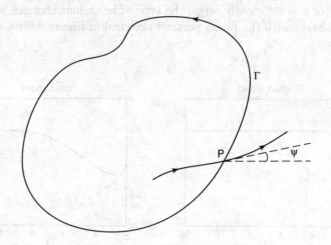

Fig. 9.11. The definition of the Poincaré index.

We can now deduce some properties of the Poincaré index.

(i) If Γ is continuously distorted without passing through any equilibrium points, n_Γ does not change.
Proof Since n_Γ must vary continuously under the action of a continuous distortion, but must also be an integer, the Poincaré index cannot change from its initial value.

(ii) If Γ does not enclose any equilibrium points, $n_\Gamma = 0$.
Proof Using the previous result, we can continuously shrink Γ down to an arbitrarily small circle without changing n_Γ. On this circle, $\psi(P)$ is almost constant, and therefore $n_\Gamma = 0$.

(iii) If Γ is the sum of two closed curves, then Γ_1 and Γ_2, $n_\Gamma = n_{\Gamma_1} + n_{\Gamma_2}$.
Proof Consider the curves shown in Figure 9.12. On the curve where Γ_1 and Γ_2 intersect, the amount by which ψ varies on traversing Γ_1 is equal and opposite to the amount by which ψ varies on traversing Γ_2, so these contributions cancel and $n_\Gamma = n_{\Gamma_1} + n_{\Gamma_2}$.

(iv) If Γ is a limit cycle then $n_\Gamma = 1$.
Proof This result is obvious.

(v) If integral paths either all enter the region enclosed by Γ or all leave this region, $n_\Gamma = 1$.
Proof This result is also obvious.

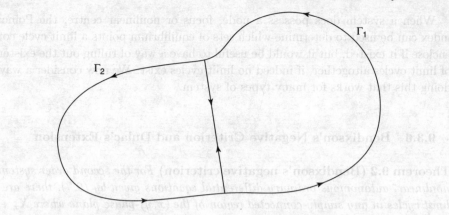

Fig. 9.12. The sum of two curves.

(vi) If Γ encloses a single node, focus or nonlinear centre, $n_\Gamma = 1$.
Proof For a node or focus, Γ can be continuously deformed into a small circle surrounding the equilibrium point. The linearized solution then shows that integral paths either all enter or all leave the region enclosed by Γ, and the previous result shows that $n_\Gamma = 1$. For a nonlinear centre, Γ can be continuously deformed into one of the limit cycles that encloses the equilibrium point, and result (iv) shows that $n_\Gamma = 1$.

(vii) If Γ encloses a single saddle point, $n_\Gamma = -1$.
Proof Γ can be continuously deformed into a small circle surrounding the saddle point, where the linearized solution gives the phase portrait shown in Figure 9.5. The direction of the integral paths makes a single clockwise rotation as P traverses Γ, and hence $n_\Gamma = -1$.

These results show that we can define the **Poincaré index of an equilibrium point** to be the Poincaré index of any curve that encloses the equilibrium point and no others. In particular, a node, focus or nonlinear centre has Poincaré index $n = 1$, whilst a saddle has Poincaré index $n = -1$. Nonhyperbolic equilibrium points can have other Poincaré indices, but we will not consider these here. Now result (iii) shows that n_Γ is simply the sum of the Poincaré indices of all the equilibrium points enclosed by Γ. Finally, and this is what all of this has been leading up to, result (iv) shows that *the sum of the Poincaré indices of the equilibrium points enclosed by a limit cycle must be 1*. A corollary of this is that a limit cycle must enclose at least one node, focus or centre.

Now let's return to our example population dynamical system, (9.20). If a limit cycle exists, it cannot cross the coordinate axes, since these are integral paths. However, there is no node, focus or nonlinear centre in the positive quadrant of the phase plane, and we conclude that no limit cycle exists.

240   PHASE PLANE METHODS

When a system does possess a node, focus or nonlinear centre, the Poincaré index can be used to determine which sets of equilibrium points a limit cycle could enclose if it existed, but it would be useful to have a way of ruling out the existence of limit cycles altogether, if indeed no limit cycles exist. We now consider a way of doing this that works for many types of system.

9.3.6   Bendixson's Negative Criterion and Dulac's Extension

Theorem 9.2 (Bendixson's negative criterion) *For the second order system of nonlinear, autonomous ordinary differential equations given by (9.7), there are no limit cycles in any simply-connected region of the (x,y)-phase plane where $X_x + Y_y$ does not change sign.*

Proof Let Γ be a limit cycle. For the vector field $(X(x,y), Y(x,y))$, Stokes' theorem states that

$$\int_D \left(\frac{\partial X}{\partial x} + \frac{\partial Y}{\partial y}\right) dx\, dy = \int_\Gamma (X\, dy - Y\, dx),$$

where D is the region enclosed by Γ. However, we can write the right hand side of this as

$$\int_\Gamma \left(X\frac{dy}{dt} - Y\frac{dx}{dt}\right) dt = \int_\Gamma (XY - YX)\, dt = 0,$$

and hence

$$\int_D \left(\frac{\partial X}{\partial x} + \frac{\partial Y}{\partial y}\right) dx\, dy = 0.$$

If $X_x + Y_y$ does not change sign in D, then the integral is either strictly positive or strictly negative†, which is a contradiction, and hence the result is proved.   □

Note that the restriction of this result to simply-connected regions (regions without holes) is crucial.

If we now apply this theorem to the system (9.20), we find that $X_x + Y_x = 5 - 6x - 4y$. Although this tells us that any limit cycle solution must be intersected by the line $6x + 4y = 5$, we cannot rule out the possibility of the existence of limit cycles in this way. We need a more powerful version of Bendixson's negative criterion, which is provided by the following theorem.

Theorem 9.3 (Dulac's extension to Bendixson's negative criterion) *If $\rho(x,y)$ is continuously differentiable in some simply-connected region D in the (x,y)-phase plane, there are no limit cycle solutions if $(\rho X)_x + (\rho Y)_y$ does not change sign.*

---

† ignoring the degenerate case where $X_x + Y_y \equiv 0$ in D.

Proof Since

$$\iint_D \left(\frac{\partial (\rho X)}{\partial x} + \frac{\partial (\rho Y)}{\partial y} \right) dx \, dy = \int_\Gamma \rho (X \, dy - Y \, dx),$$

the proof is as for Bendixson's negative criterion. □

Returning again to (9.20), we can choose $\rho = 1/xy$ and consider the positive quadrant where $\rho(x, y)$ is continuously differentiable. We find that

$$\frac{\partial (\rho X)}{\partial x} + \frac{\partial (\rho Y)}{\partial y} = -\frac{1}{x} - \frac{2}{y} < 0,$$

and hence that there can be no limit cycle solutions in the positive quadrant, as expected.

Now that we know how to try to rule out the existence of limit cycles, how can we go about showing that limit cycles *do* exist for appropriate systems? We will consider this in the next section using the Poincaré–Bendixson theorem. This will also allow us to prove that the stable separatrices of P_4, shown in Figure 9.9, do indeed behave as sketched.

9.3.7 The Poincaré–Bendixson Theorem

We begin with a definition. Consider the second order system (9.7). If I^+ is a closed subset of the phase plane and any integral path that lies in I^+ when $t = 0$ remains in I^+ for all $t \geqslant 0$, we say that I^+ is a **positively invariant set**. Similarly, a **negatively invariant set** is a closed subset, I^-, of the phase plane and all integral paths in I^- when $t = 0$ remain there when $t < 0$. For example, an equilibrium point is both a positively and a negatively invariant set, as is a limit cycle. As a less prosaic example, any subset, S, of the phase plane with outward unit normal $\mathbf{n}$ that has $(X, Y) \cdot \mathbf{n} \leqslant 0$ on its boundary, so that all integral paths enter S, is a positively invariant set.

Theorem 9.4 (Poincaré–Bendixson) *If there exists a bounded, invariant region, I, of the phase plane, and I contains no equilibrium points, then I contains at least one limit cycle.*

Note that it is crucial that the region I be bounded, but, in contrast to Bendixson's negative criterion, I does not need to be simply-connected. As we shall see, if I contains a limit cycle, it *cannot* be simply-connected. The Poincaré–Bendixson theorem says that the integral paths in I cannot wander around for ever without asymptoting to a limit cycle. Although this seems obvious, we shall see in Chapter 15 that the Poincaré–Bendixson theorem does *not* hold for third and higher order systems, in which integral paths that represent chaotic solutions can indeed wander around indefinitely without being attracted to a periodic solution.

The details of the proof are rather technical, and can be omitted on first reading.

PHASE PLANE METHODS

Proof of the Poincaré–Bendixson Theorem

We will assume that I is a positively invariant region. The proof when I is negatively invariant is identical, but with time reversed. We now need two further definitions.

Let $\mathbf{x}(t; \mathbf{x}_0)$ be the solution of (9.7) with $\mathbf{x} = \mathbf{x}_0$ when $t = 0$. We say that $\mathbf{x}_1$ is an ω-**limit point** of $\mathbf{x}_0$ if there exists a sequence, $\{t_i\}$, such that $t_i \to \infty$ and $\mathbf{x}(t_i; \mathbf{x}_0) \to \mathbf{x}_1$ as $i \to \infty$. For example, all the points on an integral path that enters a stable node or focus at $\mathbf{x} = \mathbf{x}_e$ have $\mathbf{x}_e$ as their only ω-limit point. Similarly, all points on an integral path that asymptotes to a stable limit cycle as $t \to \infty$ have all the points that lie on the limit cycle as ω-limit points. We call the set of all ω-limit points of $\mathbf{x}_0$ the ω-**limit set of** $\mathbf{x}_0$, which is denoted by $\omega(\mathbf{x}_0)$.

Let L be a finite curve such that all integral paths that meet L cross it in the same direction. We say that L is a **transversal**. If $\mathbf{x}_0$ is not an equilibrium point, it is always possible to construct a transversal through $\mathbf{x}_0$, since the slope of the integral paths is well-defined in the neighbourhood of $\mathbf{x}_0$.

Lemma 9.1 *Let I be a positively invariant region in the phase plane, and $L \subset I$ a transversal through $\mathbf{x}_0 \in I$. The integral path, $\mathbf{x}(t; \mathbf{x}_0)$, that passes through $\mathbf{x}_0$ when $t = 0$ intersects L in a monotone sequence. In other words, if $\mathbf{x}_i$ is the point where $\mathbf{x}(t; \mathbf{x}_0)$ meets L for the ith time, then $\mathbf{x}_i$ lies in the segment of L between $\mathbf{x}_{i-1}$ and $\mathbf{x}_{i+1}$.*

Proof Consider the invariant region D bounded by the integral path from $\mathbf{x}_{i-1}$ to $\mathbf{x}_i$ and the segment of L between $\mathbf{x}_{i-1}$ and $\mathbf{x}_i$. There are two possibilities. Firstly, the integral path through $\mathbf{x}_i$ enters D and remains there (see Figure 9.13(a)). In this case, $\mathbf{x}_{i+1}$, if it exists, lies in D, and $\mathbf{x}_i$ therefore lies in the segment of L between $\mathbf{x}_{i-1}$ and $\mathbf{x}_{i+1}$. Secondly, the integral path through $\mathbf{x}_{i-1}$ originates in D (see Figure 9.13(b)). In this case, $\mathbf{x}_{i-2}$, if it exists, lies in D, and $\mathbf{x}_{i-1}$ therefore lies in the segment of L between $\mathbf{x}_{i-2}$ and $\mathbf{x}_i$. □

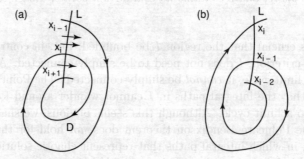

Fig. 9.13. The intersection of an integral path in an invariant region with a transversal.

9.3 SECOND ORDER EQUATIONS

Note that Lemma 9.1 does not hold in systems of higher dimension, since it relies crucially on the fact that a closed curve separates an inside from an outside (the Jordan curve theorem).

Lemma 9.2 *Consider the region I, transverse curve L and point $\mathbf{x}_0 \in I$ defined in Lemma 9.1. The ω-limit set of $\mathbf{x}_0$, $\omega(\mathbf{x}_0)$, intersects L at most once.*

Proof Suppose that $\omega(\mathbf{x}_0)$ intersects L twice, at $\hat{\mathbf{x}}$ and $\bar{\mathbf{x}}$. We can therefore find sequences of points, $\{\hat{\mathbf{x}}_i\}$ and $\{\bar{\mathbf{x}}_i\}$, lying in L such that $\{\hat{\mathbf{x}}_i\} \to \hat{\mathbf{x}}$ and $\{\bar{\mathbf{x}}_i\} \to \bar{\mathbf{x}}$ as $i \to \infty$. However, this cannot occur, since Lemma 9.1 states that these intersections must be monotonic. Hence the result is proved by contradiction. □

Lemma 9.3 *If $\mathbf{x}_1 \in \omega(\mathbf{x}_0)$ is not an equilibrium point and lies on $\mathbf{x}(t; \mathbf{x}_0)$, the integral path through $\mathbf{x}_0$, then $\mathbf{x}(t; \mathbf{x}_1)$, the integral path through $\mathbf{x}_1$, is a closed curve, also passing through $\mathbf{x}_0$.*

Proof Since $\mathbf{x}_1$ lies in $\mathbf{x}(t; \mathbf{x}_0)$, $\omega(\mathbf{x}_1) = \omega(\mathbf{x}_0)$, and hence $\mathbf{x}_1 \in \omega(\mathbf{x}_1)$. Let L be a curve through $\mathbf{x}_1$ transverse to the integral paths. By Lemma 9.2, $\mathbf{x}(t; \mathbf{x}_1)$ can only meet L once, and hence is a closed curve. □

We can now finally prove Theorem 9.4, the Poincaré–Bendixson theorem.

Proof Let $\mathbf{x}_0$ be a point in I, and hence $\mathbf{x}(t; \mathbf{x}_0) \subset I$ and $\omega(\mathbf{x}_0) \subset I$. Choose $\mathbf{x}_1 \in \omega(\mathbf{x}_0)$.

If $\mathbf{x}_1 \in \mathbf{x}(t; \mathbf{x}_0)$, then, by Lemma 9.3, the integral path through $\mathbf{x}_1$ is a closed curve and, since there are no equilibrium points in I, must be a limit cycle.

If $\mathbf{x}_1 \notin \mathbf{x}(t; \mathbf{x}_0)$, let $\mathbf{x}_2 \in \omega(\mathbf{x}_1)$. Let L be a transversal through $\mathbf{x}_2$. Since $\mathbf{x}_2 \in \omega(\mathbf{x}_1)$, the integral path though $\mathbf{x}_1$ must intersect L at a monotonic sequence of points $\mathbf{x}_{1i}$, such that $\mathbf{x}_{1i} \to \mathbf{x}_2$ as $i \to \infty$. But $\mathbf{x}_{1i} \in \omega(\mathbf{x}_0)$, so the integral path through $\mathbf{x}_0$ must pass arbitrarily close to each of the points $\mathbf{x}_{1i}$ as $t \to \infty$. However, the intersections of the integral path through $\mathbf{x}_0$ with L should be monotonic, by Lemma 9.1, and we conclude that $\mathbf{x}_{1i} = \mathbf{x}_2$, and hence that $\omega(\mathbf{x}_1)$ is the closed limit cycle through $\mathbf{x}_1$. □

Example

Consider the system

$$\dot{x} = x - y - 2x(x^2 + y^2), \quad \dot{y} = x + y - y(x^2 + y^2). \tag{9.21}$$

In order to analyze these equations, it is convenient to write them in terms of polar coordinates, (r, θ). From the definitions, $r = (x^2 + y^2)^{1/2}$ and $x = r\cos\theta$, $y = r\sin\theta$, the chain rule gives

$$\dot{r} = (x\dot{x} + y\dot{y})(x^2 + y^2)^{-1/2} = \cos\theta\, \dot{x} + \sin\theta\, \dot{y}. \tag{9.22}$$

For (9.21),

$$\dot{r} = \cos\theta\,(r\cos\theta - r\sin\theta - 2r^3\cos\theta) + \sin\theta\,(r\cos\theta + r\sin\theta - r^3\sin\theta)$$

$$= r - r^3 \left(1 + \cos^2\theta\right).$$

Since $0 \leqslant \cos^2\theta \leqslant 1$, we conclude that

$$r(1 - 2r^2) \leqslant \dot{r} \leqslant r(1 - r^2).$$

Therefore $\dot{r} > 0$ for $0 < r < 1/\sqrt{2}$ and $\dot{r} < 0$ for $r > 1$. Remembering that if $\dot{r} > 0$, r is an increasing function of t, and therefore integral paths are directed away from the origin, we can define a closed, bounded, annular region $D = \{(r, \theta) \mid r_0 \leqslant r \leqslant r_1\}$ with $0 < r_0 < 1/\sqrt{2}$ and $r_1 > 1$, such that all integral paths enter the region I, as shown in Figure 9.14. If we can show that there are no equilibrium points in I, the Poincaré–Bendixson theorem shows that at least one limit cycle exists in I.

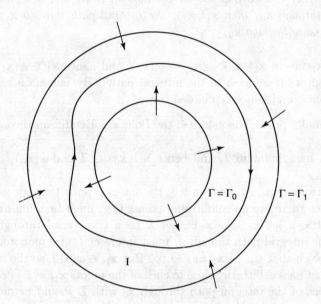

Fig. 9.14. The region I that contains a limit cycle for the system (9.21).

From the definition of $\theta = \tan^{-1}(y/x)$,

$$\dot{\theta} = \frac{1}{1 + (y/x)^2} \frac{x\dot{y} - y\dot{x}}{x^2} = \frac{1}{r}\left(\cos\theta\,\dot{y} - \sin\theta\,\dot{x}\right). \tag{9.23}$$

For (9.21),

$$\dot{\theta} = \frac{1}{r}\left\{\cos\theta\left(r\cos\theta + r\sin\theta - r^3\sin\theta\right) - \sin\theta\left(r\cos\theta - r\sin\theta - 2r^3\sin\theta\right)\right\}$$

$$= 1 + \frac{1}{2}r^2\sin 2\theta.$$

Since $-1 \leqslant \sin 2\theta \leqslant 1$, we conclude that $\dot\theta > 0$ provided $r < \sqrt{2}$, and hence that there are no equilibrium points for $0 < r < \sqrt{2}$. Therefore, provided that $1 < r_1 < \sqrt{2}$, there are no equilibrium points in I, and hence, by the Poincaré–Bendixson theorem, there is at least one limit cycle in I. Note that we know that a limit cycle must enclose at least one node, focus or centre. In this example, there is an unstable focus at $r = 0$, which is enclosed by any limit cycles, but which does not lie within I. In general, if we can construct a region in the phase plane that traps a limit cycle in this way, it cannot be simply-connected.

A corollary of the Poincaré–Bendixson theorem is that, if I is a closed, bounded, invariant region of the phase plane, and I contains no limit cycles, we can deduce that all of the integral paths that meet the boundary of I terminate at an equilibrium point contained within I. We can use similar ideas to determine the positions of the stable separatrices of P_4 for our example population dynamical system, (9.20).

Consider the region
$$R_1 = \{(x,y) \mid x \geqslant 0,\ y \geqslant 0,\ 2x+y \leqslant 2,\ 2x+2y \leqslant 3\},$$
shown in Figure 9.15. The separatrix S_1 lies in R_1 in the neighbourhood of the saddle point, P_4. No integral path enters the region R_1 through its boundaries, and we conclude that S_1 must originate at the equilibrium point at the origin, P_1.

Now consider the region
$$R_2 = \{(x,y) \mid x \geqslant 0,\ y \geqslant 0,\ 2x+y \geqslant 2,\ 2x+2y \geqslant 3,\ x^2+y^2 \leqslant r_0^2\},$$
with $r_0 > 3$, as shown in Figure 9.15. The separatrix S_2 lies in R_2 in the neighbourhood of the saddle point, P_4. No integral path enters the region R_2 through any of its straight boundaries, and we conclude that S_1 must enter R_2 through its curved boundary, $x^2+y^2 = r_0^2$. Since we can make r_0 arbitrarily large, we conclude that S_2 originates at infinity.

9.3.8 The Phase Portrait at Infinity

In order to study the behaviour of a second order system far from the origin, it is often useful to map the phase plane to a new coordinate system where the point at infinity is mapped to a finite point. There are several ways of doing this, but the most useful is the **Poincaré projection**. Consider the plane that passes through the line $x = 1$, perpendicular to the (x,y)-plane, as shown in Figure 9.16. We can set up a Cartesian coordinate system, (u,v), in this plane, with origin a unit distance above $x = 1$, $y = 0$, u-axis in the same direction as the y-axis and v-axis vertically upwards. We label the origin of the (x,y)-plane as O, and denote the point a unit distance above the origin as O'. We now consider a point B in the (x,y)-plane. The Poincaré projection maps the point B to the point C where the line $O'B$ intersects the (u,v)-plane. In particular, as $x \to \infty$, C approaches the u-axis, so that this projection is useful for studying the behaviour of a second order system for $x \gg 1$.

In order to relate u and v to x and y, we firstly consider the triangles OAB and

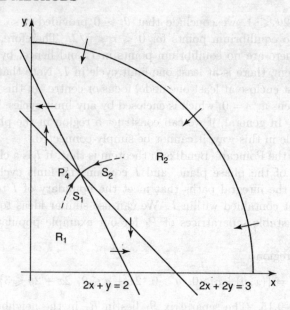

Fig. 9.15. The regions R_1 and R_2 for the population dynamical system (9.20).

$OA'B'$, shown in Figure 9.16, which are similar. This shows that $u = y/x$ and $\sqrt{x^2 + y^2} - m = \sqrt{x^2 + y^2}/x$. Secondly, we consider the similar triangles $BB'C$ and BOO', which show that $(1+v)/m = 1/\sqrt{x^2 + y^2}$. By eliminating m, we find that $v = -1/x$. The simple change of variables $u = y/x$, $v = -1/x$, and hence

$$\dot{u} = uv\dot{x} - v\dot{y}, \quad \dot{v} = v^2\dot{x}, \tag{9.24}$$

is therefore a Poincaré projection.

If we apply this to the population dynamics example (9.20), we find that

$$\dot{u} = -u\left(1 + \frac{u}{v}\right), \quad \dot{v} = -(2u + 3v + 2). \tag{9.25}$$

This has finite equilibrium points at $u = 0$, $v = -2/3$, which corresponds to P_3, and $u = 2$, $v = -2$, which corresponds to P_4. The nature of these equilibrium points remains the same after the projection, so that P_3 is a stable node and P_4 is a saddle point. In order to determine the behaviour of (9.20) for $x \gg 1$, we need to consider (9.25) close to the u-axis for $v < 0$ and $u > 0$. When $-v \ll 1$, $\dot{u} \gg 1$, whilst $\dot{v} < 0$. Integral paths close to the u-axis therefore start parallel to it, but then move away. Figure 9.17 is a sketch of the phase portrait in the (u, v)-plane. We conclude that the integral paths of (9.20) for $x \gg 1$ lead into the finite (x, y)-plane as sketched in Figure 9.9(b). The analogous transformation, $u = x/y$, $v = -1/y$ is also a Poincaré projection, and can be used to examine the behaviour for $y \gg 1$.

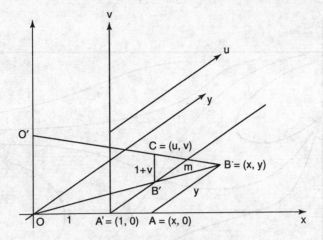

Fig. 9.16. A Poincaré projection.

We have now answered all of the questions concerning the phase portrait of (9.20) that we asked at the end of Section 9.3.4.

9.3.9 A Final Example: Hamiltonian Systems

Let Ω be a region of the (x,y)-plane and $H : \Omega \to \mathbb{R}$ be a real-valued, continuously differentiable function defined on Ω. The two-dimensional system of ordinary differential equations,

$$\dot{x} = H_y(x,y), \quad \dot{y} = -H_x(x,y), \tag{9.26}$$

is called a **Hamiltonian system** and $H(x,y)$ is called the **Hamiltonian**. Such systems occur frequently in mechanics. One example is the simple pendulum, which we studied earlier. As we have seen, this has $\dot{x} = y = H_y$, $\dot{y} = -\omega^2 \sin x = -H_x$, so that $H = \frac{1}{2}y^2 - \omega^2 \cos x$, the total energy, is a Hamiltonian for the system.

Hamiltonian systems have several general properties, which we will now investigate.

Theorem 9.5 *The integral paths of a Hamiltonian system are given by $H(x,y) =$ constant.*

Proof On an integral path $(x(t), y(t))$, $H = H(x(t), y(t))$ and

$$\frac{dH}{dt} = H_x \dot{x} + H_y \dot{y} = H_x H_y - H_y H_x = 0,$$

so that H is constant. □

248 PHASE PLANE METHODS

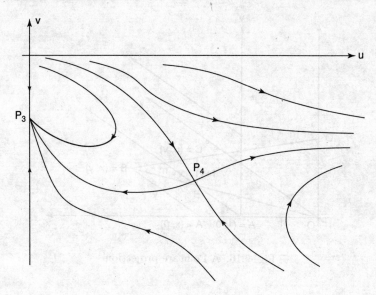

Fig. 9.17. A sketch of the phase portrait of the system of equations (9.25).

For the simple pendulum, the integral paths are therefore given by the curves $y^2 - 2\omega^2 \cos x = $ constant.

Theorem 9.6 *If* $\mathbf{x} = \mathbf{x}_e$ *is an equilibrium point of a Hamiltonian system with nonzero eigenvalues, then it is either a saddle point or a centre.*

Proof The Jacobian of (9.26) at $\mathbf{x} = \mathbf{x}_e$ is

$$J = \begin{pmatrix} H_{yx} & H_{yy} \\ -H_{xx} & -H_{xy} \end{pmatrix}\bigg|_{\mathbf{x}=\mathbf{x}_e}$$

Since $H_{xy} = H_{yx}$, the eigenvalues of J satisfy

$$\lambda^2 = H_{xy}^2(x_e, y_e) - H_{xx}(x_e, y_e)H_{yy}(x_e, y_e).$$

Note that $H_{xy}^2 \neq H_{xx}H_{yy}$ at $\mathbf{x} = \mathbf{x}_e$ since the eigenvalues are nonzero. If $H_{xy}^2 > H_{xx}H_{yy}$ at $\mathbf{x} = \mathbf{x}_e$, there is one positive and one negative eigenvalue, so $\mathbf{x}_e$ is a saddle point. If $H_{xy}^2 < H_{xx}H_{yy}$ at $\mathbf{x} = \mathbf{x}_e$, there are two complex conjugate imaginary eigenvalues. Since the integral paths are given by $H = $ constant and the conditions $H_x = H_y = 0$ and $H_{xx}H_{yy} > H_{xy}^2$ are those for a local maximum or minimum at $\mathbf{x} = \mathbf{x}_e$, the level curves of H are closed and surround $\mathbf{x} = \mathbf{x}_e$. We conclude that the equilibrium point is a nonlinear centre. □

Theorem 9.7 (Liouville's theorem) *Hamiltonian systems are area-preserving.*

Proof Consider a small triangle with one vertex at $\mathbf{x}_0$, the other two vertices at $\mathbf{x}_0 + \delta\mathbf{x}_1$ and $\mathbf{x}_0 + \delta\mathbf{x}_2$, and $|\delta\mathbf{x}_1|, |\delta\mathbf{x}_2| \ll 1$. The area of this triangle is $|\mathbf{A}|$, where

$$\mathbf{A} = \frac{1}{2} (\delta\mathbf{x}_1 \times \delta\mathbf{x}_2).$$

If all of the vertices move along integral paths of the Hamiltonian system (9.26), a Taylor expansion shows that

$$\frac{d}{dt}\delta\mathbf{x}_i \sim (\delta x_i H_{yx}(\mathbf{x}_0) + \delta y_i H_{yy}(\mathbf{x}_0), -\delta x_i H_{x\dot{x}}(\mathbf{x}_0) - \delta y_i H_{xy}(\mathbf{x}_0)),$$

for $i = 1, 2$, where $\delta\mathbf{x}_i = (\delta x_i, \delta y_i)$. Up to $O(|\delta\mathbf{x}_i|)$, we therefore have

$$2\frac{d\mathbf{A}}{dt} = \frac{d}{dt}\delta\mathbf{x}_1 \times \delta\mathbf{x}_2 + \delta\mathbf{x}_1 \times \frac{d}{dt}\delta\mathbf{x}_2$$

$$= \delta y_2 \{\delta x_1 H_{y\dot{x}}(\mathbf{x}_0) + \delta y_1 H_{yy}(\mathbf{x}_0)\} + \delta x_2 \{\delta x_1 H_{xx}(\mathbf{x}_0) + \delta y_1 H_{xy}(\mathbf{x}_0)\}$$

$$- \delta y_1 \{\delta x_2 H_{yx}(\mathbf{x}_0) + \delta y_2 H_{yy}(\mathbf{x}_0)\} - \delta x_1 \{\delta x_2 H_{xx}(\mathbf{x}_0) + \delta y_2 H_{xy}(\mathbf{x}_0)\} = 0,$$

so the area of the triangle is unchanged under the action of a Hamiltonian system. Since any area in the phase plane can be broken up into infinitesimal triangles, the Hamiltonian system is area-preserving. □

9.4 Third Order Autonomous Nonlinear Ordinary Differential Equations

The solutions of the third order system

$$\dot{x} = X(x, y, z), \quad \dot{y} = Y(x, y, z), \quad \dot{z} = Z(x, y, z), \tag{9.27}$$

can be analyzed in terms of integral paths in a three-dimensional, (x, y, z)-**phase space**. However, most of the useful results that hold for the phase plane do not hold in three or more dimensions. The crucial difference is that, in three or more dimensions, closed curves no longer divide the phase space into two distinct regions, inside and outside the curve. For example, integral paths inside a limit cycle in the phase plane are trapped there, but this is not the case in a three-dimensional phase space. There is no analogue of the Poincaré index or Bendixson's negative criterion, nor, as we noted earlier, is there an analogue of the Poincaré–Bendixson theorem. In third or higher order systems, integral paths can be attracted to **strange** or **chaotic attractors**, which have **fractal** or **noninteger dimensions**, and represent **chaotic solutions**. A simple way to get a grasp of this is to remember that cars, which drive about on a plane, often hit each other, but aircraft, which have an extra dimension to use, do so more rarely. We will examine some elementary techniques for studying chaotic solutions in Chapter 15. There are also some interesting, and useful, conservative third order systems for which a more straightforward analysis is possible (see also Exercise 13.4).

PHASE PLANE METHODS

Example

Consider the third order system

$$\dot{x} = yz, \quad \dot{y} = -xz, \quad \dot{z} = -k^2 xy. \tag{9.28}$$

In this rather degenerate system, any point on the x-, y- or z-axis is an equilibrium point. We can also see that

$$\frac{d}{dt}(x^2 + y^2) = 2x\dot{x} + 2y\dot{y} = 2xyz - 2yxz = 0.$$

Therefore, $x^2 + y^2$ is constant on any integral path, and hence all integral paths lie on the surface of a cylinder with its axis pointing in the z-direction.

Consider the integral paths on the surface of the cylinder $x^2 + y^2 = 1$. This surface is two-dimensional, so we should be able to analyze the behaviour of integral paths on it using phase plane techniques. There are equilibrium points at $(0, \pm 1, 0)$ and $(\pm 1, 0, 0)$ and the Jacobian matrix is

$$J = \begin{pmatrix} X_x & X_y & X_z \\ Y_x & Y_y & Y_z \\ Z_x & Z_y & Z_z \end{pmatrix} = \begin{pmatrix} 0 & z & y \\ -z & 0 & -x \\ -k^2 y & -k^2 x & 0 \end{pmatrix}.$$

This gives

$$J(\pm 1, 0, 0) = \begin{pmatrix} 0 & 0 & 0 \\ 0 & 0 & \mp 1 \\ 0 & \mp k^2 & 0 \end{pmatrix}, \quad J(0, \pm 1, 0) = \begin{pmatrix} 0 & 0 & \pm 1 \\ 0 & 0 & 0 \\ \mp k^2 & 0 & 0 \end{pmatrix}.$$

The points $(\pm 1, 0, 0)$ each have eigenvalues $\lambda = 0, \pm k$. The zero eigenvalue, with eigenvector $(0, 1, 0)^T$, corresponds to the fact that the y-axis is completely made up of equilibrium points. The remaining two eigenvalues are real and of opposite sign, and control the dynamics on the cylinder $x^2 + y^2 = 1$, where the equilibrium points are saddles. Similarly, the points $(0, \pm 1, 0)$ each have eigenvalues $\lambda = 0, \pm ik$, and are therefore linear centres on $x^2 + y^2 = 1$. These remain centres when nonlinear terms are taken into account, using the argument that we described earlier for the simple pendulum, since the system is unchanged by the transformation $z \mapsto -z$, $t \mapsto -t$. The phase portrait is sketched in Figure 9.18.

We can confirm that this phase portrait is correct by noting that this system actually has two other conserved quantities. From (9.28),

$$\frac{d}{dt}(k^2 y^2 - z^2) = \frac{d}{dt}(k^2 x^2 + z^2) = 0,$$

and hence $k^2 y^2 - z^2$ and $k^2 x^2 + z^2$ are constant on any integral path. Integral paths therefore lie on the intersection of the circular cylinder $x^2 + y^2 = $ constant, the hyperboloidal cylinder $k^2 y^2 - z^2 = $ constant, and the elliptical cylinder $k^2 x^2 + z^2 = $ constant. This is precisely what the phase portrait in Figure 9.18 shows.

Finally, consider the integral path with $x = 0$, $y = z = 1$ when $t = 0$. On this integral path, $k^2 x^2 + z^2 = 1$ and $x^2 + y^2 = 1$, so that

$$\dot{x} = \sqrt{1 - x^2}\sqrt{1 - k^2 x^2},$$

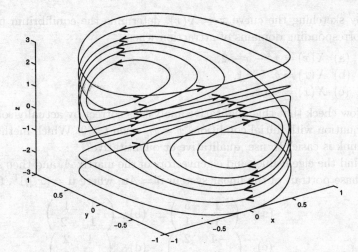

Fig. 9.18. The phase portrait of the third order system (9.28) on the surface of the cylinder $x^2 + y^2 = 1$ when $k = 2$.

which gives†

$$t = \int_0^x \frac{ds}{\sqrt{1-s^2}\sqrt{1-k^2 s^2}}.$$

This is the definition of the **Jacobian elliptic function** $\text{sn}(t; k)$. On this integral path y and z are also Jacobian elliptic functions, $y = \text{cn}(t; k)$ and $z = \text{dn}(t; k)$. The phase portrait that we have just determined now allows us to see qualitatively that these elliptic functions are periodic with t, provided that $k \neq 1$. In Sections 12.2.3 and 12.2.4 we will develop asymptotic expansions for $\text{sn}(t; k)$, firstly when k is close to unity, and secondly when $k \ll 1$. The Jacobian elliptic functions will also prove to be useful in Section 12.2.5.

Exercises

9.1   Consider the second order, autonomous ordinary differential equation

$$\ddot{x} = 3x^2 - 1,$$

where a dot represents d/dt. By integrating this equation once, obtain a relation between $\dot{x}$ and x. Sketch the phase portrait in the $(x, \dot{x})$-phase plane. Determine the coordinates of the two equilibrium points and show that there is a homoclinic orbit associated with one of them. What types of behaviour occur inside and outside the homoclinic orbit?

† Note that the properties of the particular integral path with $x = 0$, $y = z = 1$ when $t = 0$ ensure that the arguments of the square roots remain positive.

9.2 By sketching the curve $\dot{x} = X(x)$, determine the equilibrium points and corresponding domains of attraction when

(a) $X(x) = x^2 - x - 2$,
(b) $X(x) = e^{-x} - 1$,
(c) $X(x) = \sin x$.

Now check that this qualitative analysis is correct by actually solving each equation with initial conditions $x = x_0$ when $t = 0$. Which method do you think is easier to use, qualitative or quantitative?

9.3 Find the eigenvalues and eigenvectors of the matrix A, and then sketch the phase portrait of the linear system $\dot{\mathbf{u}} = A\mathbf{u}$, where $\mathbf{u} = (x, y)^T$, for $A =$

(a) $\begin{pmatrix} 1 & -5 \\ 1 & -1 \end{pmatrix}$, (b) $\begin{pmatrix} 1 & 1 \\ 1 & -2 \end{pmatrix}$,

(c) $\begin{pmatrix} -4 & 2 \\ 3 & -2 \end{pmatrix}$, (d) $\begin{pmatrix} 1 & 2 \\ 2 & 2 \end{pmatrix}$,

(e) $\begin{pmatrix} 4 & -2 \\ 3 & -1 \end{pmatrix}$, (f) $\begin{pmatrix} 2 & 1 \\ -1 & 1 \end{pmatrix}$.

9.4 Consider a second order linear system $\dot{\mathbf{u}} = A\mathbf{u}$, when the constant matrix A has two equal eigenvalues, λ. By using the Cayley–Hamilton theorem, show that there must exist a linear transformation that takes A to either

$$\Lambda_1 = \begin{pmatrix} \lambda & 0 \\ 0 & \lambda \end{pmatrix} \text{ or } \Lambda_2 = \begin{pmatrix} \lambda & 1 \\ 0 & \lambda \end{pmatrix}.$$

Solve the linear system of equations $\dot{\mathbf{v}} = \Lambda_j \mathbf{v}$ for $j = 1$ and $j = 2$, and hence sketch the phase portrait in each case. Note that in the case $j = 1$, the equilibrium point at the origin is known as a **star**, whilst when $j = 2$ it is an **improper node**.

9.5 A certain second order autonomous system has exactly two equilibrium points, both of which are saddles. Sketch a phase portrait in which (a) a separatrix connects the saddle points, (b) no separatrix connects the saddle points.

9.6 The weight at the end of a simple pendulum experiences a frictional force proportional to its velocity. Determine the equation of motion and write it as a pair of first order equations. Show that the equilibrium points are either stable points or saddle points. Sketch the phase portrait. What happens after a long time?

9.7 Find all of the equilibrium points of each of the following systems, and determine their type. Sketch the phase portrait in each case.

(a) $\dot{x} = x - y$, $\dot{y} = x + y - 2xy^2$,
(b) $\dot{x} = -3y + xy - 10$, $\dot{y} = y^2 - x^2$,
(c) $\dot{x} = y^2 - 1$, $\dot{y} = \sin x$.

9.8 Consider the system of ordinary differential equations

$$\frac{dx}{dt} = x\left(x^2 y^2 + 1\right), \quad \frac{dy}{dt} = y\left(y - 2 - x\right).$$

Determine the position and type of each equilibrium point in the (x, y)-plane. Show that the coordinate axes are integral paths. Sketch the phase portrait, taking care to ensure that your sketch is consistent with the position of the horizontal nullcline. If $x = -1$ and $y = -1$ when $t = 0$, how does the solution behave as $t \to \infty$?

9.9 The second order system

$$\frac{dr}{dt} = r(3 - r - s), \qquad \frac{ds}{dt} = s(2 - r - s),$$

can be used to model the population of sheep (s) and rabbits (r) in a closed ecosystem. Determine the position and type of all the equilibrium points. Find the directions of the separatrices close to any saddle points. Assuming that there are no limit cycles, sketch the phase portrait for $r > 0$ and $s > 0$. Which animal becomes extinct?

9.10 Explain how the system

$$\dot{x} = x(-A + b_1 y), \qquad \dot{y} = y(B - b_2 x),$$

with the constants A, B, b_1 and b_2 positive, models the populations of a carnivorous, predator species and its herbivorous prey in a closed ecosystem. Which variable is the predator population, x or y? Determine the type of each of the equilibrium points. Determine dy/dx as a function of x and y, and integrate once to obtain an equation of the form $E(x, y) =$ constant. Show that $E(x, y)$ has a local minimum at one of the equilibrium points, and hence deduce that it is a nonlinear centre. Sketch the phase portrait. What happens to the populations?

9.11 Use the concept of the Poincaré index to determine which of the following can be surrounded by a limit cycle in the phase portrait of a second order system.

(a) an unstable node,
(b) a saddle point,
(c) two saddle points, a stable node and an unstable focus,
(d) a saddle point, an unstable focus and a stable node.

Sketch a possible phase portrait in each of the cases where a limit cycle can surround the equilibrium points.

9.12 Consider the system

$$\frac{dx}{dt} = x(x^2 + y^2 - 1), \qquad \frac{dy}{dt} = y(x^2 + y^2 - 2).$$

Show that the x- and y-axes are integral paths. Show that there are no limit cycle solutions using

(a) Dulac's extension to Bendixson's negative criterion with auxiliary function $\rho(x, y) = 1/xy$,
(b) the Poincaré index.

Sketch the phase portrait.

9.13 Use the Poincaré index, Bendixson's negative criterion or Dulac's extension as appropriate to show that the following systems have no limit cycle solutions.

(a) $\dot{x} = y$, $\quad \dot{y} = 1 + x^2 - (1-x)y$,
(b) $\dot{x} = -(1-x)^3 + xy^2$, $\quad \dot{y} = y + y^3$,
(c) $\dot{x} = 2xy + x^3$, $\quad \dot{y} = -x^2 + y - y^2 + y^3$,
(d) $\dot{x} = x$, $\quad \dot{y} = 1 + x + y^2$,
(e) $\dot{x} = 1 - x^3 + y^2$, $\quad \dot{y} = 2xy$,
(f) $\dot{x} = y + x^2$, $\quad \dot{y} = -x - y + x^2 + y^2$.

(*Hint:* For (f), use Dulac's extension to Bendixson's negative criterion with auxiliary function $\rho(x, y) = e^{ax+by}$.)

9.14 Write each of the following systems in terms of polar coordinates (r, θ), and use the Poincaré–Bendixson theorem to show that at least one limit cycle solution exists.

(a) $\dot{x} = 2x + 2y - x(2x^2 + y^2)$, $\quad \dot{y} = -2x + 2y - y(2x^2 + y^2)$,
(b) $\dot{x} = x - y - x(x^2 + \frac{3}{2}y^2)$, $\quad \dot{y} = x + y - y(x^2 + \frac{1}{2}y^2)$.

9.15 (a) Write the system
$$\frac{dx}{dt} = x - y - (2x^2 + y^2)x, \qquad \frac{dy}{dt} = x + y - (x^2 + 2y^2)y$$
in terms of polar coordinates, and then use the Poincaré–Bendixson theorem to show that there is at least one limit cycle solution.

(b) Use Dulac's extension to Bendixson's negative criterion (with an auxiliary function of the form e^{ax+by} for some suitable constants a and b) to show that there is no limit cycle solution of the system with
$$\frac{dx}{dt} = y, \qquad \frac{dy}{dt} = -x - y + x^2 + y^2.$$

9.16 (a) Write the system of ordinary differential equations
$$\frac{dx}{dt} = x - y - y^3 - 2x^5 - 2x^3y^2 - xy^4,$$
$$\frac{dy}{dt} = x + y + xy^2 - 2yx^4 - 2y^3x^2 - y^5$$
in terms of polar coordinates (r, θ), and use the Poincaré–Bendixson theorem to show that at least one limit cycle solution exists.

(b) Write the system of ordinary differential equations
$$\frac{dx}{dt} = xy - x^2y + y^3, \qquad \frac{dy}{dt} = y^2 + x^3 - xy^2$$
in terms of polar coordinates, (r, θ). Show that there is a single equilibrium point at the origin and that it is nonhyperbolic. Show that the lines $\theta = \pm\pi/4$ and $\theta = \pm 3\pi/4$ are integral paths. Show that $dr/dt = 0$ when $\theta = 0$ or π. Sketch the phase portrait.

9.17 Consider the second order system

$$\frac{dp}{ds} = -pq, \quad \frac{dq}{ds} = p + q - 2,$$

which arises in a thermal ignition problem (see Section 12.2.3). Show that there are just two finite equilibrium points, one of which is a saddle. Consider S, the stable separatrix of the saddle that lies in $p > 0$. Show that this separatrix asymptotes to the other equilibrium point as $s \to -\infty$. (Hint: First show that S must meet the p-axis with $p > 2$. Next show that S must go on to meet the p-axis with $p < 2$. Finally, use the coordinate axes and S up to its second intersection with the p-axis to construct a trapping region for S.)

9.18 **Project** A particle of mass m moves under the action of an attractive central force of magnitude $\gamma m/r^\alpha$, where (r, θ) are polar coordinates and γ and α are positive constants. By using Newton's second law of motion in polar form, show that $u = 1/r$ satisfies the equation

$$\frac{d^2 u}{d\theta^2} + u = \frac{\gamma}{h^2} u^{\alpha-2}, \tag{E9.1}$$

where h is the angular momentum of the particle.

(a) Find the equilibrium points in the $(u, du/d\theta)$-phase plane, and classify them. What feature of the linear approximation will carry over to the solutions of the full, nonlinear system?

(b) If the particle moves at relativistic speeds, it can be shown that (E9.1) is modified, in the case of the inverse square law, $\alpha = 2$, appropriate to a gravitational attraction, to

$$\frac{d^2 u}{d\theta^2} + u = \frac{\gamma}{h^2} + \epsilon u^2, \tag{E9.2}$$

where ϵ is a small positive constant, and the term ϵu^2 is called **Einstein's correction**. Find the equilibrium point that corresponds to a small perturbation of the Newtonian case ($\epsilon = 0$), and show that it is a centre.

(c) Use MATLAB to solve (E9.2) numerically, and hence draw the phase portrait for the values of ϵ, γ and h appropriate to each of the three planets nearest to the Sun (you'll need to find an appropriate astronomy book in your library), and relate what you obtain to part (b) above.

CHAPTER TEN

Group Theoretical Methods

In this chapter we will develop an approach to the solution of differential equations based on finding a **group invariant**. In order to introduce the general idea, let's begin by considering some simple, first order ordinary differential equations.

The solution of the separable equation,
$$\frac{dy}{dx} = f(x)g(y),$$
is
$$\int \frac{dy}{g(y)} = \int f(x)dx + \text{constant}.$$

Another simple class of equations, often referred to as exact equations, takes the form
$$\frac{\partial \phi}{\partial x} + \frac{\partial \phi}{\partial y}\frac{dy}{dx} = 0.$$

In order to stress the equal role played by the independent variables in this equation, we will usually write
$$\frac{\partial \phi}{\partial x}dx + \frac{\partial \phi}{\partial y}dy = 0.$$

This has the solution $\phi(x,y) = \text{constant}$.

Let's now consider the equation
$$\frac{dy}{dx} = f\left(\frac{y}{x}\right).$$

In general, this is neither separable nor exact, and we are stuck unless we can use some other property of the equation. An inspection reveals that the substitution $x = \lambda \hat{x}$, $y = \lambda \hat{y}$, where λ is any real constant, leaves the form of the equation unchanged, since
$$\frac{d\hat{y}}{d\hat{x}} = f\left(\frac{\hat{y}}{\hat{x}}\right).$$

We say that the equation is **invariant** under the transformation $x \mapsto \lambda \hat{x}$, $y \mapsto \lambda \hat{y}$. The quantity $y/x \mapsto \hat{y}/\hat{x}$ is also invariant under the transformation. If we *use this invariant quantity*, $v = y/x$, *as a new dependent variable*, the equation becomes
$$v + x\frac{dv}{dx} = f(v),$$

which is separable, with solution

$$\int \frac{dv}{f(v) - v} = \int \frac{dx}{x} + \text{constant}.$$

If we regard the parameter λ as a continuous variable, the set of transformations $x \mapsto \lambda \hat{x}$, $y \mapsto \lambda \hat{y}$ forms a **group** under composition of transformations. More specifically, this is an example of a **Lie group**, named after the mathematician Sophus Lie. We will discuss exactly what we mean by a group and a Lie group below. It is the invariance of the differential equation under the action of this group that allows us to find a solution in closed form:

In the following sections, we begin by developing as much of the theory of Lie groups as we will need, and then show how this can be used as a practical tool for the solution of differential equations.

10.1 Lie Groups

Let D be a subset of $\mathbb{R}^2$ on which $x \mapsto x_1 = f(x, y; \epsilon)$, $y \mapsto y_1 = g(x, y; \epsilon)$ is a well-defined transformation from D into $\mathbb{R}^2$. We also assume that x_1 and y_1 vary continuously with the **parameter** ϵ. This set of transformations forms a **one-parameter group**, or **Lie group**, if

(i) the transformation with $\epsilon = 0$ is the **identity** transformation, so that

$$f(x, y; 0) = x, \quad g(x, y; 0) = y,$$

(ii) the transformation with $-\epsilon$ gives the **inverse** transformation, so that, if $x_1 = f(x, y; \epsilon)$ and $y_1 = g(x, y; \epsilon)$, then $x = f(x_1, y_1; -\epsilon)$ and $y = g(x_1, y_1; -\epsilon)$,

(iii) the composition of two transformations is also a member of the set of transformations, so that if $x_1 = f(x, y; \epsilon)$, $y_1 = g(x, y; \epsilon)$, $x_2 = f(x_1, y_1; \delta)$ and $y_2 = g(x_1, y_1; \delta)$, then $x_2 = f(x, y; \epsilon + \delta)$ and $y_2 = g(x, y; \epsilon + \delta)$.

Some simple one-parameter groups are:

(a) Horizontal translation, H(ϵ): $x_1 = x + \epsilon$, $y_1 = y$,
(b) Vertical translation, V(ϵ): $x_1 = x$, $y_1 = y + \epsilon$,
(c) Magnification, M(ϵ): $x_1 = e^\epsilon x$, $y_1 = e^\epsilon y$,
(d) Rotation, R(ϵ): $x_1 = x \cos \epsilon - y \sin \epsilon$, $y_1 = x \sin \epsilon + y \cos \epsilon$.

For example, to show that the set of transformations M(ϵ) forms a group, firstly note that when $\epsilon = 0$, $x_1 = x$ and $y_1 = y$. Secondly, a simple rearrangement gives $x = e^{-\epsilon} x_1$, $y = e^{-\epsilon} y_1$, so that the inverse transformation is given by M($-\epsilon$). Finally, if $x_2 = e^\delta x_1$ and $y_2 = e^\delta y_1$ then $x_2 = e^\delta \cdot e^\epsilon x = e^{\delta + \epsilon} x$, and similarly with y_2.

10.1.1 The Infinitesimal Transformation

By defining a Lie group via the transformation $x_1 = f(x,y;\epsilon)$, $y_1 = g(x,y;\epsilon)$, we are giving the **finite form** of the group. Consider what happens when $\epsilon \ll 1$. Since $\epsilon = 0$ gives the identity transformation, we can Taylor expand to obtain

$$x_1 = x + \epsilon \left(\frac{dx_1}{d\epsilon}\right)_{\epsilon=0} + \cdots, \quad y_1 = y + \epsilon \left(\frac{dy_1}{d\epsilon}\right)_{\epsilon=0} + \cdots.$$

If we now introduce the functions

$$\xi(x,y) = \left(\frac{dx_1}{d\epsilon}\right)_{\epsilon=0}, \quad \eta(x,y) = \left(\frac{dy_1}{d\epsilon}\right)_{\epsilon=0}, \tag{10.1}$$

and just retain the first two terms in the Taylor series expansions, we obtain $x_1 = x + \epsilon \xi(x,y)$, $y_1 = y + \epsilon \eta(x,y)$. This is called the **infinitesimal form** of the group. We will show later that every one-parameter group is associated with a unique infinitesimal group.

For example, the transformation $x_1 = x\cos\epsilon - y\sin\epsilon$, $y_1 = x\sin\epsilon + y\cos\epsilon$ forms the rotation group R(ϵ). When $\epsilon = 0$ this gives the identity transformation. Using the approximations $\cos\epsilon = 1 + \cdots$, $\sin\epsilon = \epsilon + \cdots$ for $\epsilon \ll 1$, we obtain the infinitesimal rotation group as $x_1 \sim x - \epsilon y$, $y_1 \sim y + \epsilon x$, and hence $\xi(x,y) = -y$ and $\eta(x,y) = x$. The transformation $x_1 = e^\epsilon x$, $y_1 = e^\epsilon y$ forms the magnification group M(ϵ). Using $e^\epsilon = 1 + \epsilon + \cdots$ for $\epsilon \ll 1$, we obtain the infinitesimal magnification group as $x_1 \sim (1+\epsilon)x$, $y_1 \sim (1+\epsilon)y$, so that $\xi(x,y) = x$ and $\eta(x,y) = y$.

We will now show that every infinitesimal transformation group is **similar**, or **isomorphic**, to a translation group. This means that, by using a change of variables, we can make any infinitesimal transformation group look like H(ϵ) or V(ϵ), which we defined earlier. Consider the equations that define ξ and η and write them in the form

$$\frac{dx_1}{d\epsilon} = \xi(x_1, y_1), \quad \frac{dy_1}{d\epsilon} = \eta(x_1, y_1),$$

a result that is correct at leading order by virtue of the infinitesimal nature of the transformation, and which we shall soon see is exact. We can also write this in the form

$$\frac{dx_1}{\xi} = \frac{dy_1}{\eta} = d\epsilon.$$

Integration of this gives solutions that are, in principle, expressible in the form $F_1(x_1, y_1) = C_1$ and $F_2(x_1, y_1) = C_2 + \epsilon$ for some constants C_1 and C_2. Since $\epsilon = 0$ corresponds to the identity transformation, we can deduce that $F_1(x_1, y_1) = F_1(x, y)$ and $F_2(x_1, y_1) = F_2(x, y) + \epsilon$. This means that if we define $u = F_1(x, y)$ and $v = F_2(x, y)$ as new variables, then the group can be represented by $u_1 = u$ and $v_1 = v + \epsilon$, so that the original group is isomorphic to the translation group V(ϵ).

10.1.2 Infinitesimal Generators and the Lie Series

Consider the change, $\delta\phi$, that occurs in a given smooth function $\phi(x,y)$ under an infinitesimal transformation. We find that

$$\delta\phi = \phi(x_1, y_1) - \phi(x, y) = \phi(x + \epsilon\xi, y + \epsilon\eta) - \phi(x, y) = \epsilon\left(\xi\frac{\partial\phi}{\partial x} + \eta\frac{\partial\phi}{\partial y}\right) + \cdots.$$

If we retain just this single term, which is consistent with the way we derived the infinitesimal transformation, we can see that $\delta\phi$ can be written in terms of the quantity

$$U\phi = \xi\frac{\partial\phi}{\partial x} + \eta\frac{\partial\phi}{\partial y}, \qquad (10.2)$$

or, in operator notation,

$$U \equiv \xi\frac{\partial}{\partial x} + \eta\frac{\partial}{\partial y}.$$

This is called the **infinitesimal generator** of the group. Any infinitesimal transformation is completely specified by $U\phi$. For example, if

$$U\phi = -y\frac{\partial\phi}{\partial x} + x\frac{\partial\phi}{\partial y},$$

$\xi(x, y) = -y$ and $\eta(x, y) = x$, so that the transformation is given by $x_1 = x - \epsilon y$, $y_1 = y + \epsilon x$. From the definition (10.2), $Ux = \xi$ and $Uy = \eta$, so that

$$U\phi = Ux\frac{\partial\phi}{\partial x} + Uy\frac{\partial\phi}{\partial y},$$

and if a group acts on (x, y) to produce new values (x_1, y_1) then

$$U\phi(x_1, y_1) = Ux_1\frac{\partial\phi}{\partial x_1} + Uy_1\frac{\partial\phi}{\partial y_1}.$$

Let's now consider a group defined in finite form by $x_1 = f(x, y; \epsilon)$, $y_1 = g(x, y; \epsilon)$ and a function $\phi = \phi(x, y)$. If we regard $\phi(x_1, y_1; \epsilon)$ as a function of ϵ, with a prime denoting $d/d\epsilon$, we find that

$$\phi(x_1, y_1; \epsilon) = \phi(x_1, y_1; 0) + \epsilon\phi'(x_1, y_1; 0) + \frac{1}{2}\epsilon^2\phi''(x_1, y_1; 0) + \cdots.$$

Since

$$\phi(x_1, y_1; 0) = \phi(x, y),$$

$$\phi'(x_1, y_1; 0) = \left.\frac{d\phi}{d\epsilon}\right|_{\epsilon=0} = \left(\frac{\partial\phi}{\partial x_1}\frac{dx_1}{d\epsilon} + \frac{\partial\phi}{\partial y_1}\frac{dy_1}{d\epsilon}\right)_{\epsilon=0}$$

$$= \left(\xi\left.\frac{\partial\phi}{\partial x_1}\right|_{\epsilon=0} + \eta\left.\frac{\partial\phi}{\partial y_1}\right|_{\epsilon=0}\right) = \xi\frac{\partial\phi}{\partial x} + \eta\frac{\partial\phi}{\partial y} = U\phi,$$

and

$$\phi''(x_1, y_1; 0) = \left.\frac{d}{d\epsilon}\left(\frac{d\phi}{d\epsilon}\right)\right|_{\epsilon=0} = U^2\phi,$$

we have
$$\phi(x_1, y_1; \epsilon) = \phi(x, y, 0) + \epsilon U\phi + \frac{1}{2}\epsilon^2 U^2 \phi + \cdots .$$

This is known as a **Lie series**, and can be written more compactly in operator form as
$$\phi(x_1, y_1; \epsilon) = e^{\epsilon U} \phi(x, y).$$

In particular, if we take $\phi(x, y, 0) = x$,
$$x_1 = x + \epsilon U x + \frac{1}{2}\epsilon^2 U^2 x + \cdots = x + \epsilon \xi + \frac{1}{2}\epsilon^2 U \xi + \cdots$$
$$= x + \epsilon \xi + \frac{1}{2}\epsilon^2 \left(\xi \frac{\partial \xi}{\partial x} + \eta \frac{\partial \xi}{\partial y} \right) + \cdots .$$

Similarly,
$$y_1 = y + \epsilon \eta + \frac{1}{2}\epsilon^2 \left(\xi \frac{\partial \eta}{\partial x} + \eta \frac{\partial \eta}{\partial y} \right) + \cdots .$$

These two relations are a representation of the group in finite form. It should now be clear that we can calculate the finite form of the group from the infinitesimal group (via the Lie series) and the infinitesimal group from the finite form of the group (via expansions for small ϵ). For example, if an infinitesimal group is represented by
$$U\phi = x \frac{\partial \phi}{\partial x} + y \frac{\partial \phi}{\partial y},$$
then
$$x_1 = x + \epsilon x + \frac{1}{2!}\epsilon^2 x + \frac{1}{3!}\epsilon^3 x + \cdots = xe^\epsilon,$$
$$y_1 = y + \epsilon y + \frac{1}{2!}\epsilon^2 y + \frac{1}{3!}\epsilon^3 y + \cdots = ye^\epsilon.$$

The finite form is therefore M(ϵ), the magnification group.

As a further example, if an infinitesimal group is represented by
$$U\phi = -y \frac{\partial \phi}{\partial x} + x \frac{\partial \phi}{\partial y},$$
then
$$Ux = -y, \quad Uy = x, \quad U^2 x = -x, \quad U^2 y = -y,$$
$$U^3 x = y, \quad U^3 y = -x, \quad U^4 x = x, \quad U^4 y = y.$$

U is therefore a cyclic operation with period 4 and the equations of the finite form of the group are
$$x_1 = x - \epsilon y - \frac{1}{2!}\epsilon^2 x + \frac{1}{3!}\epsilon^3 y + \frac{1}{4!}\epsilon^4 x + \cdots$$

$$= x\left(1 - \frac{1}{2!}\epsilon^2 + \frac{1}{4!}\epsilon^4 - \cdots\right) - y\left(\epsilon - \frac{1}{3!}\epsilon^3 + \cdots\right) = x\cos\epsilon - y\sin\epsilon.$$

Similarly,
$$y_1 = y + \epsilon x - \frac{1}{2!}\epsilon^3 y - \frac{1}{3!}\epsilon^3 x + \frac{1}{4!}\epsilon^4 y + \cdots$$

$$= y\left(1 - \frac{1}{2!}\epsilon^2 + \frac{1}{4!}\epsilon^4 - \cdots\right) + x\left(\epsilon - \frac{1}{3!}\epsilon^3 + \cdots\right) = y\cos\epsilon + x\sin\epsilon,$$

and we have the rotation group $R(\epsilon)$.

There is a rather more concise way of doing this, using the fact that
$$\frac{dx_1}{d\epsilon} = \xi(x_1, y_1), \quad \frac{dy_1}{d\epsilon} = \eta(x_1, y_1), \quad \text{subject to } x_1 = x \text{ and } y_1 = y \text{ at } \epsilon = 0$$

is an *exact* relationship according to (10.1). For the first example above, this gives $dx_1/d\epsilon = x_1$ and $dy_1/d\epsilon = y_1$, with $x = x_1$ and $y = y_1$ at $\epsilon = 0$. This first order system can readily be integrated to give $x_1 = xe^\epsilon$ and $y_1 = ye^\epsilon$.

10.2 Invariants Under Group Action

Let $x_1 = f(x, y; \epsilon)$, $y_1 = g(x, y; \epsilon)$ be the finite form of a group and let the infinitesimal transformation associated with the group have infinitesimal generator
$$U\phi = \xi\frac{\partial\phi}{\partial x} + \eta\frac{\partial\phi}{\partial y}.$$

A function $\Omega(x, y)$ is said to be **invariant** under the action of this group if, when x_1 and y_1 are derived from x and y by the operations of the group, $\Omega(x_1, y_1) = \Omega(x, y)$. Using the Lie series, we can write

$$\Omega(x_1, y_1) = \Omega(x, y) + \epsilon U\Omega + \frac{1}{2!}\epsilon^2 U^2\Omega + \cdots = \Omega(x, y) + \epsilon U\Omega + \frac{1}{2!}\epsilon^2 U(U\Omega) + \cdots,$$

so that a necessary and sufficient condition for invariance is $U\Omega = 0$, and hence that
$$\xi\frac{\partial\Omega}{\partial x} + \eta\frac{\partial\Omega}{\partial y} = 0.$$

This is a partial differential equation for Ω, whose solution is $\Omega(x, y) = C$, a constant, on the curve
$$\frac{dx}{\xi} = \frac{dy}{\eta}. \tag{10.3}$$

Since this equation has only one solution, it follows that a one-parameter group has only one invariant.

Now let's take a point (x_0, y_0) and apply the infinitesimal transformation to it, so that it is mapped to $(x_0 + \epsilon\xi, y_0 + \epsilon\eta)$. If we repeat this procedure infinitely often, we can obtain a curve that is an integral of the differential system given by (10.3). By varying the initial point (x_0, y_0) we then obtain a family of curves, all of which are solutions of (10.3), which we denote by $\Omega_f(x, y) = C$. This family of curves is invariant under the action of the group in the sense that each curve in the family is transformed into another curve of the same family under the action of the group. Suppose (x, y) becomes (x_1, y_1) under the action of the group. This means that $\Omega_f(x_1, y_1) = $ constant must represent the same family of curves. Using the Lie series we can write

$$\Omega_f(x_1, y_1) = \Omega_f(x, y) + \epsilon U\Omega_f + \frac{1}{2!}\epsilon^2 U^2 \Omega_f + \cdots.$$

The most general condition that forces the first two terms to be constant is that $U\Omega_f = $ constant should represent one family of curves, because

$$U^n \Omega_f = U^{n-1}(U\Omega_f) = U^{n-1}(\text{constant}) = 0.$$

This is conveniently written as $U\Omega_f = F(\Omega_f)$ for some arbitrary nonzero function F.

For example, the rotation group is represented in infinitesimal form by

$$U\phi = -y\frac{\partial \phi}{\partial x} + x\frac{\partial \phi}{\partial y}.$$

The equation for the invariants of this group is

$$-\frac{dx}{y} = \frac{dy}{x},$$

which can be easily integrated to give $\Omega = x^2 + y^2 = C$. This gives the intuitively obvious result that circles are invariant under rotation!

10.3 The Extended Group

If $x_1 = f(x, y; \epsilon)$ and $y_1 = g(x, y; \epsilon)$ form a group of transformations in the usual way, we can **extend** the group by regarding the differential coefficient $p = dy/dx$ as a third independent variable. Under the transformation this becomes

$$p_1 = \frac{dy_1}{dx_1} = \frac{g_x + pg_y}{f_x + pf_y} = h(x, y, p; \epsilon).$$

It can easily be verified that the triple given by (x, y, p) forms a group under the transformations above. This is known as the **extended group** of the given group. This extended group also has an infinitesimal form associated with it. If we write $x_1 = x + \epsilon\xi(x, y)$, $y_1 = y + \epsilon\eta(x, y)$, then

$$p_1 = \frac{\epsilon\eta_x + p(1 + \epsilon\eta_y)}{1 + \epsilon\xi_x + p\epsilon\xi_y}.$$

Expanding this using the binomial theorem for small ϵ, we find that

$$p_1 = p + \epsilon\left\{\eta_x + (\eta_y - \xi_x)p - \xi_y p^2\right\} = p + \epsilon\zeta. \tag{10.4}$$

The infinitesimal generator associated with this three-element group is

$$U'\phi = \xi\phi_x + \eta\phi_y + \zeta\phi_p.$$

10.4 INTEGRATION OF A FIRST ORDER EQUATION

It is of course possible, though algebraically more messy, to form further extensions of a given group by including second and higher derivatives.

We are now at the stage where we can use the theory that we have developed in this chapter to solve first order differential equations. Prior to this, it is helpful to outline two generic situations that may occur when you are confronted with a differential equation that needs to be solved.

  (i) It is straightforward to spot the group of transformations under which the equation is invariant. In this case we can give a recipe for using this invariance to solve the equation.
  (ii) No obvious group of transformations can be spotted and we need a more systematic approach to construct the group. This is called **Lie's fundamental problem** and is considerably more difficult than (i). Indeed, no general solution is known to Lie's fundamental problem for first order equations.

10.4 Integration of a First Order Equation with a Known Group Invariant

To show more explicitly that this group invariance property will lead to a more tractable differential equation than the original, let's consider a general first order ordinary differential equation, $F(x,y,p) = 0$, that is invariant under the extended group

$$U'\phi = \xi \frac{\partial \phi}{\partial x} + \eta \frac{\partial \phi}{\partial y} + \zeta \frac{\partial \phi}{\partial p}$$

derived from

$$U\phi = \xi \frac{\partial \phi}{\partial x} + \eta \frac{\partial \phi}{\partial y}.$$

We have seen that a sufficient condition for the invariance property is that $U'\phi = 0$, so we are faced with solving

$$\xi \frac{\partial \phi}{\partial x} + \eta \frac{\partial \phi}{\partial y} + \zeta \frac{\partial \phi}{\partial p} = 0.$$

The solution curves of this partial differential equation, where ϕ is constant, are the two independent solutions of the simultaneous system

$$\frac{dx}{\xi} = \frac{dy}{\eta} = \frac{dp}{\zeta}.$$

Let $u(x,y) = \alpha$ be a solution of

$$\frac{dx}{\xi} = \frac{dy}{\eta},$$

and $v(p,x,y) = \beta$ be the other independent solution.

We now show that if we know U, finding v is simply a matter of integration. To do this, recall the earlier result that any group with one parameter is similar to the translation group. Let the change of variables from (x, y) to (x_1, y_1) reduce $U\phi$ to the group of translations parallel to the y_1 axis, and call the infinitesimal generator of this group $U_1 f$. Then

$$U_1 f = U x_1 \frac{\partial f}{\partial x_1} + U y_1 \frac{\partial f}{\partial y_1} = \frac{\partial f}{\partial y_1},$$

from which we see that $U x_1 = 0$ and $U y_1 = 1$, or more explicitly,

$$\xi \frac{\partial x_1}{\partial x} + \eta \frac{\partial x_1}{\partial y} = 0, \quad \xi \frac{\partial y_1}{\partial x} + \eta \frac{\partial y_1}{\partial y} = 1.$$

The first of these equations has the solution $x_1 = u(x, y)$ and the second is equivalent to the simultaneous system

$$\frac{dx}{\xi} = \frac{dy}{\eta} = \frac{dy_1}{1}.$$

Again, one solution of this system is $u(x, y) = \alpha$. This can be used to eliminate x from the second independent solution, given by

$$\frac{dy_1}{dy} = \frac{1}{\eta(x, y)},$$

so that by a simple integration we can obtain y_1 as a function of x and y. As the extended group of translations, $U_1' f$, is identical to $U_1 f$, the most general differential equation invariant under U_1' in the new x_1, y_1 variables will therefore be a solution of the simultaneous system

$$\frac{dx_1}{0} = \frac{dy_1}{1} = \frac{dp_1}{0}.$$

This particularly simple system has solutions $x_1 = $ constant and $p_1 = $ constant, so that the differential equation can be put in the form $p_1 = F(x_1)$ for some calculable function F. In principle, it is straightforward to solve equations of this form, as they are *separable*. The solution of the original equation can then be obtained by returning to the (x, y) variables.

Example

The differential equation

$$\frac{dy}{dx} = \frac{1}{\sqrt{x + y^2}}$$

is invariant under the transformation $x = e^{-2\epsilon} x_1$, $y = e^{-\epsilon} y_1$. The infinitesimal transformation associated with this is $x_1 = x + 2\epsilon x$, $y_1 = y + \epsilon y$, so that $\xi(x, y) = 2x$ and $\eta(x, y) = y$. If we solve the system

$$\frac{dx}{\xi} = \frac{dy}{\eta},$$

we find that $y/x^{1/2} = e^C$, so that $x_1 = y/x^{1/2}$. Solving

$$\frac{dy_1}{dy} = \frac{1}{\eta} = \frac{1}{y}$$

gives $y_1 = \log y$. Some simple calculus then shows that the original differential equation transforms to

$$\frac{\frac{1}{2}x_1^3 \frac{dy_1}{dx_1}}{x_1 \frac{dy_1}{dx_1} - 1} = \frac{x_1}{\sqrt{x_1^2 + 1}},$$

which, on rearranging, gives

$$\frac{dy_1}{dx_1} = \frac{1}{x_1\left(1 - \frac{1}{2}x_1\sqrt{x_1^2+1}\right)}.$$

This is the separable equation that we are promised by the theory we have developed. The final integration of the equation can be achieved by the successive substitutions $z = \log x_1$ and $t^2 = 1 + e^{-2z}$.

10.5 Towards the Systematic Determination of Groups Under Which a First Order Equation is Invariant

If we consider a differential equation in the form

$$\frac{dy}{dx} = F(x, y)$$

and an infinitesimal transformation of the form $x_1 \sim x + \epsilon\xi(x,y)$, $y_1 \sim y + \epsilon\eta(x,y)$, (10.4) shows that

$$\frac{dy_1}{dx_1} = \frac{dy}{dx} + \epsilon\left\{\eta_x + (\eta_y - \xi_x)\frac{dy}{dx} - \xi_y\left(\frac{dy}{dx}\right)^2\right\} + \cdots.$$

Using the differential equation to eliminate dy/dx, we find that the equation will be invariant under the action of the group provided that

$$\xi\frac{\partial F}{\partial x} + \eta\frac{\partial F}{\partial y} = \eta_x + (\eta_y - \xi_x)F - \xi_y F^2. \tag{10.5}$$

So, given the function F, the fundamental problem is to determine two functions ξ and η that satisfy this first order partial differential equation. Of course, this is an underdetermined problem and has no unique solution. However, by choosing a special form for *either* ξ *or* η there are occasions when the process works, as the following example shows.

Example

Consider the equation

$$\frac{dy}{dx} = \frac{y}{x + x^2 + y^2}. \tag{10.6}$$

In this case,

$$\frac{\partial F}{\partial x} = \frac{-(1 + 2x)y}{(x + x^2 + y^2)^2}, \quad \frac{\partial F}{\partial y} = \frac{x + x^2 - y^2}{(x + x^2 + y^2)^2},$$

and (10.5) takes the form
$$(x + x^2 - y^2)\eta - (1 + 2x)y\xi = (x + x^2 + y^2)^2 \eta_x + (\eta_x - \xi_x)y(x + x^2 + y^2) - y^2 \xi_y.$$
If now we choose $\eta = 1$, this reduces to
$$(x + x^2 - y^2) + (1 + 2x)y\xi = y(x + x^2 + y^2)\xi_x + y^2 \xi_y.$$
It is not easy to solve even this equation in general, but after some trial and error, we can find the solution $\xi = x/y$. The infinitesimal transformation in this case is $x_1 = x + \epsilon x/y$, $y_1 = y + \epsilon$ and, following the procedure outlined in the last section, we now solve
$$\frac{dx}{x/y} = \frac{dy}{1}$$
to obtain $y/x =$ constant. We therefore take $x_1 = y/x$ and hence $y_1 = y$ as our new variables. In terms of these variables, (10.6) becomes
$$\frac{dy_1}{dx_1} = -\frac{1}{(1 + x_1^2)},$$
with solution $y_1 = -\tan^{-1} x_1 + C$. The solution of our original differential equation can therefore be written in the form $y + \tan^{-1}(y/x) = C$.

10.6 Invariants for Second Order Differential Equations

First order differential equations can be invariant under an infinite number of one-parameter groups. Second order differential equations can only be invariant under at most eight groups. To see where this figure comes from, let's consider the simplest form of a variable coefficient, linear, second order, differential equation,
$$\frac{d^2y}{dx^2} + q(x)y = 0. \tag{10.7}$$
Writing $x_1 = x + \epsilon\xi$ and $y_1 = y + \epsilon\eta$ we have already shown that
$$\frac{dy_1}{dx_1} = \frac{dy}{dx} + \epsilon\left\{\eta_x + (\eta_y - \xi_x)\frac{dy}{dx} - \xi_y\left(\frac{dy}{dx}\right)^2\right\}$$
$$= \frac{dy}{dx} + \epsilon\Pi\left(x, y, \frac{dy}{dx}\right) = p + \epsilon\Pi(x, y, p).$$
Now
$$\frac{d^2y_1}{dx_1^2} = \frac{d}{dx_1}\left(\frac{dy_1}{dx_1}\right) = \frac{d}{dx}\left(\frac{dy_1}{dx_1}\right)\frac{dx}{dx_1}$$
and
$$\frac{d}{dx}\left(\frac{dy_1}{dx_1}\right) = \frac{dp}{dx} + \epsilon\left(\Pi_x + \Pi_y p + \Pi_p p_x\right) + \cdots$$
$$= 1 \bigg/ \frac{dx_1}{dx} = \frac{1}{1 + \epsilon(\xi_x + \xi_y p)} + \cdots,$$

10.6 INVARIANTS FOR SECOND ORDER DIFFERENTIAL EQUATIONS

so that
$$\frac{d^2y_1}{dx_1^2} = \frac{d^2y}{dx^2} + \epsilon\left\{\Pi_x + \Pi_y\frac{dy}{dx} + \Pi_p\frac{d^2y}{dx^2} - \frac{d^2y}{dx^2}\left(\xi_x + \xi_y\frac{dy}{dx}\right)\right\} + \cdots.$$

The condition for invariance is therefore
$$\Pi_x + \Pi_y\frac{dy}{dx} + \frac{d^2y}{dx^2}\Pi_p - \frac{d^2y}{dx^2}\left(\xi_x + \xi_y\frac{dy}{dx}\right) + \xi q'(x_1)y_1 + \eta q(x_1) = 0.$$

Some simple calculation leads to
$$\{\eta_{xx} + (2\xi_x - \eta_y)q(x)y + \xi q'(x)y + q(x)\eta\} + \{2\eta_{xy} - \xi_{xx} + 3q(x)y\xi_y\}\frac{dy}{dx}$$
$$+ \{\eta_{yy} - 2\xi_{xy}\}\left(\frac{dy}{dx}\right)^2 - \xi_{yy}\left(\frac{dy}{dx}\right)^3 = 0.$$

If we now set the coefficients of powers of dy/dx to zero, we obtain
$$\xi_{yy} = 0, \quad \eta_{yy} - 2\xi_{xy} = 0, \quad 2\eta_{xy} - \xi_{xx} + 3q(x)y\xi_y = 0, \tag{10.8}$$

$$\eta_{xx} + (2\xi_x - \eta_y)q(x)y + \xi q'(x)y + q(x)\eta = 0. \tag{10.9}$$

Equation $(10.8)_1$ can be integrated to give
$$\xi = \rho(x)y + \tilde{\xi}(x).$$

Substitution of this into $(10.8)_2$ gives $\eta_{yy} = 2\rho'(x)$, which can be integrated to give
$$\eta = \rho'(x)y^2 + \tilde{\eta}(x)y + \zeta(x).$$

From $(10.8)_3$,
$$3\rho(x)q(x)y + 3\rho''(x)y + 2\tilde{\eta}'(x) - \tilde{\xi}''(x) = 0. \tag{10.10}$$

Finally, (10.9) gives
$$\{\rho'''(x)y^2 + \tilde{\eta}''(x)y + \zeta''(x)\} - q(x)y\{\tilde{\eta}(x) - 2\tilde{\xi}'(x)\} + \{\rho(x)y + \tilde{\xi}(x)\}q'(x)y$$
$$+ \{\rho'(x)y^2 + \tilde{\eta}(x)y + \zeta(x)\}q(x) = 0. \tag{10.11}$$

We can find a solution of (10.10) and (10.11) by noting that the coefficient of each power of y must be zero, which gives us four independent equations,
$$\zeta''(x) + q(x)\zeta(x) = 0, \quad \rho''(x) + q(x)\rho(x) = 0,$$

$$\tilde{\eta}''(x) - q(x)(\tilde{\eta}(x) - 2\tilde{\xi}'(x)) + \tilde{\xi}(x)q'(x) + \tilde{\eta}(x)q(x) = 0,$$

$$\tilde{\xi}''(x) = 2\tilde{\eta}'(x).$$

At this stage notice that $\rho(x)$ and $\zeta(x)$ satisfy the original second order equation, (10.7), the solution of which will involve four constants. There are also second order equations for $\tilde{\xi}(x)$ and $\tilde{\eta}(x)$, the solution of which gives rise to a further four constants. Each of these constants will generate a one-parameter group that leaves the original equation invariant, so the original equation is invariant under at most

268 GROUP THEORETICAL METHODS

eight one-parameter groups. Some of these groups are obvious. For example, since (10.7) is invariant under magnification in y, it must be invariant under $x_1 = x$, $y_1 = e^\epsilon y$.

Let's consider the group generated by $\rho(x)$ in more detail. In the usual way, we have to integrate

$$\frac{dx}{\xi(x,y)} = \frac{dy}{\eta(x,y)},$$

which leads to the equation

$$\frac{dy}{dx} = \frac{y\{\rho'(x)y + \frac{1}{2}\xi'(x) + \zeta(x)\}}{\rho(x)y + \xi(x)}.$$

This has a solution of the form $y = C\rho(x)$ provided that $\xi(x) = C\rho^2(x)$ and $\zeta(x) = 0$ for some constant C. Note that $\tilde{\eta}(x) = \frac{1}{2}\xi'(x)$ by direct integration. This means that, following the ideas of the previous section, we should define $x_1 = y/\rho(x)$. Now,

$$\frac{dy_1}{dy} = \frac{1}{\eta(x,y)} = \frac{1}{\rho'(x)y^2 + C\rho\rho'y} = \frac{1}{\rho'(x)(y^2 + C\rho y)},$$

and, since $y = C\rho(x)$,

$$\frac{dy_1}{dx} = \frac{1}{2C\rho^2},$$

which gives

$$y_1 = \frac{1}{2C}\int_{x_0}^x \frac{dx}{\rho^2} = \frac{\rho(x)}{2y}\int_{x_0}^x \frac{dt}{\rho^2(t)}.$$

We can write the differential equation in terms of these new variables by noting that

$$x_1 y_1 = \int_{x_0}^x \frac{dt}{2\rho^2(t)},$$

so that

$$\frac{d}{dx}(x_1 y_1) = \frac{1}{2\rho^2(x)}.$$

Now

$$\frac{dx_1}{d(x_1 y_1)} = \frac{dx_1}{dx}\frac{dx}{d(x_1 y_1)} = 2\rho^2\left\{\frac{1}{\rho}\frac{dy}{dx} - \frac{\rho' y}{\rho^2}\right\} = 2\left(\rho\frac{dy}{dx} - \rho' y\right),$$

which gives

$$\frac{d}{dx}\left(\frac{dx_1}{d(x_1 y_1)}\right) = 2\left\{\rho\frac{d^2y}{dx^2} + \rho'\frac{dy}{dx} - \rho'' y - \rho'\frac{dy}{dx}\right\} = 2\rho\left\{\frac{d^2y}{dx^2} + q(x)y\right\} = 0.$$

Integrating this expression gives

$$\frac{dx_1}{d(x_1 y_1)} = C_1,$$

10.6 INVARIANTS FOR SECOND ORDER DIFFERENTIAL EQUATIONS

so that $x_1 = C_1 x_1 y_1 + C_2$ or, in terms of the original variables,

$$y = C_1 \rho(x) \int_{x_0}^{x} \frac{dt}{\rho^2(t)} + C_2 \rho(x).$$

This is just the reduction of order formula, which we derived in Chapter 1. Although the invariance property produces this via a route that is rather different from that taken in Chapter 1, it is not particularly useful to us in finding solutions.

Let's now consider the invariance $x_1 = x$ and $y_1 = e^\epsilon y$, which has $\xi = 0$ and $\eta = y$. This means that

$$\frac{dx}{0} = \frac{dy}{y},$$

and we can see that this suggests using the new variables $x_1 = x$ and $y_1 = \log y$. Since $y'' = e^{y_1}(y_1'' + y_1'^2)$ under this change of variables, we obtain $y_1'' + y_1'^2 + q(x) = 0$. Putting $Y_1 = y_1'$ leads to $Y_1' + Y_1^2 + q(x) = 0$. This is a first order equation, so the group invariance property has allowed us to **reduce the order** of the original equation. As an example of this, consider the equation

$$y'' + \frac{1}{4x^2} y = 0.$$

In this case, we obtain

$$Y_1' + Y_1^2 + \frac{1}{4x^2} = 0.$$

This is a form of Ricatti's equation, which in general is difficult to solve. The exception to this is if we can spot a solution, when the equation will linearize. For this example, we can see that $Y_1 = 1/2x$ is a solution. Writing $Y_1 = 1/2x + V^{-1}$ linearizes the equation and it is straightforward to show that the general solution is $y = C_1 x^{1/2} + C_2 x^{1/2} \log x$.

At this stage, you could of course argue that you could have guessed the solution $u_1(x) = x^{1/2}$ and then reduced the order of the equation to obtain the second solution $u_2(x) = x^{1/2} \log x$. The counter argument to this is that the group theoretical method gives both the technique of reduction of order *and* an algorithm for reducing the original equation to a first order differential equation. The point can perhaps be reinforced by considering the nonautonomous equation

$$y'' + \frac{1}{x} y' + e^y = 0.$$

Using the methods derived in this chapter, we first introduce a new dependent variable, Y, defined by $y = -2 \log x + Y$. This gives us

$$Y'' + \frac{1}{x} Y' + \frac{e^Y}{x^2} = 0.$$

The invariance of this under x-magnification suggests introducing $z = \log x$ and leads to

$$\frac{d^2 Y}{dz^2} + e^Y = 0,$$

which we can immediately integrate to

$$\frac{1}{2}\left(\frac{dY}{dz}\right)^2 = C - e^Y.$$

The analysis of this equation is of course much simpler than the original one.

10.7 Partial Differential Equations

The ideas given in this chapter can be considerably extended, particularly to the area of partial differential equations. As a simple example, consider the equation for the diffusion of heat that we derived in Chapter 2, namely

$$\frac{\partial T}{\partial t} = D\frac{\partial^2 T}{\partial x^2}. \tag{10.12}$$

This equation is invariant under the group of transformations

$$x = e^\epsilon x_1, \quad t = e^{2\epsilon} t_1,$$

so that $x/t^{1/2}$ is an invariant of the transformation (can you spot which other groups it is invariant under?). If we write $\eta = x/t^{1/2}$ (here η is known as a **similarity variable**), this reduces the partial differential equation to the ordinary differential equation

$$D\frac{d^2 T}{d\eta^2} + \frac{1}{2}\eta\frac{dT}{d\eta} = 0.$$

For the initial condition $T(x,0) = H(-x)$, which is also invariant under the transformation, it is readily established from the ordinary differential equation that the solution is

$$T(x,t) = \frac{1}{\sqrt{\pi}}\int_{x/\sqrt{4Dt}}^{\infty} e^{-s^2}\,ds = 2\operatorname{erfc}\left(\frac{x}{\sqrt{4Dt}}\right), \tag{10.13}$$

which is shown in Figure 10.1.

Finally, if the initial condition is $T(x,0) = \delta(x)$, we can use the fact that (10.12) is also invariant under the **two-parameter group**

$$x = e^\epsilon x_1, \quad t = e^{2\epsilon} t_1, \quad T = e^\mu T_1.$$

The initial condition transforms to $e^\mu T_1(x_1,0) = e^{-\epsilon}\delta(x_1)$, since $\delta(ax) = \delta(x)/a$, so that the choice $\mu = -\epsilon$ makes both differential equation and initial condition invariant. As before, $x/t^{1/2}$ is an invariant, and now $t^{1/2}T$ is invariant as well. If we therefore look for a solution of the form $T(x,t) = t^{-1/2}F(x/t^{1/2})$, we find that

$$DF_{\eta\eta} + \frac{1}{2}\eta F_\eta + \frac{1}{2}F = 0.$$

A suitable solution of this is $Ae^{-\eta^2/4D}$ (for example, using the method of Frobenius), and hence

$$T(x,t) = At^{-1/2}e^{-x^2/4Dt} \tag{10.14}$$

10.7 PARTIAL DIFFERENTIAL EQUATIONS

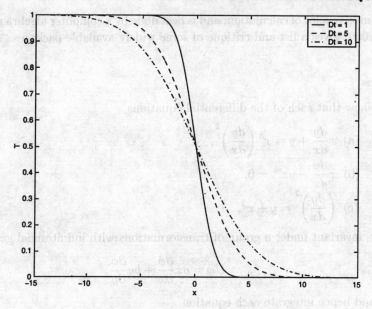

Fig. 10.1. The solution, (10.13), of the diffusion equation.

To find the constant A, we can integrate (10.12) to obtain the **integral conservation law**,

$$\frac{d}{dt}\int_{-\infty}^{\infty} T\, dx = 0.$$

Since

$$\int_{-\infty}^{\infty} T(x,0)\, dx = \int_{-\infty}^{\infty} \delta(x)\, dx = 1,$$

we must have $\int_{-\infty}^{\infty} T(x,t)\, dx = 1$. Then, from (10.14), $A = 1/\sqrt{4\pi D}$, and hence

$$T(x,t) = \frac{1}{\sqrt{4\pi Dt}} e^{-x^2/4Dt}. \tag{10.15}$$

This is known as the **point source solution** of the diffusion equation. Notice the similarity between this solution and the sequence (5.27), which we can use to *define* the Dirac delta function, as shown in Figure 5.4.

Further details of extensions to the basic method can be found in Bluman and Cole (1974) and Hydon (2000). We end this chapter with a recommendation. Given an ordinary or partial differential equation, try to find a simple invariant, such as translation, magnification or rotation. If you find one, this will lead to a reduction of order for an ordinary differential equation, or the transformation of a partial differential equation into an ordinary differential equation. If you cannot find an invariant by inspection, then you need to make a systematic search for invariants.

GROUP THEORETICAL METHODS

This can involve a lot of calculation, and is best done by a computer algebra package. Hydon (2000) gives a list and critique of some widely available packages.

Exercises

10.1 Show that each of the differential equations

(a) $x\dfrac{dy}{dx} + y = x^4 \left(\dfrac{dy}{dx}\right)^2$,

(b) $x\dfrac{dy}{dx} - yx^m = 0$,

(c) $\left(\dfrac{dy}{dx}\right)^2 = y + x^2$,

is invariant under a group of transformations with infinitesimal generator

$$U\phi = ax\dfrac{\partial \phi}{\partial x} + by\dfrac{\partial \phi}{\partial y},$$

and hence integrate each equation.

10.2 Find the general differential equation of first order, invariant under the groups

(a) $U\phi = \dfrac{\partial \phi}{\partial x} + x\dfrac{\partial \phi}{\partial y}$,

(b) $U\phi = x\dfrac{\partial \phi}{\partial x} + ay\dfrac{\partial \phi}{\partial y}$,

(c) $U\phi = y\dfrac{\partial \phi}{\partial x} + \dfrac{\partial \phi}{\partial y}$.

10.3 If the differential equation

$$\dfrac{dx}{P(x,y)} = \dfrac{dy}{Q(x,y)}$$

is invariant under a group with infinitesimal generators ξ and η, show that it has a solution in integrating factor form,

$$\int \dfrac{P\,dy - Q\,dx}{P\eta - Q\xi} = \text{constant}.$$

Hence find the solution of

$$\dfrac{dy}{dx} = \dfrac{y^4 - 2x^3 y}{2xy^3 - x^4}.$$

10.4 Show that the differential equation

$$\dfrac{d^2 y}{dx^2} + p(x) y^2 = 0$$

is invariant under at most six one-parameter groups. For the case $p(x) = x^m$, reduce the order of the equation.

10.5 Find the most general first order differential equation invariant under the group with infinitesimal generator

$$U\phi = \exp\left\{\int p(x)dx\right\} \frac{\partial \phi}{\partial y}.$$

10.6 Derive the integrating factor for $y' + P(x)y = Q(x)$ by group theoretical methods.

10.7 Find a similarity reduction of the **porous medium equation**, $u_t = (uu_x)_x$. Can you find any solutions of the resulting differential equation?

10.8 The equation $u_t = u_{xx} - u^p$, with $p \in \mathbb{R}^+$, arises in mathematical models of tumour growth, and other areas. Find some invariants of this equation, and the corresponding ordinary differential equations. Can the point source problem be resolved for this sink-like equation?

CHAPTER ELEVEN

Asymptotic Methods: Basic Ideas

The vast majority of differential equations that arise as models for real physical systems cannot be solved directly by analytical methods. Often, the only way to proceed is to use a computer to calculate an approximate, numerical solution. However, if one or more small, dimensionless parameters appear in the differential equation, it may be possible to use an **asymptotic method** to obtain an approximate solution. Moreover, the presence of a small parameter often leads to a **singular perturbation problem**, which can be difficult, if not impossible, to solve numerically.

Small, dimensionless parameters usually arise when one physical process occurs much more slowly than another, or when one geometrical length in the problem is much shorter than another. Examples occur in many different areas of applied mathematics, and we will meet several in Chapter 12. As we shall see, dimensionless parameters arise naturally when we use dimensionless variables, which we discussed at the beginning of Chapter 5. Some other examples are:

— Waves on the surface of a body of fluid or an elastic solid, with amplitude a and wavelength λ, are said to be of small amplitude if $\epsilon = a/\lambda \ll 1$. A simplification of the governing equations based on the fact that $\epsilon \ll 1$ leads to a system of linear partial differential equations (see, for example, Billingham and King, 2001). This is an example of a **regular perturbation problem**, where the problem is simplified throughout the domain of solution.

— In the high speed flow of a viscous fluid past a flat plate of length L, pressure changes due to accelerations are much greater than those due to viscous stresses, as expressed by $\mathrm{Re} = \rho U L/\mu \gg 1$. Here, Re is the Reynolds number, a dimensionless parameter that measures the ratio of acceleration to viscous forces, ρ is the fluid density, U the fluid velocity away from the plate and μ the fluid viscosity. The solution for $\mathrm{Re}^{-1} \ll 1$ has the fluid velocity equal to U everywhere except for a small neighbourhood of the plate, known as a **boundary layer**, where viscosity becomes important (see, for example, Acheson, 1990). This is an example of a singular perturbation problem, where an apparently negligible physical effect, here viscosity, becomes important in a small region.

— In aerodynamics, it is crucial to be able to calculate the lift force on the cross-section of a wing due to the flow of an inviscid fluid around it. These cross-sections are usually long and thin with aspect ratio $\epsilon = l/L \ll 1$, where l is a typical vertical width, and L a typical horizontal length. **Thin aerofoil theory** exploits the size of ϵ to greatly simplify the calculation of the lift force (see, for example, Milne-Thompson, 1960, and Van Dyke, 1964).

11.1 Asymptotic Expansions

In this chapter, we will study how to approximate various integrals as series expansions, whilst in Chapter 12, we will look at series solutions of differential equations. These series are **asymptotic expansions**. The crucial difference between asymptotic expansions and the power series that we met earlier in the book is that asymptotic expansions need not be convergent in the usual sense. We can illustrate this using an example, but firstly, there is some useful notation for comparing the sizes of functions that we will introduce here and use extensively later.

11.1.1 Gauge Functions
(i) If
$$\lim_{\epsilon \to 0} \frac{f(\epsilon)}{g(\epsilon)} = A,$$

for some nonzero constant A, we write $f(\epsilon) = O(g(\epsilon))$ for $\epsilon \ll 1$. We say that f is of order g for small ϵ. Here $g(\epsilon)$ is a **gauge function**, since it is used to gauge the size of $f(\epsilon)$. For example, when $\epsilon \ll 1$,

$$\sin \epsilon = O(\epsilon), \quad \cos \epsilon = O(1), \quad e^{-\epsilon} = O(1), \quad \cos \epsilon - 1 = O(\epsilon^2),$$

all of which can be found from the Taylor series expansions of these functions. This notation tells us nothing about the constant A. For example, $10^{10} = O(1)$. The order notation only tells us how functions behave as $\epsilon \to 0$. It is not meant to be used for comparing constants, which are all, by definition, of $O(1)$.

(ii) We also have a notation available that displays more information about the behaviour of the functions. If
$$\lim_{\epsilon \to 0} \frac{f(\epsilon)}{g(\epsilon)} = 1,$$

we write $f(\epsilon) \sim g(\epsilon)$, and say that $f(\epsilon)$ is asymptotic to $g(\epsilon)$ as $\epsilon \to 0$. For example,

$$\sin \epsilon \sim \epsilon, \quad 1 - \cos \epsilon \sim \frac{1}{2}\epsilon^2, \quad e^\epsilon - 1 \sim \epsilon, \quad \text{as } \epsilon \to 0.$$

(iii) If
$$\lim_{\epsilon \to 0} \frac{f(\epsilon)}{g(\epsilon)} = 0,$$

we write $f(\epsilon) = o(g(\epsilon))$, and say that f is much less than g. For example

$$\sin \epsilon = o(1), \quad \cos \epsilon = o(\epsilon^{-1}), \quad e^{-\epsilon} = o(\log \epsilon) \quad \text{for } \epsilon \ll 1.$$

11.1.2 Example: Series Expansions of the Exponential Integral, Ei(x)

Let's consider the exponential integral

$$\text{Ei}(x) = \int_x^\infty \frac{e^{-t}}{t} \, dt, \tag{11.1}$$

for $x > 0$. We can integrate by parts to obtain

$$\text{Ei}(x) = \left[-\frac{e^{-t}}{t}\right]_x^\infty - \int_x^\infty \frac{e^{-t}}{t^2} \, dt = \frac{e^{-x}}{x} - \int_x^\infty \frac{e^{-t}}{t^2} \, dt.$$

Doing this repeatedly leads to

$$\text{Ei}(x) = e^{-x} \sum_{m=1}^N (-1)^{m-1} \frac{(m-1)!}{x^m} + R_N, \tag{11.2}$$

where

$$R_N = (-1)^N N! \int_x^\infty \frac{e^{-t}}{t^{N+1}} \, dt. \tag{11.3}$$

This result is exact. Now, let's consider how big R_N is for $x \gg 1$. Using the fact that $x^{-(N+1)} > t^{-(N+1)}$ for $t > x$,

$$|R_N| = N! \int_x^\infty \frac{e^{-t}}{t^{N+1}} \, dt \leq \frac{N!}{x^{N+1}} \int_x^\infty e^{-t} \, dt = \frac{N! e^{-x}}{x^{N+1}},$$

and hence

$$|R_N| = O\left(e^{-x} x^{-(N+1)}\right) \quad \text{for } x \gg 1.$$

Therefore, if we truncate the series expansion (11.2) at a fixed value of N by neglecting R_N, it converges to $\text{Ei}(x)$ as $x \to \infty$. This is our first example of an **asymptotic expansion**. In common with most useful asymptotic expansions, (11.2) does *not* converge in the usual sense. The ratio of the $(N+1)^{\text{th}}$ and N^{th} terms is

$$\frac{(-1)^N N!}{x^{N+1}} \frac{x^N}{(-1)^{N-1}(N-1)!} = -\frac{N}{x},$$

which is unbounded as $N \to \infty$, so the series diverges, by the ratio test. However, for x even moderately large, (11.2) provides an extremely efficient method of calculating $\text{Ei}(x)$, as we shall see.

In order to develop the power series representation of $\text{Ei}(x)$ about $x = 0$, we would like to use the series expansion of e^{-t}. However, term by term integration is not possible as things stand, since the integrals will not be convergent. Moreover, the integral for $\text{Ei}(x)$ does not converge as $x \to 0$. We therefore have to be a little more cunning in order to obtain our power series. Firstly, note that

$$\text{Ei}(x) = \int_x^\infty \frac{1}{t}\left(e^{-t} - \frac{1}{1+t}\right) dt + \int_x^\infty \frac{1}{t(1+t)} \, dt$$

11.1 ASYMPTOTIC EXPANSIONS

$$= \int_x^\infty \frac{1}{t}\left(e^{-t} - \frac{1}{1+t}\right) dt - \log\left(\frac{x}{1+x}\right).$$

With this rearrangement, we can take the limit $x \to 0$ in the first integral, with the second integral giving $\mathrm{Ei}(x) \sim -\log x$ as $x \to 0$. A further rearrangement then gives

$$\mathrm{Ei}(x) = \int_0^\infty \frac{1}{t}\left(e^{-t} - \frac{1}{1+t}\right) dt - \int_0^x \frac{1}{t}\left(e^{-t} - 1\right) dt$$

$$- \int_0^x \frac{1}{1+t} dt - \log\left(\frac{x}{1+x}\right)$$

$$= \int_0^\infty \frac{1}{t}\left(e^{-t} - \frac{1}{1+t}\right) dt - \log x - \int_0^x \frac{1}{t}\left(e^{-t} - 1\right) dt.$$

The first term is just a constant, which we could evaluate numerically, but can be shown to be equal to minus Euler's constant, $\gamma = 0.5772\ldots$, although we will not prove this here. We can now use the power series representation of e^{-t} to give

$$\mathrm{Ei}(x) = -\gamma - \log x - \int_0^x \frac{1}{t} \sum_{n=1}^\infty \frac{(-t)^n}{n!} dt.$$

Since this power series converges uniformly for all t, we can interchange the order of summation and integration, and finally arrive at

$$\mathrm{Ei}(x) = -\gamma - \log x - \sum_{n=1}^\infty \frac{(-x)^n}{n \cdot n!}. \tag{11.4}$$

This representation of $\mathrm{Ei}(x)$ is convergent for all $x > 0$.

The philosophy that we have used earlier in the book is that we can now use (11.4) to calculate the function for any given x. There are, however, practical problems associated with this. Firstly, the series converges very slowly. For example, if we take $x = 5$ we need 20 terms of the series in order to get three-figure accuracy. Secondly, even if we take enough terms in the series to get an accurate answer, the result is the difference of many large terms. For example, the largest term in the series for $\mathrm{Ei}(20)$ is approximately 2.3×10^6, whilst for $\mathrm{Ei}(40)$ this rises to approximately 3.8×10^{14}. Unless the computer used to sum the series stores many significant figures, there will be a large roundoff error involved in calculating the $\mathrm{Ei}(x)$ as the small difference of many large terms. These difficulties do not arise for the asymptotic expansion, (11.2). Consider the two-term expansion

$$\mathrm{Ei}(x) \sim \frac{e^{-x}}{x}\left(1 - \frac{1}{x}\right).$$

Figure 11.1 shows the relative error using this expansion compared with that using the first 25 terms of the convergent series representation, (11.4). The advantages of the asymptotic series over the convergent series are immediately apparent. Note that the convergent series (11.4) also provides an asymptotic expansion valid for $x \ll 1$. We conclude that asymptotic expansions can also be convergent series.

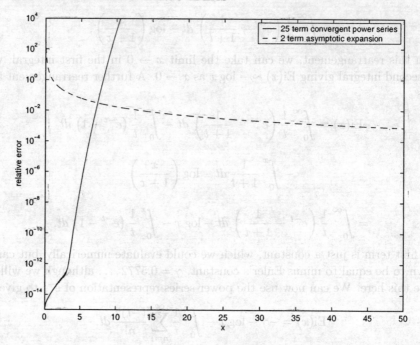

Fig. 11.1. The relative error in calculating Ei(x) using the first 25 terms of the convergent series expansion (11.4), solid line, and the two-term asymptotic expansion (11.2), dashed line.

To summarize, let

$$S_N(x) = \sum_{n=0}^{N} f_n(x).$$

— If S_N is a **convergent series** representation of $S(x)$, $S_N \to S(x)$ as $N \to \infty$ for fixed x within the radius of convergence.
— If S_N is an **asymptotic series** representation of $S(x)$, $S_N(x) \sim S(x)$ as $x \to \infty$ (or whatever limit is appropriate) for any fixed $N \geq 0$.

11.1.3 Asymptotic Sequences of Gauge Functions

Let $\delta_n(\epsilon)$, $n = 0, 1, 2, \ldots$ be a sequence of functions such that $\delta_n = o(\delta_{n-1})$ for $\epsilon \ll 1$. Such a sequence is called an **asymptotic sequence** of **gauge functions**. For example, $\delta_n = \epsilon^n$, $\delta_n = \epsilon^{n/2}$, $\delta_n = (\cot \epsilon)^{-n}$, $\delta_n = (-\log \epsilon)^{-n}$. If we can write

$$f(\epsilon) \sim \sum_{n=0}^{N} a_n \delta_n(\epsilon) \quad \text{as } \epsilon \to 0,$$

for some sequence of constants a_n, we say that $f(\epsilon)$ has an asymptotic expansion relative to the asymptotic sequence $\delta_n(\epsilon)$ for $\epsilon \ll 1$. For example, Ei(x) has an asymptotic expansion, (11.2), relative to the asymptotic sequence $x^{-n}e^{-x}$ for $x^{-1} \ll 1$. The asymptotic expansion of a function relative to a given sequence of gauge functions is unique, since each a_n can be calculated in turn from

$$a_0 = \lim_{\epsilon \to 0} \frac{f(\epsilon)}{\delta_0(\epsilon)}, \quad a_n = \lim_{\epsilon \to 0} \left\{ f(\epsilon) - \sum_{m=0}^{n-1} a_n \delta_m(\epsilon) \right\} / \delta_n(\epsilon),$$

for $n = 1, 2, \ldots$. However, a given function can have different asymptotic expansions with respect to different sequences of gauge functions. For example,

$$\text{if } \delta_n(\epsilon) = \epsilon^n, \quad \sin \epsilon = \sum_{n=1}^{\infty} (-1)^{n+1} \frac{\epsilon^{2n-1}}{(2n-1)!},$$

$$\text{if } \delta_n(\epsilon) = (1-e^{\epsilon})^n, \quad \sin \epsilon = \delta_0 + \frac{1}{2}\delta_1 + \frac{1}{3}\delta_2 + \frac{5}{4}\delta_3 + \cdots.$$

Some sequences of gauge functions are clearly easier to use than others. We also note that different functions may have the same asymptotic expansion when we consider only a finite number of terms. For example,

$$e^{\epsilon} \sim \sum_{n=0}^{N} \frac{\epsilon^n}{n!} \quad \text{as } \epsilon \to 0^+, \text{ for any } N > 0,$$

$$e^{\epsilon} + e^{-1/\epsilon} \sim \sum_{n=0}^{N} \frac{\epsilon^n}{n!} \quad \text{as } \epsilon \to 0^+, \text{ for any } N > 0.$$

Since $e^{-1/\epsilon}$ is exponentially small, it will only appear in the asymptotic expansion after all of the algebraic terms.

Now let's consider functions of a single variable x and a parameter ϵ, $f(x; \epsilon)$. We can think of this as a typical solution of an ordinary differential equation with independent variable x and a small parameter, ϵ. If f has an asymptotic expansion,

$$f(x; \epsilon) \sim \sum_{n=0}^{N} f_n(x) \delta_n(\epsilon) \quad \text{as } \epsilon \to 0,$$

that is valid for all x in some domain R, we say that the expansion is **uniformly valid** in R. If this is not the case, we say that the expansion becomes **nonuniform** in some subdomain. For example,

$$\sin(x+\epsilon) = \sin x + \epsilon \cos x - \frac{1}{2!}\epsilon^2 \sin x - \frac{1}{3!}\epsilon^3 \cos x + O(\epsilon^4) \quad \text{as } \epsilon \to 0,$$

for all x, so the expansion is uniformly valid. Now consider

$$\sqrt{x+\epsilon} = \sqrt{x}\left(1+\frac{\epsilon}{x}\right)^{1/2} \sim \sqrt{x}\left\{1 + \frac{\epsilon}{2x} - \frac{\epsilon^2}{8x^2} + O(\epsilon^3)\right\} \quad \text{as } \epsilon \to 0,$$

provided $x \gg \epsilon$. We say that the expansion becomes nonuniform as $x \to 0$, when $x = O(\epsilon)$. Note that each successive term is smaller by a factor of ϵ/x, and therefore the expansion fails to be asymptotic when $x = O(\epsilon)$. To determine an asymptotic expansion valid when $x = O(\epsilon)$, we define a new, **scaled variable**, a procedure that we will use again and again later, given by $x = \epsilon X$, with $X = O(1)$ as $\epsilon \to 0$. In terms of X,

$$\sqrt{x + \epsilon} = \epsilon^{1/2}\sqrt{X + 1}.$$

This is the trivial asymptotic expansion valid for $X = O(1)$†.

11.2 The Asymptotic Evaluation of Integrals

We have already seen in Chapters 5 and 6 that the solution of a differential equation can often be found in closed form in terms of an integral. Instead of trying to develop the asymptotic solution of a differential equation as some parameter or variable becomes small, it is often more convenient to find an integral expression for the solution and then seek an asymptotic expansion of the integral. Let's consider integrals of the type

$$I(\lambda) = \int_C e^{\lambda f(t)} g(t)\, dt, \qquad (11.5)$$

with f and g analytic functions of t, λ real and positive and C a contour that joins two distinct points in the complex t-plane, one or both of which may be at infinity. Such integrals arise very commonly in this context. For example, the Fourier transform, which we studied in Chapter 5, takes this form with $f(t) = it$, whilst the Laplace inversion integral, which we discussed in Section 6.4, has $f(t) = t$.

Another common example is the solution of Airy's equation, $y'' - xy = 0$, which we wrote in terms of Bessel functions in Section 3.8. The solution $y = \text{Ai}(x)$ can also be written in integral form as

$$\text{Ai}(x) = \frac{1}{2\pi i} \int_C e^{xt - \frac{1}{3}t^3}\, dt. \qquad (11.6)$$

To see that this is a solution of Airy's equation, note that

$$\text{Ai}'' - x\text{Ai} = \frac{1}{2\pi i} \int_C (t^2 - x) e^{xt - \frac{1}{3}t^3}\, dt = -\frac{1}{2\pi i} \int_C \frac{d}{dt}\left(e^{xt - \frac{1}{3}t^3}\right) dt = 0,$$

provided that $e^{xt - \frac{1}{3}t^3} \to 0$ as $|t| \to \infty$ on the contour C. For $|t| \gg 1$, $xt - \frac{1}{3}t^3 \sim -\frac{1}{3}t^3$, so we need $\text{Re}(t^3) > 0$ on the contour C, and therefore $-\frac{\pi}{6} < \arg(t) < \frac{\pi}{6}$, $\frac{\pi}{2} < \arg(t) < \frac{5\pi}{6}$ or $-\frac{\pi}{2} > \arg(t) > -\frac{5\pi}{6}$. As we shall see later, $\text{Ai}(x)$ is distinguished from $\text{Bi}(x)$ in that $\text{Ai}(x) \to 0$ as $x \to \infty$, in particular, with $\text{Ai}(x) \sim x^{-1/4} \exp\left(-\frac{2}{3}x^{3/2}\right)/2\sqrt{\pi}$. This means that an appropriate contour C originates with $\arg(t) = -\frac{2\pi}{3}$ and terminates with $\arg(t) = \frac{2\pi}{3}$ (for example, the contour C in Figure 11.10).

Rather than diving straight in and trying to evaluate $I(\lambda)$ in (11.5), we can proceed by considering two simpler cases first.

† and of course valid for all X in this simple example.

11.2 THE ASYMPTOTIC EVALUATION OF INTEGRALS

11.2.1 Laplace's Method

Consider the complex integral along the real line

$$I_1(\lambda) = \int_a^b e^{\lambda f(t)} g(t)\, dt \quad \text{with } f(t) \text{ real and } \lambda \gg 1. \tag{11.7}$$

In this case, the integrand is largest at the point where $f(t)$ is largest. In fact, we can approximate the integral by simply considering the contribution from the neighbourhood of the point where $f(t)$ takes its maximum value. This is known as

> **Laplace's Method**
> Integrals of the form (11.7) can be estimated by considering the contribution from the neighbourhood of the maximum value of $f(t)$ alone.

We will see how we can justify this rigorously later. For the moment, let's consider some examples.

Example 1

Consider the integral representation

$$K_\nu(x) = \int_0^\infty e^{-x \cosh t} \cosh \nu t\, dt,$$

of the modified Bessel function, $K_\nu(x)$. When $x \gg 1$, this is in the form (11.7) with $f(t) = -\cosh t$ and $g(t) = \cosh \nu t$. The maximum value of $f(t) = -\cosh t$ is at $t = 0$, where $f(t) = g(t) = 1$. For $t \ll 1$, $\cosh t \sim 1 + \frac{1}{2}t^2$, and we can therefore use Laplace's method to approximate $K_\nu(x)$ as

$$K_\nu(x) \sim \int_0^\infty e^{-x\left(1+\frac{1}{2}t^2\right)} dt = e^{-x} \int_0^\infty e^{-\frac{1}{2}xt^2}\, dt.$$

After making the substitution $t = \hat{t}\sqrt{2/x}$, this becomes

$$K_\nu(x) \sim \sqrt{\frac{2}{x}} e^{-x} \int_0^\infty e^{-\hat{t}^2}\, d\hat{t} = \sqrt{\frac{\pi}{2x}} e^{-x}, \quad \text{for } x \gg 1,$$

using (3.5).

Example 2

Consider the definition,

$$\Gamma(1 + \lambda) = \int_0^\infty t^\lambda e^{-t}\, dt = \int_0^\infty e^{\lambda \log t - t}\, dt,$$

of the gamma function, which we met in Section 3.1. Can we find the asymptotic form of the gamma function when $\lambda \gg 1$? Since the definition is not quite in the form that we require for Laplace's method, we need to make a transformation. If we let $t = \lambda \tau$, we find that

$$\Gamma(1 + \lambda) = \lambda^{1+\lambda} \int_0^\infty e^{\lambda(\log \tau - \tau)}\, d\tau.$$

This is in the form of (11.7) with $f(\tau) = \log \tau - \tau$. Since $f'(\tau) = 1/\tau - 1$ and $f''(\tau) = -1/\tau^2 < 0$, f has a local maximum at $\tau = 1$. Laplace's method states that $\Gamma(1+\lambda)$ is dominated by the contribution to the integral from the neighbourhood of $\tau = 1$. We use a Taylor series expansion,

$$\log \tau - \tau = \log\{1 + (\tau - 1)\} - 1 - (\tau - 1) = -1 - \frac{1}{2}(\tau - 1)^2 + \cdots \text{ for } |\tau - 1| \ll 1,$$

and extend the range of integration, to obtain

$$\Gamma(1+\lambda) \sim \lambda^{1+\lambda} e^{-\lambda} \int_{-\infty}^{\infty} e^{-\frac{1}{2}\lambda T^2} \, dT \text{ as } \lambda \to \infty.$$

If we now let $T = \hat{T}/\lambda^{1/2}$, we find that

$$\Gamma(1+\lambda) \sim \lambda^{\lambda+\frac{1}{2}} e^{-\lambda} \int_{-\infty}^{\infty} e^{-\frac{1}{2}\hat{T}^2} \, d\hat{T},$$

and hence

$$\lambda! = \Gamma(1+\lambda) \sim \sqrt{2\pi} \lambda^{\lambda+\frac{1}{2}} e^{-\lambda} \text{ as } \lambda \to \infty. \qquad (11.8)$$

This is known as **Stirling's formula**, and provides an excellent approximation to the gamma function, for $\lambda > 2$, as shown in Figure 11.2.

Example 3

Consider the integral

$$I(\lambda) = \int_0^{10} \frac{e^{-\lambda t}}{1+t} \, dt.$$

In this case, $f = -t$ and $g = 1/(1+t)$. Since $f'(t) = -1 \neq 0$ and the maximum value of f occurs at $t = 0$ for $t \in [0, 10]$,

$$I(\lambda) \sim \int_0^{10} e^{-\lambda t} \, dt = \frac{1}{\lambda}\left(1 - e^{-10\lambda}\right) \sim \frac{1}{\lambda} \text{ as } \lambda \to \infty.$$

In fact we can use the binomial expansion

$$(1+t)^{-1} = 1 - t + t^2 - t^3 + \cdots + (-1)^n t^n + \cdots,$$

even though this is only convergent for $|t| < 1$, since the integrand is exponentially small away from $t = 0$. We can also extend the range of integration to infinity without affecting the result, to give

$$I(\lambda) \sim \sum_{n=0}^{N} \int_0^{\infty} (-1)^n t^n e^{-\lambda t} \, dt = \sum_{n=0}^{N} \frac{(-1)^n \Gamma(n+1)}{\lambda^{n+1}} = \sum_{n=0}^{N} \frac{(-1)^n n!}{\lambda^{n+1}} \text{ as } \lambda \to \infty,$$

since, using the substitution $\tau = \lambda t$,

$$\int_0^{\infty} t^n e^{-\lambda t} \, dt = \frac{1}{\lambda^{n+1}} \int_0^{\infty} \tau^n e^{-\tau} \, d\tau = \frac{1}{\lambda^{n+1}} \Gamma(n+1) = \frac{n!}{\lambda^{n+1}}.$$

Note that, as we found for Ei(x), this is an asymptotic, rather than convergent, series representation of $I(x)$. We can justify this procedure using the following lemma.

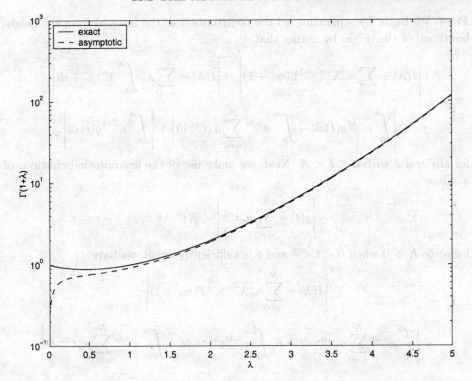

Fig. 11.2. The gamma function, $\Gamma(1+\lambda) = \lambda!$, and its asymptotic approximation using Stirling's formula, (11.8).

Lemma 11.1 (Watson's lemma) *If*

$$I(\lambda) = \int_0^A e^{-\lambda t} g(t)\, dt,$$

for $A > 0$, with g either bounded or having only integrable singularities,

$$g(t) \sim \sum_{n=0}^{N} a_n t^{\alpha_n} \quad \text{as } t \to 0,$$

and $-1 < \alpha_0 < \alpha_1 < \cdots < \alpha_N$ (which ensures that $e^{-\lambda t} g(t)$ has at worst an integrable singularity at $t = 0$), then

$$I(\lambda) \sim \int_0^\infty e^{-\lambda t} \sum_{n=0}^{N} a_n t^{\alpha_n}\, dt = \sum_{n=0}^{N} a_n \lambda^{-\alpha_n - 1} \Gamma(\alpha_n + 1) \quad \text{as } \lambda \to \infty.$$

Note that this lemma simply says that the integral is dominated by the contribution in the neighbourhood of $t = 0$, so that we can replace g with its asymptotic expansion and extend the range of integration to infinity.

ASYMPTOTIC METHODS: BASIC IDEAS

Proof We begin by separating off the contribution to the integral from the neighbourhood of the origin by noting that

$$\left| I(\lambda) - \sum_{n=0}^{N} a_n \lambda^{-\alpha_n - 1} \Gamma(\alpha_n + 1) \right| = \left| I(\lambda) - \sum_{n=0}^{N} a_n \int_0^\infty t^{\alpha_n} e^{-\lambda t} dt \right|$$

$$\leq \left| \int_0^\delta e^{-\lambda t} g(t) dt - \int_0^\infty e^{-\lambda t} \sum_{n=0}^{N} a_n t^{\alpha_n} dt \right| + \left| \int_\delta^A e^{-\lambda t} g(t) dt \right|$$

for any real δ with $0 < \delta < A$. Next, we make use of the asymptotic behaviour of g. Since

$$\left| g(t) - \sum_{n=0}^{N} a_n t^{\alpha_n} \right| < K t^{\alpha_{n+1}}$$

for some $K > 0$ when $0 < t < \delta$ and δ is sufficiently small, we have

$$\left| I(\lambda) - \sum_{n=0}^{N} a_n \lambda^{-\alpha_n - 1} \Gamma(\alpha_n + 1) \right|$$

$$< \left| \int_0^\delta e^{-\lambda t} \sum_{n=0}^{N} a_n t^{\alpha_n} dt + K \int_0^\delta e^{-\lambda t} t^{\alpha_{n+1}} dt - \int_0^\infty e^{-\lambda t} \sum_{n=0}^{N} a_n t^{\alpha_n} dt \right|$$

$$+ \left| \int_\delta^A e^{-\lambda t} g(t) dt \right|$$

$$< \left| -\sum_{n=0}^{N} a_n \int_\delta^\infty t^{\alpha_n} e^{-\lambda t} dt + K \int_0^\infty e^{-\lambda t} t^{\alpha_{n+1}} dt - K \int_\delta^\infty e^{-\lambda t} t^{\alpha_{n+1}} dt \right|$$

$$+ \left| \int_\delta^A e^{-\lambda t} g(t) dt \right|$$

$$< \sum_{n=0}^{N} |a_n| \int_\delta^\infty t^{\alpha_n} e^{-\lambda t} dt + K \lambda^{-\alpha_{n+1} - 1} \Gamma(\alpha_{n+1} + 1)$$

$$+ K \int_\delta^\infty e^{-\lambda t} t^{\alpha_{n+1}} dt + \int_\delta^A e^{-\lambda t} |g(t)| dt.$$

Now, since $e^{-\lambda t} < e^{-(\lambda - 1)\delta} e^{-t}$ for $t > \delta$,

$$\int_\delta^\infty t^{\alpha_n} e^{-\lambda t} dt < e^{-(\lambda - 1)\delta} \int_\delta^\infty t^{\alpha_n} e^{-t} dt < e^{-(\lambda - 1)\delta} \Gamma(\alpha_n + 1),$$

and also

$$\int_\delta^A e^{-\lambda t} |g(t)| dt < G e^{-\lambda \delta},$$

11.2 THE ASYMPTOTIC EVALUATION OF INTEGRALS

for $t \in [\delta, A]$, where $G = \int_\delta^A |g(t)|\, dt$. This finally shows that

$$\left| I(\lambda) - \sum_{n=0}^N a_n \lambda^{-\alpha_n - 1} \Gamma(\alpha_n + 1) \right| < K\lambda^{-\alpha_{n+1}-1} \Gamma(\alpha_{n+1} + 1) + Ge^{-\lambda\delta}$$

$$+ e^{-(\lambda-1)\delta} \left\{ \sum_{n=0}^N |a_n| \int_\delta^\infty t^{\alpha_n} e^{-t}\, dt + K \int_\delta^\infty t^{\alpha_{n+1}} e^{-t}\, dt \right\} = O(\lambda^{-\alpha_{n+1}-1}),$$

and hence the result is proved. □

A simple modification of this proof can be used to justify the use of Laplace's method in general.

11.2.2 The Method of Stationary Phase

Let's now consider an integral, again along the real line, of the form

$$I_2(\lambda) = \int_a^b e^{i\lambda F(t)} g(t)\, dt, \quad \text{with } F(t) \text{ real and } \lambda \gg 1. \tag{11.9}$$

Integrals of this type arise when Fourier transforms are used to solve differential equations (see Chapter 5). Points where $F'(t) = 0$ are called **points of stationary phase**, and the integral can be evaluated by considering the sum of the contributions from each of these points. We will consider this more carefully in the next section. For the moment, we can illustrate why this should be so by considering, as an example, the case $F(t) = (1-t)^2$, $g(t) = t$, which has $F'(t) = 0$ when $t = 1$, and

$$g(t)e^{i\lambda F(t)} = t\cos\lambda(1-t)^2 + it\sin\lambda(1-t)^2.$$

The rapid oscillations of this integrand lead to almost complete cancellation, except in the neighbourhood of the point of stationary phase, as can be seen in Figure 11.3.

Let's consider the situation when there is a single point of stationary phase at $t = t_0$. Then, since $F'(t_0) = 0$,

$$F(t) \sim F(t_0) + \frac{1}{2}(t - t_0)^2 F''(t_0) + O\left((t - t_0)^3\right) \quad \text{for } |t - t_0| \ll 1,$$

provided that $F''(t_0) \neq 0$ (see Exercise 11.11). If we assume that the integral is dominated by the contribution from the neighbourhood of the point of stationary phase,

$$I_2(\lambda) \sim \int_{t_0 - \delta}^{t_0 + \delta} g(t_0) \exp\left[i\lambda \left\{ F(t_0) + \frac{1}{2} F''(t_0)(t - t_0)^2 \right\} \right] dt$$

$$= e^{i\lambda F(t_0)} g(t_0) \int_{t_0-\delta}^{t_0+\delta} \exp\left\{ \frac{1}{2} i\lambda F''(t_0)(t - t_0)^2 \right\} dt$$

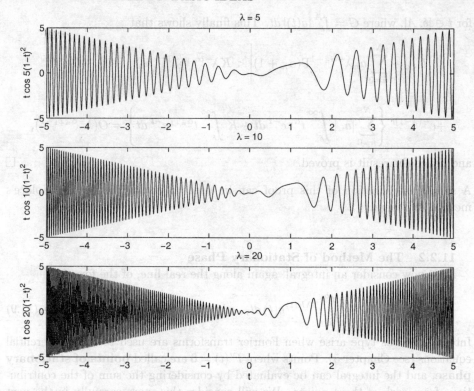

Fig. 11.3. The function $t\cos\lambda(1-t)^2$ for $\lambda = 5$, 10 and 20.

for some $\delta \ll 1$. If we now let

$$T = (t - t_0)\sqrt{\frac{1}{2}\lambda|F''(t_0)|},$$

this becomes

$$I_2(\lambda) \sim e^{i\lambda F(t_0)} g(t_0) \sqrt{\frac{2}{\lambda|F''(t_0)|}} \int_{-\delta\sqrt{\frac{1}{2}\lambda|F''(t_0)|}}^{\delta\sqrt{\frac{1}{2}\lambda|F''(t_0)|}} e^{i\,\text{sgn}\{F''(t_0)\}T^2}\,dT,$$

and, at leading order as $\lambda \to \infty$,

$$I_2(\lambda) \sim e^{i\lambda F(t_0)} g(t_0) \sqrt{\frac{2}{\lambda|F''(t_0)|}} \int_{-\infty}^{\infty} e^{\pm iT^2}\,dT.$$

We now just need to calculate

$$J = \int_0^\infty e^{iT^2}\,dT = \frac{1}{2}\int_{-\infty}^\infty e^{iT^2}\,dT.$$

To do this, consider the contours C_1, C_2 and C_3 in the complex T-plane, illustrated in Figure 11.4. Considering the contour C_2 first,

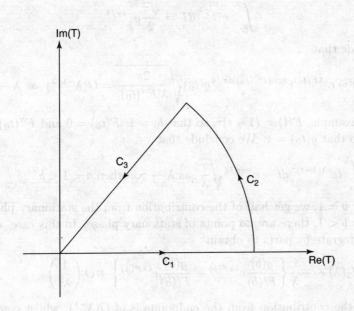

Fig. 11.4. The contours C_1, C_2 and C_3 in the complex T-plane.

$$\left| \int_{C_2} e^{iT^2} dT \right| = \left| \int_0^{\pi/4} e^{iR^2(\cos 2\theta + i \sin 2\theta)} iRe^{i\theta} d\theta \right|$$

$$\leqslant R \int_0^{\pi/4} e^{-R^2 \sin 2\theta} \, \theta \to 0 \quad \text{as } R \to \infty,$$

by Jordan's lemma. On the remaining contours,

$$\int_{C_3} e^{iT^2} dT = \int_R^0 e^{-\hat{T}^2} e^{i\pi/4} d\hat{T} \to -\frac{\sqrt{\pi}}{2} e^{i\pi/4} \quad \text{as } R \to \infty,$$

and

$$\int_{C_1} e^{iT^2} dT = \int_0^R e^{iT^2} dT \to \int_0^\infty e^{iT^2} dT \quad \text{as } R \to \infty.$$

By Cauchy's theorem

$$\int_{C_1+C_2+C_3} e^{iT^2} dT = 0,$$

and hence

$$\int_0^\infty e^{iT^2} dT = \frac{\sqrt{\pi}}{2} e^{i\pi/4}.$$

ASYMPTOTIC METHODS: BASIC IDEAS

Similarly,
$$\int_0^\infty e^{-iT^2}\,dT = \frac{\sqrt{\pi}}{2} e^{-i\pi/4}.$$

We conclude that
$$I_2(\lambda) \sim e^{i\lambda F(t_0)} e^{\mathrm{sgn}(F''(t_0))i\pi/4} g(t_0) \sqrt{\frac{2\pi}{\lambda |F''(t_0)|}} = O(\lambda^{-1/2}) \text{ as } \lambda \to \infty.$$

In our example, $F(t) = (1-t)^2$, so that $t_0 = 1$, $F(t_0) = 0$ and $F''(t_0) = 2$, and $g(t) = t$, so that $g(t_0) = 1$. We conclude that

$$\int_a^b t e^{i\lambda(1-t)^2}\,dt \sim e^{-i\pi/4}\sqrt{\frac{\pi}{\lambda}} \text{ as } \lambda \to \infty, \text{ when } a < 1 < b. \qquad (11.10)$$

If $a = 1$ or $b = 1$, we get half of the contribution from the stationary phase point. If $a > 1$ or $b < 1$, there are no points of stationary phase. In this case, we can, in general, integrate by parts to obtain

$$I_2(\lambda) \sim -\frac{i}{\lambda}\left\{\frac{g(b)}{F'(b)} e^{i\lambda F(b)} - \frac{g(a)}{F'(a)} e^{i\lambda F(a)}\right\} + O\left(\frac{1}{\lambda^2}\right). \qquad (11.11)$$

Note that the contribution from the endpoints is of $O(\lambda^{-1})$, whilst contributions from points of stationary phase are larger, of $O(\lambda^{-1/2})$.

Example

Consider (5.40), the solution of an initial value problem for the wave equation that we considered in Section 5.5.2. Let's analyze this solution when $t \gg 1$ at a point $\mathbf{x} = \mathbf{v}t$ that moves with constant velocity, $\mathbf{v}$. Since (5.40) is written in a form independent of the orientation of the coordinate axes, we can assume that $\mathbf{v} = v(0,0,1)$. Consider

$$I_+ = \int_{-\infty}^\infty \int_{-\infty}^\infty \int_{-\infty}^\infty g(\mathbf{k}) e^{it F_+(\mathbf{k})}\,dk_x\,dk_y\,dk_z,$$

where
$$g(\mathbf{k}) = \frac{\tilde{f}(\mathbf{k})}{16\pi^3 ick}, \quad F_+(\mathbf{k}) = c\sqrt{k_x^2 + k_y^2 + k_z^2} - vk_z$$

and $\mathbf{k} = (k_x, k_y, k_z)$. Starting with the k_x integral, we can see that

$$\frac{\partial F_+}{\partial k_x} = \frac{ck_x}{\sqrt{k_x^2 + k_y^2 + k_z^2}},$$

which is zero when $k_x = 0$. This is, therefore, a unique point of stationary phase and, noting that

$$\left.\frac{\partial^2 F_+}{\partial k_x^2}\right|_{k_x=0} = \frac{c}{\sqrt{k_y^2 + k_z^2}},$$

11.2 THE ASYMPTOTIC EVALUATION OF INTEGRALS

the method of stationary phase shows that

$$I_+ \sim e^{i\pi/4}\sqrt{\frac{2\pi}{ct}} \int_{-\infty}^{\infty}\int_{-\infty}^{\infty} (k_y^2 + k_z^2)^{1/4}\, \hat{g}(k_y,k_z) e^{it\hat{F}_+(k_y,k_z)}\, dk_y\, dk_z,$$

where

$$\hat{g}(k_y,k_z) = \frac{\tilde{f}(0,k_y,k_z)}{16\pi^3 ick}, \quad \hat{F}_+(k_y,k_z) = c\sqrt{k_y^2 + k_z^2} - vk_z.$$

Similarly, we can use the method of stationary phase on the k_y integral, and arrive at

$$I_+ \sim \frac{1}{8\pi^2 c^2 t}\int_{-\infty}^{\infty} \tilde{f}(0,0;k_z) e^{i(c-v)tk_z}\, dk_z.$$

When $v \neq c$, a simple change of variable, $\hat{k}_z = tk_z$, shows that $I_+ = O(1/t^2)$. However, when $v = c$, I_+ is much larger, of $O(1/t)$. This is consistent with the fact that disturbances propagate at speed c. We therefore assume that $v = c$, so that we are considering a point moving in the z-direction with constant speed c, which gives us

$$I_+ \sim \frac{1}{8\pi^2 c^2 t}\int_{-\infty}^{\infty} \tilde{f}(0,0,k_z)\, dk_z.$$

If we now define

$$I_- = \int_{-\infty}^{\infty}\int_{-\infty}^{\infty}\int_{-\infty}^{\infty} g(\mathbf{k}) e^{itF_-(\mathbf{k})}\, dk_x\, dk_y\, dk_z,$$

with

$$F_-(\mathbf{k}) = -c\sqrt{k_x^2 + k_y^2 + k_z^2} - vk_z,$$

and follow the analysis through in the same way, we find that

$$I_- \sim \frac{1}{8\pi^2 c^2 t}\int_{-\infty}^{\infty} \tilde{f}(0,0,k_z) e^{-i(c+v)tk_z}\, dk_z.$$

Since we have already decided to consider the case $v = c$, $I_- = O(1/t^2)$ for $t \gg 1$, and therefore

$$u(\mathbf{v}t) \sim I_+ \sim \frac{1}{8\pi^2 c^2 t}\int_{-\infty}^{\infty} \tilde{f}(0,0,k_z)\, dk_z,$$

where $\mathbf{v} = c(0,0,1)$. Since the z-direction can be chosen arbitrarily in this problem, we conclude that the large time solution is small, of $O(1/t^2)$, except on the surface of a sphere of radius ct, where the solution is given by

$$u(ct\mathbf{e}) \sim \frac{1}{8\pi^2 c^2 t}\int_{-\infty}^{\infty} \tilde{f}(s\mathbf{e})\, ds = O\left(\frac{1}{t}\right), \tag{11.12}$$

with $\mathbf{e}$ an arbitrary unit vector. The amplitude of the solution therefore decays like $1/t$, and also depends upon the total frequency content of the initial conditions, (5.38), in the direction of $\mathbf{e}$, as given by the integral of their Fourier transform, $\int_{-\infty}^{\infty} \tilde{f}(s\mathbf{e})\, ds$.

In summary,

> **The Method of Stationary Phase**
> Integrals of the form (11.9) can be estimated by considering the contribution from the neighbourhood of each point of stationary phase, where $F'(t) = 0$. In the absence of such points, the integral is dominated by the contributions from the endpoints of the range of integration.

11.2.3 The Method of Steepest Descents

Let's return to consider the more general case,

$$I(\lambda) = \int_C e^{\lambda f(t)} g(t)\, dt \qquad (11.13)$$

for $\lambda \gg 1$, f and g analytic functions of $t = x + iy$ and C a contour in the complex t-plane. Laplace's method and the method of stationary phase are special cases of this. Since $f(t)$ and $g(t)$ are analytic, we can deform the contour C without changing $I(\lambda)$. If we deform C onto a contour C_1, on which the imaginary part of $f(t)$ is a constant, $f(t) = \phi(t) + i\psi_0$, then (11.13) becomes

$$I(\lambda) = e^{i\lambda\psi_0} \int_{C_1} e^{\lambda\phi(t)} g(t)\, dt,$$

and we can simply use Laplace's method. This is what we really want to do, because we have shown rigorously, in Lemma 11.1, that this sort of integral is dominated by the neighbourhood of the point on C_1 where the real-valued function $\phi(t)$ is largest.

In order to take this further, we need to know what the curves $\phi = \text{constant}$ and $\psi = \text{constant}$ look like for an analytic function $f(t) = \phi(x, y) + i\psi(x, y)$. The Cauchy–Riemann equations for the analytic function $f(t)$ are (see Appendix 6)

$$\frac{\partial \phi}{\partial x} = \frac{\partial \psi}{\partial y}, \quad \frac{\partial \phi}{\partial y} = -\frac{\partial \psi}{\partial x},$$

which show that

$$\nabla\phi \cdot \nabla\psi = \frac{\partial \phi}{\partial x}\frac{\partial \psi}{\partial x} + \frac{\partial \phi}{\partial y}\frac{\partial \psi}{\partial y} = -\frac{\partial \phi}{\partial x}\frac{\partial \phi}{\partial y} + \frac{\partial \phi}{\partial y}\frac{\partial \phi}{\partial x} = 0,$$

and hence that $\nabla\phi$ is perpendicular to $\nabla\psi$. Recall that $\nabla\phi$ is normal to lines of constant ϕ, from which we conclude that *the lines of constant ϕ are perpendicular to the lines of constant ψ*. Moreover, ϕ changes most rapidly in the direction of $\nabla\phi$, in other words on the lines of constant ψ. We say that the lines where ψ is constant are **contours of steepest descent and ascent** for ϕ. Our strategy for evaluating the integral (11.13) is therefore to deform the contour of integration into one of steepest descent. Figure 11.5 illustrates these ideas for the function $f(t) = t^2$.

11.2 THE ASYMPTOTIC EVALUATION OF INTEGRALS

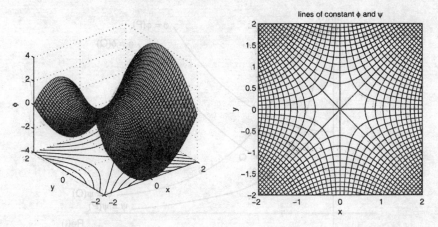

Fig. 11.5. The real part, $\phi(x,y) = x^2 - y^2$, of $f(t) = t^2$, and the lines of steepest ascent and descent for the analytic function $f(t) = t^2$.

Example: $f(t) = t^2$

When $f(t) = t^2$,

$$I(\lambda) = \int_C g(t) e^{\lambda t^2} \, dt,$$

with C a contour that joins two points, P and Q, in the complex t-plane. In this case, $f(t) = t^2 = (x+iy)^2 = x^2 - y^2 + 2ixy$, and hence $\phi = x^2 - y^2$ and $\psi = 2xy$. Therefore, ϕ is constant on the hyperbolas $x^2 - y^2 = \phi$ and ψ is constant on the hyperbolas $xy = \frac{1}{2}\psi$, as shown in Figure 11.5.

Case 1: $\psi(P) > 0$, $\psi(Q) > 0$ and $\phi(P) \neq \phi(Q)$

In this case, the ends of the contour C lie in the upper half plane, and without loss of generality, we take $\phi(P) > \phi(Q)$. We can deform C into the contour $C_1 + C_2$, with C_1 the steepest descent contour on which $\psi = \psi(P)$ and C_2 the contour on which $\phi = \phi(Q)$, as shown in Figure 11.6. On C_1 we can therefore make the change of variable

$$f(t) = f(P) - \tau, \text{ with } 0 \leqslant \tau \leqslant \phi(P) - \phi(Q).$$

The real part of f varies from $\phi(P)$ to $\phi(Q)$ as τ varies from zero to $\phi(P) - \phi(Q)$, whilst $\psi = \psi(P)$ is constant. Since $d\tau/dt = -f'(t)$,

$$\int_{C_1} g(t) e^{\lambda f(t)} \, dt = -e^{\lambda f(P)} \int_0^{\phi(P) - \phi(Q)} \frac{g(t)}{f'(t)} e^{-\lambda \tau} \, d\tau$$

$$\sim -e^{\lambda f(P)} \frac{g(P)}{f'(P)} \frac{1}{\lambda} \text{ as } \lambda \to \infty,$$

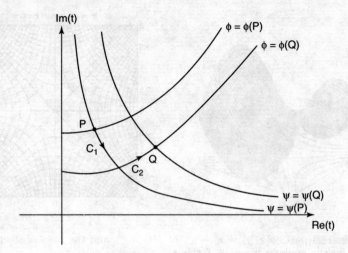

Fig. 11.6. The contours C_1 and C_2 in case 1.

using Watson's lemma, provided that $g(P)$ is bounded and nonzero. On C_2, $\phi = \phi(Q)$ and the imaginary part, ψ, varies. We can write

$$\left|\int_{C_2} g(t)e^{\lambda f(t)}\,dt\right| = \left|e^{\lambda\phi(Q)}\int_{C_2} g(t)e^{i\lambda\psi(t)}\,dt\right| \leqslant GLe^{\lambda\phi(Q)} \ll \int_{C_1} g(t)e^{\lambda f(t)}\,dt,$$

since $\phi(Q) < \phi(P)$, where $G = \max_{C_2}|g(t)|$ and $L = $ length of C_2. We conclude that

$$I(\lambda) \sim -e^{\lambda f(P)}\frac{g(P)}{f'(P)}\frac{1}{\lambda} \quad \text{as } \lambda \to \infty.$$

Case 2: $\psi(P) > 0$, $\psi(Q) > 0$ and $\phi(P) = \phi(Q)$

In this case, we must deform C into $C_1 + C_2 + C_3$, with ψ constant on C_1 and C_3 and $\phi < \phi(P)$ constant on C_3, as shown in Figure 11.7. The exact choice of contour C_2 is not important, since, as in case 1, its contribution is exponentially smaller than the contributions from C_1 and C_3. This is very similar to case 1, except that we now also have, using Laplace's method, a contribution at leading order from C_3. We find that

$$I(\lambda) \sim \left\{-e^{\lambda f(P)}\frac{g(P)}{f'(P)} + e^{\lambda f(Q)}\frac{g(Q)}{f'(Q)}\right\}\frac{1}{\lambda} \quad \text{as } \lambda \to \infty.$$

Let's consider a more interesting example.

11.2 THE ASYMPTOTIC EVALUATION OF INTEGRALS

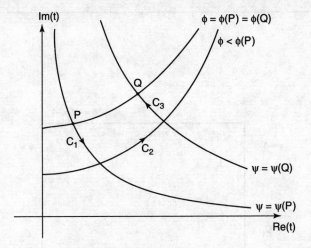

Fig. 11.7. The contours C_1, C_2 and C_3 in case 2.

Example: The Bessel function of order zero with large argument

We saw in Chapter 3 that the Bessel function $J_n(\lambda)$ has an integral representation, (3.18). Setting $n = 0$ and making the change of variable $t = \sin\theta$ leads to

$$J_0(\lambda) = \frac{1}{\pi} \int_{-1}^{1} \frac{e^{i\lambda t}}{\sqrt{1-t^2}}\, dt. \tag{11.14}$$

This is of the form (11.13) with $f(t) = it$, and hence $\phi = -y$, $\psi = x$. In this case, $P = -1$ and $Q = 1$, so that $\phi(P) = \phi(Q) = 0$. The contours of steepest descent through P and Q are just the straight lines $\psi = x = -1$ and $\psi = x = 1$ respectively. We therefore deform the contour C, which is the portion of the real axis $-1 \leqslant x \leqslant 1$, into $C_1 + C_2 + C_3$, as shown in Figure 11.8. The contribution from C_2, $y = Y > 0$, is exponentially small whatever the choice of Y ($Y = 1$ in Figure 11.8), whilst the contributions from C_1 and C_3 are dominated by the neighbourhoods of the endpoints on the real axis. On C_1 we make the change of variable $t = -1 + iy$, so that

$$\frac{1}{\pi} \int_{C_1} \frac{e^{i\lambda t}}{\sqrt{1-t^2}}\, dt = \frac{1}{\pi} \int_0^Y \frac{e^{-i\lambda} e^{-\lambda y}}{\sqrt{2iy+y^2}} i\, dy$$

$$\sim \frac{e^{-i\lambda} i}{\pi\sqrt{2i}} \int_0^\infty y^{-1/2} e^{-\lambda y}\, dy \quad \text{as } \lambda \to \infty,$$

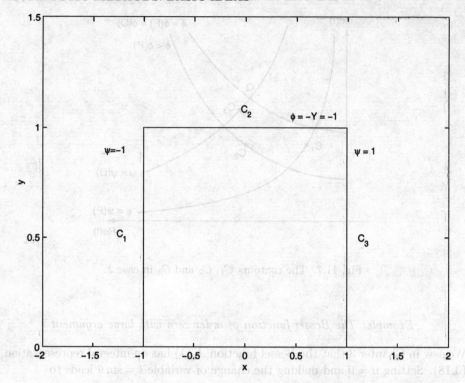

Fig. 11.8. The contours C_1, C_2 and C_3 for evaluating the asymptotic behaviour of $J_0(\lambda)$.

using Laplace's method. By making the change of variable $\hat{y} = \lambda y$, we can write this integral in terms of a gamma function, $\Gamma(1/2) = \sqrt{\pi}$, and arrive at

$$\frac{1}{\pi} \int_{C_1} \frac{e^{i\lambda t}}{\sqrt{1-t^2}} dt \sim \frac{e^{-i\lambda + i\pi/4}}{\sqrt{2\pi\lambda}} \quad \text{as } \lambda \to \infty.$$

Similarly,

$$\frac{1}{\pi} \int_{C_3} \frac{e^{i\lambda t}}{\sqrt{1-t^2}} dt \sim \frac{e^{i\lambda - i\pi/4}}{\sqrt{2\pi\lambda}} \quad \text{as } \lambda \to \infty,$$

and hence

$$J_0(\lambda) \sim \sqrt{\frac{2}{\lambda\pi}} \cos\left(\lambda - \frac{\pi}{4}\right) \quad \text{as } \lambda \to \infty.$$

So, is this the whole story? Let's return to our original example, $f(t) = t^2$. Since $f'(t) = 2t$, the real and imaginary parts of f are stationary at $t = 0$. However, $t = 0$ is neither a local maximum nor a local minimum. It is a **saddle point**. In

11.2 THE ASYMPTOTIC EVALUATION OF INTEGRALS

fact, no analytic function can have a local maximum or minimum, since its real and imaginary parts are harmonic (see Section 7.1.4 and Appendix 6). The real part of $f(t) = t^2$ has a ridge along the line $y = 0$ and a valley along the line $x = 0$, as can be seen in Figure 11.5. In cases 1 and 2, which we studied above, we were able to deform the contour C within one of the valleys.

Case 3: $\psi(P) > 0$ *and* $\psi(Q) < 0$

In this case, P and Q lie in different valleys of the function $f(t) = t^2$, and we must deform the contour so that it runs through the line of steepest descent at the saddle point, since this is the only line with ψ constant that connects the two valleys, as shown in Figure 11.9. As usual, the integrals on C_2 and C_4 are

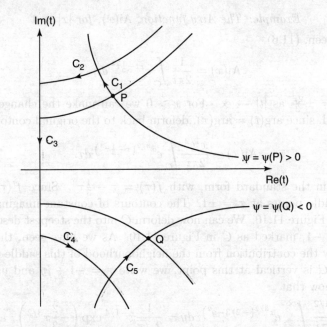

Fig. 11.9. The contours C_1, C_2, C_3, C_4 and C_5 in case 3.

exponentially small, and, using Laplace's method as in cases 1 and 2,

$$\int_{C_1} g(t)e^{\lambda t^2}\, dt \sim -e^{\lambda f(P)} \frac{g(P)}{f'(P)} \frac{1}{\lambda}, \quad \int_{C_5} g(t)e^{\lambda t^2}\, dt \sim e^{\lambda f(Q)} \frac{g(Q)}{f'(Q)} \frac{1}{\lambda}.$$

On the steepest descent contour C_3, we can make the simple change of variable $t = iy$, to arrive at

$$\int_{C_3} g(t)e^{\lambda t^2}\, dt = i \int_{\psi(P)}^{\psi(Q)} g(iy)e^{-\lambda y^2}\, dy.$$

This integral is dominated by the behaviour close to the saddle point, $y = 0$, and, if $g(0)$ is bounded and nonzero, Laplace's method shows that

$$\int_{C_3} g(t) e^{\lambda t^2} dt \sim -ig(0) \int_{-\infty}^{\infty} e^{-\lambda y^2} dy = -ig(0) \sqrt{\frac{\pi}{\lambda}}. \qquad (11.15)$$

If $\phi(P) \leq 0$ and $\phi(Q) \leq 0$, this is the dominant contribution.

Finally, what if the contour of integration, C, extends to infinity? For such an integral to converge, it must extend into the valleys of $f(t)$ as $|t| \to \infty$. In our example, if we let P and Q head off to infinity along the two different valleys, the integral is dominated by the contribution from the steepest descent contour through the saddle point, given by (11.15).

Example: The Airy function, $\mathrm{Ai}(x)$, *for* $|x| \gg 1$

As we have seen, (11.6),

$$\mathrm{Ai}(x) = \frac{1}{2\pi i} \int_C e^{xt - \frac{1}{3}t^3} dt, \qquad (11.16)$$

with $\arg(t) = \pm \frac{2\pi}{3}$ as $|t| \to \infty$. For $x > 0$ we can make the change of variable $t = x^{1/2}\tau$ and, since $\arg(t) = \arg(\tau)$, deform back to the original contour to obtain

$$\mathrm{Ai}(x) = \frac{x^{1/2}}{2\pi i} \int_C e^{x^{3/2}(\tau - \frac{1}{3}\tau^3)} d\tau. \qquad (11.17)$$

This is now in the standard form, with $f(\tau) = \tau - \frac{1}{3}\tau^3$. Since $f'(\tau) = 1 - \tau^2$, there are saddle points at $\tau = \pm 1$. The contours of constant imaginary part, ψ, are shown in Figure 11.10. We can now deform C into the steepest descent contour through $\tau = -1$, marked as C in Figure 11.10. As we have seen, the integral is dominated by the contribution from the neighbourhood of this saddle point. Since the contour C is vertical at this point, we write $\tau = -1 + iy$ and use Laplace's method to show that

$$\mathrm{Ai}(x) \sim \frac{x^{1/2}}{2\pi i} \int_{-\infty}^{\infty} e^{x^{3/2}(-2/3-y^2)} i\, dy = \frac{1}{2\sqrt{\pi}} x^{-1/4} \exp\left(-\frac{2}{3} x^{3/2}\right) \quad \text{as } x \to \infty. \qquad (11.18)$$

Similarly, when $x < 0$, we can make the transformation $t = (-x)^{1/2} T$, so that

$$\mathrm{Ai}(x) = \frac{(-x)^{1/2}}{2\pi i} \int_C e^{(-x)^{3/2}(-T - \frac{1}{3}T^3)} dT. \qquad (11.19)$$

In this case, $f(T) = -T - \frac{1}{3}T^3$ and $f'(T) = -1 - T^2$, so there are saddle points at $T = \pm i$. The contours of constant imaginary part, ψ, are shown in Figure 11.11. By deforming C into the contour $C_1 + C_2 + C_3$, we can see that the integral is dominated by the contributions from the two saddle points. The integral along C_2 is exponentially smaller. In order to calculate the leading order contribution from C_1, note that this contour passes through $T = -i$ in the direction of $1 + i$. We

11.2 THE ASYMPTOTIC EVALUATION OF INTEGRALS

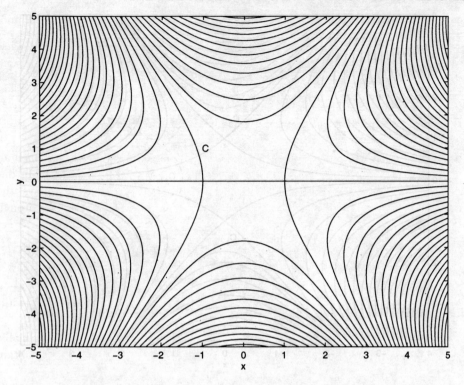

Fig. 11.10. Contours where the imaginary part of $f(\tau) = \tau - \frac{1}{3}\tau^3$ is constant.

therefore make the substitution $T = -i + (1+i)s$ and use Laplace's method to arrive at

$$\frac{(-x)^{1/2}}{2\pi i} \int_{C_1} e^{(-x)^{3/2}\left(-T-\frac{1}{3}T^3\right)} dt \sim \frac{(-x)^{1/2}}{2\pi i} e^{\frac{2}{3}i(-x)^{3/2}} (1+i) \int_{-\infty}^{\infty} e^{-2(-x)^{3/2}s^2} ds$$

$$= \frac{(-x)^{-1/4}(1+i)}{2\pi i} \sqrt{\frac{\pi}{2}} e^{\frac{2}{3}i(-x)^{3/2}}.$$

Similarly,

$$\frac{(-x)^{1/2}}{2\pi i} \int_{C_3} e^{(-x)^{3/2}\left(-T-\frac{1}{3}T^3\right)} dt \sim \frac{(-x)^{-1/4}(-1+i)}{2\pi i} \sqrt{\frac{\pi}{2}} e^{-\frac{2}{3}i(-x)^{3/2}},$$

and hence

$$\text{Ai}(x) \sim \frac{(-x)^{-1/4}}{\sqrt{\pi}} \sin\left\{\frac{2}{3}(-x)^{3/2} + \frac{\pi}{4}\right\} \quad \text{as } x \to -\infty. \quad (11.20)$$

The transition from exponential decay as $x \to \infty$ to decaying oscillations as $x \to -\infty$ is shown in Figure 11.12. Also shown is the behaviour predicted by our analysis

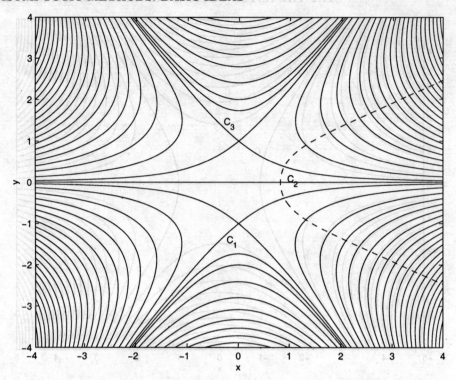

Fig. 11.11. Contours where the imaginary part of $f(T) = -T - \frac{1}{3}T^3$ is constant, along with a single contour, C_2, where the real part of f is constant.

for large $|x|$, which can be seen to be in excellent agreement with Ai(x), even for moderate values of $|x|$.

To end this section, we give a justification for the success of the method of stationary phase. Consider the example that we looked at earlier,

$$I_2(\lambda) = \int_a^b t e^{i\lambda(1-t)^2}\, dt,$$

with $a < 1 < b$. Using Cauchy's theorem on the analytic function $te^{i\lambda(1-t)^2}$, we can deform the contour C, the real axis with $a \leqslant x \leqslant b$, into $C_1 + C_2 + C_3 + C_4 + C_5$, as shown in Figure 11.13. The same arguments as those that we used above show that the largest contribution comes from the neighbourhood of the saddle point on the steepest descent contour, C_3. This is, however, just the neighbourhood of the single point of stationary phase, and, even though the contour is different, because the integrand is analytic, we obtain the same result, (11.10).

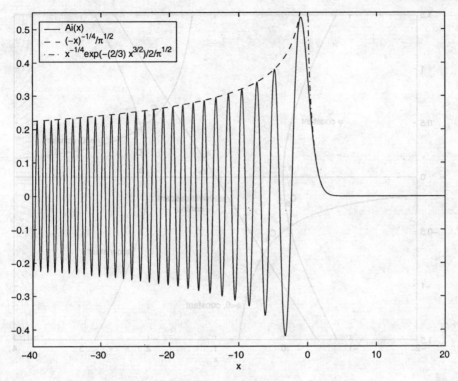

Fig. 11.12. The Airy function Ai(t).

Exercises

11.1 Calculate the order of magnitude of the following functions in terms of the simplest function of ϵ possible, as $\epsilon \to 0^+$. (a) $\sinh \epsilon$, (b) $\tan \epsilon$, (c) $\sinh(1/\epsilon)$, (d) $e^{-\epsilon}$, (e) $\cot \epsilon$, (f) $\log(1+\epsilon)$, (g) $(1-\cos \epsilon)/(1+\cos \epsilon)$, (h) $\exp\{-\cosh(1/\epsilon)\}$.

11.2 Show that $e^{-1/\epsilon} = o(\epsilon^n)$ as $\epsilon \to 0$ for all real n, provided that the complex variable ϵ is restricted to a domain that you should determine.

11.3 Consider the integral

$$I(x) = e^{-x} \int_1^x \frac{e^t}{t} dt \quad \text{as } x \to \infty.$$

By integrating by parts repeatedly, develop a series expansion of the form

$$I(x) = \frac{1}{x} + \frac{1}{x^2} + \frac{2}{x^3} + \frac{3!}{x^4} - (1+1+2+3!)\,e^{1-x} + \cdots.$$

By considering the error in terminating the expansion after the term in x^{-n}, show that the series is asymptotic as $x \to \infty$.

11.4 Show that the following are asymptotic sequences.

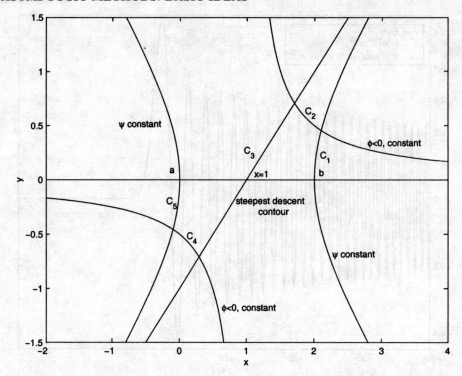

Fig. 11.13. The contours C_1, C_2, C_3, C_4 and C_5 for the method of stationary phase.

(a) $\left(\sin\dfrac{1}{\epsilon}\right)^n$, $n = 0, 1, \ldots$, as $\epsilon \to \infty$,

(b) $\log(1 + \epsilon^n)$, $n = 0, 1, \ldots$, as $\epsilon \to 0$.

11.5 Use Watson's lemma to show that

$$\int_0^\infty (t + t^2)^{-1/2} e^{-\lambda t} dt \sim \sqrt{\dfrac{\pi}{\lambda}}\left(1 - \dfrac{1}{4\lambda} + \dfrac{9}{32\lambda^2}\right) \text{ as } \lambda \to \infty.$$

11.6 After making the substitution $t = u\lambda$, use Laplace's method to show that

$$\int_0^\infty \dfrac{e^{4\lambda^{3/2} t^{1/2} - t^2}}{1 + (t/\lambda)^2} dt \sim \sqrt{\dfrac{\pi}{6}} e^{3\lambda^2} \text{ as } \lambda \to \infty.$$

11.7 Use Laplace's method to show that the modified Bessel function $K_\nu(z)$, which has an integral representation

$$K_\nu(z) = \dfrac{1}{2}\int_{-\infty}^\infty e^{\nu t - z \cosh t} dt,$$

can be approximated as $\nu \to \infty$, with $z = O(1)$ and positive, using

$$K_\nu(z) \sim \sqrt{\frac{\pi}{2\nu}} e^{-\nu} \left(\frac{2\nu}{z}\right)^\nu.$$

You may assume that the integral is dominated by the contribution from the neighbourhood of the maximum value of $\nu t - z \cosh t$.

11.8 Use the method of stationary phase to show that

$$\int_0^\infty \frac{e^{i\lambda(\frac{1}{3}t^3 - t)}}{1+t} dt \sim \frac{1}{2}\sqrt{\frac{\pi}{\lambda}} e^{-2i\lambda/3 + i\pi/4} \quad \text{as } \lambda \to \infty.$$

11.9 Use the method of stationary phase to show that

$$\int_0^{\pi/2} e^{i\lambda(2t - \sin 2t)} dt \sim \frac{1}{2}\Gamma\left(\frac{4}{3}\right)\left(\frac{6}{\lambda}\right)^{1/3} e^{i\pi/6} \quad \text{as } \lambda \to \infty.$$

11.10 Show that as $\lambda \to \infty$

(a) $\displaystyle\int_4^{1+5i} e^{i\lambda t^2} dt \sim i\frac{e^{16i\lambda}}{8}\left(\frac{1}{\lambda} - \frac{i}{32\lambda^2}\right),$

(b) $\displaystyle\int_{-5-i}^{1+5i} e^{i\lambda t^2} dt \sim \sqrt{\frac{\pi}{\lambda}} e^{i\pi/4}.$

11.11 Consider the integral (11.9) when $F(t)$ has a single point of stationary phase at $t = t_0$ with $a < t_0 < b$, $F''(t_0) = 0$ and $F'''(t_0) \ne 0$. Use the method of stationary phase to show that

$$I_2 \sim \frac{2}{3} g(t_0) e^{i\lambda F(t_0) + i\pi \operatorname{sgn}(F'''(t_0))/6} \Gamma\left(\frac{1}{3}\right)\left(\frac{6}{\lambda |F'''(t_0)|}\right)^{1/3} \quad \text{for } \lambda \gg 1.$$

11.12 Consider the integral

$$I(\lambda) = \int_P^Q e^{\lambda(\frac{1}{2}t^2 + it)} dt,$$

where P and Q are points in the complex t plane. Sketch the contours of constant real and imaginary parts of $\frac{1}{2}t^2 + it$. Show that if $P = -\frac{1}{2}$ and $Q = 2$,

$$I(\lambda) \sim \frac{e^{(2+2i)\lambda}}{(2+i)\lambda} \quad \text{as } \lambda \to \infty.$$

Show that if $P = -\frac{1}{2} + i$ and $Q = 1 - 3i$,

$$I(\lambda) \sim -i\sqrt{\frac{2\pi e^\lambda}{\lambda}} \quad \text{as } \lambda \to \infty.$$

11.13 Show that Stirling's formula for $\Gamma(z+1)$ when $z \gg 1$ holds for complex z provided $|z| \gg 1$ and $-\pi < \arg(z) < \pi$. (*Hint:* Let $z = Re^{i\theta}$.) Determine the next term in the asymptotic expansion.

11.14 From the integral representation,
$$I_\nu(x) = \frac{1}{\pi}\int_0^\pi e^{x\cos\theta}\cos\nu\theta\,d\theta - \frac{\sin\nu\pi}{\pi}\int_0^\infty e^{-x\cosh t-\nu t}\,dt,$$
of the modified Bessel function of order ν, show that
$$I_\nu(x) \sim \frac{e^x}{\sqrt{2\pi x}},$$
as $x \to \infty$ for real x and fixed ν.

11.15 The **parabolic cylinder functions** are defined by
$$D_{-2m}(x) = \frac{1}{(m-1)!}xe^{-x^2/4}\int_0^\infty e^{-s}s^{m-1}(x^2+2s)^{-m-1/2}\,ds,$$

$$D_{-2m+1}(x) = \frac{1}{(m-1)!}xe^{-x^2/4}\int_0^\infty e^{-s}s^{m-1}(x^2+2s)^{-m+1/2}\,ds,$$

for m a positive integer. Show that for real x and fixed m,
$$D_{-2m}(x) \sim x^{-2m}e^{-x^2/4}, \quad D_{-2m+1} \sim x^{-2m+1}e^{-x^2/4},$$
as $x \to \infty$.

CHAPTER TWELVE

Asymptotic Methods: Differential Equations

In this chapter we will apply the ideas that we developed in the previous chapter to the solution of ordinary and partial differential equations.

12.1 An Instructive Analogy: Algebraic Equations

Many of the essential ideas that we will need in order to solve differential equations using asymptotic methods can be illustrated using algebraic equations. These are much more straightforward to deal with than differential equations, and the ideas that we will use are far more transparent. We will consider two illustrative examples.

12.1.1 Example: A Regular Perturbation
Consider the cubic equation

$$x^3 - x + \epsilon = 0. \tag{12.1}$$

Although there is an explicit formula for the solution of cubic equations, it is rather cumbersome to use. Let's suppose instead that we only need to find the solutions for $\epsilon \ll 1$. If we simply set $\epsilon = 0$, we get $x^3 - x = 0$, and hence $x = -1$, 0 or 1. These are called the **leading order solutions** of the equation. These solutions are obviously not exact when ϵ is small but nonzero, so let's try to improve the accuracy of our approximation by seeking an asymptotic expansion of the solution (or more succinctly, an asymptotic solution) of the form

$$x = x_0 + \epsilon x_1 + \epsilon^2 x_2 + O(\epsilon^3). \tag{12.2}$$

We can now substitute this into (12.1) and equate powers of ϵ. This gives us

$$(x_0 + \epsilon x_1 + \epsilon^2 x_2)^3 - (x_0 + \epsilon x_1 + \epsilon^2 x_2) + \epsilon + O(\epsilon^3) = 0,$$

which we can rearrange into a hierarchy of powers of ϵ in the form

$$\{x_0^3 - x_0\} + \epsilon \{(3x_0 - 1)x_1 + 1\} + \epsilon^2 \{3x_0 x_1^2 + (3x_0^2 - 1)x_2\} + O(\epsilon^3) = 0.$$

At leading order we obviously get $x_0^3 - x_0 = 0$, and hence $x_0 = -1$, 0 or 1. We will concentrate on the solution with $x_0 = 1$. At $O(\epsilon)$, $(3x_0 - 1)x_1 + 1 = 2x_1 + 1 = 0$, and hence $x_1 = -\frac{1}{2}$. At $O(\epsilon^2)$, $3x_0 x_1^2 + (3x_0^2 - 1)x_2 = \frac{3}{4} + 2x_2 = 0$, and hence

$x_2 = -\frac{3}{8}$. We could, of course, continue to higher order if necessary. This shows that

$$x = 1 - \frac{1}{2}\epsilon - \frac{3}{8}\epsilon^2 + O(\epsilon^3) \quad \text{for } \epsilon \ll 1.$$

Similar expansions can be found for the other two solutions of (12.1). This is a regular perturbation problem, since we have found asymptotic expansions for all three roots of the cubic equation using the simple expansion (12.2). Figure 12.1 shows that the function $x^3 - x + \epsilon$ is qualitatively similar for $\epsilon = 0$ and $0 < \epsilon \ll 1$.

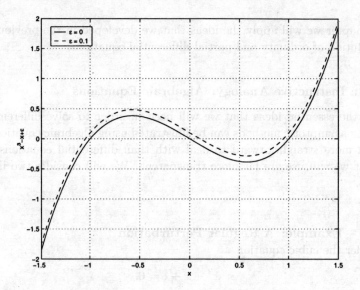

Fig. 12.1. The function $x^3 - x + \epsilon$ for $\epsilon = 0$, solid line, and $\epsilon = 0.1$, broken line.

12.1.2 Example: A Singular Perturbation

Consider the cubic equation

$$\epsilon x^3 + x^2 - 1 = 0. \tag{12.3}$$

At leading order for $\epsilon \ll 1$, $x^2 - 1 = 0$, and hence $x \doteq \pm 1$. However, we know that a cubic equation is meant to have three solutions. What's happened to the other solution? This is an example of a singular perturbation problem, where the solution for $\epsilon = 0$ is qualitatively different to the solution when $0 < \epsilon \ll 1$. The equation changes from quadratic to cubic, and the number of solutions goes from two to three. The key point is that we have implicitly assumed that $x = O(1)$. However, the term ϵx^3, which we neglect at leading order when $x = O(1)$, becomes comparable to the term x^2 for sufficiently large x, specifically when $x = O(\epsilon^{-1})$. Figure 12.2 shows how the function $\epsilon x^3 + x^2 - 1$ changes qualitatively for $\epsilon = 0$ and $0 < \epsilon \ll 1$.

12.1 AN INSTRUCTIVE ANALOGY: ALGEBRAIC EQUATIONS

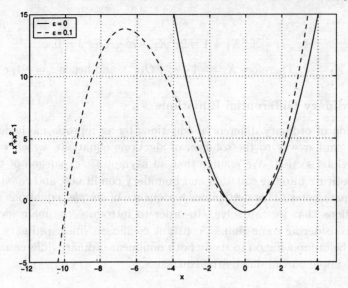

Fig. 12.2. The function $\epsilon x^3 + x^2 - 1$ for $\epsilon = 0$, solid line, and $\epsilon = 0.1$, broken line.

So, how can we proceed in a systematic way? If we expand $x = x_0 + \epsilon x_1 + O(\epsilon^2)$, we can construct the two $O(1)$ solutions, $x = \pm 1 + O(\epsilon)$, in the same manner as we did for the previous example. Since we know that there must be three solutions, we conclude that the other solution cannot have $x = O(1)$, and assume that $x = O(\epsilon^\alpha)$, with α to be determined. If we define a scaled variable, $x = \epsilon^\alpha X$, with $X = O(1)$ for $\epsilon \ll 1$, (12.3) becomes

$$\epsilon^{3\alpha+1} X^3 + \epsilon^{2\alpha} X^2 - 1 = 0. \tag{12.4}$$

We must choose α in order to obtain an **asymptotic balance** between two of the terms in (12.4). If $\alpha > 0$, the first two terms are small and cannot balance the third term, which is of $O(1)$. If $\alpha < 0$, the first two terms are large, and we can choose α so that they are of the same asymptotic order. This requires that $\epsilon^{3\alpha+1} X^3 = O(\epsilon^{2\alpha} X^2)$, and hence $\epsilon^{3\alpha+1} = O(\epsilon^{2\alpha})$. This gives $3\alpha + 1 = 2\alpha$, and hence $\alpha = -1$. This means that $x = \epsilon^{-1} X = O(\epsilon^{-1})$, as expected. Equation (12.4) now becomes

$$X^3 + X^2 - \epsilon^2 = 0. \tag{12.5}$$

Since only ϵ^2 and not ϵ appears in this rescaled equation, we expand $X = X_0 + \epsilon^2 X_1 + O(\epsilon^4)$. At leading order, $X_0^3 + X_0^2 = 0$, and hence $X_0 = -1$ or 0. Of course, $X_0 = 0$ will just give us the two solutions with $x = O(1)$ that we have already considered, so we take $X_0 = -1$. At $O(\epsilon^2)$,

$$(-1 + \epsilon^2 X_1)^3 + (-1 + \epsilon^2 X_1)^2 - \epsilon^2 + O(\epsilon^4) = 0$$

gives
$$-1 + 3\epsilon^2 X_1 + 1 - 2\epsilon^2 X_1 - \epsilon^2 + O(\epsilon^4) = 0,$$
and hence $X_1 = 1$. Therefore $X = -1 + \epsilon^2 + O(\epsilon^4)$, and hence $x = -1/\epsilon + \epsilon + O(\epsilon^3)$.

12.2 Ordinary Differential Equations

The solution of ordinary differential equations by asymptotic methods often proceeds in a similar way to the solution of algebraic equations, which we discussed in the previous section. We assume that an asymptotic expansion of the solution exists, substitute into the equation and boundary conditions, and equate powers of the small parameter. This determines a sequence of simpler equations and boundary conditions that we can solve. In order to introduce the main ideas, we will begin by considering some simple, constant coefficient linear ordinary differential equations before moving on to study both nonlinear ordinary differential equations and some simple partial differential equations.

12.2.1 Regular Perturbations

Consider the ordinary differential equation
$$y'' + 2\epsilon y' - y = 0, \tag{12.6}$$
to be solved for $0 \leqslant x \leqslant 1$, subject to the boundary conditions
$$y(0) = 0, \quad y(1) = 1. \tag{12.7}$$
Of course, we could solve this constant coefficient ordinary differential equation analytically using the method described in Appendix 5, but it is instructive to try to construct the asymptotic solution when $\epsilon \ll 1$. We seek a solution of the form
$$y(x) = y_0(x) + \epsilon y_1(x) + O(\epsilon^2).$$
At leading order, $y_0'' - y_0 = 0$, subject to $y_0(0) = 0$ and $y_0(1) = 1$, which has solution
$$y_0(x) = \frac{\sinh x}{\sinh 1}.$$
If we now substitute the expansion for y into (12.6) and retain terms up to $O(\epsilon)$, we obtain
$$(y_0 + \epsilon y_1)'' + 2\epsilon(y_0 + \epsilon y_1)' - (y_0 + \epsilon y_1) + O(\epsilon^2)$$
$$= y_0'' + \epsilon y_1'' + 2\epsilon y_0' - y_0 - \epsilon y_1 + O(\epsilon^2) = 0,$$
and hence
$$y_1'' - y_1 = -2y_0' = -2\frac{\cosh x}{\sinh 1}. \tag{12.8}$$
Similarly, the boundary conditions (12.7) show that
$$y_0(0) + \epsilon y_1(0) + O(\epsilon^2) = 0, \quad y_0(1) + \epsilon y_1(1) + O(\epsilon^2) = 1,$$

and hence
$$y_1(0) = 0, \quad y_1(0) = 0. \tag{12.9}$$

By seeking a particular integral solution of (12.8) in the form $y_{1p} = kx \sinh x$, and using the constants in the homogeneous solution, $y_{1h} = A \sinh x + B \cosh x$, to satisfy the boundary conditions (12.9), we arrive at

$$y_1 = (1-x)\frac{\sinh x}{\cosh 1},$$

and hence
$$y(x) = \frac{\sinh x}{\sinh 1} + \epsilon(1-x)\frac{\sinh x}{\cosh 1} + O(\epsilon^2), \tag{12.10}$$

for $\epsilon \ll 1$. The ratio of the second to the first term in this expansion is $\epsilon(1-x)\tanh 1$, which is uniformly small for $0 \leqslant x \leqslant 1$. This leads us to believe that the asymptotic solution (12.10) is uniformly valid in the region of solution. The situation is analogous to the example that we studied in Section 12.1.1.

One subtle point is that, although we believe that the next term in the asymptotic expansion of the solution, which we write as $O(\epsilon^2)$ in (12.10), is uniformly smaller than the two retained terms for ϵ sufficiently small, we have not proved this. We do not have a rigorous estimate for the size of the neglected term in the way that we did when we considered the exponential integral, where we were able to find an upper bound for R_N, given by (11.3). Although, for this simple, constant coefficient ordinary differential equation, we could write down the exact solution and prove that the remainder is of $O(\epsilon^2)$, in general, and in particular for nonlinear problems, this is not possible, and an act of faith is involved in trusting that our asymptotic solution provides a valid representation of the exact solution. This faith can be bolstered in a number of ways, for example, by comparing asymptotic solutions with numerical solutions, and by checking that the asymptotic solution makes sense in terms of the underlying physics of the system that we are modelling. The sensible applied mathematician always has a small, nagging doubt at the back of their mind about the validity of an asymptotic solution. For (12.6), our faith is justified, as can be seen in Figure 12.3.

12.2.2 The Method of Matched Asymptotic Expansions

Consider the ordinary differential equation

$$\epsilon y'' + 2y' - y = 0, \tag{12.11}$$

to be solved for $0 \leqslant x \leqslant 1$, subject to the boundary conditions

$$y(0) = 0, \quad y(1) = 1. \tag{12.12}$$

The observant reader will notice that this is the same as the previous example, but with the small parameter ϵ moved to the highest derivative term. We again seek an asymptotic solution of the form

$$y(x) = y_0(x) + \epsilon y_1(x) + O(\epsilon^2).$$

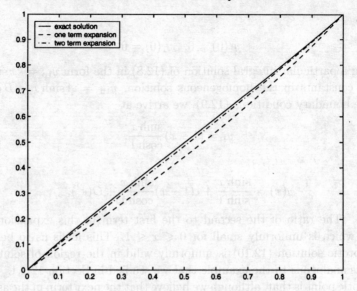

Fig. 12.3. Exact and asymptotic solutions of (12.6) when $\epsilon = 0.25$.

At leading order, $2y_0' - y_0 = 0$, which has the solution $y_0 = Ae^{x/2}$ for some constant A. However, the boundary conditions require that $y_0(0) = 0$ and $y_0(1) = 1$. How can we satisfy two boundary conditions using only one constant? Well, of course we can't. The problem is that for $\epsilon = 0$, the equation is of first order, and therefore qualitatively different from the full, second order equation. This is a singular perturbation, and is analogous to the example we studied in Section 12.1.2.

Let's proceed by satisfying the boundary condition $y_0(1) = 1$, which gives

$$y_0(x) = e^{(x-1)/2}.$$

At $O(\epsilon)$ we have

$$2y_1' - y_1 = -y_0'' = -\frac{1}{4}e^{(x-1)/2},$$

to be solved subject to $y_1(0) = 0$ and $y_1(1) = 0$. This equation can be solved using an integrating factor (see Section A5.2), which gives

$$y_1(x) = -\frac{1}{2}xe^{(x-1)/2} + ke^{(x-1)/2},$$

for some constant k. Again, we cannot satisfy both boundary conditions, and we just use $y_1(1) = 0$, which gives

$$y_1(x) = \frac{1}{2}(1-x)e^{(x-1)/2}.$$

Finally, this gives

$$y = e^{(x-1)/2}\left\{1 + \frac{1}{2}(1-x)\epsilon\right\} + O(\epsilon^2), \qquad (12.13)$$

for $\epsilon \ll 1$. This suggests that $y \to e^{-1/2}(1 + \frac{1}{2}\epsilon)$ as $x \to 0$, which clearly does not satisfy the boundary condition $y(0) = 0$. We must therefore introduce a **boundary layer** at $x = 0$, across which y adjusts to satisfy the boundary condition. The idea is that, in some small neighbourhood of $x = 0$, the term $\epsilon y''$, which we neglected at leading order, becomes important because y varies rapidly.

We rescale by defining $x = \epsilon^\alpha X$, with $\alpha > 0$ (so that $x \ll 1$) and $X = O(1)$ as $\epsilon \to 0$, and write $y(x) = Y(X)$ for $X = O(1)$. Substituting this into (12.11) gives

$$\epsilon^{1-2\alpha}\frac{d^2Y}{dX^2} + 2\epsilon^{-\alpha}\frac{dY}{dX} - Y = 0.$$

Since $\alpha > 0$, the second term in this equation is large, and to obtain an asymptotic balance at leading order we must have $\epsilon^{1-2\alpha} = O(\epsilon^{-\alpha})$, which means that $1 - 2\alpha = -\alpha$, and hence $\alpha = 1$. So $x = \epsilon X$,

$$\frac{d^2Y}{dX^2} + 2\frac{dY}{dX} - \epsilon Y = 0, \qquad (12.14)$$

and $Y(0) = 0$. It is usual to refer to the region where $\epsilon \ll x \leq 1$ as the **outer region**, with **outer solution** $y(x)$, and the boundary layer region where $x = O(\epsilon)$ as the **inner region** with **inner solution** $Y(X)$. The other boundary condition is to be applied at $x = 1$. However, $x = 1$ does not lie in the inner region, where $x = O(\epsilon)$. In order to fix a second boundary condition for (12.14), we will have to make sure that the solution in the inner region is consistent, in a sense that we will make clear below, with the solution in the outer region, which does satisfy $y(1) = 1$.

We now expand

$$Y(X) = Y_0(X) + \epsilon Y_1(X) + O(\epsilon^2).$$

At leading order, $Y_0'' + 2Y_0' = 0$, to be solved subject to $Y_0(0) = 0$. The solution is

$$Y_0 = A(1 - e^{-2X}),$$

for some constant A. At leading order, we now know that

$$y \sim e^{(x-1)/2} \qquad \text{for } \epsilon \ll x \leq 1 \qquad \text{(the outer expansion)},$$
$$Y \sim A(1 - e^{-2X}) \quad \text{for } X = O(1),\ x = O(\epsilon) \quad \text{(the inner expansion)}.$$

For these two expansions to be consistent with each other, we must have

$$\lim_{X \to \infty} Y(X) = \lim_{x \to 0} y(x), \qquad (12.15)$$

which gives $A = e^{-1/2}$. We will see below, where we make this vague notion of "consistency" more precise, that this is correct.

At $O(\epsilon)$ we obtain the equation for $Y_1(X)$ as

$$Y_1'' + 2Y_1' = Y_0 = A(1 - e^{-2X}).$$

Integrating this once gives
$$Y_1' + 2Y_1 = A\left(X + \frac{1}{2}e^{-2X}\right) + c_1,$$

for some constant c_1. This can now be solved using an integrating factor, and the solution that satisfies $Y_1(0) = 0$ is
$$Y_1 = \frac{1}{2}AX(1 + e^{-2X}) - c_2(1 - e^{-2X}),$$

for some constant c_2, which we need to determine. To summarize, the two-term asymptotic expansions are
$$y \sim e^{(x-1)/2} + \frac{1}{2}\epsilon(1-x)e^{(x-1)/2} \quad \text{for } \epsilon \ll x \leqslant 1,$$

$$Y \sim A(1 - e^{-2X}) + \epsilon\left\{\frac{1}{2}AX(1 + e^{-2X}) - c_2(1 - e^{-2X})\right\} \quad \text{for } X = O(1),\ x = O(\epsilon).$$

We can determine the constants A and c_2 by forcing the two expansions to be consistent in the sense that they should be equal in an **intermediate region** or **overlap region**, where $\epsilon \ll x \ll 1$. The point is that in such a region we expect *both* expansions to be valid.

We define $x = \epsilon^\beta \hat{X}$ with $0 < \beta < 1$, and write $y = \hat{Y}(\hat{X})$. In terms of the **intermediate variable**, $\hat{X}$, the outer expansion becomes
$$\hat{Y} \sim e^{-1/2}\exp\left(\frac{1}{2}\epsilon^\beta \hat{X}\right) + \frac{1}{2}\epsilon(1 - \epsilon^\beta \hat{X})e^{-1/2}\exp\left(\frac{1}{2}\epsilon^\beta \hat{X}\right).$$

When $\hat{X} = O(1)$, we can expand the exponentials as Taylor series, and find that
$$\hat{Y} = e^{-1/2}\left(1 + \frac{1}{2}\epsilon^\beta \hat{X}\right) + \frac{1}{2}e^{-1/2}\epsilon + o(\epsilon). \tag{12.16}$$

Since $x = \epsilon X = \epsilon^\beta \hat{X}$ gives $X = \epsilon^{\beta-1}\hat{X}$, the inner expansion is
$$\hat{Y} \sim A\left\{1 - \exp\left(-2\epsilon^{\beta-1}\hat{X}\right)\right\}$$
$$+ \epsilon\left[\frac{1}{2}A\epsilon^{\beta-1}\hat{X}\left\{1 + \exp\left(-2\epsilon^{\beta-1}\hat{X}\right)\right\} - c_2\left\{1 - \exp\left(-2\epsilon^{\beta-1}\hat{X}\right)\right\}\right].$$

Now, since $\exp(-2\epsilon^{\beta-1}\hat{X}) = o(\epsilon^n)$ for all $n > 0$ (it is exponentially small for $\beta < 1$), we have
$$\hat{Y} = A + \frac{1}{2}A\epsilon^\beta \hat{X} - c_2\epsilon + o(\epsilon). \tag{12.17}$$

Since (12.16) and (12.17) must be identical, we need $A = e^{-1/2}$, consistent with the crude assumption, (12.15), that we made above, and also $c_2 = -\frac{1}{2}e^{-1/2}$. This process, whereby we make the inner and outer expansions consistent, is known as **asymptotic matching**, and the inner and outer expansions are known as **matched asymptotic expansions**. A comparison between the one-term inner and outer solutions and the exact solution is given in Figure 12.4. It should be clear that the inner expansion is a poor approximation in the outer region and vice versa. A little later, we will show how to construct, using the inner and outer expansions, a composite expansion that is uniformly valid for $0 \leqslant x \leqslant 1$.

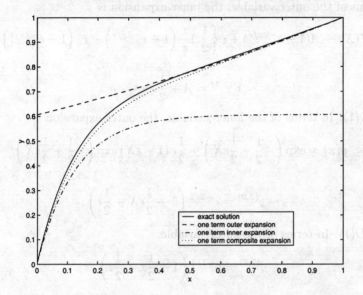

Fig. 12.4. Exact and asymptotic solutions of (12.11) when $\epsilon = 0.25$.

Van Dyke's Matching Principle

The use of an intermediate variable in an overlap region can get *very* tedious in more complicated problems. A method that works most, but not all, of the time, and is much easier to use, is **Van Dyke's matching principle**. This principle is much easier to use than to explain, but let's start with the explanation.

Let

$$y(x) \sim \sum_{n=0}^{N} \phi_n(\epsilon) y_n(x)$$

be the outer expansion and

$$Y(X) \sim \sum_{n=0}^{N} \psi_n(\epsilon) Y_n(X)$$

be the inner expansion with respect to the asymptotic sequences $\phi_n(\epsilon)$ and $\psi_n(\epsilon)$, with $x = f(\epsilon)X$. In order to analyze how the outer expansion behaves in the inner region, we can write $y(x)$ in terms of $X = x/f(\epsilon)$, and retain M terms in the resulting asymptotic expansion. We denote this by $y^{(N,M)}$, the Mth order inner approximation of the outer expansion. Similarly, we can write $Y(X)$ in terms of x, and retain M terms in the resulting expansion. We denote this by $Y^{(N,M)}$, the Mth order outer approximation of the inner expansion. Van Dyke's matching principle states that $y^{(N,M)} = Y^{(M,N)}$. Let's see how this works for our example.

In terms of the outer variable, the inner expansion is

$$Y(X) \sim A(1 - e^{-2x/\epsilon}) + \epsilon \left\{ \frac{1}{2} A \frac{x}{\epsilon} \left(1 + e^{-2x/\epsilon}\right) - c_2 \left(1 - e^{-2x/\epsilon}\right) \right\}$$

$$\sim Y^{(2,2)} = A + \frac{1}{2} Ax - c_2 \epsilon,$$

for $x = O(1)$. In terms of the inner variable, the outer expansion is

$$y(x) \sim \exp\left(-\frac{1}{2} + \frac{1}{2}\epsilon X\right) + \frac{1}{2}\epsilon (1 - \epsilon X) \exp\left(-\frac{1}{2} + \frac{1}{2}\epsilon X\right)$$

$$\sim y^{(2,2)} = e^{-1/2} \left(1 + \frac{1}{2}\epsilon X + \frac{1}{2}\epsilon\right),$$

for $X = O(1)$. In terms of the outer variable,

$$y^{(2,2)} = e^{-1/2} \left(1 + \frac{1}{2}x + \frac{1}{2}\epsilon\right).$$

Van Dyke's matching principle states that $Y^{(2,2)} = y^{(2,2)}$, and therefore gives $A = e^{-1/2}$ and $c_2 = -\frac{1}{2}e^{-1/2}$ rather more painlessly than before.

Composite Expansions

Although we now know how to construct inner and outer solutions, valid in the inner and outer regions, it would be more useful to have an asymptotic solution valid uniformly across the whole domain of solution, $0 \leqslant x \leqslant 1$ in the example. We can construct such a uniformly valid **composite expansion** by using the inner and outer expansions. We simply need to add the two expansions and subtract the expression that gives their overlap. The overlap is just the expression that appears to the appropriate order in the intermediate region, (12.17) or (12.16), or equivalently the matched part, $y^{(2,2)}$ or $Y^{(2,2)}$. For our example problem, the one-term composite expansion is

$$y \sim y_0 + Y_0 - y^{(1,1)} = e^{(x-1)/2} + e^{-1/2}(1 - e^{-2X}) - e^{-1/2}$$

$$= e^{(x-1)/2} - e^{-1/2 - 2x/\epsilon} \text{ for } 0 \leqslant x \leqslant 1 \text{ as } \epsilon \to 0.$$

This composite expansion is shown in Figure 12.4, and shows good agreement with the exact solution across the whole domain, as expected. Note that, in terms of Van Dyke's matching principle, we can write the composite solution of any order as

$$y \sim y_c^{(M,N)} = \sum_{n=0}^{M} y_n(x) + \sum_{n=0}^{N} Y_n(X) - y^{(M,N)}.$$

12.2 ORDINARY DIFFERENTIAL EQUATIONS

The Location of the Boundary Layer

In our example, when we constructed the outer solution, we chose to satisfy the boundary condition at $x = 1$ and assume that there was a boundary layer at $x = 0$. Why is this correct? Let's see what happens if we assume that there is a boundary layer at $x = x_0$. Strictly speaking, if $x_0 \neq 0$ and $x_0 \neq 1$ this is an **interior layer**. We define scaled variables $y(x) = Y(X)$ and $x = x_0 + \epsilon^\alpha X$, with $\alpha > 0$ and Y, $X = O(1)$ for $\epsilon \ll 1$. As before, we find that we can only obtain an asymptotic balance at leading order by taking $\alpha = 1$, so that $x = x_0 + \epsilon X$ and

$$Y'' + 2Y' - \epsilon = 0.$$

At leading order, as before, $Y_0 = A_0 + B_0 e^{-2X}$. As $X \to -\infty$, Y_0 becomes exponentially large, and cannot be matched to the outer solution. This forces us to take $x_0 = 0$, since then this solution is only needed for $X \geqslant 0$, and, as we have seen, we can construct an asymptotic solution.

Interior Layers

Singular perturbations of ordinary differential equations need not always result in a boundary layer. As an example, consider

$$\epsilon y'' + 2xy' + 2x = 0, \tag{12.18}$$

to be solved for $-1 < x < 1$, subject to the boundary conditions

$$y(-1) = 2, \quad y(1) = 3. \tag{12.19}$$

We will try to construct the leading order solution for $\epsilon \ll 1$. The leading order outer solution satisfies $2x(y' + 1) = 0$, and hence $y = k - x$ for some constant k. If $y(-1) = 2$ we need $y = 1-x$, whilst if $y(1) = 3$ we need $y = 4-x$. We clearly cannot satisfy both boundary conditions with the same outer solution, so let's look for a boundary or interior layer at $x = x_0$ by defining $y(x) = Y(X)$ and $x = x_0 + \epsilon^\alpha X$, with $Y, X = O(1)$. In terms of these scaled variables, (12.18) becomes

$$\epsilon^{1-2\alpha} Y_{XX} + 2(x_0 + \epsilon^\alpha X)(\epsilon^{-\alpha} Y_X + 1) = 0.$$

If $x_0 \neq 0$, for a leading order balance we need $\epsilon^{1-2\alpha} = O(\epsilon^{-\alpha})$, and hence $\alpha = 1$. In this case, at leading order,

$$Y_{XX} + 2x_0 Y_X = 0,$$

and hence $Y = A + B e^{-2x_0 X}$. For $x_0 > 0$ this grows exponentially as $X \to -\infty$, whilst for $x_0 < 0$ this grows exponentially as $X \to \infty$. In either case, we cannot match these exponentially growing terms with the outer solution. This suggests that we need $x_0 = 0$, when

$$\epsilon^{1-2\alpha} Y_{XX} + 2X Y_X + 2\epsilon^\alpha X = 0.$$

For a leading order asymptotic balance we need $\alpha = 1/2$, and hence a boundary layer with width of $O(\epsilon^{1/2})$. At leading order,

$$Y_{XX} + 2X Y_X = 0,$$

which, after multiplying by the integrating factor, e^{X^2}, gives

$$\frac{d}{dX}\left(e^{X^2} Y_X\right) = 0,$$

and hence

$$Y = B + A \int_{-\infty}^{X} e^{-s^2} ds.$$

Now, since the interior layer is at $x = 0$, the outer solution must be

$$y = \begin{cases} 1 - x & \text{for } -1 \leqslant x < O(\epsilon^{1/2}), \\ 4 - x & \text{for } O(\epsilon^{1/2}) < x \leqslant 1. \end{cases}$$

Since $y \to 4$ as $x \to 0^+$ and $y \to 1$ as $x \to 0^-$, we must have $Y \to 4$ as $X \to \infty$ and $Y \to 1$ as $X \to -\infty$. This allows us to fix the constants A and B and find that

$$Y(X) = 1 + \frac{3}{\sqrt{\pi}} \int_{-\infty}^{X} e^{-s^2} ds = \frac{1}{2}(5 + 3\,\mathrm{erf}(x)),$$

which leads to the one-term composite solution

$$y \sim y_c = -x + \frac{1}{2}\left\{5 + 3\,\mathrm{erf}\left(\frac{x}{\sqrt{\epsilon}}\right)\right\} \quad \text{for } -1 \leqslant x \leqslant 1 \text{ and } \epsilon \ll 1. \tag{12.20}$$

This is illustrated in Figure 12.5 for various values of ϵ. Note that the boundary conditions at $x = \pm 1$ are only satisfied by the composite expansion at leading order.

12.2.3 Nonlinear Problems

Example 1: Elliptic functions of large period

As we have already seen in Section 9.4, the Jacobian elliptic function $x = \mathrm{sn}(t; k)$ satisfies the equation

$$\frac{dx}{dt} = \sqrt{1 - x^2}\sqrt{1 - k^2 x^2}, \tag{12.21}$$

subject to $x = 0$ when $t = 0$, and has periodic solutions for $k \neq 1$. When $k = 1$, the solution that corresponds to $\mathrm{sn}(t; k)$ is a heteroclinic path that connects the equilibrium points $(\pm 1, 0, 0)$ in the phase portrait shown in Figure 9.18, and hence the period tends to infinity as $k \to 1$. When k is close to unity, it seems reasonable to assume that the period of the solution is large but finite. Can we quantify this? Let's assume that $k^2 = 1 - \delta^2$, with $\delta \ll 1$, and seek an asymptotic solution for the first quarter period of $x(t)$, with $0 \leqslant x \leqslant 1$. Figure 12.6 shows $\mathrm{sn}(t; k)$ for various values of δ, and we can see that the period does increase as δ decreases and k approaches unity. The function $\mathrm{sn}(t; k)$ is available in MATLAB as `ellipj`. The quarter period is simply the value of t when $\mathrm{sn}(t; k)$ reaches its first maximum.

We seek an asymptotic solution of the form

$$x = x_0 + \delta^2 x_1 + \delta^4 x_2 + O\left(\delta^6\right).$$

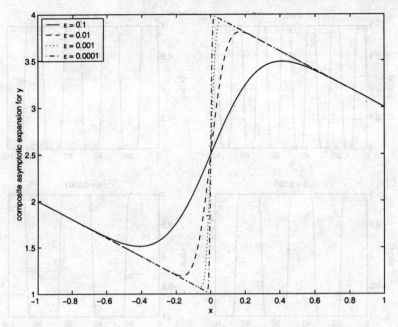

Fig. 12.5. The composite asymptotic solution, (12.20), of (12.18).

Using a binomial expansion, (12.21) is

$$\frac{dx}{dt} = (1-x^2)\left(1+\delta^2\frac{x^2}{1-x^2}\right)^{1/2} = 1 - x^2 + \frac{1}{2}\delta^2 x^2 + \frac{1}{8}\delta^4 \frac{x^4}{1-x^2} + O(\delta^6).$$

This binomial expansion is only valid when x is not too close unity, so we should expect any asymptotic expansion that we develop to become nonuniform as $x \to 1$, and we treat this as the outer expansion.

At leading order,

$$\frac{dx_0}{dt} = 1 - x_0^2, \quad \text{subject to } x_0(0) = 0,$$

which has solution $x_0 = \tanh t$. At $O(\delta^2)$,

$$\frac{dx_1}{dt} = -2x_0 x_1 + \frac{1}{2}x_0^2 = -2\tanh t\, x_1 + \frac{1}{2}\tanh^2 t, \quad \text{subject to } x_1(0) = 0.$$

Using the integrating factor $\cosh^2 t$, we can find the solution

$$x_1 = \frac{1}{4}\left(\tanh t - t\,\text{sech}^2 t\right).$$

We can now see that

$$x \sim 1 - 2e^{-2t} + \frac{1}{4}\delta^2 + O(\delta^4) \quad \text{as } t \to \infty.$$

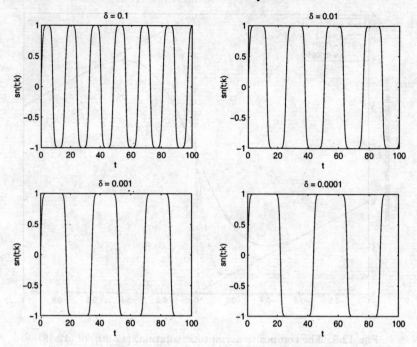

Fig. 12.6. The Jacobian elliptic function sn($t; k$) for various values of δ.

Although x approaches unity as $t \to \infty$, there is no nonuniformity in this expansion, so we need to go to $O(\delta^4)$. At this order,

$$\frac{dx_2}{dt} = -x_1^2 - 2x_0 x_2 + x_0 x_1 - \frac{1}{8}\left(\frac{x_0^4}{1 - x_0^2}\right) \quad \text{subject to } x_2(0) = 0.$$

Solving this problem would be extremely tedious. Fortunately, we don't really want to know the exact expression for x_2, just its behaviour as $t \to \infty$. Using the known behaviour of the various hyperbolic functions, we find that

$$\frac{dx_2}{dt} + 2x_2 \sim -\frac{1}{32} e^{2t} \quad \text{as } t \to \infty,$$

and hence from solving this simple linear equation,

$$x_2 \sim -\frac{1}{128} e^{2t} \quad \text{as } t \to \infty.$$

This shows that

$$x \sim 1 - 2e^{-2t} + \frac{1}{4}\delta^2 - \frac{1}{128}\delta^4 e^{2t} + O(\delta^6) \quad \text{as } t \to \infty. \tag{12.22}$$

We can now see that the fourth term in this expansion becomes comparable to the third when $\delta^4 e^{2t} = O(\delta^2)$, and hence as $t \to \log(1/\delta)$, when $x = 1 + O(\delta^2)$.

12.2 ORDINARY DIFFERENTIAL EQUATIONS

We therefore define new variables for an inner region close to $x = 1$ as

$$x = 1 - \delta^2 X, \quad t = \log\left(\frac{1}{\delta}\right) + T.$$

On substituting these inner variables into (12.21), we find that, at leading order,

$$\frac{dX}{dT} = -\sqrt{2X(2X+1)}.$$

Using the substitution $X = \bar{X} - \frac{1}{4}$ brings this separable equation into a standard form, and the solution is

$$X = \frac{1}{4}\{\cosh(K - 2T) - 1\}. \qquad (12.23)$$

We now need to determine the constant K by matching the inner solution, (12.23), with the outer solution, whose behaviour as $x \to 1$ is given by (12.22). Writing the inner expansion in terms of the outer variables and retaining terms up to $O(\delta^2)$ leads to

$$x = 1 - \frac{1}{8}e^K e^{-2t} + \frac{1}{4}\delta^2 + O(\delta^4),$$

for $t = O(1)$. Comparing this with (12.22) shows that we need $\frac{1}{8}e^K e^{-2t} = 2e^{-2t}$, and hence $K = 4\log 2$, which gives

$$x = 1 - \frac{1}{4}\delta^2\{\cosh(4\log 2 - 2T) - 1\} + O(\delta^4) \qquad (12.24)$$

when $T = O(1)$. From this leading order approximation, $x = 1$ when $T = T_0 = 2\log 2 + O(\delta^2)$. This is the quarter period of the solution, so the period τ satisfies

$$\frac{1}{4}\tau = \log\left(\frac{1}{\delta}\right) + T_0,$$

and hence

$$\tau = 4\log\left(\frac{4}{\delta}\right) + O(\delta^2),$$

for $\delta \ll 1$. We conclude that the period of the solution does indeed tend to infinity as $\delta \to 0$, $k \to 1^-$, but only logarithmically fast. Figure 12.7 shows a comparison between the exact and analytical solutions. The agreement is very good for all $\delta \leqslant 1$. We produced this figure using the MATLAB script

```
Texact = []; d = 10.^(-7:0.25:0); Tasymp = 4*log(4./d);
options = optimset('Display','off','TolX', 10^-10);
for del = d
    k = sqrt(1-del^2); T2 = 2*log(4/del);
    Texact = [Texact 2*fzero(@ellipj,T2,options,k)];
end
plot(log10(d),Texact,log10(d),Tasymp, '--')
xlabel('log_{10}\delta'), ylabel('T')
legend('exact','asymptotic')
```

This uses the MATLAB function `fzero` to find where the elliptic function is zero, using the asymptotic expression as an initial estimate. Note that the function `optimset` allows us to create a variable `options` that we can pass to `fzero` as a parameter, which controls the details of its execution. In this case, we turn off the output of intermediate results, and set the convergence tolerance to 10^{-10}.

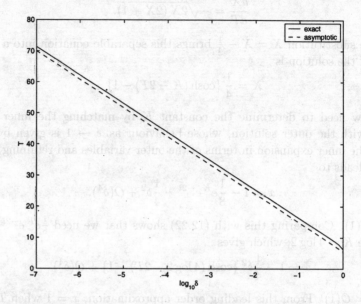

Fig. 12.7. A comparison of the exact period of the elliptic function $\text{sn}(t; k)$ for $k = \sqrt{1-\delta^2}$.

Finally, by adding the solutions in the inner and outer regions and subtracting the matched part, $1 + \frac{1}{4}\delta^2 - 2e^{-2t}$, we can obtain a composite expansion, uniformly valid for $0 \leqslant t \leqslant \frac{1}{4}\tau = \log(4/\delta) + O(\delta^2)$, as

$$x = \tanh t + 2e^{-2t} + \frac{1}{4}\delta^2 \left[\tanh t - t\,\text{sech}^2 t - \cosh\left\{\log\left(\frac{16}{\delta^2}\right) - 2t\right\}\right] + O(\delta^4).$$

Example 2: A thermal ignition problem

Many materials decompose to produce heat. This decomposition is usually more rapid the higher the temperature. This leads to the possibility of thermal ignition. As a material gets hotter, it releases heat more rapidly, which heats it more rapidly, and so on. This positive feedback mechanism can lead to the unwanted, and potentially disastrous, combustion of many common materials, ranging from packets of powdered milk to haystacks. The most common physical mechanism that can break this feedback loop is the diffusion of heat through the material and out through its surface. The rate of heat production due to decomposition is proportional to the volume of the material and the rate of heat loss from its surface proportional to surface area. For a sufficiently large volume of material, heat production dominates heat loss, and the material ignites. Determining the critical temperature below which it is safe to store a potentially combustible material is an important and difficult problem.†

We now want to develop a mathematical model of this problem. In Section 2.6.1, we showed how to derive the diffusion equation, (2.12), which governs the flow of heat in a body. We now need to include the effect of a chemical reaction that produces $R(x, y, z, t)$ units of heat, per unit volume, per unit time, by adding a term $R\,\delta t\,\delta x\,\delta y\,\delta z$ to the right hand side of (2.11). On taking the limit δt, δx, δy, $\delta z \to 0$, we arrive at

$$\rho c \frac{\partial T}{\partial t} = -\nabla \cdot \mathbf{Q} + R,$$

and hence, for a steady state solution ($\partial/\partial t = 0$), and since Fourier's law of heat conduction is $\mathbf{Q} = -k\nabla T$,

$$k\nabla^2 T + R = 0. \qquad (12.25)$$

The rate of combustion of the material can be modelled using the **Arrhenius law**, $R = Ae^{-T_a/T}$, where A is a constant and T_a is the **activation temperature**, also a constant. It is important to note that T is the **absolute temperature** here. The Arrhenius law can be derived from first principles using statistical mechanics, although we will not attempt this here (see, for example, Flowers and Mendoza, 1970). Figure 12.8 shows that the reaction rate is zero at absolute zero ($T = 0$), and remains small until T approaches the activation temperature, T_a, when it increases, with $R \to A$ as $T \to \infty$. After defining $u = T/T_a$ and rescaling distances

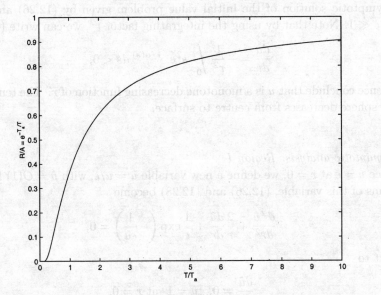

Fig. 12.8. The Arrhenius reaction rate law.

† For more background on combustion theory, see Buckmaster and Ludford (1982).

ASYMPTOTIC METHODS: DIFFERENTIAL EQUATIONS

with $\sqrt{k/A}$, we arrive at the nonlinear partial differential equation

$$\nabla^2 u + e^{-1/u} = 0.$$

For a uniform sphere of combustible material with a spherically-symmetric temperature distribution, this becomes

$$\frac{d^2 u}{dr^2} + \frac{2}{r}\frac{du}{dr} + e^{-1/u} = 0, \qquad (12.26)$$

subject to

$$\frac{du}{dr} = 0 \text{ at } r = 0, \quad u = u_a \text{ at } r = r_a. \qquad (12.27)$$

Here, u_a is the dimensionless absolute temperature of the surroundings and r_a the dimensionless radius of the sphere. Note that, as we discussed earlier, we would expect that the larger r_a, the smaller u_a must be to prevent the ignition of the sphere.† A positive solution of this boundary value problem represents a possible steady state solution in which these two physical processes are in balance. If no such steady state solution exists, we can conclude that the material will ignite. A small trick that makes the study of this problem easier is to replace (12.27) with

$$\frac{du}{dr} = 0, \quad u = \epsilon \text{ at } r = 0. \qquad (12.28)$$

We can then solve the initial value problem given by (12.26) and (12.28) for a given value of ϵ, the dimensionless temperature at the centre of the sphere, and then determine the corresponding value of $u_a = u(r_a)$. Our task is therefore to construct an asymptotic solution of the initial value problem given by (12.26) and (12.28) when $\epsilon \ll 1$. Note that by using the integrating factor r^2, we can write (12.26) as

$$\frac{du}{dr} = -\frac{1}{r^2}\int_0^r s^2 e^{-1/u(s)}\, ds < 0,$$

and hence conclude that u is a monotone decreasing function of r. The temperature of the sphere decreases from centre to surface.

Asymptotic analysis: Region I

Since $u = \epsilon$ at $r = 0$, we define a new variable $\hat{u} = u/\epsilon$, with $\hat{u} = O(1)$ for $\epsilon \ll 1$. In terms of this variable, (12.26) and (12.28) become

$$\frac{d^2 \hat{u}}{dr^2} + \frac{2}{r}\frac{d\hat{u}}{dr} + \frac{1}{\epsilon}\exp\left(-\frac{1}{\epsilon \hat{u}}\right) = 0,$$

subject to

$$\frac{d\hat{u}}{dr} = 0, \quad \hat{u} = 1 \text{ at } r = 0.$$

† For all the technical details of this problem, which was first studied by Gel'fand (1963), see Billingham (2000).

12.2 ORDINARY DIFFERENTIAL EQUATIONS

At leading order,
$$\frac{d^2\hat{u}}{dr^2} + \frac{2}{r}\frac{d\hat{u}}{dr} = 0, \qquad (12.29)$$

which has the trivial solution $\hat{u} = 1$. We're going to need more than this to be able to proceed, so let's look for an asymptotic expansion of the form

$$\hat{u} = 1 + \phi_1(\epsilon)\hat{u}_1 + \phi_2(\epsilon)\hat{u}_2 + \cdots,$$

where $\phi_2 \ll \phi_1 \ll 1$ are to be determined. As we shall see, we need a three-term asymptotic expansion to be able to determine the scalings for the next asymptotic region.

Firstly, note that

$$\frac{1}{\epsilon}e^{-1/\epsilon\hat{u}} \sim \frac{1}{\epsilon}\exp\left\{-\frac{1}{\epsilon}(1+\phi_1\hat{u}_1)^{-1}\right\} \sim \frac{1}{\epsilon}\exp\left\{-\frac{1}{\epsilon}(1-\phi_1\hat{u}_1)\right\}$$

$$\sim \frac{1}{\epsilon}e^{-1/\epsilon}\exp\left(\frac{\phi_1}{\epsilon}\hat{u}_1\right) \sim \frac{1}{\epsilon}e^{-1/\epsilon}\left(1 + \frac{\phi_1}{\epsilon}\hat{u}_1\right),$$

provided that $\phi_1 \ll \epsilon$, which we can check later. Equation (12.26) then shows that

$$\phi_1\left(\frac{d^2\hat{u}_1}{dr^2} + \frac{2}{r}\frac{d\hat{u}_1}{dr}\right) + \phi_2\left(\frac{d^2\hat{u}_2}{dr^2} + \frac{2}{r}\frac{d\hat{u}_2}{dr}\right) \sim -\frac{1}{\epsilon}e^{-1/\epsilon}\left(1 + \frac{\phi_1}{\epsilon}\hat{u}_1\right). \qquad (12.30)$$

In order to obtain a balance of terms, we therefore take

$$\phi_1 = \frac{1}{\epsilon}e^{-1/\epsilon} \ll \epsilon, \quad \phi_2 = \frac{1}{\epsilon^2}e^{-1/\epsilon}\phi_1 = \frac{1}{\epsilon^3}e^{-2/\epsilon},$$

and hence expand

$$\hat{u} = 1 + \frac{1}{\epsilon}e^{-1/\epsilon}\hat{u}_1 + \frac{1}{\epsilon^3}e^{-2/\epsilon}\hat{u}_2 + \cdots.$$

Now, using (12.30),
$$\frac{d^2\hat{u}_1}{dr^2} + \frac{2}{r}\frac{d\hat{u}_1}{dr} = -1, \qquad (12.31)$$

subject to
$$\frac{d\hat{u}_1}{dr} = \hat{u}_1 = 0 \text{ at } r = 0.$$

Using the integrating factor r^2, we find that the solution is $\hat{u}_1 = -\frac{1}{6}r^2$. Similarly,

$$\frac{d^2\hat{u}_2}{dr^2} + \frac{2}{r}\frac{d\hat{u}_2}{dr} = \frac{1}{6}r^2,$$

subject to
$$\frac{d\hat{u}_2}{dr} = \hat{u}_2 = 0 \text{ at } r = 0,$$

and hence $\hat{u}_2 = \frac{1}{120}r^4$. This means that

$$\hat{u} \sim 1 - \frac{1}{6\epsilon}e^{-1/\epsilon}r^2 + \frac{1}{120\epsilon^3}e^{-2/\epsilon}r^4 \qquad (12.32)$$

for $\epsilon \ll 1$. This expansion is not uniformly valid, since for sufficiently large r the second term is comparable to the first, when $r = O(\epsilon^{1/2}e^{1/2\epsilon})$, and the third is comparable to the second, when $r = O(\epsilon e^{1/2\epsilon}) \ll \epsilon^{1/2}e^{1/2\epsilon}$. As r increases, the first nonuniformity to occur is therefore when $r = O(\epsilon e^{1/2\epsilon})$ and $u = \epsilon + O(\epsilon^2)$. Note that this is why we needed a three-term expansion to work out the scalings for the next asymptotic region.

What would have happened if we had taken only a two-term expansion and looked for a new asymptotic region with $r = O(\epsilon^{1/2}e^{1/2\epsilon})$ and $u = O(\epsilon)$? If you try this, you will find that it is impossible to match the solutions in the new region to the solutions in region I. After some thought, you notice that the equations at $O(1)$ and $O(\frac{1}{\epsilon}e^{-1/\epsilon})$ in region I, (12.29) and (12.31), do not depend at all on the functional form of the term $e^{-1/\epsilon\hat{u}}$, which could be replaced by $e^{-1/\epsilon}$ without affecting (12.29) or (12.31). This is a sign that we need another term in the expansion in order to capture the effect of the only nonlinearity in the problem.

Asymptotic analysis: Region II

In this region we define scaled variables $u = \epsilon + \epsilon^2 U$, $r = \epsilon e^{1/2\epsilon}R$, with $U = O(1)$ and $R = O(1)$ for $\epsilon \ll 1$. At leading order, (12.26) becomes

$$\frac{d^2U}{dR^2} + \frac{2}{R}\frac{dU}{dR} + e^U = 0, \qquad (12.33)$$

to be solved subject to the matching condition

$$U \sim -\frac{1}{6}R^2 \quad \text{as } R \to 0. \qquad (12.34)$$

Equation (12.33) is nonlinear and nonautonomous, which usually means that we must resort to finding a numerical solution. However, we have seen in Chapter 10 that we can often make progress using group theoretical methods. Equation (12.33) is invariant under the transformation $U \mapsto U + k$, $R \mapsto e^{-k/2}R$. We can therefore make the nonlinear transformation

$$p(s) = e^{-2s}e^U, \quad q(s) = 2 + e^{-s}\frac{dU}{dR}, \quad R = e^{-s},$$

after which (12.33) and (12.34) become

$$\frac{dp}{ds} = -pq, \quad \frac{dq}{ds} = p + q - 2, \qquad (12.35)$$

subject to

$$p \sim e^{-2s}, \quad q \sim 2 - \frac{1}{3}e^{-2s} \quad \text{as } s \to \infty. \qquad (12.36)$$

The problem is still nonlinear, but is now autonomous, so we can use the phase plane methods that we studied in Chapter 9.

There are two finite equilibrium points, at $P_1 = (0,2)$ and $P_2 = (2,0)$ in the (p,q) phase plane. After determining the Jacobian at each of these points and calculating the eigenvalues in the usual way, we find that P_1 is a saddle point and P_2 is an unstable, anticlockwise spiral. Since (12.36) shows that we are interested in a solution that asymptotes to P_1 as $s \to \infty$, this solution must be represented by one of the stable separatrices of P_1. Furthermore, since $p = e^{-2s}e^U > 0$, the unique integral path that represents the solution is the stable separatrix of P_1 that lies in the half plane $p > 0$. What happens to this separatrix as $s \to -\infty$? A sensible, and correct, guess would be that it asymptotes to the unstable spiral at P_2, as shown in Figure 12.9, for which we calculated the solution numerically using MATLAB (see Section 9.3.4). The proof that this is the case is Exercise 9.17, which comes with some hints.

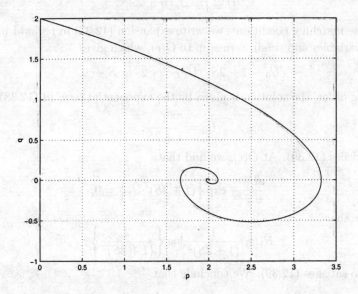

Fig. 12.9. The behaviour of the solution of (12.35) subject to (12.36) in the (p,q)-phase plane.

Since the solution asymptotes to P_2, we can determine its behaviour as $s \to -\infty$ by considering the local solution there. The eigenvalues of P_2 are $\frac{1}{2}(1 \pm i\sqrt{7})$, and therefore

$$p \sim 2 + Ae^{s/2} \sin\left(\frac{\sqrt{7}}{2}s + B\right) \quad \text{as } s \to -\infty,$$

for some constants A and B. Since $U = 2s + \log p$ and $s = -\log R$, this shows that

$$U \sim -2\log R + \log 2 - \frac{A}{2\sqrt{R}} \sin\left(\frac{\sqrt{7}}{2}\log R - B\right) \quad \text{as } R \to \infty. \quad (12.37)$$

We conclude that

$$u \sim \epsilon + \epsilon^2(\log 2 - 2\log R) \quad \text{as } R \to \infty,$$

for $\epsilon \ll 1$. When $\log R = O(\epsilon^{-1})$, the second term is comparable to the first term, and we have a further nonuniformity.

Asymptotic analysis: Region III

We define scaled variables $u = \epsilon \bar{U}$ and $S = \epsilon \log R = -\frac{1}{2} - \epsilon \log \epsilon + \epsilon \log r$, $r = \epsilon e^{1/2\epsilon} e^{S/\epsilon}$, with $U = O(1)$ and $S = O(1)$ for $\epsilon \ll 1$. In terms of these variables, (12.26) becomes

$$\epsilon \frac{d^2 \bar{U}}{dS^2} + \frac{d\bar{U}}{dS} + \exp\left\{\frac{1}{\epsilon}\left(-\frac{1}{\bar{U}} + 1 + 2S\right)\right\} = 0. \tag{12.38}$$

We expand $\bar{U}$ as

$$\bar{U} = \bar{U}_0 + \epsilon \bar{U}_1 + \cdots.$$

To find the matching conditions, we write expansion (12.37) in region I in terms of the new variables and retain terms up to $O(\epsilon)$, which gives

$$\bar{U}_0 \sim 1 - 2S, \quad \bar{U}_1 \sim \log 2 \quad \text{as } S \to 0. \tag{12.39}$$

At leading order, the solution is given by the exponential term in (12.38) as

$$\bar{U}_0 = \frac{1}{1 + 2S},$$

which satisfies (12.39). At $O(\epsilon)$, we find that

$$\frac{d\bar{U}_0}{dS} + \exp\left\{(1 + 2S)^2 \bar{U}_1\right\} = 0,$$

and hence that

$$\bar{U}_1 = \frac{1}{(1 + 2S)^2} \log\left\{\frac{2}{(1 + 2S)^2}\right\},$$

which also satisfies (12.39). We conclude that

$$u = \frac{\epsilon}{1 + 2S} + \frac{\epsilon^2}{(1 + 2S)^2} \log\left\{\frac{2}{(1 + 2S)^2}\right\} + O(\epsilon^3),$$

for $S = O(1)$ and $\epsilon \ll 1$. This expansion remains uniform, with $u \to 0$ as $S \to \infty$, and hence $R \to \infty$, and the solution is complete.

We can now determine $u_a = u(r_a)$. If $r_a \ll \epsilon e^{1/2\epsilon}$, $r = r_a$ lies in region I, so that $u_a \sim \epsilon - \frac{1}{6} e^{-1/\epsilon} r_a^2 \sim \epsilon$. In other words, for u_a sufficiently small that $r_a \ll u_a e^{1/2u_a}$, a steady state solution is always available, and we predict that the sphere will not ignite.

If $r_a = O(\epsilon e^{1/2\epsilon})$, we need to consider the solution in region II. The oscillations in p lead to oscillations in u_a as a function of ϵ, as shown in Figure 12.10. In particular, the fact that $p < p_{\max} \approx 3.322$, as can be seen in Figure 12.9, leads to a maximum value of u_a for which a steady state solution is available, namely

$$u_{a\max} \sim \epsilon + \epsilon^2 \log\left(\frac{p_{\max} \epsilon^2 e^{1/\epsilon}}{r_a^2}\right). \tag{12.40}$$

Finally, if $r_a = O(\epsilon e^{3/2\epsilon})$, the solution in region III shows that

$$u_a \sim \frac{\epsilon}{\frac{1}{2} + 2\epsilon \log(r_a/\epsilon)} < u_{a\max}.$$

We conclude that $u_{a\mathrm{max}}$ (the **critical ignition temperature** or **critical storage temperature**) gives an asymptotic approximation to the hottest ambient temperature at which a steady state solution exists, and hence at which the material will not ignite.

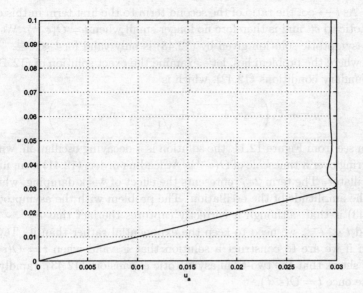

Fig. 12.10. The ambient temperature, u_a, as a function of ϵ in region II when $r_\mathrm{a} = 10^9$.

12.2.4 The Method of Multiple Scales

Let's now return to solving a simple, linear, constant coefficient ordinary differential equation that, at first sight, seems like a regular perturbation problem. Consider

$$\ddot{y} + 2\epsilon \dot{y} + y = 0, \qquad (12.41)$$

to be solved subject to

$$y(0) = 1, \quad \dot{y}(0) = 0, \qquad (12.42)$$

for $t \geqslant 0$, where a dot denotes d/dt. Since this is an initial value problem, we can think of y developing as a function of time, t. As usual, we expand

$$y = y_0(t) + \epsilon y_1(t) + O(\epsilon^2)$$

for $\epsilon \ll 1$. At leading order, $\ddot{y}_0 + y_0 = 0$, subject to $y_0(0) = 1$, $\dot{y}_0(0) = 0$, which has solution $y_0 = \cos t$. At $O(\epsilon)$,

$$\ddot{y}_1 + y_1 = -2\dot{y}_0 = 2\sin t,$$

subject to $y_1(0) = 0$, $\dot{y}_1(0) = 0$. After noting that the particular integral solution of this equation takes the form $y_{1p} = kt\cos t$ for some constant k, we find that $y_1 = -t\cos t + \sin t$. This means that

$$y(t) \sim \cos t + \epsilon(-t\cos t + \sin t) \tag{12.43}$$

as $\epsilon \to 0$. As $t \to \infty$, the ratio of the second term to the first term in this expansion is asymptotic to ϵt, and is therefore no longer small when $t = O(\epsilon^{-1})$. We conclude that the asymptotic solution given by (12.43) is only valid for $t \ll \epsilon^{-1}$.

To see where the problem lies, let's examine the exact solution of (12.41) subject to the boundary conditions (12.42), which is

$$y = e^{-\epsilon t}\left(\cos\sqrt{1-\epsilon^2}\,t + \frac{\epsilon}{\sqrt{1-\epsilon^2}}\sin\sqrt{1-\epsilon^2}\,t\right). \tag{12.44}$$

As we can see from Figure 12.11, the solution is a decaying oscillation, with the decay occurring over a timescale of $O(\epsilon^{-1})$. At leading order, (12.41) is an undamped, linear oscillator. The term $2\epsilon\dot{y}$ represents the effect of weak damping, which slowly reduces the amplitude of the oscillation. The problem with the asymptotic expansion (12.43) is that, although it correctly captures the fact that $e^{-\epsilon t} \sim 1 - \epsilon t$ for $\epsilon \ll 1$ and $t \ll \epsilon^{-1}$, we need to keep the exponential rather than its Taylor series expansion if we are to construct a solution that is valid when $t = O(\epsilon^{-1})$. Figure 12.11 shows that the two-term asymptotic expansion, (12.43), rapidly becomes inaccurate once $t = O(\epsilon^{-1})$.

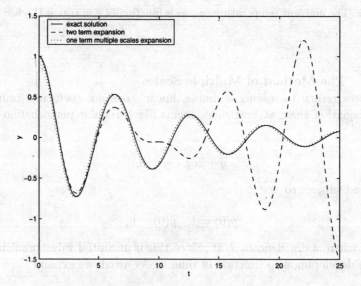

Fig. 12.11. The exact solution, (12.44), two-term asymptotic solution, (12.43), and one-term multiple scales solution (12.50) of (12.41) when $\epsilon = 0.1$.

The **method of multiple scales**, in its most basic form, consists of defining a new **slow time variable**, $T = \epsilon t$, so that when $t = O(1/\epsilon)$, $T = O(1)$, and the slow decay can therefore be accounted for. We then look for an asymptotic solution

$$y = y_0(t,T) + \epsilon y_1(t,T) + O(\epsilon^2),$$

with each term a function of both t, to capture the oscillation, and T, to capture the slow decay. After noting that

$$\frac{d}{dt} = \frac{\partial}{\partial t} + \epsilon \frac{\partial}{\partial T},$$

(12.41) becomes

$$\frac{\partial^2 y}{\partial t^2} + 2\epsilon \frac{\partial^2 y}{\partial t \partial T} + \epsilon^2 \frac{\partial^2 y}{\partial T^2} + 2\epsilon \frac{\partial y}{\partial t} + 2\epsilon^2 \frac{\partial y}{\partial T} + y = 0, \qquad (12.45)$$

to be solved subject to

$$y(0,0) = 1, \quad \frac{\partial y}{\partial t}(0,0) + \epsilon \frac{\partial y}{\partial T}(0,0) = 0. \qquad (12.46)$$

At leading order,

$$\frac{\partial^2 y_0}{\partial t^2} + y_0 = 0,$$

subject to

$$y_0(0,0) = 1, \quad \frac{\partial y_0}{\partial t}(0,0) = 0.$$

Although this is a partial differential equation, the only derivatives are with respect to t, so we can solve as if it were an ordinary differential equation in t. However, we must remember that the 'constants' of integration can actually be functions of the other variable, T. This means that

$$y_0 = A_0(T)\cos t + B_0(T)\sin t,$$

and the boundary conditions show that

$$A_0(0) = 1, \quad B_0(0) = 0. \qquad (12.47)$$

The functions $A_0(T)$ and $B_0(T)$ are, as yet, undetermined.

At $O(\epsilon)$,

$$\frac{\partial^2 y_1}{\partial t^2} + y_1 = -2\frac{\partial^2 y_0}{\partial t \partial T} - 2\frac{\partial y_0}{\partial t} = 2\left(\frac{dA_0}{dT} + A_0\right)\sin t - 2\left(\frac{dB_0}{dT} + B_0\right)\cos t. \qquad (12.48)$$

Because of the appearance on the right hand side of the equation of the terms $\cos t$ and $\sin t$, which are themselves solutions of the homogeneous version of the equation, the particular integral solution will involve the terms $t\sin t$ and $t\cos t$. As we have seen, it is precisely terms of this type that lead to a nonuniformity in the asymptotic expansion when $t = O(\epsilon^{-1})$. The terms proportional to $\sin t$ and $\cos t$ in (12.48) are known as **secular terms** and, to keep the asymptotic expansion

uniform, we must eliminate them by choosing A_0 and B_0 appropriately. In this example, we need
$$\frac{dA_0}{dT} + A_0 = 0, \quad \frac{dB_0}{dT} + B_0 = 0, \tag{12.49}$$
and hence
$$y_1 = A_1(T)\cos t + B_1(T)\sin t.$$
The initial conditions for (12.49) are given by (12.47), so the solutions are $A_0 = e^{-T}$ and $B_0 = 0$. We conclude that
$$y \sim e^{-T}\cos t = e^{-\epsilon t}\cos t \tag{12.50}$$
for $\epsilon \ll 1$. This asymptotic solution is consistent with the exact solution, (12.44), and remains valid when $t = O(\epsilon^{-1})$, as can be seen in Figure 12.11. In fact, we will see below that this solution is only valid for $t \ll \epsilon^{-2}$.

In order to proceed to find more terms in the asymptotic expansion using the method of multiple scales, we can take the exact solution, (12.44), as a guide. We know that
$$y \sim e^{-\epsilon t}\left\{\cos\left(1 - \frac{1}{2}\epsilon^2\right)t + \epsilon\sin\left(1 - \frac{1}{2}\epsilon^2\right)t\right\}, \tag{12.51}$$
for $\epsilon \ll 1$. This shows that the phase of the oscillation changes by an $O(1)$ amount when $t = O(\epsilon^{-2})$. In order to capture this, we seek a solution that is a function of the two timescales
$$T = \epsilon t, \quad \tau = \left(1 + a\epsilon^2 + b\epsilon^3 + \cdots\right)t,$$
with the constants $a, b, \ldots$ to be determined. Although this looks like a bit of a cheat, since we are only doing this because we know the exact solution, this approach works for a wide range of problems, including nonlinear differential equations.

In order to develop a one-term multiple scale expansion, we needed to consider the solution up to $O(\epsilon)$. This suggests that we will need to expand up to $O(\epsilon^2)$ to construct a two-term multiple scales expansion, with
$$y = y_0(\tau, T) + \epsilon y_1(\tau, T) + \epsilon^2 y_2(\tau, T) + O(\epsilon^3).$$
After noting that
$$\frac{d}{dt} = \epsilon\frac{\partial}{\partial T} + \left(1 + a\epsilon^2 + b\epsilon^3 + \cdots\right)\frac{\partial}{\partial \tau},$$
equation (12.41) becomes
$$\frac{\partial^2 y}{\partial \tau^2} + 2a\epsilon^2\frac{\partial^2 y}{\partial \tau^2} + 2\epsilon\frac{\partial^2 y}{\partial \tau \partial T} + \epsilon^2\frac{\partial^2 y}{\partial T^2} + 2\epsilon\frac{\partial y}{\partial \tau} + 2\epsilon^2\frac{\partial y}{\partial T} + y + O(\epsilon^3) = 0, \tag{12.52}$$
to be solved subject to
$$y(0,0) = 1, \quad \left(1 + a\epsilon^2 + b\epsilon^3 + \cdots\right)\frac{\partial y}{\partial \tau}(0,0) + \epsilon\frac{\partial y}{\partial T}(0,0) = 0. \tag{12.53}$$
We already know that
$$y_0 = e^{-T}\cos\tau, \quad y_1 = A_1(T)\cos\tau + B_1(T)\sin\tau.$$

At $O(\epsilon^2)$,

$$\frac{\partial^2 y_2}{\partial \tau^2} + y_2 = -2a\frac{\partial^2 y_0}{\partial \tau^2} - 2\frac{\partial^2 y_1}{\partial \tau \partial T} + \frac{\partial^2 y_0}{\partial T^2} - 2\frac{\partial y_1}{\partial \tau} - 2\frac{\partial y_0}{\partial T}$$

$$= 2\left(\frac{dA_1}{dt} + A_1\right)\sin\tau - 2\left\{\frac{dB_1}{dt} + B_1 - \left(a + \frac{1}{2}\right)e^{-T}\right\}\cos\tau.$$

In order to remove the secular terms we need

$$\frac{dA_1}{dt} + A_1 = 0, \quad \frac{dB_1}{dt} + B_1 = \left(a + \frac{1}{2}\right)e^{-T}.$$

At $O(\epsilon)$ the boundary conditions are

$$y_1(0,0) = A_1(0) = 0, \quad \frac{\partial y_1}{\partial \tau}(0,0) + \frac{\partial y_0}{\partial T}(0,0) = B_1(0) - 1 = 0,$$

and hence

$$A_1 = 0, \quad B_1 = \left(a + \frac{1}{2}\right)Te^{-T} + e^{-T}.$$

This gives us

$$y \sim e^{-T}\cos\tau + \epsilon\left\{\left(a + \frac{1}{2}\right)Te^{-T} + e^{-T}\right\}\sin\tau.$$

However, the part of the $O(\epsilon)$ term proportional to T will lead to a nonuniformity in the expansion when $T = O(\epsilon^{-1})$, and we must therefore remove it by choosing $a = -1/2$. We could have deduced this directly from the differential equation for B_1, since the term proportional to e^{-T} is secular. We conclude that

$$\tau = \left(1 - \frac{1}{2}\epsilon^2 + \cdots\right)t,$$

and hence obtain (12.51), as expected.

Example 1: The van der Pol Oscillator

The governing equation for the **van der Pol oscillator** is

$$\frac{d^2y}{dt^2} + \epsilon(y^2 - 1)\frac{dy}{dt} + y = 0, \qquad (12.54)$$

for $t \geq 0$. For $\epsilon \ll 1$ this is a linear oscillator with a weak nonlinear term, $\epsilon(y^2-1)\dot{y}$. For $|y| < 1$ this term tends to drive the oscillations to a greater amplitude, whilst for $|y| > 1$, this term damps the oscillations. It is straightforward (at least if you're an electrical engineer!) to build an electronic circuit from resistors and capacitors whose output is governed by (12.54). It was in this context that this system was first studied extensively as a prototypical nonlinear oscillator. It is also straightforward to construct a forced van der Pol oscillator, which leads to a nonzero right hand side in (12.54), and study the chaotic response of the circuit.

Since the damping is zero when $y = 1$, a reasonable guess at the behaviour of the solution for $\epsilon \ll 1$ would be that there is an oscillatory solution with unit amplitude. Let's investigate this plausible, but incorrect, guess by considering the solution with initial conditions

$$y(0) = 1, \quad \frac{dy}{dt}(0) = 0. \tag{12.55}$$

We will use the method of multiple scales, and define a slow time scale, $T = \epsilon t$. We seek an asymptotic solution of the form

$$y = y_0(t, T) + \epsilon y_1(t, T) + O(\epsilon^2).$$

As for our linear example, we find that

$$y_0 = A_0(T) \cos\{t + \phi_0(T)\}.$$

For this problem, it is more convenient to write the solution in terms of an amplitude, $A_0(T)$, and phase, $\phi_0(T)$. The boundary conditions show that

$$A_0(0) = 1, \quad \phi_0(0) = 0. \tag{12.56}$$

At $O(\epsilon)$,

$$\frac{\partial^2 y_1}{\partial t^2} + y_1 = -2 \frac{\partial^2 y_0}{\partial t \partial T} - (y_0^2 - 1) \frac{\partial y_0}{\partial t}$$

$$= 2 \frac{dA_0}{dT} \sin(t + \phi_0) + 2 A_0 \frac{d\phi_0}{dT} \cos(t + \phi_0) + A_0 \sin(t + \phi_0) \left\{ A_0^2 \cos^2(t + \phi_0) - 1 \right\}.$$

In order to pick out the secular terms on the right hand side, we note that†

$$\sin\theta \cos^2\theta = \sin\theta - \sin^3\theta = \sin\theta - \frac{3}{4}\sin\theta + \frac{1}{4}\sin 3\theta = \frac{1}{4}\sin\theta + \frac{1}{4}\sin 3\theta.$$

This means that

$$\frac{\partial^2 y_1}{\partial t^2} + y_1 = \left\{ 2\frac{dA_0}{dT} + \frac{1}{4}A_0^3 - A_0 \right\} \sin(t + \phi_0) + 2A_0 \frac{d\phi_0}{dT} \cos(t + \phi_0) + \frac{1}{4}A_0^3 \sin 3(t + \phi_0).$$

To suppress the secular terms we therefore need

$$\frac{d\phi_0}{dT} = 0, \quad \frac{dA_0}{dT} = \frac{1}{8}A_0(4 - A_0^2).$$

Subject to the boundary conditions (12.56), the solutions are

$$\phi_0 = 0, \quad A_0 = 2(1 + 3e^{-T})^{-1/2}.$$

Therefore

$$y \sim 2(1 + 3e^{-T})^{-1/2} \cos t$$

for $\epsilon \ll 1$, and we conclude that the amplitude of the oscillation actually tends to 2 as $t \to \infty$, as shown in Figure 12.12.

† To get $\cos^n \theta$ in terms of $\cos m\theta$ for $m = 1, 2, \ldots, n$, use $e^{in\theta} = \cos n\theta + i \sin n\theta = (e^{i\theta})^n = (\cos\theta + i\sin\theta)^n$ and equate real and imaginary parts.

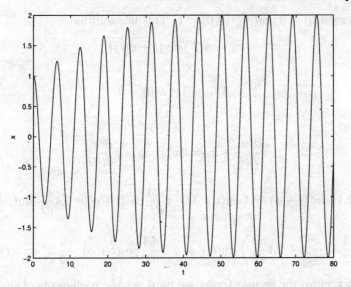

Fig. 12.12. The leading order solution of the van der Pol equation, (12.54), subject to $y(0) = 1$, $\dot{y}(0) = 0$, when $\epsilon = 0.1$.

Example 2: Jacobian elliptic functions with almost simple harmonic behaviour

Let's again turn our attention to the Jacobian elliptic function $x = \text{sn}(t; k)$, which satisfies (12.21). When $k \ll 1$, this function is oscillatory and, at leading order, performs simple harmonic motion. We can see this more clearly by differentiating (12.21) and eliminating dx/dt to obtain

$$\frac{d^2x}{dt^2} + (1+k^2)x = 2k^2x^3. \quad (12.57)$$

The initial conditions, of which there must be two for this second order equation, are $x = 0$ and, from (12.21), $dx/dt = 1$ when $t = 0$. Let's now use the method of multiple scales to see how this small perturbation affects the oscillation after a long time. As usual, we define $T = k^2 t$ and $x = x(t, T)$, in terms of which (12.57) becomes

$$\frac{\partial^2 x}{\partial t^2} + 2k^2 \frac{\partial^2 x}{\partial t \partial T} + k^4 \frac{\partial^2 x}{\partial^2 T} + (1+k^2)x = 2k^2 x^3. \quad (12.58)$$

We seek an asymptotic solution

$$x = x_0(t, T) + k^2 x_1(t, T) + O(k^4).$$

At leading order

$$\frac{\partial^2 x_0}{\partial t^2} + x_0 = 0,$$

subject to $x_0(0,0) = 0$ and $\frac{\partial x_0}{\partial t}(0,0) = 1$. This has solution

$$x_0 = A(T)\sin\{t + \phi(t)\},$$

and

$$A(0) = 1, \quad \phi(0) = 0. \tag{12.59}$$

At $O(k^2)$,

$$\frac{\partial^2 x_1}{\partial t^2} + x_1 = 2x_0^3 - x_0 - 2\frac{\partial^2 x_0}{\partial t \partial T}$$

$$= 2A^3\sin^3(t+\phi) - A\sin(t+\phi) - 2\frac{dA}{dT}\cos(t+\phi) + 2A\frac{d\phi}{dT}\sin(t+\phi)$$

$$= \left(\frac{3}{2}A^3 - A + 2A\frac{d\phi}{dT}\right)\sin(t+\phi) - 2\frac{dA}{dT}\cos(t+\phi) - \frac{1}{2}A^3\sin 3(t+\phi).$$

In order to remove the secular terms, we must set the coefficients of $\sin(t+\phi)$ and $\cos(t+\phi)$ to zero. This gives us two simple ordinary differential equations to be solved subject to (12.59), which gives

$$A = 1, \quad \phi = -\frac{1}{4}T,$$

and hence

$$x = \sin\left\{\left(1 - \frac{1}{4}k^2\right)t\right\} + O(k^2),$$

for $k \ll 1$. We can see that, as we would expect from the analysis given in Section 9.4, the leading order amplitude of the oscillation does not change with t, in contrast to the solution of the van der Pol equation that we studied earlier. However, the period of the oscillation changes by $O(k^2)$ even at this lowest order. If we take the analysis to $O(k^4)$, we find that the amplitude of the oscillation is also dependent on k (see King, 1988).

12.2.5 Slowly Damped Nonlinear Oscillations: Kuzmak's Method

The method of multiple scales, as we have described it above, is appropriate for weakly perturbed *linear* oscillators. Can we make any progress if the leading order problem is *nonlinear*? We will concentrate on the simple example,

$$\frac{d^2y}{dt^2} + 2\epsilon\frac{dy}{dt} + y - y^3 = 0, \tag{12.60}$$

subject to

$$y(0) = 0, \quad \frac{dy}{dt}(0) = v_0 > 0, \tag{12.61}$$

with ϵ small and positive. Let's begin by considering the leading order problem, with $\epsilon = 0$. As we saw in Chapter 9, since dy/dt does not appear in (12.60) when

$\epsilon = 0$, we can integrate once to obtain

$$\frac{dy}{dt} = \pm\sqrt{E - y^2 + \frac{1}{2}y^4}, \qquad (12.62)$$

with $E = v_0^2$ from the initial conditions, (12.61). If we now assume that $v_0^2 < 1/2$†, and scale y and t using

$$y = \left(1 - \sqrt{1 - 2E}\right)^{1/2} \hat{y}, \quad t = \left(\frac{1 - \sqrt{1 - 2E}}{E}\right)^{1/2} \hat{t},$$

(12.62) becomes

$$\frac{d\hat{y}}{d\hat{t}} = \pm\sqrt{1 - \hat{y}^2}\sqrt{1 - k^2\hat{y}^2}, \qquad (12.63)$$

where

$$k = \left(\frac{1 - \sqrt{1 - 2E}}{1 + \sqrt{1 - 2E}}\right)^{1/2}. \qquad (12.64)$$

If we compare (12.63) with the system that we studied in Section 9.4, we find that

$$y = \left(1 - \sqrt{1 - 2E}\right)^{1/2} \operatorname{sn}\left\{\left(\frac{E}{1 - \sqrt{1 - 2E}}\right)^{1/2} t; k\right\}. \qquad (12.65)$$

In the absence of any damping ($\epsilon = 0$), y varies periodically in a manner described by the Jacobian elliptic function sn. In addition, Example 2 in the previous section shows that $y \sim v_0 \sin t$ when $v_0 \ll 1$. This is to be expected, since the nonlinear term in (12.60) is negligible when v_0, and hence y, is small.

For ϵ small and positive, but nonzero, by analogy with what we found using the method of multiple scales in the previous section, we expect that weak damping leads to a slow decrease in the amplitude of the oscillation, and possibly some change in its phase. In order to construct an asymptotic solution valid for $t = O(\epsilon^{-1})$, when the amplitude and phase of the oscillation have changed significantly, we begin in the usual way by defining a slow time scale, $T = \epsilon t$. However, for a nonlinear oscillator, the frequency of the leading order solution depends upon the amplitude of the oscillation, so it is now convenient to define

$$\psi = \frac{\theta(T)}{\epsilon} + \phi(T), \quad \theta(0) = 0,$$

and seek a solution $y \equiv y(\psi, T)$. This was first done by Kuzmak (1959), although not in full generality.

Since

$$\frac{dy}{dt} = (\theta' + \epsilon\phi')\frac{\partial y}{\partial \psi} + \epsilon\frac{\partial y}{\partial T},$$

where a prime denotes d/dT, we can see that $\theta'(T) \equiv \omega(T)$ is the frequency of the oscillation at leading order and $\phi(T)$ the change in phase, both of which we must

† The usual graphical argument shows that this is a sufficient condition for the solution to be periodic in t.

determine as part of the solution. In terms of these new variables, (12.60) and (12.61) become

$$(\omega + \epsilon\phi')^2 \frac{\partial^2 y}{\partial \psi^2} + 2\epsilon(\omega + \epsilon\phi') \frac{\partial^2 y}{\partial T \partial \psi} + \epsilon^2 \frac{\partial^2 y}{\partial T^2} + \epsilon(\omega' + \epsilon\phi'') \frac{\partial y}{\partial \psi}$$

$$+ 2\epsilon(\omega + \epsilon\phi') \frac{\partial y}{\partial \psi} + 2\epsilon^2 \frac{\partial y}{\partial T} + y - y^3 = 0, \qquad (12.66)$$

subject to

$$y(0,0) = 0, \quad (\omega(0) + \epsilon\phi'(0)) \frac{\partial y}{\partial \psi}(0,0) + \epsilon \frac{\partial y}{\partial T}(0,0) = v_0. \qquad (12.67)$$

We now expand

$$y = y_0(\psi, T) + \epsilon y_1(\psi, T) + O(\epsilon^2),$$

and substitute into (12.66) and (12.67). At leading order, we obtain

$$\omega^2(T) \frac{\partial^2 y_0}{\partial \psi^2} + y_0 - y_0^3 = 0, \qquad (12.68)$$

subject to

$$y_0(0,0) = 0, \quad \omega(0) \frac{\partial y_0}{\partial \psi}(0,0) = v_0. \qquad (12.69)$$

As we have seen, this has solution

$$y_0 = \left(1 - \sqrt{1 - 2E(T)}\right)^{1/2} \operatorname{sn}\left\{ \left(\frac{E(T)}{1 - \sqrt{1 - 2E(T)}}\right)^{1/2} \frac{\psi}{\omega(T)}; k(T) \right\}, \qquad (12.70)$$

where k is given by (12.64). The initial conditions, (12.69), show that

$$E(0) = v_0^2, \quad \phi(0) = 0. \qquad (12.71)$$

The period of the oscillation, $P(T)$, can be determined by noting that the quarter period is

$$\frac{1}{4} P(T) = \omega(T) \int_0^{\left(1 - \sqrt{1-2E(T)}\right)^{1/2}} \frac{ds}{\sqrt{E(T) - s^2 + \frac{1}{2}s^4}},$$

which, after a simple rescaling, shows that

$$P(T) = 4\omega(T) \left(\frac{1 - \sqrt{1 - 2E(T)}}{E(T)}\right)^{1/2} K(k(E)), \qquad (12.72)$$

where

$$K(k) = \int_0^1 \frac{ds}{\sqrt{1 - s^2}\sqrt{1 - k^2 s^2}} \qquad (12.73)$$

is the **complete elliptic integral of the first kind**.

As in the method of multiple scales, we need to go to $O(\epsilon)$ to determine the

12.2 ORDINARY DIFFERENTIAL EQUATIONS

behaviour of the solution on the slow time scale, T. Firstly, note that we have three unknowns to determine, $E(T)$, $\phi(T)$ and $\omega(T)$, whilst for the method of multiple scales, we had just two, equivalent to $E(T)$ and $\phi(T)$. Since we introduced $\theta(T)$ simply to account for the fact that the period of the oscillation changes with the amplitude, we have the freedom to choose this new time scale so that the period of the oscillation is constant. For convenience, we take $P = 1$, so that

$$\frac{d\theta}{dT} = \omega \equiv \omega(E) = \frac{1}{4K(k(E))} \left(\frac{E}{1 - \sqrt{1 - 2E}} \right)^{1/2}. \qquad (12.74)$$

We also need to note for future reference the parity with respect to ψ of y_0 and its derivatives. Both y_0 and $\partial^2 y_0 / \partial \psi^2$ are odd functions, whilst $\partial y_0 / \partial \psi$ is an even function of ψ. In addition, we can now treat y_0 as a function of ψ and E, with

$$y_0(\psi, E) = \left(1 - \sqrt{1 - 2E}\right)^{1/2} \operatorname{sn}\left\{4K(k(E))\psi;\, k(E)\right\}. \qquad (12.75)$$

At $O(\epsilon)$ we obtain

$$\omega^2 \frac{\partial^2 y_1}{\partial \psi^2} + (1 - 3y_0^2)\, y_1 = -2\omega\phi' \frac{\partial^2 y_0}{\partial \psi^2} - 2\omega \frac{\partial^2 y_0}{\partial \psi \partial E} \frac{dE}{dT} - \omega' \frac{\partial y_0}{\partial \psi} - 2\omega \frac{\partial y_0}{\partial \psi},$$

which we write as

$$L(y_1) = R_{1\mathrm{odd}} + R_{1\mathrm{even}}, \qquad (12.76)$$

where

$$L = \omega^2 \frac{\partial^2}{\partial \psi^2} + 1 - 3y_0^2, \qquad (12.77)$$

and

$$R_{1\mathrm{odd}} = -2\omega\phi' \frac{\partial^2 y_0}{\partial \psi^2}, \quad R_{1\mathrm{even}} = -2\omega \frac{\partial^2 y_0}{\partial \psi \partial E} \frac{dE}{dT} - \omega' \frac{\partial y_0}{\partial \psi} - 2\omega \frac{\partial y_0}{\partial \psi}. \qquad (12.78)$$

Now, by differentiating (12.68) with respect to ψ, we find that

$$L\left(\frac{\partial y_0}{\partial \psi}\right) = 0,$$

so that $\partial y_0 / \partial \psi$ is a solution of the homogeneous version of (12.76). For the solution of (12.76) to be periodic, the right hand side must be orthogonal to the solution of the homogeneous equation, and therefore orthogonal to $\partial y_0 / \partial \psi$.† This is equivalent to the elimination of secular terms in the method of multiple scales. Since $\partial y_0 / \partial \psi$ is even in ψ, this means that

$$\int_0^1 \frac{\partial y_0}{\partial \psi} R_{1\mathrm{even}}\, d\psi = 0,$$

which is the equivalent of the secularity condition that determines the amplitude

---

† Strictly speaking, this is a result that arises from **Floquet theory**, the theory of linear ordinary differential equations with periodic coefficients.

of the oscillation in the method of multiple scales. Using (12.78), we can write this as†

$$\frac{d}{dE}\left\{\omega \int_0^1 \left(\frac{\partial y_0}{\partial \psi}\right)^2 d\psi\right\} \frac{dE}{dT} + 2\omega \int_0^1 \left(\frac{\partial y_0}{\partial \psi}\right)^2 d\psi = 0. \qquad (12.79)$$

To proceed, we firstly note that

$$\int_0^1 \left(\frac{\partial y_0}{\partial \psi}\right)^2 d\psi = 4 \int_0^{(1-\sqrt{1-2E})^{1/2}} \frac{\partial y_0}{\partial \psi} dy_0$$

$$= \frac{4}{\omega} \int_0^{(1-\sqrt{1-2E})^{1/2}} \sqrt{E - y_0^2 + \frac{1}{2}y_0^4}\, dy_0,$$

and hence that

$$\frac{\partial}{\partial E}\left\{\omega \int_0^1 \left(\frac{\partial y_0}{\partial \psi}\right)^2 d\psi\right\} = 2 \int_0^{(1-\sqrt{1-2E})^{1/2}} \frac{dy_0}{\sqrt{E - y_0^2 + \frac{1}{2}y_0^4}} = \frac{1}{2\omega}. \qquad (12.80)$$

Secondly, using the results of Section 9.4, we find that

$$\int_0^1 \left(\frac{\partial y_0}{\partial \psi}\right)^2 d\psi$$

$$= 16K^2(E)\left(1 - \sqrt{1-2E}\right) \int_0^1 \left\{1 - \text{sn}^2\left(4K\psi; k\right)\right\}\left\{1 - k^2\text{sn}^2\left(4K\psi; k\right)\right\} d\psi$$

$$= 16K(E)\left(1 - \sqrt{1-2E}\right) \int_0^K \left\{1 - \text{sn}^2\left(\psi; k\right)\right\}\left\{1 - k^2\text{sn}^2\left(\psi; k\right)\right\} d\psi.$$

We now need a standard result on elliptic functions‡, namely that

$$\int_0^K \left\{1 - \text{sn}^2\left(\psi; k\right)\right\}\left\{1 - k^2\text{sn}^2\left(\psi; k\right)\right\} d\psi$$

$$= \frac{1}{3k^2}\left\{\left(1 + k^2\right) L(k) - \left(1 - k^2\right) K(k)\right\},$$

where

$$L(k) = \int_0^1 \sqrt{\frac{1 - k^2 s^2}{1 - s^2}}\, ds,$$

is the **complete elliptic integral of the second kind**§. Equation (12.79) and the definition of ω, (12.74), then show that

$$\frac{dE}{dT} = -\frac{4E}{3k^2 K}\left\{\left(1 + k^2\right) L(k) - \left(1 - k^2\right) K(k)\right\}. \qquad (12.81)$$

† Note that the quantity in the curly brackets in (12.79) is often referred to as the **action**.
‡ See, for example, Byrd (1971).
§ Although the usual notation for the complete elliptic integral of the second kind is $E(k)$, the symbol E is already spoken for in this analysis.

12.2 ORDINARY DIFFERENTIAL EQUATIONS

This equation, along with the initial condition, (12.71), determines $E(T)$. For $v_0 \ll 1$, this reduces to the multiple scales result, $dE/dT = -2E$ (see Exercise 12.10). It is straightforward to integrate (12.81) using MATLAB, since the complete elliptic integrals of the first and second kinds can be calculated using the built-in function `ellipke`. Note that we can simultaneously calculate $\theta(T)$ numerically by solving (12.74) subject to $\theta(0) = 0$. Figure 12.13 shows the function $E(T)$ calculated for $E(0) = 0.45$, and also the corresponding result using multiple scales on the linearized problem, $E = E(0)e^{-2T}$.

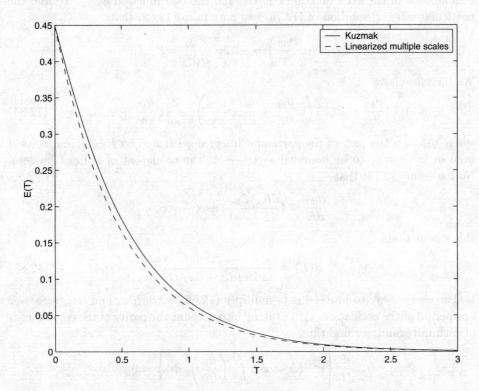

Fig. 12.13. The solution of (12.81) when $E(0) = 0.45$, and the corresponding multiple scales solution of the linearized problem, $E(T) = E(0)e^{-2T}$.

We now need to find an equation that determines the phase, $\phi(T)$. Unfortunately, unlike the method of multiple scales, we need to determine the solution at $O(\epsilon)$ in order to make progress. By differentiating (12.68) with respect to E, we find that

$$L\left(\frac{\partial y_0}{\partial E}\right) = -2\omega \frac{d\omega}{dE} \frac{\partial^2 y_0}{\partial \psi^2}.$$

We also note that

$$L\left(\psi \frac{\partial y_0}{\partial \psi}\right) = 2\omega^2 \frac{\partial^2 y_0}{\partial \psi^2}, \qquad (12.82)$$

and hence that
$$L(y_{0\text{odd}}) = 0,$$
where
$$y_{0\text{odd}} = \omega \frac{\partial y_0}{\partial E} + \frac{d\omega}{dE} \psi \frac{\partial y_0}{\partial \psi}. \qquad (12.83)$$

The general solution of the homogeneous version of (12.76) is therefore a linear combination of the even function $\partial y_0/\partial \psi$ and the odd function $y_{0\text{odd}}$. To find the particular integral solution of (12.76), we note from (12.82) that
$$L\left(-\frac{\phi'}{\omega}\psi\frac{\partial y_0}{\partial \psi}\right) = -2\omega\phi'\frac{\partial^2 y_0}{\partial \psi^2} = R_{1\text{odd}}.$$

We therefore have
$$y_1 = A(T)\frac{\partial y_0}{\partial \psi} + B(T)\left(\omega\frac{\partial y_0}{\partial E} + \frac{d\omega}{dE}\psi\frac{\partial y_0}{\partial \psi}\right) - \frac{\phi'}{\omega}\psi\frac{\partial y_0}{\partial \psi} + y_{1\text{even}}, \qquad (12.84)$$

where $y_{1\text{even}}$ is the part of the particular integral generated by $R_{1\text{even}}$, and is itself even in ψ. For y_1 to be bounded as $\psi \to \infty$, the coefficient of ψ must be zero. Noting from (12.75) that
$$\frac{\partial y_0}{\partial E} \sim 4\frac{dK}{dE}\psi\frac{\partial y_0}{\partial \psi} \quad \text{as } \psi \to \infty,$$
this means that
$$B(T) = \frac{\phi'/\omega}{4\omega dK/dE + d\omega/dE}. \qquad (12.85)$$

The easiest way to proceed is to multiply (12.66) by $\partial y/\partial \psi$, and integrate over one period of the oscillation. After taking into account the parity of the components of each integrand, we find that
$$\frac{d}{dT}\left\{e^{2T}(\omega+\epsilon\phi')\int_0^1\left(\frac{\partial y}{\partial \psi}\right)^2 d\psi\right\} = 0.$$

At leading order, this reproduces (12.79), whilst at $O(\epsilon)$ we find that
$$\frac{d}{dT}\left\{e^{2T}\left(2\omega\int_0^1\frac{\partial y_0}{\partial \psi}\frac{\partial y_1}{\partial \psi}d\psi + \phi'\int_0^1\left(\frac{\partial y_0}{\partial \psi}\right)^2 d\psi\right)\right\} = 0. \qquad (12.86)$$

Now, using what we know about the parity of the various components of y_1, we can show that
$$\int_0^1\frac{\partial y_0}{\partial \psi}\frac{\partial y_1}{\partial \psi}d\psi = \frac{\phi'}{d\omega/dE}\int_0^1\frac{\partial y_0}{\partial \psi}\frac{\partial^2 y_0}{\partial \psi \partial E}d\psi = \frac{\phi'}{2d\omega/dE}\frac{\partial}{\partial E}\int_0^1\left(\frac{\partial y_0}{\partial \psi}\right)^2 d\psi,$$
and hence from (12.86) that
$$\frac{d}{dT}\left[e^{2T}\frac{\phi'}{d\omega/dE}\frac{\partial}{\partial E}\left\{\omega\int_0^1\left(\frac{\partial y_0}{\partial \psi}\right)^2 d\psi\right\}\right] = 0.$$

Using (12.80), this becomes

$$\frac{d}{dT}\left(\frac{e^{2T}\phi'}{\omega d\omega/dE}\right) = 0. \tag{12.87}$$

This is a second order equation for $\phi(T)$. Although we know that $\phi(0) = 0$, to be able to solve (12.87) we also need to know $\phi'(0)$.

At $O(\epsilon)$, the initial conditions, (12.67), are

$$y_1(0,0) = 0, \quad \omega(E(0))\frac{\partial y_1}{\partial \psi}(0,0) = -E'(0)\frac{\partial y_0}{\partial E}(0,E(0)) - \phi'(0)\frac{\partial y_0}{\partial \psi}(0,E(0)). \tag{12.88}$$

By substituting the solution (12.84) for y_1 into the second of these, we find that

$$\omega(E(0))\left\{A(0)\frac{\partial^2 y_0}{\partial \psi^2}(0,E(0)) + \dot{B}(0)\omega(E(0))\frac{\partial^2 y_0}{\partial \psi \partial E}(0,E(0))\right.$$

$$\left. + \left(B(0)\frac{d\omega}{dE}(E(0)) - \frac{\phi'(0)}{\omega(E(0))}\right)\frac{\partial y_0}{\partial \psi}(0,E(0)) + \frac{\partial y_{1\text{even}}}{\partial \psi}(0,0)\right\}$$

$$= -E'(0)\frac{\partial y_0}{\partial E}(0,E(0)) - \phi'(0)\frac{\partial y_0}{\partial \psi}(0,E(0)).$$

Using (12.85), all of the terms that do not involve $\phi'(0)$ are odd in ψ, and therefore vanish when $\psi = 0$. We conclude that $\phi'(0) = 0$, and hence from (12.87) that $\phi(T) = 0$.

Figure 12.14 shows a comparison between the leading order solution computed using Kuzmak's method, the leading order multiple scales solution of the linearized problem, $y = \sqrt{v_0}e^{-T}\sin t$, and the numerical solution of the full problem when $\epsilon = 0.01$. The numerical and Kuzmak solutions are indistinguishable. Although the multiple scales solution gives a good estimate of $E(T)$, as shown in figure 12.13, it does not give an accurate solution of the full problem.

To see how the method works in general for damped nonlinear oscillators, the interested reader is referred to Bourland and Haberman (1988).

12.2.6 The Effect of Fine Scale Structure on Reaction–Diffusion Processes

Consider the two-point boundary value problem

$$\frac{d}{dx}\left\{D\left(x,\frac{x}{\epsilon}\right)\frac{d\theta}{dx}\right\} + R\left(\theta,x,\frac{x}{\epsilon}\right) = 0 \text{ for } 0 < x < 1, \tag{12.89}$$

subject to the boundary conditions

$$\frac{d\theta}{dx} = 0 \text{ at } x = 0 \text{ and } x = 1. \tag{12.90}$$

We can think of $\theta(x)$ as a dimensionless temperature, so that this boundary value problem models the steady distribution of heat in a conducting material. When $\epsilon \ll 1$, this material has a fine scale structure varying on a length scale of $O(\epsilon)$, and

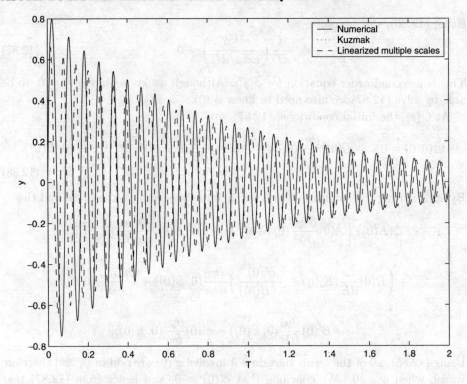

Fig. 12.14. The numerical solution of (12.60) subject to (12.61) when $v_0^2 = 0.45$ and $\epsilon = 0.01$ compared to asymptotic solutions calculated using the leading order Kuzmak solution and the leading order multiple scales solution of the linearized problem.

a coarse, background structure varying on a length scale of $O(1)$. The steady state temperature distribution represents a balance between the diffusion of heat, $(D\theta_x)_x$ (where subscripts denote derivatives), and its production by some chemical reaction, R (see Section 12.2.3, Example 2 for a derivation of this type of balance law). If we integrate (12.89) over $0 \leqslant x \leqslant 1$ and apply the boundary conditions (12.90), we find that the chemical reaction term must satisfy the solvability condition

$$\int_0^1 R\left(\theta(x), x, \frac{x}{\epsilon}\right) dx = 0 \qquad (12.91)$$

for a solution to exist. Physically, since (12.90) states that no heat can escape from the ends of the material, (12.91) simply says that the overall rate at which heat is produced must be zero for a steady state to be possible, with sources of heat in some regions of the material balanced by heat sinks in other regions.

In order to use asymptotic methods to determine the solution at leading order when $\epsilon \ll 1$, we begin by introducing the **fast variable**, $\hat{x} = x/\epsilon$. In the usual way (see Section 12.2.4), (12.89) and (12.90) become

$$(D\theta_{\hat{x}})_{\hat{x}} + \epsilon\{(D\theta_{\hat{x}})_x + (D\theta_x)_{\hat{x}}\} + \epsilon^2\{(D\theta_x)_x + R\} = 0, \qquad (12.92)$$

subject to

$$\theta_{\hat{x}} + \epsilon\theta_x = 0 \text{ at } x = \hat{x} = 0 \text{ and } x = 1,\, \hat{x} = 1/\epsilon, \qquad (12.93)$$

with $R \equiv R(\theta(x,\hat{x}), x, \hat{x})$, $D \equiv D(x, \hat{x})$. It is quite awkward to apply a boundary condition at $\hat{x} = 1/\epsilon \gg 1$, but we shall see that we can determine the equation satisfied by θ at leading order without using this, and we will not consider it below. We now expand θ as

$$\theta = \theta_0(x,\hat{x}) + \epsilon\theta_1(x,\hat{x}) + \epsilon^2\theta_2(x,\hat{x}) + O(\epsilon^3).$$

At leading order,

$$(D\theta_{0\hat{x}})_{\hat{x}} = 0,$$

subject to

$$\theta_{0\hat{x}} = 0 \text{ at } x = \hat{x} = 0.$$

We can integrate this once to obtain $D\theta_{0\hat{x}} = \alpha(x)$, with $\alpha(0) = 0$, and then once more, which gives

$$\theta_0 = \alpha(x)\int_0^{\hat{x}} \frac{ds}{D(x,s)} + f_0(x).$$

At $O(\epsilon)$, we find that

$$(D\theta_{1\hat{x}})_{\hat{x}} = -(D\theta_{0\hat{x}})_x - (D\theta_{0x})_{\hat{x}},$$

which we can write as

$$(D\theta_{1\hat{x}} + D\theta_{0x})_{\hat{x}} = -\alpha'(x). \qquad (12.94)$$

We can integrate (12.94) to obtain

$$D\theta_{1\hat{x}} + D\theta_{0x} = -\alpha'(x)\hat{x} + \beta(x).$$

Substituting for θ_0 and integrating again shows that

$$\theta_1 = f_1(x) - f_0'(x)\hat{x} - \alpha'(x)\hat{x}\int_0^{\hat{x}} \frac{ds}{D(x,s)}$$

$$-\alpha(x)\frac{\partial}{\partial x}\int_0^{\hat{x}} \frac{\hat{x}-s}{D(x,s)}ds + \beta(x)\int_0^{\hat{x}} \frac{ds}{D(x,s)}. \qquad (12.95)$$

When $\hat{x}$ is large, there are terms in this expression that are of $O(\hat{x}^2)$. These are secular, and become comparable with the leading order term in the expansion for θ when $\hat{x} = 1/\epsilon$. We must eliminate this secularity by taking

$$\lim_{\epsilon \to 0}\left\{\alpha'(x)\epsilon\int_0^{1/\epsilon} \frac{ds}{D(x,s)} - \alpha(x)\epsilon^2\frac{\partial}{\partial x}\int_0^{1/\epsilon} \frac{1/\epsilon - s}{D(x,s)}ds\right\} = 0. \qquad (12.96)$$

This is a first order ordinary differential equation for $\alpha(x)$. Since $\alpha(0) = 0$, we conclude that $\alpha \equiv 0$. Note that each of the terms in (12.96) is of $O(1)$ for $\epsilon \ll 1$. For example, $\epsilon \int_0^{1/\epsilon} \frac{ds}{D(x,s)}$ is the mean value of $1/D$ over the fine spatial scale. We now have simply $\theta_0 = f_0(x)$. This means that, at leading order, θ varies only on the coarse, $O(1)$ length scale. This does not mean that the fine scale structure has no effect, as we shall see.

Since we now know that $\theta_0 = O(1)$ when $\hat{x} = 1/\epsilon$, we must also eliminate the secular terms that are of $O(\hat{x})$ when $\hat{x}$ is large in (12.95). We therefore require that

$$f_0' = \frac{d\theta_0}{dx} = \lim_{\epsilon \to 0} \left\{ \epsilon \beta(x) \int_0^{1/\epsilon} \frac{ds}{D(x,s)} \right\}. \tag{12.97}$$

In order to determine β, and hence an equation satisfied by θ_0, we must consider one further order in the expansion.

At $O(\epsilon^2)$, (12.92) gives

$$(D\theta_{2\hat{x}} + D\theta_{1x})_{\hat{x}} = -(D\theta_{1\hat{x}})_x - (D\theta_{0x})_x - R(\theta_0(x), x, \hat{x})$$

$$= -\beta'(x) - R(\theta_0(x), x, \hat{x}), \tag{12.98}$$

which we can integrate twice to obtain

$$\theta_2 = f_2(x) - f_1'(x)\hat{x} + \frac{1}{2}\frac{d^2\theta_0}{dx^2}\hat{x}^2 - \frac{\partial}{\partial x}\left\{\beta(x)\int_0^{\hat{x}} \frac{\hat{x}-s}{D(x,s)} ds\right\} + \gamma(x) \int_0^{\hat{x}} \frac{ds}{D(x,s)}$$

$$-\beta'(x)\int_0^{\hat{x}} \frac{s}{D(x,s)} ds - \int_0^{\hat{x}} \frac{1}{D(x,s)} \int_0^s R(\theta_0(x), x, u) \, du \, ds. \tag{12.99}$$

In order to eliminate the secular terms of $O(\hat{x}^2)$, we therefore require that

$$\lim_{\epsilon \to 0} \left[\frac{1}{2}\frac{d^2\theta_0}{dx^2} - \epsilon^2 \frac{\partial}{\partial x}\left\{\beta(x) \int_0^{1/\epsilon} \frac{1/\epsilon - s}{D(x,s)} ds\right\} \right.$$

$$\left. -\beta'(x)\epsilon^2 \int_0^{1/\epsilon} \frac{s}{D(x,s)} ds - \epsilon^2 \int_0^{1/\epsilon} \frac{1}{D(x,s)} \int_0^s R(\theta_0(x), x, u) \, du \, ds \right] = 0.$$

If we now use (12.97) to eliminate $\beta(x)$, we arrive at

$$\lim_{\epsilon \to 0} \left\{ \frac{d^2\theta_0}{dx^2} - \frac{2\frac{d}{dx}\left(\epsilon^2 \int_0^{1/\epsilon} \frac{s}{D(x,s)} ds\right)}{\epsilon \int_0^{1/\epsilon} \frac{ds}{D(x,s)}} \frac{d\theta_0}{dx} \right.$$

$$\left. + 2\epsilon^2 \int_0^{1/\epsilon} \frac{1}{D(x,s)} \int_0^s R(\theta_0(x), x, u) \, du \, ds \right\} = 0.$$

After multiplying through by a suitable integrating factor, we can see that θ_0 satisfies the reaction–diffusion equation

$$\left\{\bar{D}(x)\theta_{0x}\right\}_x + \bar{R}(\theta_0(x), x) = 0, \tag{12.100}$$

on the coarse scale, where

$$\bar{D}(x) = \lim_{\epsilon \to 0} \left[\exp\left\{ -2 \int_0^x \frac{\frac{d}{dX}\left(\epsilon^2 \int_0^{1/\epsilon} \frac{s}{D(X,s)} ds\right)}{\epsilon \int_0^{1/\epsilon} \frac{ds}{D(X,s)}} dX \right\} \right], \qquad (12.101)$$

$$\bar{R}(\theta_0(x), x) = \lim_{\epsilon \to 0} \left\{ 2\bar{D}(x)\epsilon^2 \int_0^{1/\epsilon} \frac{1}{D(x,s)} \int_0^s R(\theta_0(x), x, u)\, du\, ds \right\} \qquad (12.102)$$

are the **fine scale averages**.

This asymptotic analysis, which is called **homogenization**, shows that the leading order temperature does not vary on the fine scale, and satisfies a standard reaction–diffusion equation on the coarse scale. However, the fine scale structure modifies the reaction term and diffusion coefficient through (12.101) and (12.102), with $\bar{D}$ the **homogenized diffusion coefficient** and $\bar{R}$ the **homogenized reaction term**. If we were to seek higher order corrections, we would find that there are variations in the temperature on the fine scale, but that these are at most of $O(\epsilon)$.

One case where $\bar{D}$ and $\bar{R}$ take a particularly simple form is when $D(x, \hat{x}) = D_0(x)\hat{D}(\hat{x})$ and $R(\theta(x), x, \hat{x}) = R_0(\theta(x), x)\hat{R}(\hat{x})$. On substituting these into (12.101) and (12.102), we find, after cancelling a constant common factor between $\bar{D}$ and $\bar{R}$, that we can use $\bar{D}(x) = D_0(x)$ and $\bar{R}(\theta, x) = KR_0(\theta, x)$, where

$$K = \lim_{\epsilon \to 0} \left\{ 2\epsilon \int_0^{1/\epsilon} \frac{1}{\hat{D}(s)} \int_0^s \hat{R}(u)\, du\, ds \right\}.$$

The homogenized diffusion coefficient and reaction term are simply given by the terms $D_0(x)$ and $R_0(\theta, x)$, modified by a measure of the mean value of the ratio of the fine scale variation of each, given by the constant K. In particular, when $\hat{D}(\hat{x}) = 1/(1 + A_1 \sin k_1 \hat{x})$ and $\hat{R}(\hat{x}) = 1 + A_2 \sin k_2 \hat{x}$ for some positive constants k_1, k_2, A_1 and A_2, with $A_1 < 1$ and $A_2 < 1$, we find that $K = 1$. We can illustrate this for a simple case where it is straightforward to find the exact solution of both (12.89) and (12.100) analytically. Figure 12.15 shows a comparison between the exact and asymptotic solutions for various values of ϵ when $k_1 = k_2 = 1$, $A_1 = A_2 = 1/2$, $D_0(x) = 1/(1 + x)$ and $R_0 = 2x - 1$. The analytical solution of (12.89) that vanishes at $x = 0$ (the solution would otherwise contain an arbitrary constant in this case) is

$$\theta = \frac{1}{2}x^2 - \frac{1}{4}x^4 + \epsilon^2 \left[\frac{1}{2}(1 - 2x) \left\{ \frac{1}{8} \cos\left(\frac{2x}{\epsilon}\right) - \sin\left(\frac{x}{\epsilon}\right) \right\} - \frac{1}{4}x - \frac{1}{16} \right]$$

$$+ \epsilon^3 \left[2\cos\left(\frac{x}{\epsilon}\right) + \frac{3}{16}\sin\left(\frac{2x}{\epsilon}\right) - 2 \right],$$

whilst the corresponding solution of (12.100) correctly reproduces the leading order part of this.

Homogenization has been used successfully in many more challenging applications than this linear, steady state reaction diffusion equation, for example,

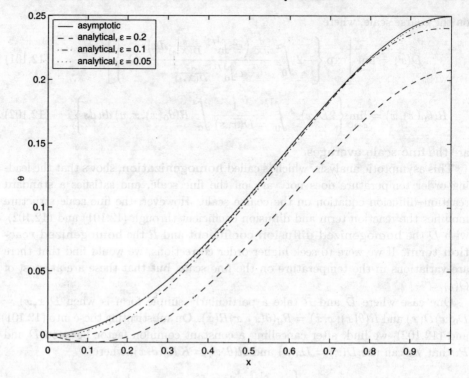

Fig. 12.15. Analytical and asymptotic solutions of (12.89) when $D = 1/(1+x)\left(1 + \frac{1}{2}\sin x\right)$ and $R = (2x - 1)\left(1 + \frac{1}{2}\sin x\right)$.

in assessing the strength of elastic media with small carbon fibre reinforcements (Bakhvalov and Panasenko, 1989).

12.2.7 The WKB Approximation

In all of the examples that we have seen so far, we have used an expansion in algebraic powers of a small parameter to develop perturbation series solutions of differential or algebraic equations. This procedure has been at its most complex when we have needed to match a slowly varying outer solution with an exponentially-rapidly varying, or dissipative, boundary layer. This procedure doesn't always work! For example, if we consider the two-point boundary value problem

$$\epsilon^2 y''(x) + \phi(x)y(x) = 0 \quad \text{subject to } y(0) = 0 \text{ and } y(1) = 1, \tag{12.103}$$

the procedure fails as there are no terms to balance with the leading order term, $\phi(x)y(x)$. If there were a first derivative term in this problem, the procedure would work, although we would have a singular perturbation problem. However, a first derivative term can always be removed. Suppose that we have a differential equation

of the form
$$w'' + p(x)w' + q(x)w = 0.$$

By writing $w = Wu$, we can easily show that
$$W'' + \left(\frac{2u'}{u} + p\right)W' + \left(\frac{u''}{u} + p\frac{u'}{u} + q\right)W = 0.$$

By choosing $\frac{2u'}{u} + p = 0$ and hence $u = \exp\{-\frac{1}{2}\int^x p(t)\,dt\}$, we can remove the first derivative term. Because of this, there is a sense in which $\epsilon^2 y'' + \phi y = 0$ is a generic second order ordinary differential equation, and we need to develop a perturbation method that can deal with it.

The Basic Expansion

A suitable method was proposed by Wentzel, Kramers and Brillioun (and perhaps others as well) in the 1920s. The appropriate asymptotic development is of the form
$$y = \exp\left\{\frac{\psi_0(x)}{\epsilon} + \psi_1(x) + O(\epsilon)\right\}.$$

Differentiating this gives
$$y' = \exp\left\{\frac{\psi_0}{\epsilon} + \psi_1\right\}\left\{\frac{\psi_0'}{\epsilon} + \psi_1' + O(\epsilon)\right\},$$

and
$$y'' = \exp\left\{\frac{\psi_0}{\epsilon} + \psi_1 + O(\epsilon)\right\}\left\{\frac{\psi_0''}{\epsilon} + \psi_1''(x) + O(\epsilon)\right\}$$
$$+ \exp\left\{\frac{\psi_0}{\epsilon} + \psi_1 + O(\epsilon)\right\}\left\{\frac{\psi_0'}{\epsilon} + \psi_1' + O(\epsilon)\right\}^2$$
$$= \exp\left\{\frac{\psi_0}{\epsilon} + \psi_1\right\}\left\{\frac{(\psi_0')^2}{\epsilon^2} + \frac{1}{\epsilon}(2\psi_0'\psi_1' + \psi_0'') + O(1)\right\}.$$

If we substitute these into (12.103), we obtain
$$(\psi_0')^2 + \epsilon(2\psi_0'\psi_1' + \psi_0'') + \phi(x) + O(\epsilon^2) = 0,$$

and hence
$$(\psi_0')^2 = -\phi(x), \quad \psi_1' = -\frac{\psi_0''}{2\psi_0'} = -\frac{1}{2}\frac{d}{dx}(\log \psi_0'). \qquad (12.104)$$

If $\phi(x) > 0$, say for $x > 0$, we can simply integrate these equations to obtain
$$\psi_0 = \pm i \int^x \phi^{1/2}(t)\,dt + \text{constant}, \quad \psi_1 = -\frac{1}{4}\log(\phi(x)) + \text{constant}.$$

All of these constants of integration can be absorbed into a single constant, which may depend upon ϵ, in front of the exponential. The leading order solution is rapidly oscillating, or **dispersive**, in this case, and can be written as

$$y = \frac{A(\epsilon)}{\phi^{1/4}(x)} \exp\left\{i\frac{\int^x \phi^{1/2}(t)\,dt}{\epsilon}\right\} + \frac{B(\epsilon)}{\phi^{1/4}(x)} \exp\left\{-i\frac{\int^x \phi^{1/2}(t)\,dt}{\epsilon}\right\} + \cdots. \tag{12.105}$$

Note that this asymptotic development has assumed that $x = O(1)$, and, in order that it remains uniform, we must have $\psi_1 \ll \psi_0/\epsilon$.

If $\phi(x) < 0$, say for $x < 0$, then some minor modifications give an exponential, or **dissipative**, solution of the form,

$$y = \frac{C(\epsilon)}{|\phi(x)|^{1/4}} \exp\left\{\frac{\int^x |\phi(t)|^{1/2}\,dt}{\epsilon}\right\} + \frac{D(\epsilon)}{|\phi(x)|^{1/4}} \exp\left\{-\frac{\int^x |\phi(t)|^{1/2}\,dt}{\epsilon}\right\} + \cdots. \tag{12.106}$$

If $\phi(x)$ is of one sign in the domain of solution, one or other of the expansions (12.105) and (12.106) will be appropriate. If, however, $\phi(x)$ changes sign, we will have an oscillatory solution on one side of the zero and an exponentially growing or decaying solution on the other. We will consider how to deal with this combination of dispersive and dissipative behaviour after studying a couple of examples.

Example 1: Bessel functions for $x \gg 1$

We saw in Chapter 3 that Bessel's equation is

$$y'' + \frac{1}{x}y' + \left(1 - \frac{\nu^2}{x^2}\right)y = 0.$$

If we make the transformation $y = x^{-1/2}Y$, we obtain the generic form of the equation,

$$Y'' + \left(1 + \frac{\frac{1}{2} - \nu^2}{x^2}\right)Y = 0. \tag{12.107}$$

Although this equation currently contains no small parameter, we can introduce one in a useful way by defining $\bar{x} = \delta x$. If x is large and positive, we can have $\bar{x} = O(1)$ by choosing δ sufficiently small. We have introduced this **artificial small parameter** as a device to help us determine how the Bessel function behaves for $x \gg 1$, and it *cannot* appear in the final result when we change variables back from $\bar{x}$ to x.

In terms of $\bar{x}$ and $\bar{Y}(\bar{x}) = Y(x)$, (12.107) becomes

$$\delta^2 \bar{Y}'' + \left(1 + \delta^2 \frac{\frac{1}{2} - \nu^2}{\bar{x}^2}\right)\bar{Y} = 0. \tag{12.108}$$

By direct comparison with the derivation of the WKB expansion above, in which we neglected terms of $O(\delta^2)$,

$$\psi_0 = \pm i \int^{\bar{x}} dt = \pm i\bar{x}, \quad \psi_1 = -\frac{1}{4}\log 1 = 0,$$

12.2 ORDINARY DIFFERENTIAL EQUATIONS

so that

$$\bar{Y} \sim A(\delta) \exp\left(\frac{i\bar{x}}{\delta}\right) + B(\delta) \exp\left(-\frac{i\bar{x}}{\delta}\right) + \cdots,$$

and hence

$$y \sim \frac{1}{x^{1/2}} \left(Ae^{ix} + Be^{-ix}\right)$$

is the required expansion. Note that, although we can clearly see that the Bessel functions are slowly decaying, oscillatory functions of x as $x \to \infty$, we cannot determine the constants A and B using this technique. As we have already seen, it is more appropriate to use the integral representation (11.14).

We can show that the WKB method is *not* restricted to ordinary differential equations with terms in just y'' and y by considering a further example.

Example 2: A boundary layer

Let's try to find a uniformly valid approximation to the solution of the two-point boundary value problem

$$\epsilon y'' + p(x)y' + q(x)y = 0 \text{ subject to } y(0) = \alpha, \ y(1) = \beta \quad (12.109)$$

when $\epsilon \ll 1$ and $p(x) > 0$. If we assume a WKB expansion,

$$y = \exp\left\{\frac{\psi_0(x)}{\epsilon} + \psi_1(x) + O(\epsilon)\right\},$$

and substitute into (12.109), we obtain at $O(1/\epsilon)$

$$\psi_0' \{\psi_0' + p(x)\} = 0, \quad (12.110)$$

and at $O(1)$,

$$2\psi_0'\psi_1' + \psi_0'' + p(x)\psi_1' + q(x) = 0. \quad (12.111)$$

Using the solution $\psi_0' = 0$ of (12.110) and substituting into (12.111) gives $\psi_1' = -q(x)/p(x)$, which generates a solution of the form

$$y_1 = \exp\left\{\frac{c_1}{\epsilon} - \int_0^x \frac{q(t)}{p(t)} + \cdots\right\} \sim C_1(\epsilon) \exp\left\{-\int_0^x \frac{q(t)}{p(t)}\right\}. \quad (12.112)$$

The second solution of (12.110) has $\psi_0' = -p(x)$ and hence

$$\psi_0 = -\int^x p(t)\, dt + c_2.$$

Equation (12.111) then gives

$$\psi_1 = -\log p(x) + \int_0^x \frac{q(t)}{p(t)}\, dt.$$

348   ASYMPTOTIC METHODS: DIFFERENTIAL EQUATIONS

The second independent solution therefore takes the form

$$y_2 = \exp\left\{\frac{c_2 - \int_0^x p(t)\,dt}{\epsilon} - \log p(x) + \int_0^x \frac{q(t)}{p(t)}\,dt + \cdots\right\}$$

$$\sim \frac{C_2}{p(x)}\exp\left\{-\frac{1}{\epsilon}\int_0^x p(t)\,dt + \int_0^x \frac{q(t)}{p(t)}\,dt\right\}. \qquad (12.113)$$

Combining (12.112) and (12.113) then gives the general asymptotic solution as

$$y = C_1 \exp\left\{-\int_0^x \frac{q(t)}{p(t)}\right\} + \frac{C_2}{p(x)}\exp\left\{-\frac{1}{\epsilon}\int_0^x p(t)\,dt + \int_0^x \frac{q(t)}{p(t)}\,dt\right\}. \qquad (12.114)$$

We can now apply the boundary conditions in (12.109) to obtain

$$\alpha = C_1 + \frac{C_2}{p(0)}, \quad \beta = C_1 \exp\left\{-\int_0^1 \frac{q(t)}{p(t)}\right\} + \frac{C_2}{p(1)}\exp\left\{-\frac{1}{\epsilon}\int_0^1 p(t)\,dt + \int_0^1 \frac{q(t)}{p(t)}\,dt\right\}.$$

The term $\exp\left\{-\frac{1}{\epsilon}\int_0^1 p(t)\,dt\right\}$ is uniformly small and can be neglected, so that the asymptotic solution can be written as

$$y \sim \beta \exp\left\{\int_x^1 \frac{q(t)}{p(t)}\,dt\right\}$$

$$+ \frac{p(0)}{p(x)}\left[\alpha - \beta \exp\left\{\int_0^1 \frac{q(t)}{p(t)}\,dt\right\}\right]\exp\left\{-\frac{1}{\epsilon}\int_0^x p(t)\,dt + \int_0^x \frac{q(t)}{p(t)}\,dt\right\}.$$

Finally, the last exponential in this solution is negligibly small unless $x = O(\epsilon)$ (the boundary layer), so we can write

$$y \sim \beta \exp\left\{\int_x^1 \frac{q(t)}{p(t)}\,dt\right\} + \frac{p(0)}{p(x)}\left[\alpha - \beta \exp\left\{\int_0^1 \frac{q(t)}{p(t)}\,dt\right\}\right]\exp\left\{-\frac{p(0)x}{\epsilon}\right\}.$$

This is precisely the composite expansion that we would have obtained if we had used the method of matched asymptotic expansions instead.

Connection Problems

Let's now consider the boundary value problem

$$\epsilon^2 y''(x) + \phi(x)y(x) = 0 \quad \text{subject to } y(0) = 1,\ y \to 0 \text{ as } x \to -\infty, \qquad (12.115)$$

with

$$\begin{aligned}\phi(x) &> 0 & \text{for } x > 0, \\ \phi(x) &\sim \phi_1 x & \text{for } |x| \ll 1,\ \phi_1 > 0, \\ \phi(x) &< 0 & \text{for } x < 0.\end{aligned} \qquad (12.116)$$

To prevent nonuniformities as $|x| \to \infty$, we will also insist that $|\phi(x)| \gg x^{-2}$ for $|x| \gg 1$. Using the expansions (12.105) and (12.106), we can immediately write

$$y = \frac{A(\epsilon)}{\phi^{1/4}(x)}\exp\left\{i\frac{\int_0^x \phi^{1/2}(t)\,dt}{\epsilon}\right\} +$$

$$\frac{B(\epsilon)}{\phi^{1/4}(x)} \exp\left\{-i\frac{\int_0^x \phi^{1/2}(t)\,dt}{\epsilon}\right\} + \cdots \quad \text{for } x > 0, \qquad (12.117)$$

$$y = C(\epsilon) \exp\left\{-\frac{1}{\epsilon}\int_x^0 |\phi(t)|^{1/2}\,dt - \frac{1}{4}\log|\phi(x)| + \cdots\right\} \quad \text{for } x < 0. \qquad (12.118)$$

The problem of determining how A and B depend upon C is known as a **connection problem**, and can be solved by considering an inner solution in the neighbourhood of the origin.

For $|x| \ll 1$, $\phi \sim \phi_1 x$, so we can estimate the sizes of the terms in the WKB expansion. For $x < 0$

$$y \sim C(\epsilon)\exp\left\{-\frac{1}{\epsilon}\int_x^0 (-\phi_1 t)^{1/2}\,dt - \frac{1}{4}\log(-\phi_1 x)\right\}$$

$$\sim C(\epsilon)\exp\left\{-\frac{2}{3\epsilon}\phi_1^{1/2}(-x)^{3/2} - \frac{1}{4}\log\phi_1 - \frac{1}{4}\log(-x)\right\}.$$

We can now see that the second term becomes comparable to the first when $-x = O(\epsilon^{2/3})$. A similar estimate of the solution for $x > 0$ also gives a nonuniformity when $x = O(\epsilon^{2/3})$. The WKB solutions will therefore be valid in two outer regions with $|x| \gg \epsilon^{2/3}$. We will need a small inner region, centred on the origin, and the inner solution must match with the outer solutions. Equation (12.115) shows that the *only* rescaling possible near to the origin is in a region where $x = O(\epsilon^{2/3})$, and $\bar{x} = x/\epsilon^{2/3} = O(1)$ for $\epsilon \ll 1$. Writing (12.117) and (12.118) in terms of $\bar{x}$ leads to the matching conditions

$$\bar{y} \sim \frac{C(\epsilon)}{\phi_1^{1/4}(-\bar{x})^{1/4}\epsilon^{1/6}} \exp\left\{-\frac{2}{3}(-\bar{x})^{3/2}\phi_1^{1/2}\right\} \quad \text{as } \bar{x} \to -\infty, \qquad (12.119)$$

$$\bar{y} \sim \frac{A(\epsilon)}{\phi_1^{1/4}\bar{x}^{1/4}\epsilon^{1/6}}\exp\left\{i\frac{2}{3}\bar{x}^{3/2}\phi_1^{1/2}\right\} + \frac{B(\epsilon)}{\phi_1^{1/4}\bar{x}^{1/4}\epsilon^{1/6}}\exp\left\{-i\frac{2}{3}\bar{x}^{3/2}\phi_1^{1/2}\right\} \quad \text{as } \bar{x} \to \infty.$$

$$(12.120)$$

If we now rewrite (12.115) in the inner region, making use of $\phi \sim \epsilon^{2/3}\phi_1\bar{x}$ at leading order, we arrive at

$$\frac{d^2\bar{y}}{d\bar{x}^2} + \phi_1\bar{x}\bar{y} = 0,$$

subject to $\bar{y}(0) = 1$, and the matching conditions (12.119) and (12.120).

$$(12.121)$$

We can write (12.121)† in terms of a standard equation by defining $t = -\phi_1^{1/3}\bar{x}$, in terms of which (12.121) becomes

$$\frac{d^2\bar{y}}{dt^2} = t\bar{y}.$$

---

† Note that equation (12.121) is valid for $|x| \ll 1$, and we would expect its solution to be valid in the same domain. Hence there is an overlap of the domains for which the inner and outer solutions are valid, namely $\epsilon^{2/3} \ll |x| \ll 1$, and we can expect the asymptotic matching process to be successful. In fact the overlap domain can be refined to $\epsilon^{2/3} \ll |x| \ll \epsilon^{2/5}$ (see Exercise 12.17).

This is Airy's equation, which we met in Section 3.8, so the solution can be written as

$$\bar{y} = a\mathrm{Ai}(t) + b\mathrm{Bi}(t) = a\mathrm{Ai}\left(-\phi_1^{1/3}\bar{x}\right) + b\mathrm{Bi}\left(-\phi_1^{1/3}\bar{x}\right). \quad (12.122)$$

In Section 11.2.3 we determined the asymptotic behaviour of $\mathrm{Ai}(t)$ for $|t| \gg 1$ using the method of steepest descents. The same technique can be used for $\mathrm{Bi}(t)$, and we find that

$$\mathrm{Ai}(t) \sim \frac{1}{2}\pi^{-1/2}t^{-1/4}\exp\left(-\frac{2}{3}t^{3/2}\right),$$

$$\mathrm{Bi}(t) \sim \pi^{-1/2}t^{-1/4}\exp\left(\frac{2}{3}t^{3/2}\right) \quad \text{as } t \to \infty, \quad (12.123)$$

$$\mathrm{Ai}(t) \sim \frac{1}{\sqrt{\pi}}(-t)^{-1/4}\sin\left\{\frac{2}{3}(-t)^{3/2} + \frac{\pi}{4}\right\},$$

$$\mathrm{Bi}(t) \sim \frac{1}{\sqrt{\pi}}(-t)^{-1/4}\cos\left\{\frac{2}{3}(-t)^{3/2} + \frac{\pi}{4}\right\} \quad \text{as } t \to -\infty. \quad (12.124)$$

Using this known behaviour to determine the behaviour of the inner solution, (12.122), shows that

$$\bar{y} \sim a\frac{1}{2}\pi^{-1/2}\left(\phi_1^{1/3}\bar{x}\right)^{-1/4}\exp\left\{-\frac{2}{3}\left(\phi_1^{1/3}\bar{x}\right)^{3/2}\right\}$$

$$+b\pi^{-1/2}\left(\phi_1^{1/3}\bar{x}\right)^{-1/4}\exp\left\{\frac{2}{3}\left(\phi_1^{1/3}\bar{x}\right)^{3/2}\right\} \quad \text{as } \bar{x} \to -\infty.$$

In order to satisfy the matching condition (12.119), we must have $b = 0$, so that only the Airy function Ai appears in the solution, and

$$C(\epsilon) = \frac{1}{2}\pi^{-1/2}\epsilon^{1/6}\phi_1^{1/6}a. \quad (12.125)$$

The boundary condition $\bar{y}(0) = 1$ then gives $a = 1/\mathrm{Ai}(0) = \Gamma\left(\frac{2}{3}\right)3^{2/3}$, and hence determines $C(\epsilon)$ through (12.125).

As $\bar{x} \to \infty$,

$$\bar{y} \sim a\frac{1}{\sqrt{\pi}}\left(\phi_1^{1/3}\bar{x}\right)^{-1/4}\sin\left\{\frac{2}{3}\left(\phi_1^{1/3}\bar{x}\right)^{3/2} + \frac{\pi}{4}\right\}$$

$$\sim \frac{a}{\sqrt{\pi}\phi_1^{1/12}\bar{x}^{1/4}}\frac{1}{2i}\left\{\exp\left(\frac{2}{3}i\phi_1^{1/2}\bar{x}^{3/2} + \frac{\pi}{4}\right) - \exp\left(-\frac{2}{3}i\phi_1^{1/2}\bar{x}^{3/2} - \frac{\pi}{4}\right)\right\}.$$

We can therefore satisfy the matching condition (12.120) by taking

$$A(\epsilon) = B^*(\epsilon) = \frac{a\phi_1^{1/6}\epsilon^{1/6}}{2i\sqrt{\pi}}.$$

Since A and B are complex conjugate, the solution is real for $x > 0$, as of course it should be. This determines all of the unknown constants, and completes the solution. The Airy function $Ai(t)$ is shown in Figure 11.12, which clearly shows the transition from dispersive oscillatory behaviour as $t \to -\infty$ to dissipative, exponential decay as $t \to \infty$.

12.3 Partial Differential Equations

Many of the asymptotic methods that we have met can also be applied to partial differential equations. As you might expect, the task is usually rather more difficult than we have found it to be for ordinary differential equations. We will proceed by considering four generic examples.

Example 1: Asymptotic solutions of the Helmholtz equation

As an example of an elliptic partial differential equation, let's consider the solution of the **Helmholtz equation**,

$$\nabla^2 \phi + \epsilon^2 \phi = 0 \quad \text{for } r \leqslant 1, \tag{12.126}$$

subject to $\phi(1,\theta) = \sin\theta$ for $\epsilon \ll 1$. This arises naturally as the equation that governs time-harmonic solutions of the wave equation,

$$\frac{1}{c^2} \frac{\partial^2 z}{\partial t^2} = \nabla^2 z,$$

which we met in Chapter 3. If we write $z = e^{i\omega t} \phi(\mathbf{x})$, we obtain (12.126), with $\epsilon = \omega/c$. Since $\phi = O(1)$ on the boundary, we expand $\phi = \phi_0 + \epsilon^2 \phi_2 + O(\epsilon^4)$, and obtain, at leading order,

$$\nabla^2 \phi_0 = 0, \quad \text{subject to } \phi_0(1,\theta) = \sin\theta.$$

If we seek a separable solution of the form $\phi_0 = f(r) \sin\theta$, we obtain

$$f'' + \frac{1}{r} f' - \frac{1}{r^2} f = 0, \quad \text{subject to } f(1) = 1.$$

This has solutions of the form $f = Ar + Br^{-1}$, so the bounded solution that satisfies the boundary condition is $f(r) = r$, and hence $\phi_0 = r \sin\theta$. At $O(\epsilon^2)$, (12.126) gives

$$\nabla^2 \phi_2 = -r \sin\theta, \quad \text{subject to } \phi_2(1,\theta) = 0.$$

If we again seek a separable solution, $\phi_2 = F(r) \sin\theta$, we arrive at

$$F'' + \frac{1}{r} F' - \frac{1}{r^2} F = -\frac{r}{a}, \quad \text{subject to } F(1) = 0.$$

Using the variation of parameters formula, this has the bounded solution

$$\phi_2 = \frac{1}{8} \left(r - r^3 \right) \sin\theta.$$

The two-term asymptotic expansion of the solution can therefore be written as

$$\phi = r \sin\theta + \frac{1}{8} \epsilon^2 \left(r - r^3 \right) \sin\theta + O(\epsilon^4),$$

which is bounded and uniformly valid throughout the circle, $r \leqslant 1$.

Let's also consider a boundary value problem for the **modified Helmholtz equation**,

$$\epsilon^2 \nabla \phi - \phi = 0 \text{ subject to } \phi(1,\theta) = 1 \text{ and } \phi \to 0 \text{ as } r \to \infty. \quad (12.127)$$

Note that $\phi = 0$ satisfies both the partial differential equation and the far field boundary condition, but not the boundary condition at $r = 1$. This suggests that we need a boundary layer near $r = 1$. If we define $r = 1 + \epsilon \bar{r}$ with $\bar{r} = O(1)$ in the boundary layer for $\epsilon \ll 1$, we obtain

$$\phi_{\bar{r}\bar{r}} - \phi = 0,$$

at leading order. This has solution

$$\phi = \bar{A}(\theta)e^{\bar{r}} + \bar{B}(\theta)e^{-\bar{r}},$$

which will match with the far field solution if $\bar{A} = 0$, and satisfy the boundary condition at $\bar{r} = 0$ if $\bar{B} = 1$. The inner solution, and also a composite solution valid at leading order for all $r \geqslant a$, is therefore

$$\phi = \exp\left(\frac{r-1}{\epsilon}\right).$$

Example 2: The small and large time solutions of a diffusion problem

Consider the initial value problem for the diffusion equation,

$$\frac{\partial c}{\partial t} = D \frac{\partial^2 c}{\partial x^2} \text{ for } -\infty < x < \infty \text{ and } t > 0, \quad (12.128)$$

to be solved subject to the initial condition

$$c(x,0) = \begin{cases} f_0(x) & \text{for } x \leqslant 0, \\ 0 & \text{for } x > 0, \end{cases} \quad (12.129)$$

with $f_0 \in C^2(\mathbb{R})$, $f_0 \to 0$ as $x \to -\infty$, $f_0(0) \neq 0$ and

$$\int_{-\infty}^{0} f_0(x)\, dx = f_{\text{tot}}. \quad (12.130)$$

We could solve this using either Laplace or Fourier transforms. The result, however, would be in the form of a convolution integral, which does not shed much light on the structure of the solution. We can gain a lot of insight by asking how the solution behaves just after the initial state begins to diffuse (the small time solution, $t \ll 1$), and after a long time ($t \gg 1$).

The small time solution, $t \ll 1$

The general effect of diffusion is to smooth out gradients in the function $c(x,t)$. It can be helpful to think of c as a distribution of heat or a chemical concentration. This smoothing is particularly pronounced at points where c is initially discontinuous, in this case at $x = 0$ only. Diffusion will also spread the initial data into $x > 0$, where $c = 0$ initially. For this reason, we anticipate that there will be three distinct asymptotic regions.

12.3 PARTIAL DIFFERENTIAL EQUATIONS

— Region I: $x < 0$. In this region we expect a gradual smoothing out of the initial data.
— Region II: $|x| \ll 1$. The major feature in this region will be an instantaneous smoothing out of the initial discontinuity.
— Region III: $x > 0$. There will be a flux from $x < 0$ into this region, so we expect an immediate change from $c = 0$ to c nonzero.

Thinking about the physics of the problem before doing any detailed calculation is usually vital to unlocking the structure of the asymptotic solution of a partial differential equation.

Let's begin our analysis in region I by posing an asymptotic expansion valid for $t \ll 1$,

$$c(x,t) = f_0(x) + tf_1(x) + t^2 f_2(x) + O(t^3).$$

Substituting this into (12.128) gives

$$f_1 = Df_0'', \quad 2f_2 = Df_1'',$$

and hence

$$c(x,t) = f_0(x) + tDf_0''(x) + \frac{1}{2}t^2 D^2 f_0''''(x) + O(t^3). \tag{12.131}$$

Note that c increases in regions where $f_0'' > 0$, and vice versa, as physical intuition would lead us to expect. Note also that as $x \to 0^-$, $c(x,t) \sim f_0(0)$. However, in $x > 0$ we would expect c to be small, and the solutions in regions I and III will not match together without a boundary layer centred on the origin, namely region II.

Before setting up this boundary layer, it is convenient to find the solution in region III, where we have noted that c is small. If we try a WKB expansion of the form†

$$c(x,t) = \exp\left\{-\frac{A(x)}{t} + B(x)\log t + C(x) + o(1)\right\},$$

and substitute into (12.128), we obtain

$$A = DA_x^2, \quad A_x B_x = 0, \quad B = -D(A_{xx} + 2A_x C_x).$$

The solutions of these equations are

$$A = \frac{x^2}{4D} + \beta x + D\beta^2, \quad B = b, \quad C = -\left(b + \frac{1}{2}\right)\log(2\beta D + x) + d,$$

where β, b and d are constants of integration. The WKB solution in region III is therefore

$$c = \exp\left\{-\frac{1}{t}\left(\frac{x^2}{4D} + \beta x + D\beta^2\right) + b\log t - \left(b + \frac{1}{2}\right)\log(2\beta D + x) + d + o(1)\right\}. \tag{12.132}$$

† Note that we are using the small time, t, in the WKB expansion that we developed in Section 12.2.7.

In the boundary layer, region II, we define a new variable, $\eta = x/t^\alpha$. In region II, $|\eta| = O(1)$, and hence $|x| = O(t^\alpha) \ll 1$, with $\alpha > 0$ to be determined. In terms of η, (12.128) becomes

$$\frac{\partial c}{\partial t} - \frac{\alpha}{t}\eta\frac{\partial c}{\partial \eta} = \frac{D}{t^{2\alpha}}\frac{\partial^2 c}{\partial \eta^2}.$$

Since $c = O(1)$ in the boundary layer (remember, c must match with the solution in region I as $\eta \to -\infty$, where $c = O(1)$), we can balance terms in this equation to find a distinguished limit when $\alpha = 1/2$. The boundary layer therefore has thickness of $O(t^{1/2})$, which is a typical diffusive length scale. If we now expand as $c = c_0(\eta) + o(1)$, we have, at leading order,

$$-\frac{1}{2}\eta c_{0\eta} = Dc_{0\eta\eta}. \tag{12.133}$$

If we now write the solutions in regions I and III, given by (12.131) and (12.132), in terms of η, we arrive at the matching conditions

$$c_0 \sim f_0(0) \quad \text{as } \eta \to -\infty, \tag{12.134}$$

$$c_0 \sim \exp\left\{-\frac{D\beta^2}{t} - \frac{\beta\eta}{t^{1/2}} - \frac{\eta^2}{4D} + b\log t - \right.$$

$$\left. \left(b+\frac{1}{2}\right)\log\left(2\beta D + t^{1/2}\eta\right) + d + o(1)\right\} \quad \text{as } \eta \to \infty. \tag{12.135}$$

The solution of (12.133) is

$$c_0(\eta) = F + G\int_{-\infty}^{\eta} e^{-s^2/4D}\,ds.$$

As $\eta \to -\infty$, $c \sim F = f_0(0)$. As $\eta \to \infty$, we can use integration by parts to show that

$$c_0 = f_0(0) + G\left\{\int_{-\infty}^{\infty} e^{-s^2/4D}\,ds - \frac{2D}{\eta}e^{-\eta^2/4D} + O\left(\frac{1}{\eta^2}e^{-\eta^2/4D}\right)\right\}.$$

In order that this is consistent with the matching condition (12.135), we need

$$f_0(0) + G\int_{-\infty}^{\infty} e^{-s^2/4D}\,ds = 0,$$

and hence $G = -f_0(0)/\sqrt{4\pi D}$. This leaves us with

$$c_0 \sim \exp\left\{-\frac{\eta^2}{4D} + \log(-2DG) - \log\eta\right\}.$$

For this to be consistent with (12.135) we need $\beta = 0$, $b = 1/2$ and $d = \log(-2DG)$.

The structure of the solution that we have constructed allows us to be rather more precise about how diffusion affects the initial data. For $|x| \gg t^{1/2}$, $x < 0$ there is a slow smoothing of the initial data that involves algebraic powers of t, given by (12.131). For $|x| \gg t^{1/2}$, $x > 0$, c is exponentially small, driven by a diffusive flux across the boundary layer. For $|x| = O(t^{1/2})$ there is a boundary layer, with the solution changing by $O(1)$ over a small length of $O(t^{1/2})$. This small time solution continues to evolve, and, when $t = O(1)$, is not calculable by asymptotic methods. When t is sufficiently large, a new asymptotic structure emerges, which we shall consider next.

The large time solution, $t \gg 1$

After a long time, diffusion will have spread out the initial data in a more or less uniform manner, and the structure of the solution is rather different from that which we discussed above for $t \ll 1$. We will start our asymptotic development where $x = O(1)$, and seek a solution of the form

$$c(x,t) = c_0(t) + c_1(x,t) + \cdots,$$

with $|c_1| \ll |c_0|$ for $t \gg 1$ to ensure that the expansion is asymptotic. If we substitute this into (12.128), we obtain

$$\dot{c}_0(t) = Dc_{1xx},$$

at leading order, which can be integrated to give

$$c_1(x,t) = \frac{\dot{c}_0(t)}{2D} x^2 + \alpha_1^{\pm}(t)x + \beta_1^{\pm}(t). \qquad (12.136)$$

The distinction between α_1^+ and α_1^-, and similarly for $\beta_1^{\pm}$, is to account for differences in the solution for $x > 0$ and $x < 0$, introduced by the linear terms. As $|x| \to \infty$, c_1 grows quadratically, which causes a nonuniformity in the expansion, specifically when $x = O\left(\sqrt{c_0(t)/|\dot{c}_0(t)|}\right)$. In order to deal with this, we introduce a scaled variable, $\eta = x/\sqrt{c_0(t)/|\dot{c}_0(t)|}$, with $\eta = O(1)$ for $t \gg 1$ in this outer region. In order to match with the solution in the inner region, where $x = O(1)$, we need

$$c(\eta, t) \to c_0(t) \quad \text{as } \eta \to 0. \qquad (12.137)$$

In terms of η, (12.128) becomes

$$\frac{\partial c}{\partial t} - \frac{1}{2}\eta \frac{|\dot{c}_0(t)|}{c_0(t)} \frac{d}{dt}\left(\frac{c_0(t)}{|\dot{c}_0(t)|}\right) \frac{\partial c}{\partial \eta} = \frac{|\dot{c}_0(t)|}{c_0(t)} D \frac{\partial^2 c}{\partial \eta^2}. \qquad (12.138)$$

Motivated by the matching condition (12.137), we will try to solve this using the expansion $c = c_0(t)F_{\pm}(\eta) + o(c_0(t))$, subject to $F_{\pm} \to 1$ as $\eta \to 0^{\pm}$ and $F_{\pm} \to 0$ as $\eta \to \pm\infty$. The superscript $\pm$ indicates whether the solution is for $\eta > 0$ or $\eta < 0$. It is straightforward to substitute this into (12.138), but this leads to some options. The first and third terms are of $O(\dot{c}_0(t))$, whilst the second term is of $O(\dot{c}_0(t)\frac{d}{dt}\left(\frac{c_0(t)}{|\dot{c}_0(t)|}\right))$. Should we include the second term in the leading order balance or not? Let's see what happens if we do decide to balance these terms to get the richest limit. We must then have $c_0(t)/\dot{c}_0(t) = O(t)$, and hence $c_0 = Ct^{-\alpha}$ for some constants C and α. This looks like a sensible gauge function.

If we proceed, (12.138) becomes, at leading order,

$$F_{\pm} - \frac{1}{2}\eta F'_{\pm} = DF''_{\pm}. \qquad (12.139)$$

This is slightly more difficult to solve than (12.133). However, if we look for a quadratic solution, we quickly find that $F_\pm = \eta^2 + 2D$ is a solution. Using the method of reduction of order, the general solution of (12.139) is

$$F_\pm = (\eta^2 + 2D)\left\{A^\pm + B^\pm \int_0^\eta \frac{e^{-s^2/2D}}{(s^2+2D)^2}\, ds\right\}. \qquad (12.140)$$

As $\eta \to 0$, by Taylor expanding the integrand, we can show that

$$F_\pm = (\eta^2 + 2D)\left(A^\pm + B^\pm \frac{\eta}{2D}\right) + O(\eta^3).$$

To match with the inner solution, we therefore require that $A^\pm = 1/2D$. As $\eta \to \pm\infty$, using integration by parts, we find that

$$F_\pm = (\eta^2 + 2D)\left\{A^\pm \pm B^\pm \int_0^\infty \frac{e^{-s^2/2D}}{(s^2+2D)^2}\, ds + O\left(\frac{1}{\eta^5}e^{-\eta^2/4D}\right)\right\},$$

so that we require

$$A^\pm \pm B^\pm \int_0^\infty \frac{e^{-s^2/2D}}{(s^2+2D)^2}\, ds = 0.$$

The outer solution can therefore be written as

$$F_\pm = \left(1 + \frac{\eta^2}{2D}\right)\left\{1 \mp \int_0^\eta \frac{e^{-s^2/2D}}{(s^2+2D)^2}\, ds \bigg/ \int_0^\infty \frac{e^{-s^2/2D}}{(s^2+2D)^2}\, ds\right\}. \qquad (12.141)$$

In order to determine α, and hence the size of $c_0(t)$, some further work is needed. Firstly, we can integrate (12.128) and apply the initial condition, to obtain

$$\int_{-\infty}^\infty c(x,t)\, dx = \int_{-\infty}^0 f_0(x)\, dx = f_{tot}. \qquad (12.142)$$

This just says that mass is conserved during the diffusion process. Secondly, we can write down the composite expansion

$$c = c_{\text{inner}} + c_{\text{outer}} - (c_{\text{inner}})_{\text{outer}} = c_0(t) F_\pm(\eta),$$

and use this in (12.142) to obtain

$$\int_{-\infty}^\infty c_0(t) F_\pm(\eta)\, dx = c_0(t)\sqrt{\frac{c_0(t)}{|\dot{c}_0(t)|}} \int_{-\infty}^\infty F_\pm(\eta)\, d\eta = f_{tot}.$$

This is now a differential equation for c_0 in the form

$$\frac{c_0^{3/2}(t)}{|\dot{c}_0(t)|^{1/2}} = \frac{f_{tot}}{\int_{-\infty}^\infty F_\pm(\eta)\, d\eta}.$$

In Exercise 12.18 we find that $\int_{-\infty}^\infty F_\pm(\eta)\, d\eta = \sqrt{2\pi D}$ and hence that $c_0(t) = $

$f_{\text{tot}}/\sqrt{4\pi Dt}$. We can now calculate that $\sqrt{c_0(t)/|\ddot{c}_0(t)|} = O(t^{1/2})$, which is the usual diffusive length scale. Our asymptotic solution therefore takes the form

$$c(x,t) \sim \begin{cases} \dfrac{f_{\text{tot}}}{\sqrt{4\pi Dt}} & \text{for } |x| = O(t^{1/2}), \\ \dfrac{f_{\text{tot}}}{\sqrt{4\pi Dt}} F_{\pm}(\eta) & \text{for } |x| \gg t^{1/2}. \end{cases} \qquad (12.143)$$

The success of this approach justifies our decision to choose $c_0(t)$ in order to obtain the richest distinguished limit. Notice that the large time solution has "forgotten" the precise details of the initial conditions. It only "remembers" the area under the initial data, at leading order.

If we consider the particular case $f_0(t) = e^x$, we find (see Exercise 12.18) that an exact solution is available, namely $c(x,t) = \frac{1}{2} e^{x+Dt} \text{erfc}\left(\frac{x}{\sqrt{4Dt}} + \sqrt{Dt}\right)$. This solution is plotted in Figure 12.16 at various times, and we can clearly see the structures that our asymptotic solutions predict emerging for both small and large times. In Figure 12.17 we plot $c(0,t)$ as a function of Dt. Our asymptotic solution predicts that $c(0,t) = \frac{1}{2} + o(1)$ for $t \ll 1$, consistent with Figure 12.17(a). In Figure 12.17(b) we can see that the asymptotic solution, $c(0,t) \sim 1/\sqrt{4\pi Dt}$ as $t \to \infty$, is in excellent agreement with the exact solution.

Example 3: The wave equation with weak damping

(i) Linear damping

Consider the equation

$$\frac{\partial^2 y}{\partial t^2} = c^2 \frac{\partial^2 y}{\partial x^2} - \epsilon \frac{\partial y}{\partial t}, \quad \text{for } t > 0 \text{ and } -\infty < x < \infty, \qquad (12.144)$$

subject to the initial conditions

$$y(x,0) = Y_0(x), \quad \frac{\partial y}{\partial t}(x,0) = 0, \qquad (12.145)$$

with $\epsilon \ll 1$. The one-dimensional wave equation, (12.144) with $\epsilon = 0$, governs the small amplitude motion of an elastic string, which we met in Section 3.9.1. The additional term, ϵy_t, represents a weak, linear damping, proportional to the velocity of the string, for example due to drag on the string as it moves through the air.

The form of the initial conditions suggests that we should consider an asymptotic expansion $y = y_0 + \epsilon y_1 + O(\epsilon^2)$. On substituting this into (12.144) and (12.145) we obtain

$$\frac{\partial^2 y_0}{\partial t^2} - c^2 \frac{\partial^2 y_0}{\partial x^2} = 0, \quad \text{subject to } y_0(x,0) = Y_0(x), \ y_{0t}(x,0) = 0, \qquad (12.146)$$

$$\frac{\partial^2 y_1}{\partial t^2} - c^2 \frac{\partial^2 y_1}{\partial x^2} = -\frac{\partial y_0}{\partial t}, \quad \text{subject to } y_1(x,0) = y_{1t}(x,0) = 0. \qquad (12.147)$$

The initial value problem given by (12.146) is one that we studied in Section 3.9.1, and has d'Alembert's solution, (3.43),

$$y_0(x,t) = \frac{1}{2}\{Y_0(x-ct) + Y_0(x+ct)\}. \qquad (12.148)$$

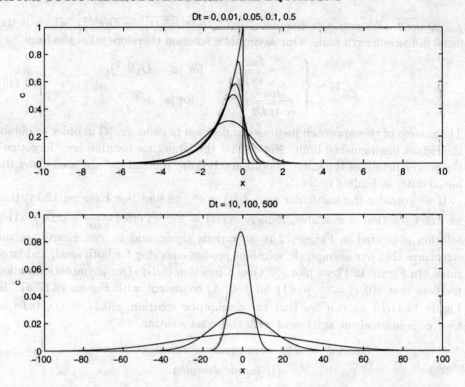

Fig. 12.16. The solution of the diffusion equation with $f_0(x) = e^x$ at various times.

This solution represents the sum of two waves, one travelling to the left and one travelling to the right, each with speed c, without change of form, and with half the initial amplitude. This is illustrated in Figure 12.18 for the initial condition $Y_0(x) = 1/(1+x^2)$. The splitting of the initial profile into left- and right-travelling waves is clearly visible.

In terms of the characteristic variables, $\xi = x - ct$, $\eta = x + ct$, (12.147) becomes

$$-4c^2 \frac{\partial^2 y_1}{\partial \xi \partial \eta} = c \left(\frac{\partial y_0}{\partial \xi} - \frac{\partial y_0}{\partial \eta} \right) = \frac{1}{2} c \{Y_0'(\xi) - Y_0'(\eta)\}.$$

Integrating this expression twice gives the solution

$$y_1 = \frac{1}{8c} \{\xi Y_0(\eta) - \eta Y_0(\xi)\} + F_1(\xi) + G_1(\eta). \tag{12.149}$$

The initial conditions show that

$$F_1(x) + G_1(x) = 0,$$

and

$$F_1'(x) - G_1'(x) = \frac{1}{4c} \{x Y_0'(x) - Y_0(x)\},$$

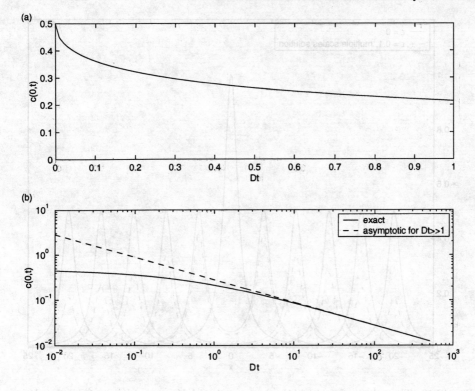

Fig. 12.17. The solution of the diffusion equation with $f_0(x) = e^x$ at $x = 0$.

which can be integrated once to give

$$F_1(x) - G_1(x) = \frac{1}{4c}\left\{xY_0(x) - 2\int_0^x Y_0(s)\,ds\right\} + b.$$

Finally,

$$F_1(x) = -G_1(x) = \frac{1}{8c}\left\{xY_0(x) - 2\int_0^x Y_0(s)\,ds\right\} + \frac{1}{2}b,$$

which, in conjunction with (12.149), shows that

$$y_1 = -\frac{1}{4}t\{Y_0(x+ct) + Y_0(x-ct)\} + \frac{1}{4c}\int_{x-ct}^{x+ct} Y_0(s)\,ds. \tag{12.150}$$

We can now see that $y_1 = O(t)$ for $t \gg 1$, and therefore that our asymptotic expansion becomes nonuniform when $t = O(\epsilon^{-1})$.

We will proceed using the method of multiple scales, defining a slow time scale $T = \epsilon t$, and looking for a solution $y = y(x, t, T)$. In terms of these new independent variables, (12.144) becomes

$$y_{tt} + 2\epsilon y_{tT} + \epsilon^2 y_{TT} = c^2 y_{xx} - \epsilon y_t - \epsilon^2 y_T. \tag{12.151}$$

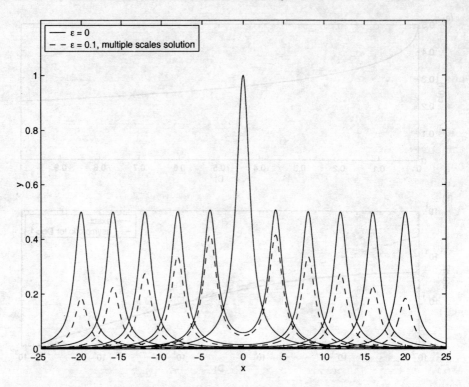

Fig. 12.18. The solution of (12.144) when $c = 1$ and $Y_0(x) = 1/(1+x^2)$ at equal time intervals, $t = 0, 4, 8, 12, 16, 20$, when $\epsilon = 0$ and 0.1.

If we now seek an asymptotic solution of the form $y = y_0(x,t,T) + \epsilon y_1(x,t,T) + O(\epsilon^2)$, at leading order we obtain (12.146), as before, but now the solution is

$$y_0(x,t,T) = F_0(\xi, T) + G_0(\eta, T), \tag{12.152}$$

with

$$F_0(\xi, 0) = \frac{1}{2} Y_0(\xi), \quad G_0(\eta, 0) = \frac{1}{2} Y_0(\eta). \tag{12.153}$$

As usual in the method of multiple scales, we need to go to $O(\epsilon)$ to determine F_0 and G_0. We find that

$$-4c y_{1\xi\eta} = F_{0\xi} - G_{0\eta} + 2\left(F_{0\xi T} - G_{0\eta T}\right). \tag{12.154}$$

On solving this equation, the presence of the terms of the right hand side causes y_1 to grow linearly with t. In order to eliminate them, we must have

$$F_{0\xi T} = -\frac{1}{2} F_{0\xi}, \quad G_{0\eta T} = -\frac{1}{2} G_{0\eta}. \tag{12.155}$$

If we solve these equations subject to the initial conditions (12.153), we obtain

$$F_0 = \frac{1}{2}e^{-T}Y_0(\xi), \quad G_0 = \frac{1}{2}e^{-T}Y_0(\eta),$$

and hence

$$y_0 = \frac{1}{2}e^{-\epsilon t/2}\{Y_0(x-ct) + Y_0(x+ct)\}. \quad (12.156)$$

This shows that the small term, ϵy_t, in (12.144) leads to an exponential decay of the amplitude of the solution over the slow time scale, $t = O(\epsilon^{-1})$, consistent with our interpretation of this as a damping term. Figure 12.18 shows how this slow exponential decay affects the solution.

(ii) Nonlinear damping

What happens if we replace the linear damping term ϵy_t with a nonlinear damping term, $\epsilon(y_t)^3$? We must then solve

$$\frac{\partial^2 y}{\partial t^2} = c^2\frac{\partial^2 y}{\partial x^2} - \epsilon\left(\frac{\partial y}{\partial t}\right)^3, \quad \text{for } t > 0 \text{ and } -\infty < x < \infty, \quad (12.157)$$

subject to the initial conditions

$$y(x,0) = Y_0(x), \quad \frac{\partial y}{\partial t}(x,0) = 0. \quad (12.158)$$

We would again expect a nonuniformity when $t = O(\epsilon^{-1})$, so let's go straight to a multiple scales expansion, $y = y_0(x,t,T) + \epsilon y_1(x,t,T)$. At leading order, as before, we have (12.152) and (12.153). At $O(\epsilon)$,

$$-4cy_{1\xi\eta} = c^2(F_{0\xi} - G_{0\eta})^3 + 2(F_{0\xi T} - G_{0\eta T}). \quad (12.159)$$

In order to see clearly which terms are secular, we integrate the expression $(F_{0\xi} - G_{0\eta})^3$ twice to obtain

$$\eta \int_0^\xi F_{0\xi}^3(s)\,ds - 3G_0(\eta)\int_0^\xi F_{0\xi}^2(s)\,ds + 3F_0(\xi)\int_0^\eta G_{0\xi}^2(s)\,ds - \xi\int_0^\eta G_{0\eta}^3(s)\,ds.$$

Assuming that $F_0(s)$ and $G_0(s)$ are integrable as $s \to \pm\infty$, we can see that the terms that become unbounded as ξ and η become large are those associated with $F_{0\xi}^3$ and $G_{0\eta}^3$. We conclude that, to eliminate secular terms in (12.159), we need

$$F_{0\xi T} = -\frac{1}{2}c^2 F_{0\xi}^3, \quad G_{0\eta T} = -\frac{1}{2}c^2 G_{0\eta}^3,$$

to be solved subject to (12.153). The solutions are

$$F_{0\xi} = \frac{Y_0'(\xi)}{\sqrt{4 + c^2 T\{Y_0'(\xi)\}^2}}, \quad G_{0\eta} = \frac{Y_0'(\eta)}{\sqrt{4 + c^2 T\{Y_0'(\eta)\}^2}},$$

and hence

$$y_0 = -\left[\int_\xi^\infty \frac{Y_0'(s)}{\sqrt{4 + c^2 T\{Y_0'(s)\}^2}}\,ds + \int_\eta^\infty \frac{Y_0'(s)}{\sqrt{4 + c^2 T\{Y_0'(s)\}^2}}\,ds\right]. \quad (12.160)$$

In general, these integrals cannot be determined analytically, but we can see that the amplitudes of the waves do decay as T increases, and are of $O(T^{-1/2})$ for $T \gg 1$.

Example 4: The measurement of oil fractions using local electrical probes

As we remarked at the beginning of the book, many problems that arise in engineering are susceptible to mathematical modelling. We can break the modelling process down into separate steps.

(i) Identify the important physical processes that are involved.
(ii) Write down the governing equations and boundary conditions.
(iii) Define dimensionless variables and identify dimensionless constants.
(iv) Solve the governing equations using either a numerical method or an asymptotic method.

Note that, although it is possible that we can find an analytical solution, this is highly unlikely when studying real world problems. As we discussed at the start of Chapter 11, when one or more of the dimensionless parameters is small, we can use an asymptotic solution technique. Let's now discuss an example of this type of situation.

For obvious reasons, oil companies are interested in how much oil is coming out of their oilwells, and often want to make this measurement at the point where oil is entering the well as droplets, rather than at the surface. One tool that can be lowered into a producing oilwell to assist with this task is a **local probe**. This is a device with a tip that senses whether it is in oil or water. The output from the probe can be time-averaged to give the local oil fraction at the tip, and an array of probes deployed to give a measurement of how the oil fraction varies across the well. We will consider a simple device that distinguishes between oil and water by measuring electrical conductivity, which is several orders of magnitude higher in saline water than in oil.

The geometry of the electrical probe, which is made from sharpening the tip of a coaxial cable like a pencil, is shown in Figure 12.19. A voltage is applied to the core of the probe, whilst the outer layer, or cladding, is earthed. A measurement of the current between the core and the cladding is then made to determine the conductivity of the surrounding medium. Although this measurement gives a straightforward way of distinguishing between oil and water when only one liquid is present, for example when dipping the probe into a beaker containing a single liquid, the difficulty lies in interpreting the change in conductivity as a droplet of oil approaches, meets, deforms around and is penetrated by the probe. If we want to understand and model this process, there is clearly a difficult fluid mechanical problem to be solved before we can begin to relate the configuration of the oil droplet to the current through the probe (see Billingham and King, 1995). We will pre-empt all of this fluid mechanical complexity by considering what happens if, in the course of the interaction of an oil droplet with a probe, a thin layer of oil forms on the surface of the probe. How thin must this oil layer become before the current through the probe is effectively equal to that of a probe in pure water?

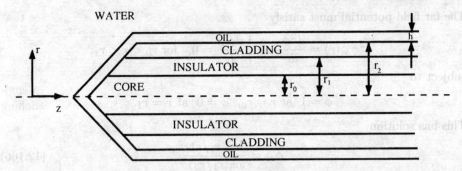

Fig. 12.19. A cross-section through an axisymmetric electrical probe.

In order to answer this question, we must solve a problem in electrostatics, since the speed at which oil–water interfaces move is much less than the speed at which electromagnetic disturbances travel (the speed of light)†. The electrostatic potential, ϕ, is an axisymmetric solution of Laplace's equation,

$$\nabla^2 \phi = 0. \qquad (12.161)$$

We will assume that the conducting parts of the probe are **perfect conductors**, so that

$$\phi = \begin{cases} 1 & \text{at the surface of the core,} \\ 0 & \text{at the earthed surface of the cladding.} \end{cases} \qquad (12.162)$$

At interfaces between different media, for example oil and water or oil and insulator, we have the jump conditions

$$[\phi] = 0, \quad \left[\sigma \frac{\partial \phi}{\partial n}\right] = 0. \qquad (12.163)$$

Square brackets indicate the change in the enclosed quantity across an interface, σ is the **conductivity**, which is different in each medium (oil, water and insulator), and $\partial/\partial n$ is the derivative in the direction normal to the interface. Equation (12.163) represents continuity of potential and continuity of current at an interface. To complete the problem, we have the far field conditions that

$$\phi \to 0 \quad \text{as } r^2 + z^2 \to \infty \text{ outside the probe}, \qquad (12.164)$$

and

$$\phi \sim \phi_\infty(r) \quad \text{as } z \to \infty \text{ for } r_0 < r < r_1, \qquad (12.165)$$

using cylindrical polar coordinates coaxial with the probe, and $r = 0$ at the tip. Here r_0 and r_1 are the inner and outer radii of the insulator, as shown in Figure 12.19.

† This, in itself, is an asymptotic approximation that can be made rigorous by defining a small parameter, the ratio of a typical fluid speed to the speed of light. Some approximations are, however, so obvious that justifying them rigorously is a little too pedantic. For a simple introduction to electromagnetism, see Billingham and King (2001).

The far field potential must satisfy

$$\nabla^2 \phi_\infty(r) = \frac{d^2\phi_\infty}{dr^2} + \frac{1}{r}\frac{d\phi_\infty}{dr} = 0, \quad \text{for } r_0 < r < r_1,$$

subject to

$$\phi = 1 \text{ at } r = r_0, \quad \phi = 0 \text{ at } r = r_1.$$

This has solution

$$\phi_\infty = \frac{\log(r_1/r)}{\log(r_1/r_0)}. \tag{12.166}$$

Finally, we will assume that the probe is surrounded by water, except for a uniform layer of oil on its surface of thickness $h \ll r_0$. Our task is to solve the boundary value problem given by (12.161) to (12.165).

This is an example of a problem where the governing equation and boundary conditions are fairly straightforward, but the geometry is complicated. Problems like this are usually best solved numerically. However, in this case we have one region where the aspect ratio is small – the thin oil film. The potential is likely to change rapidly across this film compared with its variation elsewhere, and a numerical method will have difficulty handling this. We can, however, make some progress by looking for an asymptotic solution in the thin film. The first thing to do is to set up a local coordinate system in the oil film. The quantities h and r_0 are the natural length scales with which to measure displacements across and along the film, so we let η measure displacement across the film, with $\eta = 0$ at the surface of the probe and $\eta = 1$ at the surface of the water, and let ξ measure displacement along the film, with $\xi = 0$ at the probe tip and $\xi = 1$ a distance r_0 from the tip. Away from the tip and the edge of the probe, which we will not consider for the moment, this provides us with an orthogonal coordinate system, and (12.161) becomes

$$\frac{\partial^2 \phi}{\partial \eta^2} + \delta^2 \frac{\partial^2 \phi}{\partial \xi^2} = 0,$$

where

$$\delta = \frac{h}{r_0} \ll 1.$$

At leading order, $\partial^2 \phi/\partial \eta^2 = 0$, and hence ϕ varies linearly across the film, with

$$\phi = A(\xi)\eta + B(\xi). \tag{12.167}$$

Turning our attention now to (12.163)$_2$, since we expect variations of ϕ in the water and the insulator to take place over the geometrical length scale r_0, we have

$$\delta_j \frac{\partial \phi}{\partial \eta} = \delta \frac{\partial \phi}{\partial n} \quad \text{at interfaces}, \tag{12.168}$$

where

$$\delta_j = \frac{\sigma_o}{\sigma_j},$$

with the subscripts o, w and i indicating oil, water and insulator respectively. We expect that $\delta_w \ll 1$ and $\delta_i = O(1)$, since oil and the insulator have similar conductivities, both much less than that of saline water. We conclude that, at the interface between oil and insulator, at leading order $\partial\phi/\partial\eta = 0$, and hence $A(\xi) = 0$ there. Also, from the conditions at the surface of the cladding and core, (12.162), $B(\xi) = 0$ at the cladding and $B(\xi) = 1$ at the core.

Returning now to (12.168) with $j = w$, note that we have two small parameters, δ and δ_w. Double limiting processes like this ($\delta \to 0$, $\delta_w \to 0$) have to be treated with care, as the final result usually depends on how fast one parameter tends to zero compared with the other. In this case, we obtain the richest asymptotic balance by assuming that

$$\frac{\delta}{\delta_w} = K = \frac{h\sigma_w}{r_0\sigma_o} = O(1) \text{ as } \delta \to 0,$$

and

$$\frac{\partial\phi}{\partial n} = K\frac{\partial\phi}{\partial\eta} = KA(\xi).$$

We can now combine all of the information that we have, to show that at the surface of the probe, the potential in the water satisfies

$$\frac{\partial\phi}{\partial n} = \begin{cases} K(\phi - 1) & \text{at the surface of the core,} \\ 0 & \text{at the surface of the insulator,} \\ K\phi & \text{at the surface of the cladding.} \end{cases} \quad (12.169)$$

The fact that the oil film is thin allows us to apply these conditions at the surface of the probe at leading order. The key point is that this asymptotic analysis allows us to eliminate the thin film from the geometry of the problem at leading order, and instead include its effect in the boundary conditions (12.169). The solution of (12.161) subject to (12.164), (12.165) and (12.169) in the region outside the probe is geometrically simple, and easily achieved using a computer. We will not show how to do this here, as it is outside the scope of this book. We can, however, extract one vital piece of information from our analysis. We have proceeded on the basis that $K = O(1)$. What happens if $K \gg 1$ or $K \ll 1$? If $K \gg 1$, at leading order (12.169) becomes

$$\phi = \begin{cases} 1 & \text{at the surface of the core,} \\ 0 & \text{at the surface of the cladding,} \end{cases}$$

$$\frac{\partial\phi}{\partial n} = 0 \text{ at the surface of the insulator.} \quad (12.170)$$

These are precisely the boundary conditions that would apply at leading order in the absence of an oil layer. We conclude that if $K \gg 1$, and hence $h \ll r_0\sigma_o/\sigma_w$, the film of oil is too thin to prevent a current from passing from core to cladding through the water, and the oil cannot be detected by the probe. If $K \ll 1$, at leading order (12.169) becomes $\partial\phi/\partial n = 0$ at the surface of the probe, and hence $\phi = 0$ in the water. This then shows that $\phi = 1 - \eta$ in the oil film over the core and $\phi = 0$ in the rest of the film, from which it is straightforward to calculate the current flowing from core to cladding. In this case the oil film dominates the response of the probe, effectively insulating it from the water outside.

We conclude that there is a critical oil film thickness, $h_c = r_0 \sigma_o / \sigma_w$. For $h \ll h_c$ the oil film is effectively invisible, for $h \gg h_c$ the external fluid is effectively invisible, and the current through the probe is determined by the thickness of the film, whilst for $h = O(h_c)$ both the oil and water affect the current through the boundary conditions (12.169). For a typical oil and saline water, $h_c \approx 10^{-9}$ m. This is such a small length that, in practice, any thin oil film coating a probe insulates it from the external fluid, and can lead to practical difficulties with this technique. In reality, local probes are used with alternating rather than direct current driving the core. One helpful effect of this is to increase the value of h_c, due to the way that the impedances† of oil and water change with the frequency of the driving potential.

Exercises

12.1 Determine the first two terms in the asymptotic expansion for $0 < \epsilon \ll 1$ of all the roots of each of the equations

(a) $x^3 + \epsilon x^2 - x + \epsilon = 0$,
(b) $\epsilon x^3 + x^2 - 1 = 0$,
(c) $\epsilon x^4 + (1 - 3\epsilon)x^3 - (1 + 3\epsilon)x^2 - (1 + \epsilon)x + 1 = 0$,
(d) $\epsilon x^4 + (1 - 3\epsilon)x^3 - (1 - 3\epsilon)x^2 - (1 + \epsilon)x + 1 = 0$.

In each case, sketch the left hand side of the equation for $\epsilon = 0$ and $\epsilon \ll 1$.

12.2 The function $y(x)$ satisfies the ordinary differential equation

$$\epsilon y'' + (4 + x^2)(y' + 2y) = 0, \quad \text{for } 0 \leqslant x \leqslant 1,$$

subject to $y(0) = 0$ and $y(1) = 1$, with $\epsilon \ll 1$. Show that a boundary layer is possible only at $x = 0$. Use the method of matched asymptotic expansions to determine two-term inner and outer expansions, which you should match using either Van Dyke's matching principle, or an intermediate variable. Hence show that

$$y'(0) \sim \frac{4e^2}{\epsilon} - 4e^2 + 8e^2 \tan^{-1}\left(\frac{1}{2}\right) \quad \text{as } \epsilon \to 0.$$

Construct a composite expansion, valid up to $O(\epsilon)$.

12.3 Determine the leading order outer and inner approximations to the solution of

$$\epsilon y'' + x^{1/2} y' + y = 0 \quad \text{for } 0 \leqslant x \leqslant 1,$$

subject to $y(0) = 0$ and $y(1) = 1$, when $\epsilon \ll 1$. Hence show that

$$y'(0) \sim \epsilon^{-2/3} \frac{e^2}{\Gamma\left(\frac{2}{3}\right)} \left(\frac{3}{2}\right)^{1/3}.$$

† the a.c. equivalents of the conductivities.

12.4 The function $y(x)$ satisfies the ordinary differential equation
$$\epsilon y'' + (1+x)y' - y + 1 = 0,$$
for $0 \leqslant x \leqslant 1$, subject to the boundary conditions $y(0) = y(1) = 0$, with $\epsilon \ll 1$, where a prime denotes d/dx. Determine a two-term inner expansion and a one-term outer expansion. Match the expansions using either Van Dyke's matching principle or an intermediate region. Hence show that
$$y'(0) \sim \frac{1}{2\epsilon} + 1 \text{ as } \epsilon \to 0.$$

12.5 Consider the boundary value problem
$$\epsilon(2y + y'') + 2xy' - 4x^2 = 0 \text{ for } -1 \leqslant x \leqslant 2,$$
subject to
$$y(-1) = 2, \quad y(2) = 7,$$
with $\epsilon \ll 1$. Show that it is not possible to have a boundary layer at either $x = -1$ or $x = 2$. Determine the rescaling needed for an interior layer at $x = 0$. Find the leading order outer solution away from this interior layer, and the leading order inner solution. Match these two solutions, and hence show that $y(0) \sim 2$ as $\epsilon \to 0$. Sketch the leading order solution.

Now determine the outer solutions up to $O(\epsilon)$. Show that a term of $O(\epsilon \log \epsilon)$ is required in the inner expansion. Match the two-term inner and outer expansions, and hence show that $y(0) = 2 - \frac{3}{2}\epsilon \log \epsilon + O(\epsilon)$ for $\epsilon \ll 1$.

12.6 Consider the ordinary differential equation
$$\epsilon y'' + yy' - y = 0 \text{ for } 0 \leqslant x \leqslant 1,$$
subject to $y(0) = \alpha$, $y(1) = \beta$, with α and β constants, and $\epsilon \ll 1$.
  (a) Assuming that there is a boundary layer at $x = 0$, determine the leading order inner and outer solutions when $\alpha = 0$ and $\beta = 3$.
  (b) Assuming that there is an interior layer at $x = x_0$, determine the leading order inner and outer solutions, and hence show that $x_0 = 1/2$ when $\alpha = -1$ and $\beta = 1$.

12.7 Use the method of multiple scales to determine the leading order solution, uniformly valid for $t \ll \epsilon^{-2}$, of
$$\frac{d^2 y}{dt^2} + y = \epsilon y^3 \left(\frac{dy}{dt}\right)^2,$$
subject to $y = 1$, $dy/dt = 0$ when $t = 0$, for $\epsilon \ll 1$.

12.8 Consider the ordinary differential equation
$$\ddot{y} + \epsilon \dot{y} + y + \epsilon^2 y \cos^2 t = 0,$$
for $t \geqslant 0$, subject to $y(0) = 1$, $\dot{y}(0) = 0$, where a dot denotes d/dt. Use the method of multiple scales to determine a two-term asymptotic expansion, uniformly valid for all $t \ll \epsilon^{-3}$ when $\epsilon \ll 1$.

368 ASYMPTOTIC METHODS: DIFFERENTIAL EQUATIONS

12.9 Consider the ordinary differential equation
$$\ddot{y} + \epsilon \dot{y} + y + \epsilon^2 y^3 = 0,$$
for $t \geqslant 0$, subject to $y(0) = 1$, $\dot{y}(0) = 0$, where a dot denotes d/dt. Use the method of multiple scales, with
$$T = \epsilon t, \quad \tau = t + \epsilon^2 a t, \quad y \equiv y(\tau, T),$$
to show that
$$y \sim e^{-T/2} \cos \tau \quad \text{for } \epsilon \ll 1.$$
Show further that $a = -1/8$ and determine the next term in the asymptotic expansion. (You will need to consider the first three terms in the asymptotic expansion for y.)

12.10 Show that when $v_0 \ll 1$, (12.81) becomes $dE/dT = -2E$ at leading order.

12.11 Consider the initial value problem
$$\frac{d^2 y}{dt^2} - 2\epsilon \frac{dy}{dt} - y + y^3 = 0, \tag{E12.1}$$
subject to
$$y(0) = y_i > 1, \quad \frac{dy}{dt}(0) = 0. \tag{E12.2}$$
Use a graphical argument to show that when $\epsilon = 0$, y is positive, provided that $y_i < \sqrt{2}$. Use Kuzmak's method to determine the leading order approximation to the solution when $0 < \epsilon \ll 1$ and $1 < y_i < \sqrt{2}$. You should check that your solution is consistent with the linearized solution when $y_i - 1 \ll 1$. Hence show that y first becomes negative when $t = t_0 \sim T_0/\epsilon$, where
$$T_0 = \int_{\frac{1}{2} y_i^2 (y_i^2 - 2)}^0 \frac{3K\left(\frac{1}{k}\right)}{4\sqrt{1 + 2E}\left(\sqrt{1 + 2E} + 1\right)}$$
$$\times \frac{1}{\{(2k^2 - 1) L\left(\frac{1}{k}\right) - 2(k^2 - 1) K\left(\frac{1}{k}\right)\}} dE,$$
$$k \equiv k(E) = \left(\frac{\sqrt{1 + 2E} + 1}{2\sqrt{1 + 2E}}\right)^{1/2}.$$

Hints:
(a) The leading order solution can be written in terms of the Jacobian elliptic function cn (see Section 9.4).
(b)
$$\int_{\sqrt{1 - \frac{1}{k^2}}}^1 \sqrt{1 - x^2} \sqrt{1 - k^2 + k^2 x^2} \, dx$$
$$= \frac{1}{3k} \left\{(2k^2 - 1) L\left(\frac{1}{k}\right) - 2(k^2 - 1) K\left(\frac{1}{k}\right)\right\}.$$

12.12 Show that when $D(x, \hat{x}) = D_0(x)\hat{D}(\hat{x})$ and $R(\theta(x), x, \hat{x}) = R_0(\theta(x), x)\hat{R}(\hat{x})$, the homogenized diffusion coefficient and reaction term given by (12.101) and (12.102) can be written in the simple form described in Section 12.2.6.

12.13 Give a physical example that would give rise to the initial–boundary value problem

$$\frac{\partial \theta}{\partial t} = \frac{\partial}{\partial x}\left(D\left(x, \frac{x}{\epsilon}\right)\frac{\partial \theta}{\partial x}\right) \text{ for } 0 < x < 1,$$

subject to

$$\theta(x, 0) = \theta_i(x) \text{ for } 0 \leqslant x \leqslant 1,$$

$$\frac{\partial \theta}{\partial x} = 0 \text{ at } x = 0 \text{ and } x = 1 \text{ for } t > 0.$$

When $0 < \epsilon \ll 1$, use homogenization theory to show that, at leading order, θ is a function of x and t only, and satisfies an equation of the form

$$F(x)\frac{\partial \theta}{\partial t} = \frac{\partial}{\partial x}\left(\bar{D}(x)\frac{\partial \theta}{\partial x}\right),$$

where $\bar{D}(x)$ is given by (12.101), and $F(x)$ is a function that you should determine. If $D(x, x/\epsilon) = D_0(x)\hat{D}(x/\epsilon)$, show that, at leading order, θ satisfies the diffusion equation

$$\frac{\partial \theta}{\partial \hat{t}} = \frac{\partial}{\partial x}\left(D_0(x)\frac{\partial \theta}{\partial x}\right),$$

where

$$\hat{t} = \frac{t}{\lim_{\epsilon \to 0}\left(2\epsilon^2 \int_0^{1/\epsilon} \frac{s}{\hat{D}(s)}ds\right)}.$$

12.14 Use the WKB method to find the eigensolutions of the differential equation

$$y''(x) + (\lambda - x^2)y(x) = 0,$$

subject to $y \to 0$ as $|x| \to \infty$, when $\lambda \gg 1$.

12.15 Find the first two terms in the WKB approximation to the solution of the fourth order equation

$$\epsilon y''''(x) = \{1 - \epsilon V(x)\} y(x)$$

that satisfies $V(\pm\infty) = 0$ and $y(\pm\infty) = 0$, when $\epsilon \ll 1$.

12.16 Solve the connection problem

$$\epsilon^2 y''(x) + cx^2 y(x) = 0,$$

subject to $y(0) = 1$ and $y \to 0$ as $x \to -\infty$, when $\epsilon \ll 1$.

12.17 By determining the next term in the WKB expansion, verify that the overlap domain for the inner and outer solutions of the boundary value problem (12.115) is $\epsilon^{2/3} \ll |x| \ll \epsilon^{2/5}$ (see the footnote just after (12.121)).

12.18 Use Fourier transforms to solve (12.128) subject to (12.129) with $f_0(x) = e^x$, and hence show that $c(x,t) = \frac{1}{2}e^{x+Dt}\text{erfc}\left(\frac{x}{\sqrt{4Dt}} + \sqrt{Dt}\right)$. Use this expression to show that $c(0,t) \sim 1/\sqrt{4\pi Dt}$ as $t \to \infty$. By comparing this result with the large time asymptotic solution that we derived in Section 12.3, show that $\int_{-\infty}^{\infty} F_{\pm}(\eta)\,d\eta = \sqrt{2\pi D}$.

12.19 Find the small time solution of the reaction–diffusion equation
$$u_t = Du_{xx} + \alpha u,$$
subject to the initial condition
$$u(x,0) = \begin{cases} f_0(x) & \text{for } |x| < a, \\ 0 & \text{for } |x| \geqslant a. \end{cases}$$
What feature of the solution arises as a result of the reaction term, αu?

12.20 Find the leading order solution of the partial differential equation
$$c_t = (D + \epsilon x)^2 c_{xx} \quad \text{for } x > 0,\ t > 0,$$
when $\epsilon \ll 1$, subject to the initial condition $c(x,0) = f(x)$ and the boundary condition $c(0,t) = 0$. Your solution should remain valid for large x and t.

12.21 Find a uniformly valid solution of the hyperbolic equation
$$\epsilon(u_t + u_x) + (t-1)^2 u = 1 \quad \text{for } -\infty < x < \infty,\ t > 0,$$
when $\epsilon \ll 1$, subject to the initial condition $u(x,0) = 0$.

12.22 Find the leading order asymptotic solution of
$$\epsilon(u_{xx} + u_{yy}) + u_x + \beta u_y = 0 \quad \text{for } x > 0,\ 0 < y < L,$$
when $\epsilon \ll 1$, subject to the boundary conditions $u(x,0) = f(x)$, $u(0,y) = g(y)$ and $u(x,L) = 0$. Your solution should be uniformly valid in the domain of solution, so you will need to resolve any boundary layers that are required.

12.23 Find a uniformly valid leading order asymptotic solution of
$$\epsilon(u_{tt} - c^2 u_{xx}) + u_t + \alpha u_x = 0 \quad \text{for } -\infty < x < \infty,\ t > 0,$$
when $\epsilon \ll 1$, subject to the initial conditions $u(x,0) = f(x)$, $u_t(x,0) = 0$. How does the solution for $c > |\alpha|$ differ from the solution when $c < |\alpha|$?

12.24 **Project: The triple deck**
Consider the innocuous looking two-point boundary value problem
$$\epsilon y'' + x^3 y' + (x^3 - \epsilon)y = 0, \quad \text{subject to } y(0) = \tfrac{1}{2},\ y(1) = 1, \quad \text{(E12.3)}$$
with $\epsilon \ll 1$.

(a) Show that the *outer* solution is $y = e^{1-x}$.

(b) Since the outer solution does not satisfy the boundary condition at $x = 0$, there must be a boundary layer. By writing $\bar{x} = x/\epsilon^\mu$, show that there are two possibilities, $\mu = 1$ and $\mu = \frac{1}{2}$.

As there is now a danger of confusion in our notation, we will define $\bar{x} = x/\epsilon$, and refer to the region where $\bar{x} = O(1)$ as the **lower deck**, and $x^* = x/\epsilon^{1/2}$, and refer to the region where $x^* = O(1)$ as the **middle deck**. We will refer to the region where $x = O(1)$ as the **upper deck**. This nomenclature comes from the study of boundary layers in high Reynolds number fluid flow.

(c) Show that the leading order solution in the lower deck is $\bar{y} = \frac{1}{2}e^{-\bar{x}}$, and in the middle deck $y^* = a \exp\left\{-1/2\,(x^*)^2\right\}$.

(d) Apply the simplest form of matching condition to show that $a = e$, and hence that the solution in the middle deck is

$$y^* = \exp\left\{1 - \frac{1}{2\,(x^*)^2}\right\}.$$

(e) The matching between the lower and middle decks requires that

$$\lim_{\bar{x} \to \infty} \bar{y} = \lim_{x^* \to 0} y^*,$$

which is satisfied automatically without fixing any constants. Show that the composite solution takes the form

$$y = e^{1-x} + \exp\left\{1 - \frac{1}{2\,(x^*)^2}\right\} + \frac{1}{2}e^{-\bar{x}} - e.$$

(f) Integrate (E12.3) numerically using MATLAB, and compare the numerical solution with the asymptotic solution for $\epsilon = 0.1$, 0.01 and 0.001. Can you see the structure of the triple deck?

(g) The equations for steady, high Reynolds number flow of a viscous Newtonian fluid with velocity $\mathbf{u}$ and pressure p are

$$\nabla \cdot \mathbf{u} = 0, \quad (\mathbf{u} \cdot \nabla \mathbf{u}) = -\nabla p + \frac{1}{\mathrm{Re}}\nabla^2 \mathbf{u},$$

where Re is the Reynolds number. If these are to be solved in $-\infty < x < \infty$, $y > f(x)$ subject to $\mathbf{u} = \mathbf{0}$ on $y = f(x)$ and $\mathbf{u} \sim U\mathbf{i}$ as $x^2 + y^2 \to \infty$, where would you expect a triple or double deck structure to appear?

You can find some more references and background material in Sobey (2000). From the data in this book, estimate the maximum thickness of the boundary layer on a plate 1 m long, with water flowing past it at 10 m s^{-1}.

CHAPTER THIRTEEN

Stability, Instability and Bifurcations

In this chapter, we will build upon the ideas that we introduced in Chapter 9, which was concerned with phase plane methods for nonlinear, autonomous, ordinary differential equations. We will study what happens when an equilibrium point has one or more eigenvalues with zero real part. These are nonhyperbolic equilibrium points, which we were unable to study by simple linearization in Chapter 9. Next, we will introduce the idea of Lyapunov functions, and show how they can be used to study the stability of nonhyperbolic equilibrium points. We will also consider differential equations that contain a parameter, and examine what can happen to the qualitative form of the solutions as the parameter varies. At parameter values where one or more equilibrium points is nonhyperbolic, a local bifurcation point, the qualitative nature of the solutions changes. Finally, we will look at an example of a global bifurcation.

13.1 Zero Eigenvalues and the Centre Manifold Theorem

Let's consider the structure of an equilibrium point which has one zero eigenvalue and one negative eigenvalue. After shifting the equilibrium point to the origin, writing the system in terms of coordinates with axes in the directions of the eigenvectors, and rescaling time so that the negative eigenvalue has unit magnitude, the system will have the form

$$\dot{x} = P(x,y), \quad \dot{y} = -y + Q(x,y), \tag{13.1}$$

where P and Q are nonlinear functions with $P(0,0) = Q(0,0) = 0$. The linear approximation to (13.1) is $\dot{x} = 0$, $\dot{y} = -y$, which has the solution $x = 0$, $y = y_0 e^{-t}$ that passes through the origin. This shows that points on the y-axis close to the origin approach the origin as $t \to \infty$. We say that the **local stable manifold**† in some neighbourhood, U, of the origin is

$$\omega^s(\mathbf{0}, U) = \{(x,y) \mid x = 0, \ (x,y) \in U\}.$$

This is the set of points that are, locally, attracted to the origin in the direction of the eigenvector that corresponds to the negative eigenvalue. We are going to have to work harder to determine the behaviour of the other integral paths near the origin.

† We will carefully define what we mean by a manifold in Section 13.1.2.

13.1 ZERO EIGENVALUES AND THE CENTRE MANIFOLD THEOREM

In order to work out what is going on, we define a **local centre manifold** for (13.1) to be a C^k curve $\omega^c(\mathbf{0}, U)$ in a neighbourhood U of the origin such that

(i) $\omega^c(\mathbf{0}, U)$ is an invariant set within U, so that if $\mathbf{x}(0) \in \omega^c(\mathbf{0}, U)$, then $\mathbf{x}(t) \in \omega^c(\mathbf{0}, U)$ when $\mathbf{x}(t) \in U$. In other words, a solution that initially lies on the centre manifold remains there when it lies in U.

(ii) $\omega^c(\mathbf{0}, U)$ is the graph of a C^k function that is tangent to the x-axis at the origin, so that

$$\omega^c(\mathbf{0}, U) = \left\{ (x, y) \mid y = h(x),\ h(0) = \frac{dh}{dx}(0) = 0,\ (x, y) \in U \right\}.$$

This gives us the picture of the local stable and centre manifolds shown in Figure 13.1. The main idea now is that the qualitative behaviour of the solution in

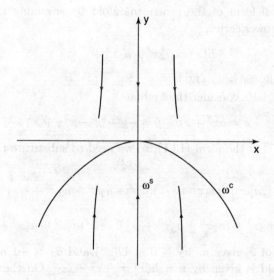

Fig. 13.1. The local stable and centre manifolds of the system (13.1).

the neighbourhood of the origin, excluding the stable manifold, is determined by the behaviour of the solution on the centre manifold. This means that the local dynamics are governed by a first order differential equation.

Theorem 13.1 (The centre manifold theorem) *The equilibrium point at the origin of the system (13.1) is stable/unstable if and only if the equilibrium point at $x = 0$ of the first order differential equation*

$$\dot{x} = P(x, h(x)), \qquad (13.2)$$

where $y = h(x)$ is the local centre manifold, is stable/unstable. Integral paths in the neighbourhood of the local centre manifold are attracted onto the centre manifold exponentially fast.

Proof This can be found in Wiggins (1990). □

13.1.1 Construction of the Centre Manifold

Now that we know that a local centre manifold exists, how can we determine its equation, $y = h(x)$? Since $\dot{y} = \dot{x}dh/dx$, (13.2) shows that

$$\frac{dh}{dx}P(x, h(x)) = -h(x) + Q(x, h(x)), \qquad (13.3)$$

subject to

$$h(0) = \frac{dh}{dx}(0) = 0. \qquad (13.4)$$

We will not usually be able to solve (13.3) analytically, but we can proceed to determine the *local* form of the centre manifold by assuming that $h(x)$ can be represented as a power series,

$$h(x) = a_2 x^2 + a_3 x^3 + \cdots, \qquad (13.5)$$

which automatically satisfies (13.4).

As an example, let's consider the system

$$\dot{x} = ax^3 + xy, \quad \dot{y} = -y + y^2 + x^2 y + x^3 \qquad (13.6)$$

with $a > 0$. This is of the form (13.1), so we need to substitute (13.5) into (13.3). This shows that

$$(2a_2 x + 3a_3 x^2 + \cdots)(ax^3 + a_2 x^3 + a_3 x^4 + \cdots)$$

$$= -a_2 x^2 - a_3 x^3 + (a_2 x^2 + a_3 x^3 + \cdots)^2 + x^2(a_2 x^2 + a_3 x^3 + \cdots) + x^3.$$

Equating powers of x gives us $a_2 = 0$ at $O(x^2)$ and $a_3 = +1$ at $O(x^3)$, so that the centre manifold is given by $y = h(x) = +x^3 + \cdots$. On the centre manifold, we therefore have $\dot{x} = ax^3 - x^4 + \cdots$. For $|x| \ll 1$ we can ignore the quartic term and just consider $\dot{x} \approx ax^3$, so that $\dot{x} > 0$ for $x > 0$ and $\dot{x} < 0$ for $x < 0$. Integral paths that begin on the local centre manifold therefore asymptote to the origin as $t \to -\infty$, and we conclude that the local phase portrait is of a nonlinear saddle, as shown in Figure 13.2. More specifically, if $x = x_0$ on the centre manifold when $t = 0$, $x \approx x_0/\sqrt{1 - 2ax_0^2 t}$, so that $x \sim x_0/\sqrt{-2ax_0^2 t}$ as $t \to -\infty$. This algebraic behaviour on the centre manifold is in contrast to the exponential behaviour that occurs on the unstable separatrix of a linear saddle point.

Example: Travelling waves in cubic autocatalysis

If two chemicals, which we label A and B, react through a mechanism known as **cubic autocatalysis**, we write

$$A + 2B \to 3B, \quad \text{rate } kab^2, \qquad (13.7)$$

13.1 ZERO EIGENVALUES AND THE CENTRE MANIFOLD THEOREM

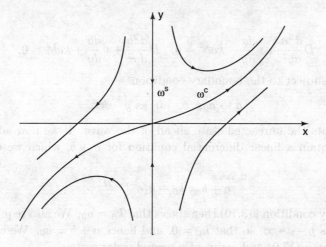

Fig. 13.2. The local phase portrait for the system (13.6).

where k is the **reaction rate constant** and a and b are the concentrations of the two chemicals, which are measured in moles m^{-3}†. The chemical B is known as the **autocatalyst**, since it catalyses its own production. The greater the concentration of B, the faster it is produced by the reaction (13.7). If these two chemicals then react in a long thin tube, so that their concentrations only vary in the x-direction, the main physical processes that act, in the absence of any underlying fluid flow, are chemical reaction and one-dimensional diffusion. We can derive the partial differential equations that govern this situation using the arguments that we described in Section 12.2.3, example 2. Specifically, since the rate of change of the amount of each chemical in a control volume is equal to the total diffusive flux through the bounding surface plus the total rate of production of that chemical within the volume, we find that

$$\frac{\partial a}{\partial t} = D\frac{\partial^2 a}{\partial x^2} - kab^2, \quad \frac{\partial b}{\partial t} = D\frac{\partial^2 b}{\partial x^2} + kab^2, \qquad (13.8)$$

where t is time and D the constant diffusivity of the chemicals. If a small amount of the autocatalyst is introduced locally into a spatially uniform expanse of A with $a = a_0$, waves of chemical reaction will propagate away from the initiation site. We can study one of these by looking for a wave that propagates from left to right at constant speed $v > 0$, represented by a solution of the form $a = a(y)$, $b = b(y)$, where $y = x - vt$. Such a solution is called a **travelling wave solution**.‡

With a and b functions of y alone, (13.8) become nonlinear ordinary differential

---

† A mole is a fixed number (Avogadro's number $\approx 6.02 \times 10^{23}$) of molecules of the substance.
‡ For more background and details on reaction–diffusion equations and travelling wave solutions, see Billingham and King (2001).

equations,
$$D\frac{d^2a}{dy^2} + v\frac{da}{dy} - kab^2 = 0, \quad D\frac{d^2b}{dy^2} + v\frac{db}{dy} + kab^2 = 0, \tag{13.9}$$

to be solved subject to the boundary condition
$$a \to a_0, \quad b \to 0 \quad \text{as } y \to \infty. \tag{13.10}$$

This represents the unreacted state ahead of the wave. If we now add equations (13.9), we obtain a linear differential equation for $a + b$, which we can solve to obtain
$$a + b = k_0 + k_1 e^{-vy/D}.$$

The boundary condition (13.10) then shows that $k_0 = a_0$. We also require that $a+b$ is bounded as $y \to -\infty$, so that $k_1 = 0$, and hence $a + b = a_0$. We can therefore eliminate a from (13.9) and arrive at a second order system
$$\frac{db}{dy} = c, \quad D\frac{dc}{dy} = -vc - kb^2(a_0 - b), \tag{13.11}$$

subject to
$$b \to 0, \quad c \to 0 \quad \text{as } y \to \infty. \tag{13.12}$$

Note that, since we require that both of the chemical concentrations should be positive for a physically meaningful solution, we need $0 \leqslant b \leqslant a_0$.

The only equilibrium points of (13.11) are $b = c = 0$ and $b = a_0, c = 0$. The second of these represents the fully reacted state behind the travelling wave, where all of the chemical A has been converted into the autocatalyst, B, so we also require that
$$b \to a_0, \quad c \to 0 \quad \text{as } y \to -\infty. \tag{13.13}$$

We can write the boundary value problem given by (13.11) to (13.13) in a more convenient form by defining the dimensionless variables
$$\beta = \frac{b}{a_0}, \quad \gamma = \sqrt{\frac{ka_0^2}{D}}\frac{c}{a_0}, \quad z = \sqrt{\frac{ka_0^2}{D}}y, \quad V = \frac{v}{\sqrt{ka_0^2 D}},$$

so that
$$\frac{d\beta}{dz} = \gamma, \quad \frac{d\gamma}{dz} = -V\gamma - \beta^2(1-\beta), \tag{13.14}$$

subject to
$$\beta \to 0, \quad \gamma \to 0 \quad \text{as } z \to \infty, \tag{13.15}$$
$$\beta \to 1, \quad \gamma \to 0 \quad \text{as } z \to -\infty, \tag{13.16}$$

and $0 \leqslant \beta \leqslant 1$ for a physically meaningful solution. The dimensionless wave speed, V, is now the only parameter.

The next step is to study the solutions of (13.14) subject to (13.15) and (13.16) in the (β, γ) phase plane and determine for what values of the wave speed, V, solutions

13.1 ZERO EIGENVALUES AND THE CENTRE MANIFOLD THEOREM

exist. As a preliminary, let's see if we can guess a solution of this system. A plausible functional form is $\gamma = k\beta(1-\beta)$, which satisfies the boundary conditions. Since

$$\frac{d\gamma}{d\beta} = -V - \frac{\beta^2(1-\beta)}{\gamma}, \tag{13.17}$$

we find that we can satisfy this equation with $k = -1/\sqrt{2}$ and $V = 1/\sqrt{2}$. This gives

$$\frac{d\beta}{dz} = -\frac{1}{\sqrt{2}}\beta(1-\beta),$$

and hence

$$\beta = \beta_e = \frac{1}{1+e^{(z-z_0)/\sqrt{2}}}, \quad \gamma = \gamma_e = -\frac{1}{\sqrt{2}}\frac{e^{(z-z_0)/\sqrt{2}}}{(1+e^{(z-z_0)/\sqrt{2}})^2}. \tag{13.18}$$

This is an exact solution for $V = 1/\sqrt{2}$ and any constant z_0, as shown in Figure 13.3. The presence of z_0 in (13.18) simply shows that the solution can be given an arbitrary displacement in the z-direction and remain a solution, as we would expect for a wave that propagates at constant speed without change of form.

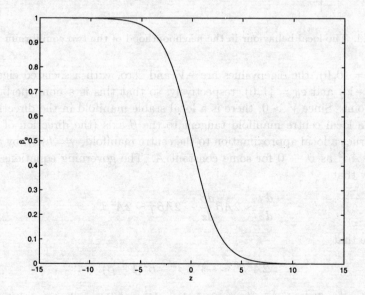

Fig. 13.3. The analytical solution of the cubic travelling wave problem, $\beta = \beta_e(z)$, with $z_0 = 0$.

So now we know that a solution exists for $V = 1/\sqrt{2}$. What about other values of V? Let's go along our usual route, and determine the nature of the two equilibrium points, $P_1 = (1, 0)$ and $P_2 = (0, 0)$. The Jacobian is

$$J = \begin{pmatrix} 0 & 1 \\ -2\beta + 3\beta^2 & -V \end{pmatrix}.$$

STABILITY, INSTABILITY AND BIFURCATIONS

At P_1, the eigenvalues are real and of opposite sign, so that P_1 is a saddle point. Boundary condition (13.16) shows that we need the integral path that represents the solution to asymptote to P_1 as $z \to -\infty$. Since the unstable separatrices of P_1 are the only integral paths that do this, the solution must be represented by the unstable separatrix of P_1 that lies in the physically meaningful region, $0 \leqslant \beta \leqslant 1$, shown in Figure 13.4 as S_1. The other boundary condition, (13.15), shows that we need S_1 to asymptote to the other equilibrium point, P_2, as $z \to \infty$, if it is to represent a solution.

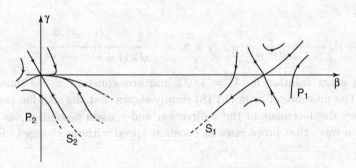

Fig. 13.4. The local behaviour in the neighbourhood of the two equilibrium points.

At $P_2 = (0,0)$, the eigenvalues are $-V$ and zero, with associated eigenvectors $\mathbf{e}_- = (1, -V)$ and $\mathbf{e}_0 = (1, 0)$, respectively, so that this is a nonhyperbolic equilibrium point. Since $V > 0$, there is a local stable manifold in the direction of $\mathbf{e}_-$ and also a local centre manifold tangent to the β-axis (the direction of $\mathbf{e}_0$). We can construct a local approximation to the centre manifold, $\gamma = h(\beta)$, by assuming that $\gamma \sim A\beta^2$ as $\beta \to 0$ for some constant A. The governing equations, (13.14), then show that

$$\frac{d\gamma}{dz} \sim 2A\beta \frac{d\beta}{dz} \sim 2A\beta\gamma \sim 2A^2\beta^3,$$

and hence that

$$2A^2\beta^3 \sim -VA\beta^2 - \beta^2(1-\beta).$$

By equating coefficients of β^2, we find that $A = -1/V$, and hence that the local centre manifold has $\gamma \sim -\beta^2/V$ as $\beta \to 0$. This means that $\gamma = d\beta/dz < 0$ on the local centre manifold. Points on the centre manifold in $\beta > 0$ are therefore attracted to P_2 as $z \to \infty$ with, from $d\beta/dz \sim -\beta^2/V$, $\beta \sim V/z$ as $z \to \infty$. In contrast, points on the centre manifold in $\beta < 0$ are swept away as z increases. This type of behaviour is characteristic of a new type of equilibrium point, known as a **saddle-node**. To the right of the stable manifold, the point behaves like a stable node, attracting integral paths onto the centre manifold and into the origin, whilst to the left of the stable manifold the point is unstable, as shown in Figure 13.4.

13.1 ZERO EIGENVALUES AND THE CENTRE MANIFOLD THEOREM

Apart from the local stable manifold, on which $\beta = O(e^{-Vz})$ as $z \to \infty$ (recall that the negative eigenvalue of P_2 is $-V$), any other integral paths that asymptote to the origin as $z \to \infty$ do so on the centre manifold, with $\beta \sim V/z$.

Finally, for any given value of V, we need to determine whether the integral path S_1 lies to the right of the stable manifold of P_2, which we label S_2, and therefore enters the origin and represents a solution, or whether S_1 lies to the left of S_2, is swept away from the origin into $\beta < 0$, and therefore does not represent a physically meaningful solution. We will return to this problem in Section 13.3.3.

13.1.2 The Stable, Unstable and Centre Manifolds

We end this section by defining more carefully what we mean by a manifold, and generalizing the definitions of the stable, unstable and centre manifolds to n^{th} order systems.

Let's begin with some definitions. A **homeomorphism** is a mapping $f : L \to N$ that is one-to-one, onto and continuous and has a continuous inverse. Here, L and N are subsets of $\mathbb{R}^n$. A C^k **diffeomorphism** is a mapping $f : L \to N$ that is one-to-one, onto and k times differentiable with a k-times differentiable inverse. A **smooth diffeomorphism** is a C^∞ diffeomorphism and a homeomorphism is a C^0 diffeomorphism. An m-**dimensional manifold** is a set $M \subset \mathbb{R}^n$ for which each $x \in M$ has a neighbourhood U in which there exists a homeomorphism $\phi : U \to \mathbb{R}^m$, where $m \leqslant n$. A manifold is said to be differentiable if there is a diffeomorphism rather than a homeomorphism $\phi : U \to \mathbb{R}^m$. For example, a smooth curve in $\mathbb{R}^3$ is a one-dimensional differentiable manifold, and the surface of a sphere is a two-dimensional differentiable manifold.

Now that we have defined these ideas, let's consider the behaviour of the solutions of

$$\frac{d\mathbf{x}}{dt} = \mathbf{f}(\mathbf{x}), \qquad (13.19)$$

where $\mathbf{x}, \mathbf{f}(\mathbf{x}) \in \mathbb{R}^n$. In the neighbourhood of an equilibrium point, $\bar{\mathbf{x}}$, of (13.19), there exist three invariant manifolds.

(i) The **local stable manifold**, $\omega_{\text{loc}}^{\text{s}}$, of dimension s, is spanned by the eigenvectors of A whose eigenvalues have real parts less than zero.

(ii) The **local unstable manifold**, $\omega_{\text{loc}}^{\text{u}}$, of dimension u, is spanned by the eigenvectors of A whose eigenvalues have real parts greater than zero.

(iii) The **local centre manifold**, $\omega_{\text{loc}}^{\text{c}}$, of dimension c, is spanned by the eigenvectors of A whose eigenvalues have zero real parts.

Note that $s + c + u = n$. Solutions lying in $\omega_{\text{loc}}^{\text{s}}$ are characterized by exponential decay and those in $\omega_{\text{loc}}^{\text{u}}$ by exponential growth. The behaviour on the centre manifold is determined by the nonlinear terms in (13.19), as we described earlier in this section. For a linear system these manifolds exist globally, whilst for a nonlinear system they exist in some neighbourhood of the equilibrium point.

Example

Let's consider the simple system

$$\dot{x} = x(2-x), \quad \dot{y} = -y + x, \tag{13.20}$$

and try to determine the equations of the local manifolds that pass through the equilibrium point at the origin. The Jacobian at the origin is

$$J = \begin{pmatrix} 2 & 0 \\ 1 & -1 \end{pmatrix},$$

which has eigenvalues 2 and -1 and associated eigenvectors of $(3,1)^T$ and $(0,1)^T$ respectively, so that ω^u_{loc} is the line $y = x/3$ and ω^s_{loc} is the y-axis. We can solve the nonlinear system directly, since the equation for x is independent of y, and the equation for y is linear. The solution is

$$x = \frac{2A}{A + 2e^{-t}}, \quad y = e^{-t}\left\{B + 2e^t - \frac{4}{A}\log(Ae^t + 2)\right\},$$

where $A \neq 0$ and B are constants. There is also the obvious solution $x = 0$, $y = Be^{-t}$, the y-axis, which gives the local stable manifold, ω^s_{loc}, and also the **global stable manifold**, $\omega^s(\bar{x})$. Points that lie in the local unstable manifold, ω^u_{loc}, have $y \to 0$ as $t \to -\infty$. Since $y \sim e^{-t}(B - 4\log 2/A)$ as $t \to -\infty$, we must have $B = A + 4/A \log 2$, so that the **global unstable manifold**, $\omega^u(\bar{x})$, is given in parametric form by

$$\left(\frac{2A}{A + 2e^{-t}}, e^{-t}\left\{A + \frac{4}{A}\log 2 + 2e^t - \frac{4}{A}\log(Ae^t + 2)\right\}\right).$$

The phase portrait is sketched in Figure 13.5.

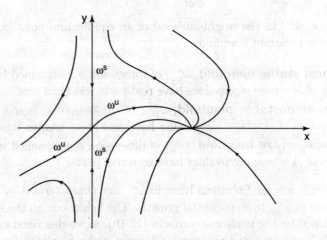

Fig. 13.5. The phase portrait of the system (13.20).

13.2 Lyapunov's Theorems

Although we are now familiar with the idea of the stability of an equilibrium point in an informal way, in order to develop the idea of a Lyapunov function, we need to consider some more formal definitions of stability. Note that, in this section, we will consider systems of autonomous ordinary differential equations, written in vector form as $\dot{\mathbf{x}} = \mathbf{f}(\mathbf{x})$, that have an isolated equilibrium point at the origin.

An equilibrium point, $\mathbf{x} = \mathbf{x}_e$, is **Lyapunov stable** if for all $\epsilon > 0$ there exists a $\delta > 0$ such that for all $\mathbf{x}(0)$ with $|\mathbf{x}(0) - \mathbf{x}_e| < \delta$, $|\mathbf{x}(t) - \mathbf{x}_e| < \epsilon$ for all $t > 0$. In other words, integral paths that start close to a Lyapunov stable equilibrium point remain close for all time, as shown in Figure 13.6.

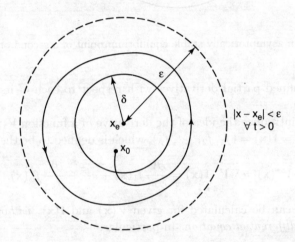

Fig. 13.6. A Lyapunov stable equilibrium point of a second order system.

An equilibrium point, $\mathbf{x} = \mathbf{x}_e$, is **asymptotically stable** if there exists a $\delta > 0$ such that for all $\mathbf{x}(0)$ with $|\mathbf{x}(0) - \mathbf{x}_e| < \delta$, $|\mathbf{x}(t) - \mathbf{x}_e| \to 0$ as $t \to \infty$. This is a stronger definition of stability than Lyapunov stability, and states that integral paths that start sufficiently close to an asymptotically stable equilibrium point are attracted into it, as illustrated in Figure 13.7. Stable nodes and spirals are both Lyapunov and asymptotically stable. It should be clear that asymptotically stable equilibrium points are also Lyapunov stable. However, Lyapunov stable equilibrium points, for example centres, are not necessarily asymptotically stable.

Now that we have formalized our notions of stability, we need one more new concept. Let $V : \Omega \to \mathbb{R}$, where $\Omega \subset \mathbb{R}^n$ and $\mathbf{0} \in \Omega$. We say that the function V is **positive definite** on Ω if and only if $V(\mathbf{0}) = 0$ and $V(\mathbf{x}) > 0$ for $\mathbf{x} \in \Omega$, $\mathbf{x} \neq \mathbf{0}$. For example, for $n = 3$, $V(\mathbf{x}) = V(x_1, x_2, x_3) = x_1^2 + x_2^2 + x_3^2$ is positive definite in $\mathbb{R}^3$, whilst $V(x_1, x_2, x_3) = x_2^2$ is not, since it is zero on the plane $x_2 = 0$, and not just at the origin. Note that if $-V(\mathbf{x})$ is a positive definite function, we say that V is a **negative definite** function. In the following, we will assume that V is continuous

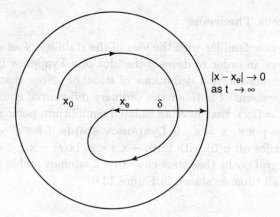

Fig. 13.7. An asymptotically stable equilibrium point of a second order system.

and has well-defined partial derivatives with respect to each of its arguments, so that $V \in C^1(\Omega)$.

We can now introduce the idea of the derivative of a function $V(\mathbf{x})$ with respect to the system $\dot{\mathbf{x}} = \mathbf{f}(\mathbf{x}) = (f_1, f_2, \ldots, f_n)$, which is defined to be the scalar product

$$V^*(\mathbf{x}) = \nabla V \cdot \mathbf{f}(\mathbf{x}) = \frac{\partial V}{\partial x_1} f_1(\mathbf{x}) + \cdots + \frac{\partial V}{\partial x_n} f_n(\mathbf{x}).$$

This derivative can be calculated for given $V(\mathbf{x})$ and $\mathbf{f}(\mathbf{x})$, *without knowing the solution of the differential equation*. In particular,

$$\frac{dV}{dt} = \frac{\partial V}{\partial x_1} \dot{x}_1 + \cdots + \frac{\partial V}{\partial x_n} \dot{x}_n = \frac{\partial V}{\partial x_1} f_1(\mathbf{x}) + \cdots + \frac{\partial V}{\partial x_n} f_n(\mathbf{x}) = V^*(\mathbf{x}),$$

so that the total derivative of V with respect to the solution of the equations coincides with our definition of the derivative with respect to the system. This allows us to prove three important theorems.

Theorem 13.2 *If, in some region $\Omega \subset \mathbb{R}^n$ that contains the origin, there exists a scalar function $V(\mathbf{x})$ that is positive definite and for which $V^*(\mathbf{x}) \leqslant 0$, then the origin is Lyapunov stable. The function $V(\mathbf{x})$ is known as a **Lyapunov function**.*

Proof Since V is positive definite in Ω, there exists a sphere of radius $r > 0$[†] contained within Ω such that

$$V(\mathbf{x}) > 0 \text{ for } \mathbf{x} \neq \mathbf{0} \text{ and } |\mathbf{x}| < r,$$

$$V^*(\mathbf{x}) \leqslant 0 \text{ for } |\mathbf{x}| \leqslant r.$$

[†] the set of points with $|\mathbf{x}| \leqslant r$ in $\mathbb{R}^n$.

Let $\mathbf{x} = \mathbf{x}(t)$ be the solution of the differential equation $\dot{\mathbf{x}} = \mathbf{f}(\mathbf{x})$ with $\mathbf{x}(0) = \mathbf{x}_0$. By the local existence theorem, Theorem 8.1, extended to higher order systems, this solution exists for $0 \leqslant t < t^*$ with $t^* > 0$. This solution can then be continued for $t \geqslant t^*$, and we denote by t_1 the largest value of t for which the solution exists. There are two possibilities, either $t_1 = \infty$ or $t_1 < \infty$. We now show that, for $|\mathbf{x}_0|$ sufficiently small, $t_1 = \infty$.

From the definition of the derivative with respect to a system,

$$\frac{dV}{dt}(\mathbf{x}(t)) = V^*(\mathbf{x}(t)) \quad \text{for } 0 \leqslant t < t_1.$$

We can integrate this equation to give

$$V(\mathbf{x}(t)) - V(\mathbf{x}_0) = \int_0^t V^*(\mathbf{x}(s)) \, ds \leqslant 0,$$

since V^* is negative definite. This means that $0 < V(\mathbf{x}(t)) \leqslant V(\mathbf{x}_0)$ for $0 \leqslant t < t_1$. Now let ϵ satisfy $0 < \epsilon \leqslant r$, and let S be the closed, spherical shell with inner and outer radii ϵ and r†. By continuity of V, and since S is closed, $\mu = \min_{\mathbf{x} \in S} V(\mathbf{x})$ exists and is strictly positive. Since $V(\mathbf{x}) \to 0$ as $|\mathbf{x}| \to 0$, we can choose δ with $0 < \delta < \mu$ such that for $|\mathbf{x}_0| \leqslant \delta$, $V(\mathbf{x}_0) < \mu$, so that $0 < V(\mathbf{x}(t)) \leqslant V(\mathbf{x}_0) < \mu$ for $0 \leqslant t < t_1$. Since μ is the minimum value of V in S, this gives $|\mathbf{x}(t)| < \epsilon$ for $0 \leqslant t < t_1$. If there exists t_2 such that $|\mathbf{x}(t_2)| = \epsilon$, then, when $t = t_2$, we also have, from the definition of μ, $\mu \leqslant V(\mathbf{x}(t_2)) \leqslant V(\mathbf{x}_0) < \mu$, which cannot hold. We conclude that $t_1 = \infty$, and that, for a given $\epsilon > 0$, there exists a $\delta > 0$ such that when $|\mathbf{x}_0| < \delta$, $|\mathbf{x}(t)| < \epsilon$ for $t \geqslant 0$, and hence that the origin is Lyapunov stable. □

The proofs of the following two theorems are rather similar, and we will not give them here.

Theorem 13.3 *If, in some region $\Omega \subset \mathbb{R}^n$ that contains the origin, there exists a scalar function $V(\mathbf{x})$ that is positive definite and for which $V^*(\mathbf{x})$ is negative definite, then the origin is asymptotically stable.*

Theorem 13.4 *If, in some region $\Omega \subset \mathbb{R}^n$ that contains the origin, there exists a scalar function $V(\mathbf{x})$ such that $V(\mathbf{0}) = 0$ and $V^*(\mathbf{x})$ is either positive definite or negative definite, and if, in every neighbourhood N of the origin with $N \subset \Omega$, there exists at least one point $\mathbf{a}$ such that $V(\mathbf{a})$ has the same sign as $V^*(\mathbf{a})$, then the origin is unstable.*

Theorems 13.2 to 13.4 are known as Lyapunov's theorems, and have a geometrical interpretation that is particularly attractive for two-dimensional systems. The equation $V(\mathbf{x}) = c$ then represents a surface in the (x, y, V)-space. By varying c (through positive values only, since V is positive definite), we can obtain a series of contour lines, with $V = 0$ at the origin a local minimum on the surface, as shown in Figure 13.8. Since $\dot{\mathbf{x}} = \mathbf{f}(\mathbf{x})$, the vector field $\mathbf{f}$ represents the direction taken by

† the set of points with $\epsilon \leqslant |\mathbf{x}| \leqslant r$ in $\mathbb{R}^n$.

an integral path at any point. The vector normal to the surface $V(\mathbf{x}) = c$ is ∇V, so that, if $V^* = \nabla V \cdot \mathbf{f} \leqslant 0$, integral paths cannot point into the exterior of the region $V(\mathbf{x}) < c$. We conclude that an integral path that starts sufficiently close to the origin, for example with $V(\mathbf{x}_0) < c_1$, cannot leave the region bounded by $V(\mathbf{x}) = c_1$, and hence that the origin is Lyapunov stable. Similarly, if $V^* < 0$, the integral paths must actually cross from the exterior to the interior of the region. Hence V will decrease monotonically to zero from its initial value when the integral path enters the region Ω, and we conclude that the origin is asymptotically stable.

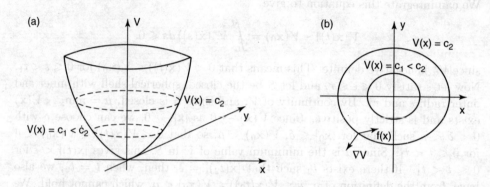

Fig. 13.8. (a) The local behaviour and (b) a contour plot of a Lyapunov function near the origin.

Although we can now see why a Lyapunov function is useful, it can take considerable ingenuity to actually construct one for a given system.

Example 1

Consider the system

$$\dot{x} = -x - 2y^2, \quad \dot{y} = xy - y^3.$$

The origin is the only equilibrium point, and the linearized system is $\dot{x} = -x$, $\dot{y} = 0$. The eigenvalues are therefore 0 and -1, so this is a nonhyperbolic equilibrium point. Let's try to construct a Lyapunov function. We start by trying $V = x^2 + \alpha y^2$. This is clearly positive definite for $\alpha > 0$, and $V(0,0) = 0$. In addition,

$$V^* = \frac{dV}{dt} = 2x(-x - 2y^2) + 2\alpha y(xy - y^3) = -2x^2 + 2(\alpha - 2)xy^2 - 2\alpha y^4.$$

If we choose $\alpha = 2$, then $dV/dt = -2x^2 - 4y^4 < 0$ for all x and y excluding the origin. From Theorems 13.2 and 13.3 we conclude that the origin is both Lyapunov and asymptotically stable. As a general guideline, it is worth looking for a homogeneous, algebraic Lyapunov function when $\mathbf{f}$ has a simple algebraic form.

13.2 LYAPUNOV'S THEOREMS

Example 2

Consider the second order differential equation $\ddot{\theta} + f(\theta) = 0$ for $-\pi \leqslant \theta \leqslant \pi$, with $\theta f(\theta)$ positive, f differentiable and $f(0) = 0$. We can write this as a second order system,
$$\dot{x}_1 = x_2, \quad \dot{x}_2 = -f(x_1),$$
where $x_1 = \theta$. The origin is an equilibrium point, but is it stable? In order to construct a Lyapunov function, it is helpful to think in terms of an equivalent physical system. By analogy with the model for a simple pendulum, which we discussed in Section 9.1, we can think of $f(\theta)$ as the restoring force and $\dot{\theta}$ as the angular velocity. The total energy of the system is the sum of the kinetic and potential energies, which we can write as $E = \frac{1}{2}\dot{\theta}^2 + \int_0^\theta f(s)\,ds$. If this energy were to decrease/not grow, we would expect the motionless, vertical state of the pendulum to be asymptotically/Lyapunov stable. Guided by this physical insight, we define
$$V = \frac{1}{2}x_2^2 + \int_0^{x_1} f(s)\,ds.$$
Clearly $V(0,0) = 0$, and, since $\int_0^{x_1} f(s)\,ds$ is positive by the assumption that $\theta f(\theta) \geqslant 0$, V is positive definite for $-\pi \leqslant x_1 \leqslant \pi$. Finally,
$$V^* = \frac{dV}{dt} = f(x_1)x_2 + x_2 \cdot -f(x_1) = 0.$$
By Theorem 13.2, V is a Lyapunov function, and the origin is Lyapunov stable.

Example 3

Consider the differential equation $\ddot{x} + \dot{x} + x + x^2 = 0$. We can write this as a second order system,
$$\dot{x}_1 = x_2, \quad \dot{x}_2 = -x_1 - x_1^2 - x_2, \tag{13.21}$$
where $x_1 = x$. This has two equilibrium points, at $(-1,0)$ and $(0,0)$. It is straightforward to determine the eigenvalues of these equilibrium points and show that both are hyperbolic, with $(-1,0)$ a saddle point and $(0,0)$ a stable, clockwise spiral. We can now construct a Lyapunov function that will give us some idea of the domain of attraction of the stable equilibrium point at the origin. Consider the function
$$V = \frac{1}{2}(x_1^2 + x_2^2) + \frac{1}{3}x_1^3.$$
This function vanishes at the origin and is positive in the region
$$\Omega = \left\{(x_1, x_2) \mid x_2^2 > -x_1^2 - \frac{2}{3}x_1^3\right\},$$
which is sketched, along with the phase portrait, in Figure 13.9.

Let's consider the curve $V = \frac{1}{6}$, which passes through the saddle point and the point $(\frac{1}{2}, 0)$, as shown in Figure 13.10. If $V = V_0 < \frac{1}{6}$, we have a curve that encloses the origin, but not the saddle point. By taking $V = V_0 < \frac{1}{6}$ arbitrarily close to $\frac{1}{6}$, we can make the curve $V = V_0$ arbitrarily close to the saddle point. As we are

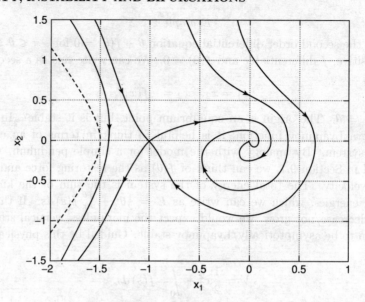

Fig. 13.9. The region Ω and the phase portrait of the system (13.21). The region Ω lies to the right of the curved, broken line, which is $V = 0$.

interested in the domain of attraction of the equilibrium point at the origin, we will focus on Ω_0, the subset of Ω given by $V < V_0 < \frac{1}{6}$, with V_0 close to $\frac{1}{6}$.

Since

$$V^* = \frac{dV}{dt} = x_2\left(x_1 + x_1^2\right) + \left(-x_1 - x_1^2 - x_2\right) x_2 = -x_2^2 \leqslant 0,$$

we immediately have from Theorem 13.3 that the origin is Lyapunov stable. To prove asymptotic stability requires more work, as $V^* = 0$ on $x_2 = 0$, which could allow trajectories to escape from the region Ω_0 through the two points where $V = V_0$ meets the x_1-axis, which are labelled as A and B in Figure 13.10. There are various ways of dealing with this. The obvious one is to choose a different Lyapunov function, which is possible, but technically difficult. We will use a phase plane analysis. Consider S_A, the integral path through the point A. All other integral paths through the boundary of Ω_0 in the neighbourhood of A enter Ω_0. The integral path S_A cannot, therefore, lie along the boundary of Ω_0. If S_A does not enter Ω_0, it must intersect the integral path that enters Ω_0 at $x_2 = 0^+$, which is not possible. We conclude that the integral path through A enters Ω_0, as shown in Figure 13.11. A similar argument holds at B.

Since the Lyapunov function, V, is monotone decreasing away from the x_1-axis, there cannot be any limit cycle solutions in Ω_0. Finally, since Ω_0 has all integral paths entering it, and contains a single, stable equilibrium point and no limit cycles,

13.2 LYAPUNOV'S THEOREMS

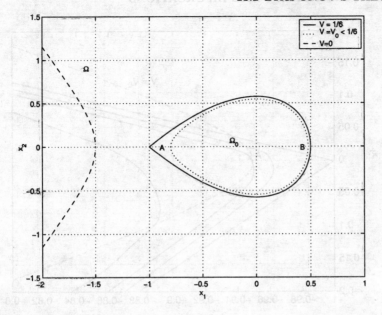

Fig. 13.10. The regions Ω and Ω^* and the curves $V = 0$, $V = \frac{1}{6}$ and $V = V_0 < \frac{1}{6}$.

we conclude from the Poincaré–Bendixson theorem, Theorem 9.4, that all integral paths in Ω_0 enter the origin, which is therefore asymptotically stable, and that Ω_0 lies within the domain of attraction of the origin. In fact, we can see from Figure 13.9 that this domain of attraction is considerably larger than Ω_0, and is bounded by the stable separatrices of the saddle point at $(-1, 0)$.

Example 4

Consider the system

$$\dot{x} = x^2 - y^2, \quad \dot{y} = -2xy.$$

This is a genuinely nonlinear system with an equilibrium point at the origin. The linear approximation at the origin is $\dot{x} = \dot{y} = 0$, so both eigenvalues are zero. Let's try a Lyapunov function of the form $V = \alpha xy^2 - x^3$, which has $V(0,0) = 0$. Since

$$V^* = \frac{dV}{dt} = (\alpha y^2 - 3x^2)(x^2 - y^2) - 4\alpha x^2 y^2 = 3(1-\alpha)x^2 y^2 - \alpha y^4 - 3x^4,$$

we can choose $\alpha = 1$, so that $V^* = -y^4 - 3x^4$, which is negative definite. We can see that $V = x(y^2 - x^2) = 0$ when $x = 0$ or $y = \pm x$, so that V changes sign six times on any circle that surrounds the origin. In particular, in every neighbourhood of the origin there is at least one point where V has the same sign as V^*, so that all of the conditions of Theorem 13.4 are satisfied by V, and hence the origin is unstable.

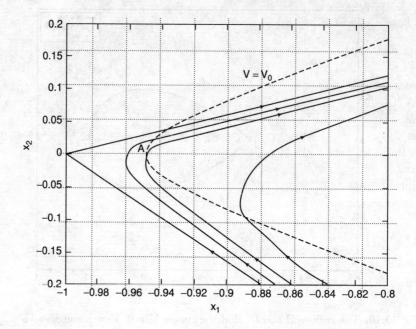

Fig. 13.11. The phase portrait in the neighbourhood of the point A and the saddle point at $(-1, 0)$.

Further refinements exist of the Lyapunov Theorems 13.2 to 13.4 that we have studied in this section, and the interested reader is referred to Coddington and Levinson (1955) for further information.

13.3 Bifurcation Theory

13.3.1 First Order Ordinary Differential Equations

Let's consider the first order ordinary differential equation, (9.5), whose hyperbolic equilibrium points we studied in Section 9.2. For a hyperbolic equilibrium point at $x = x_1$, we saw that a simple linearization about $x = x_1$ determines the local behaviour and stability. If $X'(x_1) = 0$, $x = x_1$ is a nonhyperbolic equilibrium point, and we need to retain more terms in the Taylor expansion of X, (9.6), in order to sort out what happens close to the equilibrium point. For example, if $X(x_1) = X'(x_1) = 0$ and $X''(x_1) \neq 0$,

$$\frac{d\bar{x}}{dt} \approx \frac{1}{2} X''(x_1) \bar{x}^2 \quad \text{for } \bar{x} \ll 1,$$

and hence

$$\bar{x} \approx -\frac{2}{X''(x_1)(t-t_0)},$$

for some constant t_0. The graph of $X(x_1+\bar{x})$ close to $\bar{x} = 0$ is shown in Figure 13.12 for $X''(x_1) < 0$. Focusing on this case, we can see that $\bar{x} \to 0$ $(x \to x_1)$ as $t \to \infty$ for $t_0 < 0$, whilst $\bar{x} \to -\infty$ as $t \to t_0$ for $t_0 > 0$. This nonhyperbolic equilibrium point therefore attracts solutions from $x \geqslant x_1$ and repels them from $x < x_1$. Note that the rate at which solutions asymptote to $x = x_1$ from $x > x_1$ is algebraic, in contrast to the faster, exponential approach associated with stable hyperbolic equilibrium points.

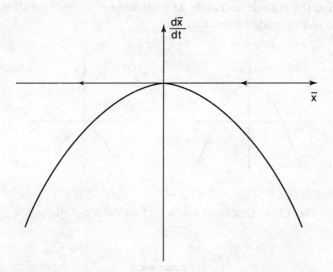

Fig. 13.12. The graph of $d\bar{x}/dt = X(x_1+\bar{x}) \approx \frac{1}{2}X''(x_1)\bar{x}^2$ for $X''(x_1) < 0$.

A system that contains one or more nonhyperbolic equilibrium points is said to be **structurally unstable**. This means that a small perturbation, not to the solution but to the model itself, for example the addition of a small extra term to $X(x)$, can lead to a qualitative difference in the structure of the set of solutions, for example, a change in the number of equilibrium points or in their stability. Consider the function shown in Figure 13.12. The addition of a small positive constant to $X(x)$ would shift the graph upwards by a small amount, and give two equilibrium solutions, whilst the addition of a small negative constant would shift the graph downwards and lead to the absence of any equilibrium solutions. Let's investigate this further.

Consider the equation

$$\dot{x} = \mu - x^2, \tag{13.22}$$

where μ is a parameter. For $\mu > 0$ there are two hyperbolic equilibrium points at $x = \pm\sqrt{\mu}$, whilst for $\mu < 0$ there are no equilibrium points, as shown in Figure 13.13. When $\mu = 0$, $X(x) = -x^2$ and $x = 0$ is a nonhyperbolic equilibrium point of the type that we analyzed above. We can now draw a **bifurcation diagram**, which shows the position of the equilibrium solutions as a function of μ, with stable equilibria as solid lines and unstable equilibria as dashed lines, as shown in Figure 13.14. The point $\mu = 0$, $x = 0$ is called a **bifurcation point**, because the qualitative nature of the phase line changes there. The bifurcation associated with $\dot{x} = \mu - x^2$ is called a **saddle–node** bifurcation, for reasons that will become clear in Section 13.3.2. Any first order system that undergoes a saddle–node bifurcation, in other words one that contains a bifurcation point where two equilibrium solutions meet and then disappear, can be written in the form $\dot{x} = \mu - x^2$ in the neighbourhood the bifurcation point. The equation $\dot{x} = \mu - x^2$ is called the **normal form** for the saddle–node bifurcation.

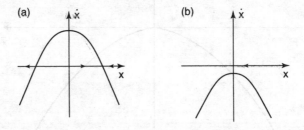

Fig. 13.13. Graphs of $\dot{x} = \mu - x^2$ for (a) $\mu > 0$, (b) $\mu < 0$.

Example

Consider the ordinary differential equation

$$\dot{y} = \lambda - 2\lambda y - y^2. \tag{13.23}$$

This has equilibrium points at $y = -\lambda \pm \sqrt{\lambda^2 + \lambda}$, so there are no real equilibrium points for $-1 < \lambda < 0$ and two equilibrium points otherwise. This suggests that there are saddle–node bifurcations at $\lambda = 0$, $y = 0$ and $\lambda = -1$, $y = 1$. Now note that, at the equilibrium points,

$$\frac{d\dot{y}}{dy} = -2\lambda - 2y = \mp\sqrt{\lambda^2 + \lambda}.$$

Using the analysis of the previous section, we can see that the equilibrium point with the larger value of y is stable, whilst the other is unstable. The bifurcation diagram is shown in Figure 13.15, and certainly looks as if it contains two saddle–node bifurcations.

For $y \ll 1$ and $\lambda \ll 1$, $\dot{y} = \lambda - 2\lambda y - y^2 \approx \lambda - y^2$, which is precisely the normal form for the saddle–node bifurcation. All of the terms on the right hand side of

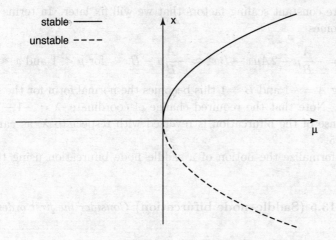

Fig. 13.14. The saddle–node bifurcation diagram.

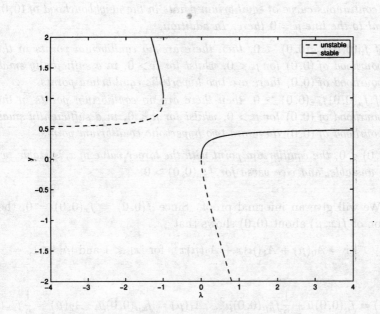

Fig. 13.15. The bifurcation diagram for $\dot{y} = \lambda - 2\lambda y - y^2$.

(13.23) are small, but the only one that is sure to be smaller than at least one of the others is the second, since $y \ll 1$ means that $\lambda y \ll \lambda$. We make no assumption about how big λ is compared with y^2. To examine the neighbourhood of the other bifurcation point, we shift the origin there using $\lambda = -1 + A\mu$, $y = 1 + Bx$, where

A and B are constant scaling factors that we will fix later. In terms of μ and x, (13.23) becomes

$$\dot{x} = -\frac{A}{B}\mu - 2A\mu x - Bx^2 \approx -\frac{A}{B}\mu - Bx^2 \quad \text{for } \mu \ll 1 \text{ and } x \ll 1.$$

By choosing $A = -1$ and $B = 1$ this becomes the normal form for the saddle–node bifurcation. Note that the required change of coordinate, $\lambda = -1 - \mu$, indicates that the sense of the bifurcation is reversed with respect to λ, as can be seen in Figure 13.15.

We can formalize the notion of a saddle–node bifurcation using the following theorem.

Theorem 13.5 (Saddle–node bifurcation) *Consider the first order differential equation*

$$\dot{x} = f(x, \mu),$$

with $f(0,0) = f_x(0,0) = 0$. Provided that $f_\mu(0,0) \neq 0$ and $f_{xx}(0,0) \neq 0$, there exists a continuous curve of equilibrium points in the neighbourhood of $(0,0)$, which is tangent to the line $\mu = 0$ there. In addition,

(i) *if $f_\mu(0,0)f_{xx}(0,0) < 0$, then there are no equilibrium points in the neighbourhood of $(0,0)$ for $\mu < 0$, whilst for $\mu > 0$, in a sufficiently small neighbourhood of $(0,0)$ there are two hyperbolic equilibrium points.*
(ii) *if $f_\mu(0,0)f_{xx}(0,0) > 0$, then there are no equilibrium points in the neighbourhood of $(0,0)$ for $\mu > 0$, whilst for $\mu < 0$, in a sufficiently small neighbourhood of $(0,0)$ there are two hyperbolic equilibrium points.*

If $f_{xx}(0,0) < 0$, the equilibrium point with the larger value of x is stable, whilst the other is unstable, and vice versa for $f_{xx}(0,0) > 0$.

Proof We will give an informal proof. Since $f(0,0) = f_x(0,0) = 0$, the Taylor expansion of $f(x, \mu)$ about $(0,0)$ shows that

$$\dot{x} \sim A_0(\mu) + A_1(\mu)x + A_2(\mu)x^2 \quad \text{for } |x| \ll 1 \text{ and } |\mu| \ll 1,$$

where

$$A_0(\mu) = f_\mu(0,0)\mu + \frac{1}{2}f_{\mu\mu}(0,0)\mu^2, \quad A_1(\mu) = f_{x\mu}(0,0)\mu, \quad A_2(\mu) = \frac{1}{2}f_{xx}(0,0).$$

There are therefore equilibrium points at

$$x = \frac{-A_1 \pm \sqrt{A_1^2 - 4A_0 A_2}}{2A_2} \sim \pm\sqrt{\frac{-2f_\mu(0,0)\mu}{f_{xx}(0,0)}} \quad \text{for } |\mu| \ll 1.$$

This shows that the location of the equilibrium points is as described in the theorem. Finally, at the equilibrium points,

$$\frac{d\dot{x}}{dx} \sim A_1(\mu) + 2A_2(\mu)x \sim f_{xx}(0,0)x,$$

since $x = O(|\mu|^{1/2})$. The stability of the equilibrium points is therefore determined by the sign of $f_{xx}(0,0)$. □

Example: The CSTR

Many industrially important chemicals are produced in bulk using a continuous flow, stirred tank reactor (CSTR). This is simply a large container, to which fresh chemicals are continuously supplied and from which the resulting reactants are continuously withdrawn, as sketched in Figure 13.16. A stirrer ensures that the chemicals in the CSTR are well mixed. Let's consider what happens when chemicals A and B that react through the cubic autocatalytic reaction step (13.7) are fed into a CSTR, and assume that the idea is to convert as much of the reactant A into the autocatalyst B as possible.

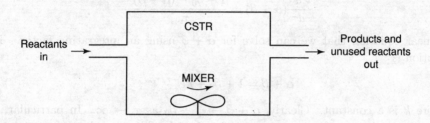

Fig. 13.16. A continuous flow, stirred tank reactor (CSTR).

Since the CSTR is well stirred, we assume that the concentrations of the chemicals are spatially uniform, and given by $a(t)$ and $b(t)$. The rate of change of the total amount of species A in the CSTR is equal to the rate at which it is produced by chemical reaction plus the rate at which it flows in minus the rate at which it flows out. If the CSTR has constant volume V, constant inlet and (by conservation of mass) outlet flowrate q and inlet concentration of A given by a_0, we have

$$\frac{d}{dt}(Va) = -Vkab^2 + q(a_0 - a),$$

and hence

$$\frac{da}{dt} = -kab^2 + \frac{a_0 - a}{t_{\text{res}}}. \tag{13.24}$$

The **residence time**, $t_{\text{res}} = V/q$, is the time it takes for a volume V of fresh reactants to flow into the CSTR, and characterizes the period for which a fluid element typically remains within the CSTR. Similarly,

$$\frac{db}{dt} = kab^2 + \frac{b_0 - b}{t_{\text{res}}}, \tag{13.25}$$

where b_0 is the inlet concentration of the autocatalyst, B.

We now define dimensionless variables

$$\alpha = \frac{a}{a_0}, \quad \beta = \frac{b}{a_0}, \quad \tau = ka_0^2 t,$$

in terms of which (13.24) and (13.25) become

$$\frac{d\alpha}{d\tau} = -\alpha\beta^2 + \frac{1-\alpha}{\tau_{\text{res}}}, \qquad (13.26)$$

$$\frac{d\beta}{d\tau} = \alpha\beta^2 + \frac{\beta_0 - \beta}{\tau_{\text{res}}}, \qquad (13.27)$$

where

$$\tau_{\text{res}} = ka_0^2 t_{\text{res}}, \quad \beta_0 = \frac{b_0}{a_0}.$$

We can now add (13.26) and (13.27) to obtain

$$\frac{d}{d\tau}(\alpha + \beta) = \frac{1 + \beta_0 - (\alpha + \beta)}{\tau_{\text{res}}},$$

a linear equation that we can solve for $\alpha + \beta$ using an integrating factor. The solution is

$$\alpha + \beta = 1 + \beta_0 + k e^{-\tau/\tau_{\text{res}}},$$

where k is a constant. Clearly, $\alpha + \beta \to 1 + \beta_0$ as $\tau \to \infty$. In particular, for $\tau \gg \tau_{\text{res}}$, and hence for $t \gg t_{\text{res}}$, $\alpha + \beta \sim 1 + \beta_0$. The term $k e^{-\tau/\tau_{\text{res}}}$ represents an initial transient, which decays to zero exponentially fast over a time scale given by the residence time. We therefore assume that $\alpha + \beta = 1 + \beta_0$, and can thereby eliminate β from (13.26). In the analysis that follows, it is more convenient to work in terms of $z = 1 - \alpha$, the extent to which the reactant A has been converted to B. This gives us a nonlinear, first order ordinary differential equation for z,

$$\frac{dz}{d\tau} = R(z) - F(z), \qquad (13.28)$$

where

$$R(z) = (1-z)(z + \beta_0)^2, \quad F(z) = \frac{z}{\tau_{\text{res}}}.$$

We can now see that a steady state, where $dz/d\tau = 0$, occurs when the rate of reaction, $R(z)$, is balanced by the rate at which A flows into the CSTR, $F(z)$, and hence when $F(z) = R(z)$. The function $R(z)$ is a cubic polynomial, whilst $F(z)$ is a straight line through the origin. The steady states are therefore given by the points of intersection of these two curves.

Case 1: No autocatalyst in the inflow, $\beta_0 = 0$

When $\beta_0 = 0$, $R(z)$ has a repeated root at $z = 0$, as shown in Figure 13.17. The straight line $F(z)$ always passes through $z = 0$, which is therefore always a steady state. The state $z = 0$ represents a CSTR that contains no autocatalyst, just the reactant A supplied by the inlet flow. A simple calculation of the steady state solutions shows that for $\tau_{\text{res}} < 4$, $z = 0$ is the only steady state, whilst for

13.3 BIFURCATION THEORY

$\tau_{\text{res}} > 4$, there are two further points of intersection between $F(z)$ and $R(z)$, and hence two other steady states, at $z = z_1$ and $z = z_2$, where

$$z_1 = \frac{1}{2}\left(1 - \sqrt{1 - \frac{4}{\tau_{\text{res}}}}\right), \quad z_2 = \frac{1}{2}\left(1 + \sqrt{1 - \frac{4}{\tau_{\text{res}}}}\right).$$

Note that $0 < z_1 < \frac{1}{2}$ and $\frac{1}{2} < z_2 < 1$, and $z_1 \to 0$, $z_2 \to 1$ as $\tau_{\text{res}} \to \infty$. The steady states are sketched in Figure 13.18. The stability of the steady states is easily calculated, and we find that $z = 0$ and $z = z_2$ are stable states, whilst $z = z_1$ is unstable. There is a saddle–node bifurcation at $\tau_{\text{res}} = 4$. This situation, with $\beta_0 = 0$, is a realistic one, since it is probably desirable to run the system with just a single species entering the CSTR. We clearly want to run the CSTR in the state $z = z_2$, where more than half of the reactant A is converted to B. However, we also want to make the residence time as small as possible, to increase the rate at which B is produced. However, we can now see that, if τ_{res} is slowly decreased, z_2 will also slowly decrease, and that when τ_{res} decreases past 4, the saddle–node bifurcation point, the situation changes dramatically. The only available steady state is $z = 0$, so that no autocatalyst remains in the CSTR, and the reaction stops. This is known as **washout**. Attempts to recover the desirable state $z = z_2$ by increasing the residence time, τ_{res}, are doomed to failure when there is no autocatalyst entering the CSTR.

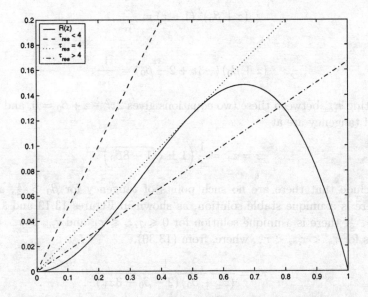

Fig. 13.17. The curves $R(z)$ and $F(z)$ when $\beta_0 = 0$.

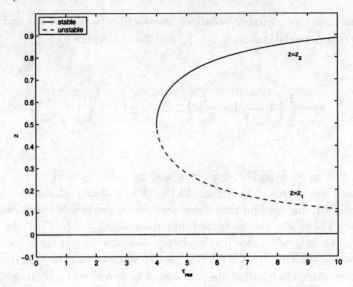

Fig. 13.18. The bifurcation diagram when $\beta_0 = 0$.

Case 2: Autocatalyst in the inflow, $\beta_0 > 0$

Let's now see how the situation is affected by including some autocatalyst in the inflow. When $\beta_0 > 0$, the cubic polynomial $R(z)$ has a single positive root at $z = 1$, and is strictly positive for $0 \leqslant z < 1$, as shown in Figure 13.19. In order to determine when $F(z)$ is tangent to $R(z)$, we must simultaneously solve $R(z) = F(z)$ and $R'(z) = F'(z)$, which gives

$$(z + \beta_0)^2 (1 - z) = \frac{z}{\tau_{\text{res}}}, \qquad (13.29)$$

$$(z + \beta_0)(-3z + 2 - \beta_0) = \frac{1}{\tau_{\text{res}}}. \qquad (13.30)$$

Eliminating τ_{res} between these two equations gives $2z^2 - z + \beta_0 = 0$, and hence the points of tangency are at

$$z = z_{\pm} = \frac{1}{4}\left(1 \pm \sqrt{1 - 8\beta_0}\right).$$

We conclude that there are no such points of tangency for $\beta_0 > \frac{1}{8}$, and hence that there is a unique stable solution, as shown in Figures 13.19 and 13.20. For $0 < \beta_0 < \frac{1}{8}$, there is a unique solution for $0 \leqslant \tau_{\text{res}} \leqslant \tau_+$ and $\tau_{\text{res}} > \tau_-$, and three solutions for $\tau_+ < \tau_{\text{res}} < \tau_-$, where, from (13.30),

$$\tau_{\pm} = \frac{1}{(z_{\pm} + \beta_0)(2 - \beta_0 - 3z_{\pm})}.$$

There are now two saddle–node bifurcations, at $\tau_{\text{res}} = \tau_{\pm}$.

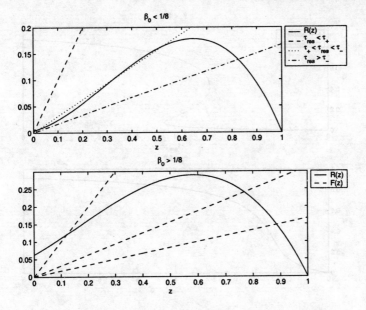

Fig. 13.19. The curves $R(z)$ and $F(z)$ when $\beta_0 > 0$.

We conclude that for $\beta_0 > \frac{1}{8}$, there is no possibility of washout, just a unique steady state solution for any given residence time, τ_{res}. For $0 < \beta < \frac{1}{8}$, we again have the possibility, if not of a washout of the autocatalyst, at least of a dramatic decrease in the concentration of B when τ_{res} falls below τ_+. However, if τ_{res} is now increased again, there will be a dramatic increase in the concentration of B as τ_{res} increases past τ_-. This change in the steady state from when a parameter is increased to when it is decreased is known as **hysteresis**.

Finally, note that when $\beta = \frac{1}{8}$ the two saddle–node points merge and disappear at $\tau_{\text{res}} = \frac{64}{27}$. This is itself a bifurcation, and is known as a **codimension two bifurcation**. Such a bifurcation can only occur in a system that has at least two parameters (here β_0 and τ_{res}).

We will study two other types of bifurcation. Consider the normal form

$$\dot{x} = \mu x - x^2. \tag{13.31}$$

This system has equilibrium points at $x = 0$ and $x = \mu$. When $\mu = 0$ we again have the nonhyperbolic equilibrium point given by $\dot{x} = -x^2$, whilst when $\mu \neq 0$ there are always two equilibrium points. In this case,

$$\frac{d\dot{x}}{dx} = \mu - 2x = \begin{cases} -\mu & \text{at } x = \mu, \\ \mu & \text{at } x = 0, \end{cases}$$

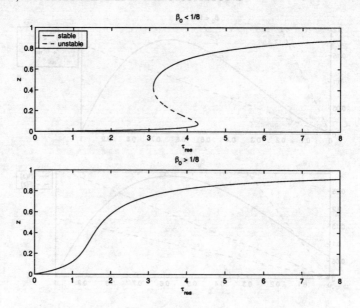

Fig. 13.20. The bifurcation diagram when $\beta_0 > 0$.

and hence

$$x = \mu \text{ is } \begin{cases} \text{stable for } \mu > 0, \\ \text{unstable for } \mu < 0, \end{cases}$$

$$x = -\mu \text{ is } \begin{cases} \text{unstable for } \mu > 0, \\ \text{stable for } \mu < 0. \end{cases}$$

The bifurcation diagram is shown in Figure 13.21. This is called a **transcritical bifurcation**. At the bifurcation point, the two equilibrium solutions pass through each other and exchange stabilities, so this sort of bifurcation is often referred to as an **exchange of stabilities**.

Theorem 13.6 (Transcritical bifurcation) *Consider the first order differential equation*

$$\dot{x} = f(x, \mu),$$

with $f(0,0) = f_x(0,0) = f_\mu(0,0) = 0$. *Provided that* $f_{xx}(0,0) \neq 0$ *and* $f_{\mu x}^2(0,0) - f_{xx}(0,0)f_{\mu\mu}(0,0) > 0$, *there exist two continuous curves of equilibrium points in the neighbourhood of* $(0,0)$. *These curves intersect transversely at* $(0,0)$. *For each* $\mu \neq 0$ *there are two hyperbolic equilibrium points in the neighbourhood of* $x = 0$. *If* $f_{xx}(0,0) < 0$, *the equilibrium point with the larger value of* x *is stable, whilst the other is unstable, and vice versa for* $f_{xx}(0,0) > 0$.

We will leave the informal proof as Exercise 13.9.

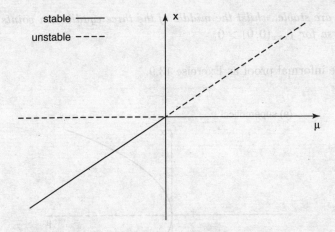

Fig. 13.21. The transcritical bifurcation.

Finally, consider the normal form

$$\dot{x} = \mu x - x^3. \qquad (13.32)$$

This has a single equilibrium point at $x = 0$ for $\mu < 0$, and three equilibrium points at $x = 0$ and $x = \pm\sqrt{\mu}$ for $\mu > 0$. The bifurcation diagram is shown in Figure 13.22(a). The bifurcation at $x = 0$, $\mu = 0$ is called a **supercritical pitchfork bifurcation**. A similar bifurcation, in which the two new equilibrium points created at the bifurcation point are unstable, is the **subcritical pitchfork bifurcation**, with normal form $\dot{x} = \mu x + x^3$, whose bifurcation diagram is shown in Figure 13.22(b).

Theorem 13.7 (Pitchfork bifurcation) *Consider the first order differential equation*

$$\dot{x} = f(x, \mu),$$

with $f(0,0) = f_x(0,0) = f_\mu(0,0) = f_{xx}(0,0) = 0$. Provided that $f_{\mu x}(0,0) \neq 0$ and $f_{xxx}(0,0) \neq 0$, there exist two continuous curves of equilibrium points in the neighbourhood of $(0,0)$. One curve passes through $(0,0)$ transverse to the line $\mu = 0$, whilst the other is tangent to $\mu = 0$ at $x = 0$. In addition,

(i) *if $f_{\mu x}(0,0) f_{xxx}(0,0) < 0$, then, close to $(0,0)$, there is a single equilibrium point for $\mu < 0$ and three equilibrium points for $\mu > 0$.*
(ii) *if $f_{\mu x}(0,0) f_{xxx}(0,0) > 0$, then, close to $(0,0)$, there is a single equilibrium point for $\mu > 0$ and three equilibrium points for $\mu < 0$.*

If $f_{xxx}(0,0) < 0$, the single equilibrium point and the outer two of the three equilib-

rium points are stable, whilst the middle of the three equilibrium points is unstable, and vice versa for $f_{xxx}(0,0) > 0$.

We leave the informal proof as Exercise 13.9.

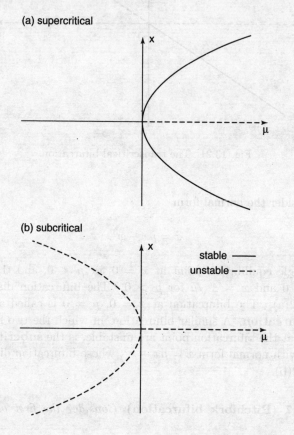

Fig. 13.22. The (a) supercritical and (b) subcritical pitchfork bifurcation.

Although these bifurcations seem rather abstract, they often describe the qualitative behaviour of complicated physical systems, as we saw for the CSTR. Another simple example is the buckling beam. Consider a straight beam of elastic material under compression from two equal and opposite forces along its axis, one applied at each end, as shown in Figure 13.23(a). Now consider the displacement, x, of the midpoint of the beam. For sufficiently small applied forces, the beam undergoes a simple axial compression and $x = 0$. As the applied forces are slowly increased, at a ⎵⎵⎵ the beam buckles, as shown in Figure 13.23(b), because the unbuckled ⎵⎵⎵ e. If the beam has perfect left–right symmetry, it is equally ⎵⎵⎵ right. When the displacement of the midpoint of the beam ⎵⎵⎵ of magnitude of the applied forces, we obtain the bifurcation diagram corresponding to a supercritical pitchfork, as in Figure 13.22(a). Of course, in practice, a beam, for example a ruler held between your forefingers, will not have perfect symmetry, and will buckle in a preferred direction. This suggests that the pitchfork bifurcation is itself structurally unstable. You can investigate this by trying Exercise 13.7.

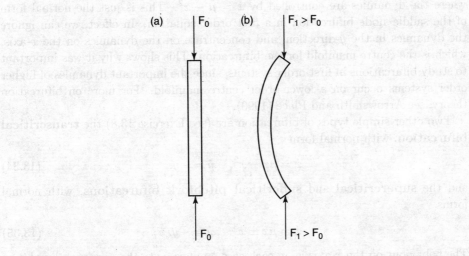

Fig. 13.23. An elastic beam under compression (a) before and (b) after buckling.

13.3.2 Second Order Ordinary Differential Equations

As for first order equations, second order systems that possess an equilibrium point with an eigenvalue with zero real part are structurally unstable. A small change in the governing equations can change the qualitative nature of the phase portrait. For example, the simple, frictionless pendulum, with phase portrait shown in Figure 9.2(b), contains a nonlinear centre. Since a centre has two eigenvalues with zero real part, we conclude that this system is structurally unstable. In terms of the physics, consider what happens if we allow for a tiny amount of friction. This will gradually reduce the energy and the amplitude of the motion, with the pendulum being brought to rest as $t \to \infty$. This is reflected in what happens to the phase portrait when a term $\mu d\theta/dt$, with $\mu > 0$, is added to the left hand side of (9.2) to model the effect of friction. No matter how small μ is, the nonlinear centre is transformed into a stable focus, and all integral paths asymptote to $\theta = \dot{\theta} = 0$ as $t \to \infty$ (see Exercise 9.6).

Now consider the second order system

$$\dot{x} = \mu - x^2, \quad \dot{y} = -y. \tag{13.33}$$

For $\mu < 0$ there are no equilibrium points, whilst for $\mu > 0$ there are two equilibrium points at $P_+ = (+\sqrt{\mu}, 0)$ and $P_- = (-\sqrt{\mu}, 0)$. The point P_+ is a stable node, whilst P_- is a saddle point. When $\mu = 0$ there is a single, nonhyperbolic equilibrium point at $(0,0)$, a saddle–node. The various phase portraits are shown in Figure 13.24. The point $\mu = 0$, $x = y = 0$ is called a **saddle–node bifurcation point**, and (13.33) is its normal form. At such a point, a saddle and a node collide and disappear. Note that, since $\dot{y} = -y$, all integral paths asymptote to the x-axis as $t \to \infty$, where the dynamics are controlled by $\dot{x} = \mu - x^2$. This is just the normal form of the saddle–node bifurcation in a first order equation. In effect, we can ignore the dynamics in the y-direction, and concentrate on the dynamics on the x-axis, which is the centre manifold for the bifurcation. This shows why it was important to study bifurcations in first order systems, since the important dynamics of higher order systems occur on a lower order centre manifold. For more on bifurcation theory, see Arrowsmith and Place (1990).

Two other simple types of bifurcation are (see Exercise 13.8) the **transcritical bifurcation**, with normal form

$$\dot{x} = \mu x - x^2, \quad \dot{y} = -y, \tag{13.34}$$

and the **supercritical** and **subcritical pitchfork bifurcations**, with normal forms

$$\dot{x} = \mu x \pm x^3, \quad \dot{y} = -y. \tag{13.35}$$

The behaviour on the x-axis is, in each case, analogous to the corresponding bifurcation in a one-dimensional system, which we studied in Section 13.3.1. In each of these examples, one eigenvalue passes through zero at the bifurcation point. We will not consider the more degenerate case where both eigenvalues pass through zero (see Guckenheimer and Holmes, 1983). However, we will consider one other type of bifurcation, which occurs when a complex conjugate pair of eigenvalues passes through the imaginary axis away from the origin. This is called the **Hopf bifurcation**, and has no counterpart in first order systems.

The normal form of the Hopf bifurcation is most easily written in terms of polar coordinates as

$$\dot{r} = \mu r + ar^3, \quad \dot{\theta} = \omega. \tag{13.36}$$

Let's assume that $a < 0$. For $\mu \leqslant 0$ there is a stable focus at the origin, and no other equilibrium points, and all trajectories spiral into the origin from infinity. For $\mu > 0$ there is an unstable focus at the origin, and no other equilibrium points, but there is also a stable limit cycle, given by $r = \sqrt{-\mu/a}$, $\dot{\theta} = \omega$. Trajectories spiral onto the limit cycle both from the origin and from infinity, as shown in Figure 13.25. When $\mu = 0$ a linear analysis shows that the origin is a centre, and hence has eigenvalues with zero real part. However, this is an example where the nonlinear terms, here $\dot{r} = ar^3$, cause the integral paths to spiral into the equilibrium point. Note that this is a **supercritical Hopf bifurcation**. The **subcritical Hopf bifurcation**, for which the limit cycle is unstable and exists for $\mu < 0$, occurs when $a > 0$. In general, for equations that model real physical systems, any limit cycle solutions

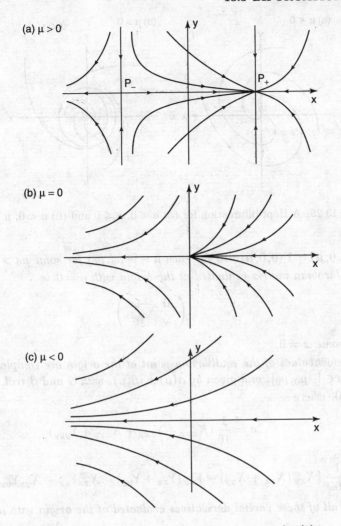

Fig. 13.24. A saddle–node bifurcation in a second order system, for (a) $\mu > 0$, (b) $\mu = 0$, (c) $\mu < 0$.

that exist are formed at Hopf bifurcations, and it is therefore crucial to determine the position of these bifurcations. In order to demonstrate that a Hopf bifurcation occurs when a pair of eigenvalues crosses the imaginary axis, we can appeal to the **Hopf bifurcation theorem**.

Theorem 13.8 (Hopf) *Consider the second order system*

$$\dot{x} = X(x, y; \mu), \quad \dot{y} = Y(x, y; \mu).$$

Suppose that

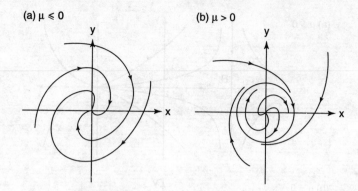

Fig. 13.25. A Hopf bifurcation for (a) $a < 0$, $\mu \leq 0$ and (b) $a < 0$, $\mu > 0$.

(i) $X(0,0;\mu) = Y(0,0;\mu) = 0$ for each $\mu \in [-\mu_0, \mu_0]$ for some $\mu_0 > 0$,
(ii) the Jacobian matrix evaluated at the origin with $\mu = 0$ is

$$\begin{pmatrix} 0 & -\omega \\ \omega & 0 \end{pmatrix}$$

for some $\omega \neq 0$,
(iii) the eigenvalues of the equilibrium point at the origin are complex conjugate for $\mu \in [-\mu_0, \mu_0]$, and given by $\alpha(\mu) \pm i\beta(\mu)$, with α and β real,
(iv) $a \neq 0$, where

$$a = \frac{1}{16}(X_{xxx} + Y_{xxy} + X_{xyy} + Y_{yyy})$$

$$+ \frac{1}{16\omega}\{X_{xy}(X_{xx} + X_{yy}) - Y_{xy}(Y_{xx} + Y_{yy}) - X_{xx}Y_{xx} + X_{yy}Y_{yy}\}, \quad (13.37)$$

with all of these partial derivatives evaluated at the origin with $\mu = 0$.

Then

(i) If $a\alpha'(0) < 0$, a unique limit cycle solution bifurcates from the origin in $\mu > 0$ as μ passes through zero. For $\mu < 0$ there exists a neighbourhood of the origin that does not contain a limit cycle solution. The stability of the limit cycle is the same as that of the origin in $\mu < 0$.
(ii) If $a\alpha'(0) > 0$, a unique limit cycle solution bifurcates from the origin in $\mu < 0$ as μ passes through zero. For $\mu > 0$ there exists a neighbourhood of the origin that does not contain a limit cycle solution. The stability of the limit cycle is the same as that of the origin in $\mu > 0$.

The amplitude of the limit cycle grows like $|\mu|^{1/2}$ and its period tends to $2\pi/|\omega|$ as $\mu \to 0$.

We will not prove this theorem here. The idea of the proof is to go through a series of algebraic transformations of the differential equations that reduce them to the

13.3 BIFURCATION THEORY

normal form, (13.36), and is algebraically very unpleasant. For as straightforward an explanation of the details as you could hope for, see Glendinning (1994). Instead, we will concentrate on a concrete example to show how the proof of the theorem works.

Example

Consider the second order system

$$\dot{x} = X(z, y; \mu) = \mu x - \omega y + x^3, \quad \dot{y} = Y(x, y; \mu) = \omega x + \mu y + y^2, \quad (13.38)$$

where μ and ω are constants. The linear part of the system is in the normal form for an equilibrium point at the origin with complex conjugate eigenvalues $\lambda = \mu + i\omega$, which we discussed in Section 9.3.2. This means that the origin is a stable focus for $\mu < 0$, a linear centre for $\mu = 0$ and an unstable focus for $\mu > 0$. The system therefore satisfies conditions (i), (ii) and (iii) of Theorem 13.8. If we now substitute this particular choice of X and Y into (13.37), we find that $a = 3/8$, and hence that $a\alpha'(0) = 3/8 > 0$. The Hopf bifurcation at $\mu = 0$ is therefore subcritical, with an unstable limit cycle emerging from the origin in $\mu < 0$.

Let's now try to write (13.38) in the normal form (13.36) for sufficiently small x, y and μ, using the same transformations as those used in the proof of the Hopf bifurcation theorem. We begin by defining $z = x + iy$, in terms of which (13.38) can be written concisely as

$$\dot{z} = \lambda z - \frac{1}{4}i(z - z^*)^2 + \frac{1}{8}(z + z^*)^3. \quad (13.39)$$

The next step is to make a quadratic **near identity transformation**,

$$z = w + a_1 w^2 + a_2 w w^* + a_3 w^{*2}. \quad (13.40)$$

The idea is that such a transformation leaves the linear part of the equation unchanged when written in terms of w, but can be used to simplify the nonlinear part by an appropriate choice of the constants a_1, a_2 and a_3. Specifically, we can eliminate quadratic terms from (13.39). For $|z| \ll 1$,

$$w = z - a_1 w^2 - a_2 w w^* - a_3 w^{*2} = z - a_1 \left(z - a_1 w^2 - a_2 w w^* - a_3 w^{*2}\right)^2$$

$$- a_2 \left(z - a_1 w^2 - a_2 w w^* - a_3 w^{*2}\right)\left(z^* - a_1^* w^{*2} - a_2^* w w^* - a_3^* w^2\right)$$

$$- a_3 \left(z^* - a_1^* w^{*2} - a_2^* w w^* - a_3^* w^2\right)^2 = z - a_1 z^2 - a_2 z z^* - a_3 z^{*2} + \left(2a_1^2 + a_2 a_3^*\right) z^3$$

$$+ \left(3a_1 a_2 + a_2 a_2^* + 2a_3^2\right) z^2 z^*$$

$$+ \left(2a_1 a_3 + a_1^* a_2 + a_2^2 + 2a_2^* a_3\right) z z^{*2} + \left(a_2 a_3 + 2a_1^* a_3\right) z^{*3} + O(|z|^4).$$

If we now take this expression, differentiate, and replace $\dot{z}$ and $\dot{z}^*$ using (13.39), we arrive, after considerable effort, at

$$\dot{w} = \lambda z - \frac{1}{4}i(z - z^*)^2 + \frac{1}{8}(z - z^*)^3$$

406 STABILITY, INSTABILITY AND BIFURCATIONS

$$-\left(2a_1z+a_2z^*\right)\left\{\lambda z-\frac{1}{4}i\left(z-z^*\right)^2\right\}-\left(a_2z+2a_3z^*\right)\left\{\lambda^*z^*+\frac{1}{4}i\left(z-z^*\right)^2\right\}$$

$$+3\left(2a_1^2+a_2a_3^*\right)\lambda z^3+\left(3a_1a_2+a_2a_2^*+2a_3^2\right)\left(2\lambda+\lambda^*\right)z^2z^*$$

$$+\left(2a_1a_3+a_1^*a_2+a_2^2+2a_2^*a_3\right)\left(\lambda+2\lambda^*\right)zz^{*2}+3\left(a_2a_3+2a_1^*a_3\right)\lambda^*z^{*3}+O(|z|^4).$$

We can now use the definition, (13.40), of w to eliminate z from the right hand side, retaining only cubic terms and larger, and arrive at

$$\dot{w}=\lambda\left(w+a_1w^2+a_2ww^*+a_3w^{*2}\right)$$

$$-\left(\frac{1}{4}i+2a_1\lambda\right)\left(w^2+2a_1w^3+2a_2w^2w^*+2a_3ww^{*2}\right)$$

$$+\left\{\frac{1}{2}i-a_2\left(\lambda+\lambda^*\right)\right\}\left\{ww^*+a_3^*w^3+\left(a_1+a_2^*\right)w^2w^*+\left(a_1^*+a_2\right)ww^{*2}+a_3w^{*3}\right\}$$

$$-\left(\frac{1}{4}i+2a_3\lambda^*\right)\left(w^{*2}+2a_1^*w^{*3}+2a_2^*w^{*2}w+2a_3^*w^*w^2\right)$$

$$+\frac{1}{4}i\left(2a_1w+a_2w^*\right)\left(w-w^*\right)^2-\frac{1}{4}i\left(a_2w+2a_3w^*\right)\left(w-w^*\right)^2+\frac{1}{8}\left(w+w^*\right)^3$$

$$+3\left(2a_1^2+a_2a_3^*\right)\lambda w^3+\left(3a_1a_2+a_2a_2^*+2a_3^2\right)\left(2\lambda+\lambda^*\right)w^2w^*$$

$$+\left(2a_1a_3+a_1^*a_2+a_2^2+2a_2^*a_3\right)\left(\lambda+2\lambda^*\right)ww^{*2}+3\left(a_2a_3+2a_1^*a_3\right)\lambda^*w^{*3}+O(|w|^4).$$

We can now see that we can eliminate all of the quadratic terms by choosing

$$a_1=-\frac{i}{4\lambda},\quad a_2=\frac{i}{2\lambda},\quad a_3=\frac{i}{4\left(\lambda-2\lambda^*\right)},$$

which leaves us with

$$\dot{w}=\lambda w+A_1w^3+A_2w^2w^*+A_3ww^{*2}+A_4w^{*3},$$

where

$$A_1=-2a_1\left(\frac{1}{4}i+2a_1\lambda\right)+a_3^*\left\{\frac{1}{2}i-a_2\left(\lambda+\lambda^*\right)\right\}+\frac{1}{8}$$

$$+\frac{1}{4}i\left(2a_1-a_2\right)+3\left(2a_1+a_2a_3^*\right)\lambda,$$

$$A_2=-2a_2\left(\frac{1}{4}i+2a_1\lambda\right)+\left(a_1+a_2^*\right)\left\{\frac{1}{2}i-a_2\left(\lambda+\lambda^*\right)\right\}+\frac{3}{8}$$

$$-2a_3^*\left(\frac{1}{4}i+2a_3\lambda^*\right)+\frac{1}{4}i\left(3a_2-2a_3-4a_1\right)+\left(3a_1a_2+a_2a_2^*+2a_3^2\right)\left(2\lambda+\lambda^*\right),$$

$$A_3=-2a_3\left(\frac{1}{4}i+2a_1\lambda\right)+\left(a_1^*+a_2\right)\left\{\frac{1}{2}i-a_2\left(\lambda+\lambda^*\right)\right\}+\frac{3}{8}$$

$$-2a_2^*\left(\frac{1}{4}i + 2a_3\lambda^*\right) + \frac{1}{4}i\left(4a_3 - 3a_2 + 2a_1\right) + \left(2a_1a_3 + a_1^*a_2 + a_2^2 + 2a_2^*a_3\right)(\lambda + 2\lambda^*),$$

$$A_4 = -2a_1^*\left(\frac{1}{4}i + 2a_3\lambda^*\right) + a_3\left\{\frac{1}{2}i - a_2(\lambda + \lambda^*)\right\} + \frac{1}{8}$$

$$+\frac{1}{4}i\left(a_2 - 2a_3\right) + 3\left(a_2a_3 + a_1^*a_3\right)\lambda^*.$$

The next step is to make another near identity transformation,

$$w = v + b_1v^3 + b_2v^2v^* + b_3vv^{*2} + b_4v^{*3}, \qquad (13.41)$$

and try to eliminate the cubic terms as well. Proceeding exactly as we did for the quadratic terms, although the algebra is now easier since we only retain cubic terms, we find that

$$\dot{v} = \lambda v + (A_1 - 2\lambda b_1)v^3 + \{A_2 - (\lambda + \lambda^*)b_2\}v^2v^* + (A_3 - 2\lambda^*b_3)vv^{*2}$$

$$+ \{A_4 + (\lambda - 3\lambda^*)b_4\}v^{*3} + O(|v|^4).$$

At first sight, it would appear that we can eliminate all of the cubic terms in the same way as we did the quadratic terms. However, the coefficient of b_2 is $\lambda + \lambda^* = 2\mu$, which is small when $\mu \ll 1$. Since we need $b_2 = O(1)$ for $|v| \ll 1$ and $\mu \ll 1$, we conclude that we cannot eliminate the v^2v^* term, and hence that the simplest normal form is

$$\dot{v} \sim \lambda v + av|v|^2,$$

using $v^2v^* = v|v|^2$, with $a = A_2(0)$. A little algebra then shows that $a = A_2(0) = 3/8$, consistent with our earlier calculation of a from (13.37). Indeed, the hard part of the proof the Hopf bifurcation is to show that a, as defined in (13.37), appears in the normal form in this way. If we now write $v = re^{i\theta}$ and separate real and imaginary parts, we recover the normal form (13.36), and we are done.

We should reiterate that this transformation and the ensuing algebra is *not* what needs to be done whenever the system you are studying contains a Hopf bifurcation point. The Hopf bifurcation theorem is far easier to use. We went through the details of this example purely to illustrate the steps involved in the proof.

Figure 13.26 shows the unstable limit cycle solution for various $\mu < 0$. We calculated these solutions numerically using MATLAB (see Section 9.3.4). For $|\mu|$ sufficiently small, we can see that the limit cycle is circular. However, as $|\mu|$ increases, the limit cycle moves away from the neighbourhood of the origin and becomes increasingly distorted as it begins to interact with a saddle point that lies close to $(-1, -1)$. In fact, when $\mu = \mu_0 \approx -0.12176$, the limit cycle collides with this saddle point, and is destroyed. This is an example of a homoclinic bifurcation, which we discuss below.

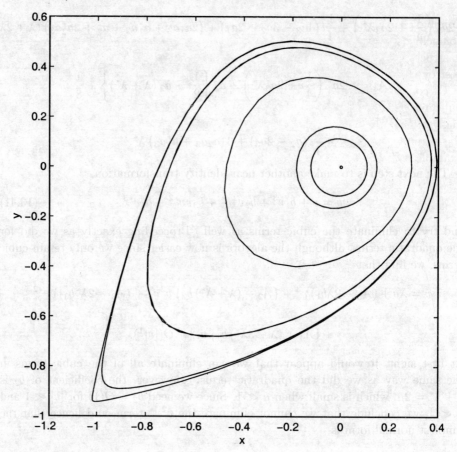

Fig. 13.26. The unstable limit cycle solution of (13.36) for $\mu = -0.002, -0.005, -0.01, -0.05, -0.1, -0.12$ and -0.12176.

13.3.3 Global Bifurcations

With the exception of the previous example, all of the bifurcations that we have studied so far have been **local bifurcations**. That is, they have arisen because of a change in the nature of an equilibrium point as a parameter passes through a critical value. A **global bifurcation** is one that occurs because the qualitative nature of the solutions changes due to the interaction of two or more distinct features of the phase portrait. For second order systems, the crucial features are usually limit cycles and the separatrices of any saddle points. Figure 13.27 illustrates an example of a **homoclinic bifurcation**, which can occur when a limit cycle interacts with a saddle point. A limit cycle is formed in a Hopf bifurcation when the bifurcation parameter, μ, is equal to μ_1. As μ increases, the amplitude of the limit cycle grows until, when $\mu = \mu_2 > \mu_1$, the limit cycle collides with the saddle point, forming a connection between the stable and unstable separatrices. This is the homoclinic bifurcation point. With the separatrices joined up like this, the system is structurally unstable, since an arbitrarily small change to the equations can destroy the connection. For $\mu > \mu_2$, there is no limit cycle.

Of course, it is one thing to describe a global bifurcation in qualitative terms, but quite another to be able to quantify when the bifurcation occurs, since a local analysis is of no help. We must usually resort to numerical methods, as we did for the example shown in Figure 13.26. We will therefore confine ourselves to a single example of a global bifurcation where analytical progress is possible.

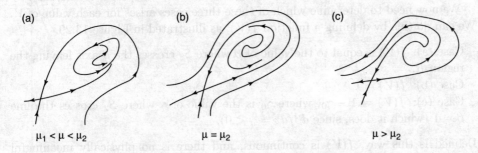

Fig. 13.27. A homoclinic bifurcation.

Example: Travelling waves in cubic autocatalysis (continued)

In Section 13.1.1 we performed a local analysis of the two equilibrium points of (13.14). In order to determine for what values of V a physically meaningful solution exists, we need to determine for what values of V the unstable separatrix of P_1, labelled S_1 in Figure 13.4, enters the origin as $z \to \infty$. This is a global problem involving the relative positions of S_1 and the stable manifold of P_2, labelled S_2 in Figure 13.4.

We begin by defining the region

$$R = \left\{ (\beta, \gamma) \mid 0 < \beta < \frac{2}{3}, \; -\frac{4}{27}V < \gamma < 0 \right\}$$

$$\bigcup \left\{ (\beta, \gamma) \mid \frac{2}{3} \leqslant \beta < 1, \; \gamma_H(\beta) < \gamma < 0 \right\},$$

where $\gamma_H(\beta) = -\beta^2(1-\beta)/V$ is the horizontal isocline. The region R is shown in Figure 13.28. Note that the point $\beta = \frac{2}{3}$, $\gamma = -\frac{4}{27}V$ is the local minimum of $\gamma_H(\beta)$. This region is constructed so that there are three qualitatively different possibilities.

— Case (a): S_2 enters R through the β-axis,

— Case (b): $S_2 = S_1$, and hence asymptotes to P_1 as $z \to -\infty$,
— Case (c): S_2 enters R through its lower boundary.

Note that S_2 cannot enter R through the γ-axis, since $d\beta/dz = \gamma < 0$ there. Cases (a), (b) and (c) are illustrated in Figure 13.28. In case (a), S_1 lies below S_2 and therefore does not enter the origin and is swept into the region $\beta < 0$. It cannot represent a physically meaningful solution in this case. In case (b), $S_2 = S_1$ represents a physically meaningful solution, and this solution enters the origin on the stable manifold. In case (c), S_1 lies above S_2 and is therefore attracted into the origin on the centre manifold. These arguments can be made more rigorous, but we will not do this here (see Billingham and Needham, 1991, for more details).

We now need to determine which of these three cases arises for each value of V. We can do this by defining a function $f(V)$, as illustrated in Figure 13.29.

— Case (a): $f(V)$ is equal to the value of β where S_2 crosses the β-axis leaving the region R,
— Case (b): $f(V) = 1$,
— Case (c): $f(V) = 1 - \gamma_0$, where γ_0 is the value of γ where S_2 crosses the line $\beta = 1$ (which it does, since $d\beta/dz = \gamma < 0$).

Defined is this way, $f(V)$ is continuous, and there is no physically meaningful solution of (13.14) when $f(V) < 1$ (case(a)), and a unique physically meaningful solution when $f(V) \geqslant 1$ (cases (b) and (c)).

Lemma 13.1 $f(V)$ *is strictly monotone increasing for* $V > 0$.

Proof When $V = V_0 > 0$, we define the region

$$D(V_0) = \{(\beta, \gamma) \mid 0 \leqslant \gamma \leqslant \gamma_S(\beta)|_{V=V_0} , 0 \leqslant \beta \leqslant 1\},$$

where $\gamma = \gamma_S(\beta)$ is the equation of S_2 within the region R, as illustrated in Figure 13.29. From (13.17),

$$\frac{\partial}{\partial V}\left(\frac{d\gamma}{d\beta}\right) = -1 < 0.$$

This means that the slope of the integral path through any fixed point is strictly monotone decreasing as V increases. In particular, when $V = V_1 > V_0$, all integral paths that meet the curved boundary of $D(V_0)$ (a boundary given by S_2 when $V = V_0$) enter $D(V_0)$. In addition, from (13.14), all integral paths that meet the straight parts of the boundary of $D(V_0)$ also enter $D(V_0)$. Finally, since S_2 is directed along the vector $\mathbf{e}_- = (1, -V)$ as it enters the origin, it lies outside $D(V_0)$ in a sufficiently small neighbourhood of the origin when $V = V_1$. We conclude that when $V = V_1 > V_0$, S_2 cannot pass through the boundary of $D(V_0)$, and therefore that $f(V_1) > f(V_0)$. □

We will leave the next stage of our argument as Exercise 13.11, in which you are asked, helped by some hints, to show that $f(V) = O(V^2)$ for $V \ll 1$, and that $f(V) \sim V$ for $V \gg 1$. Now, since $f(V) < 1$ for V sufficiently small, $f(V) > 1$ for V

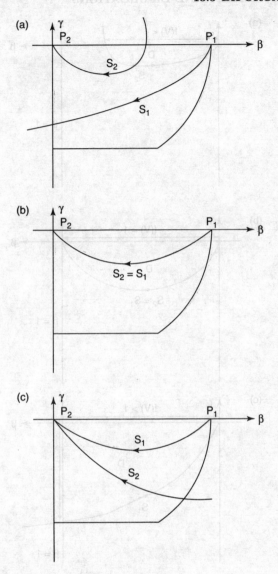

Fig. 13.28. The region R and the three possible types of global behaviour of S_1 and S_2.

sufficiently large, $f(V)$ is strictly monotone increasing by Lemma 13.1, and $f(V)$ is continuous, we conclude from the intermediate value theorem that there exists a unique value $V = V^*$, such that $f(V) \geqslant 1$ for $V \geqslant V^*$ and $f(V) < 1$ for $V < V^*$. When $V = V^*$ there is therefore a global bifurcation, since we have now shown that

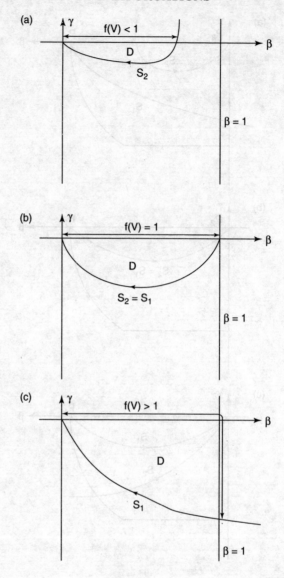

Fig. 13.29. The definition of $f(V)$ and the region $D(V)$ in each of the three cases.

case (a) occurs when $V < V^*$, case (b) occurs when $V = V^*$ and case (c) occurs when $V > V^*$. This global bifurcation comes about purely because of the relative positions of the manifolds S_1 and S_2.

In order to determine the numerical value of V^*, we would usually have to solve the governing equations (13.14) numerically. However, we do know that when $V = V^*$ the solution asymptotes to the origin on the stable manifold with $\beta = O(e^{-V^*z})$ as $z \to \infty$, whilst for $V > V^*$, the solution asymptotes to the origin on the centre manifold, with $\beta \sim V/z$ as $z \to \infty$. We were also able to show that there is an analytical solution, (13.18), when $V = 1/\sqrt{2}$, which has $\beta = O(e^{-z/\sqrt{2}})$ as $z \to \infty$. This must therefore be the unique solution that corresponds to $V = V^*$, and we conclude that the global bifurcation point is $V = V^* = 1/\sqrt{2}$.

In conclusion, we have shown that a unique travelling wave solution exists for each $V \geqslant 1/\sqrt{2}$, and that the solution with $V = 1/\sqrt{2}$ asymptotes to zero exponentially fast as $z \to \infty$, whilst this decay is only algebraic for $V > 1/\sqrt{2}$. This has implications for the selection of the speed of the waves generated in an initial value problem for equations (13.8), in particular that localized initial inputs of autocatalyst generate waves with the minimum speed, $V = 1/\sqrt{2}$. The reader is referred to Billingham and Needham (1991) for further details, and to King and Needham (1994) for a similar analysis when the diffusion coefficient is not constant.

Exercises

13.1 Sketch the phase portraits of the systems

(a) $\dot{x} = -x - 2y^2$, $\dot{y} = xy - y^3$,
(b) $\dot{x} = x^2$, $\dot{y} = -y - x^2$,
(c) $\dot{x} = y + x^2$, $\dot{y} = -y - x^2$,

in the neighbourhood of the origin, including in your sketch the unstable and centre manifolds.

13.2 Use arguments based on Lyapunov's theorems to determine the stability of the equilibrium points of

(a) $\dot{x} = x^2 - 2y^2$, $\dot{y} = -4xy$,
(b) $\dot{x} = xy^2 - x$, $\dot{y} = x^2y - y$.

13.3 Consider the second order differential equation $\ddot{\theta} + k\dot{\theta} + f(\theta) = 0$, where f is continuously differentiable on the interval $|\theta| < \alpha$, with $\alpha > 0$ a given constant, $\theta f(\theta) > 0$ for $\theta \neq 0$ and $f(0) = 0$. By considering a suitable Lyapunov function, show that the origin is an asymptotically stable equilibrium point.

13.4 **Euler's equations** for a rigid body spinning freely about a fixed point in the absence of external forces are

$$A\dot{\omega}_1 - (B-C)\omega_2\omega_3 = 0,$$
$$B\dot{\omega}_2 - (C-A)\omega_3\omega_1 = 0,$$
$$C\dot{\omega}_3 - (A-B)\omega_1\omega_2 = 0,$$

where A, B and C are the principal moments of inertia, and $\omega = (\omega_1, \omega_2, \omega_3)$ is the angular velocity of the body relative to its principal axes.

Find all of the steady states of Euler's equations. Using the ideas that we developed in Section 9.4, show that $\omega_1^2 + \omega_2^2 + \omega_3^2$ is a constant of the motion.

Sketch the phase portrait on the surface of the sphere $w_1^2 + w_2^2 + w_3^2 = w_0^2$. Deduce that the steady state with $w_1 = w_0$, $w_2 = w_3 = 0$ is unstable if $C < A < B$ or $B < A < C$, but stable otherwise.

Show that

$$V = [\{B(A-B)w_2^2 + C(A-C)w_3^2\} + Bw_2^2 + Cw_3^2 + A(w_1^2 + 2w_0w_1)]^2,$$

is a Lyapunov function for the case when A is the largest moment of inertia, so that the state $w_1 = w_0$, $w_2 = w_3 = 0$ is stable. Suggest a Lyapunov function that will establish the stability of this state when A is the smallest moment of inertia. Are these states asymptotically stable? Perform a simple experiment to verify your conclusions.

13.5   A particle of mass m lies at $\mathbf{r} = (x, y, z)$ and moves in a potential field $W(x, y, z)$, so that its equation of motion is

$$m\ddot{\mathbf{r}} = -\nabla W.$$

By writing $\dot{x} = u$, $\dot{y} = v$ and $\dot{z} = w$, express this equation of motion in terms of first derivatives only. Suppose that W has a local minimum at $\mathbf{r} = \mathbf{0}$. By using the Lyapunov function

$$V = W + \frac{1}{2}m\left(u^2 + v^2 + w^2\right),$$

show that the origin is a stable point of equilibrium for the particle. What do the level curves of V represent physically? Is the origin asymptotically stable?

If an additional, nonconservative force, $\mathbf{f}(u, v, w)$, also acts on the particle, so that

$$m\ddot{\mathbf{r}} = -\nabla W + \mathbf{f},$$

describe qualitatively how the stability of the point of equilibrium at the origin is affected.

13.6   Consider the three systems

(a) $\dot{y} = y(y-1)(y-2\lambda)$,
(b) $\dot{y} = y^2 + 4\lambda^2 - 1$,
(c) $\dot{y} = -y(4y^2 + \lambda^2 - 1)$.

Sketch the bifurcation diagram for each system, and show that each system has two bifurcation points of the same type, which you should determine. Close to each bifurcation point, write each system in the normal form appropriate to the bifurcation.

13.7   (a) Consider the system

$$\dot{x} = -\epsilon + \mu x - x^2.$$

When $\epsilon = 0$ this is the normal form for a transcritical bifurcation. Sketch the bifurcation diagram when $\epsilon \neq 0$, dealing separately with the cases $\epsilon > 0$ and $\epsilon < 0$. In one case there are two saddle–node bifurcation points, in the other there are no bifurcation points.

(b) Consider the system

$$\dot{x} = \mu x + 2\epsilon x^2 - x^3,$$

with $\epsilon \geqslant 0$. When $\epsilon = 0$ this is the normal form for a supercritical pitchfork bifurcation. Sketch the bifurcation diagram when ϵ is small, but nonzero.

What can you deduce from the answers to the two parts of this question?

13.8 Sketch the phase portraits when $\mu > 0$, $\mu = 0$ and $\mu < 0$ for the normal forms of (a) the transcritical and (b) the supercritical pitchfork bifurcation, given by (13.34) and (13.35).

13.9 Give an informal proof of the transcritical bifurcation theorem and the pitchfork bifurcation theorem.

13.10 Consider the system $\dot{x} = \mu x - \omega y + x^3$, $\dot{y} = \omega x + \mu y + y^3$. Use the Hopf bifurcation theorem to determine the nature of the Hopf bifurcation at $\mu = 0$. Use a near identity transformation to write this system in normal form, and confirm that this is consistent with the Hopf bifurcation theorem. Use MATLAB to solve this system of equations numerically. Plot how the limit cycle solution changes with μ, and determine for what range of values of μ it exists.

13.11 (a) Seek an asymptotic solution of (13.17) subject to the boundary condition (13.15), valid when $V \gg 1$, by rescaling $\gamma = V\bar{\gamma}$ and using an asymptotic expansion

$$\bar{\gamma}(\beta) = \bar{\gamma}_0(\beta) + V^{-2}\bar{\gamma}_1(\beta) + O(V^{-4}).$$

Hence show that $f(V)$, as defined in Section 13.3.3, satisfies

$$f(V) = V + 1 - \frac{1}{6}V^{-1} + O(V^{-3}) \text{ for } V \gg 1.$$

(b) Repeat part (a) when $V \ll 1$. In this case, you will need to seek a rescaling of the form $\beta = \phi(V)\hat{\beta}$, $\gamma = \psi(V)\hat{\gamma}$, and determine ϕ and ψ by seeking an asymptotic balance in (13.17) and (13.15). You should find that, at leading order,

$$\frac{d\hat{\gamma}}{d\hat{\beta}} = -1 - \frac{\hat{\beta}^2}{\hat{\gamma}}, \quad \text{subject to } \hat{\gamma} \sim -\hat{\beta} \text{ as } \hat{\beta} \to 0.$$

Integrate this equation numerically using MATLAB, and hence show that

$$f(V) \sim \beta^* V^2 \text{ as } V \to 0,$$

where β^* is a constant that you should determine numerically.

13.12 **Project** Consider the CSTR system that we studied in Section 13.3.1, but where, in addition, the autocatalyst is itself unstable, and breaks down to form the final product C through the chemical reaction

$$B \to C, \quad \text{rate } k_2 b.$$

(a) Show that the concentrations of A and B now satisfy the equations

$$\frac{d\alpha}{d\tau} = -\alpha\beta^2 + \frac{1-\alpha}{\tau_{res}}, \qquad (E13.1)$$

$$\frac{d\beta}{d\tau} = \alpha\beta^2 + \frac{\beta_0 - \beta}{\tau_{res}} - \frac{\beta}{\tau_2}, \qquad (E13.2)$$

where τ_2 is a dimensionless constant that you should determine.

(b) Using the ideas that we developed in Section 13.3.1, show that the steady state solutions are again given by the points of intersection of a cubic polynomial and a straight line through the origin. *Without making any quantitative calculations*, sketch the position of the steady states in the (τ_{res}, z)-plane, firstly when there is no autocatalyst in the inflow, and secondly when there is.

(c) Now restrict your attention to the case where there is no autocatalyst in the inflow. Determine the range of values for which there are three steady state solutions. Show that the smallest of these steady states is stable and that the middle steady state is a saddle point. Show that the largest steady state loses stability through a Hopf bifurcation, whose location you should determine. Use the Hopf bifurcation theorem to determine when this is supercritical and when it is subcritical.

(d) Use MATLAB to integrate (E13.1) and (E13.2) numerically, and hence investigate what happens to the limit cycle that forms at the Hopf bifurcation point. Draw the complete bifurcation diagram, indicating the location of any limit cycles. What advice would you give to an engineer trying to maximize the output of C from the CSTR? *Hint:* For some parameter ranges, there is more than one limit cycle.

CHAPTER FOURTEEN

Time-Optimal Control in the Phase Plane

Many physical systems that are amenable to mathematical modelling do not exist in isolation from human intervention. A good example is the British economy, for which the Treasury has a complicated mathematical model. The state of the system (the economy) is given by values of the dependent variables (for example, unemployment, foreign exchange rates, growth, consumer spending and inflation), and the government attempts to control the state of the system to a target state (low inflation, high employment, high growth) by varying several control parameters (most notably taxes and government spending). There is also a cost associated with any particular action, which the government tries to minimize (some function of, for example, government borrowing and, one would hope, the environmental cost of any government action or inaction). The optimal control leads to the economy reaching the target state with the smallest possible cost.

Another system, for which we have studied a simple mathematical model, consists of two populations of different species coexisting on an isolated island. For the case of two herbivorous species, which we studied in Chapter 9, we saw that one species or the other will eventually die out. If the island is under human management, this may well be undesirable, and we would like to know how to intervene to maintain the island close to a state of equilibrium, which we know, if left uncontrolled, is unstable. We could choose between either continually culling the more successful species, continually introducing animals of the less successful species or some combination of these two methods of control. Each of these actions has a cost associated with it.

These are examples of optimal control problems. Optimal control is a huge topic, and in this short, introductory chapter we will study just about the simplest type of problem, which involves linear, constant coefficient ordinary differential equations. These are, however, extremely important, as it is often necessary to control small deviations from a steady state, for example when steering a ship, for which linear equations are a good approximation. We will also restrict our attention to time-optimal control, for which the cost function is the time taken for the system to reach the target state. We wish to drive the system to the target state in the shortest possible time.

The crucial results that we will work towards are the properties of the controllability matrix and the application of the time-optimal maximum principle. Although we will give proofs of the various results that we need, the main thing is to know how to apply them.

TIME-OPTIMAL CONTROL IN THE PHASE PLANE

14.1 Definitions

Let's begin by formalizing some of the ideas that we introduced above. Consider the system of ordinary differential equations

$$\dot{\mathbf{x}} = \mathbf{f}(\mathbf{x}, \mathbf{u}, t), \quad \text{subject to } \mathbf{x}(0) = \mathbf{x}^0.$$

Note that we use the superscript ^{0} to indicate the initial state and the superscript ^{1} to indicate the final state. We say that $\mathbf{x}(t)$ is the **state vector**, whose components $\mathbf{x} = (x_1(t), x_2(t), \ldots, x_n(t))$ are the **state variables**, and $\mathbf{u}$ is the **control vector**, whose components $\mathbf{u} = (u_1(t), u_2(t), \ldots, u_m(t))$ are the **control variables**. We assume that $\mathbf{f}$ is continuously differentiable with respect to $\mathbf{x}$, $\mathbf{u}$ and t, but that $\mathbf{u}$ is merely integrable, so that we can allow for discontinuous changes in its components, the control variables. As we have seen, these conditions guarantee the existence and uniqueness of the solution for a given control $\mathbf{u}$.

A **control problem** takes the form of a question: if $\mathbf{x} = \mathbf{x}^0$ when $t = 0$, can we choose the control vector $\mathbf{u}(t)$ so that $\mathbf{x} = \mathbf{x}^1$, the **target state**, when $t = t_1$? In other words, can the system be controlled from $\mathbf{x}^0$ to $\mathbf{x}^1$, reaching $\mathbf{x}^1$ when $t = t_1$? If there is a **cost function**,

$$J = \int_0^{t_1} g_0(\mathbf{x}(t), \mathbf{u}(t), t)\, dt,$$

associated with the control problem, such that we seek controls $\mathbf{u}$ for which J is a minimum, we have an **optimal control problem**. If $g_0 = 1$, and hence $J = t_1$, so that we seek to minimize the time taken to reach the state $\mathbf{x}^1$, we have a **time-optimal control problem**. This is the type of problem that we will be studying. In particular, for first order, linear, constant coefficient equations,

$$\frac{dx}{dt} = Ax + Bu(t),$$

and for second order, linear, constant coefficient equations,

$$\frac{dx_1}{dt} = A_{11}x_1 + A_{12}x_2 + B_{11}u_1(t) + B_{12}u_2(t),$$

$$\frac{dx_2}{dt} = A_{21}x_1 + A_{22}x_2 + B_{21}u_1(t) + B_{22}u_2(t).$$

We can write this more concisely using matrix notation as

$$\frac{d\mathbf{x}}{dt} = A\mathbf{x} + B\mathbf{u},$$

where A and B are constant, 2×2 matrices. Note that there are at most n independent control variables for an nth order linear system of this form.

14.2 First Order Equations

A good way of introducing many of the important, basic concepts of control theory is to study one-dimensional systems, in other words, first order, linear ordinary differential equations with constant coefficients.

Example: The tightrope walker

A tightrope walker is inherently unstable. Walking a tightrope is rather like trying to balance a pencil on its tip. Although the walker has an equilibrium position, it is unstable, and she must push herself back towards the vertical to stay upright. Moreover, if her deviations from the vertical become too large, they cannot be controlled, and she falls off the tightrope. This is typical of many types of problem where the idea is to control small deviations from equilibrium as quickly as possible – linear, time-optimal control.

A *very* simple model for the tightrope walker is

$$\frac{dx}{dt} = x + u, \tag{14.1}$$

where x represents her angular deviation from the vertical and u her attempts to control her balance. Let's assume that $x = x^0 > 0$ when $t = 0$. Can this deviation from the vertical be controlled at all? If so, how quickly can it be controlled, and what should the control be? If there is no attempt to control the deviation, $u = 0$, $x = x^0 e^t$ and the tightrope walker falls off. The upright position, $x = 0$, is an unstable equilibrium state. For a general control $u(t)$, we can solve (14.1) using the integrating factor e^{-t}, so that

$$\frac{d}{dt}(e^{-t}x) = e^{-t}u(t),$$

and hence

$$x = x^0 e^t + e^t \int_0^t e^{-\tau}u(\tau)\,d\tau \tag{14.2}$$

Since we want to control the system to $x = 0$ when $t = t_1$, we must have

$$x^0 + \int_0^{t_1} e^{-\tau}u(\tau)\,d\tau = 0. \tag{14.3}$$

Clearly, since $x^0 > 0$, u must be negative for at least some of the period $0 \leq t \leq t_1$. Now, if $u(t)$ is not bounded below (the tightrope walker can apply an arbitrarily large restoring force), it is easy to choose u to satisfy (14.3), for example with $u(t) = -e^t x^0/t_1$. This is unrealistic, since we can control the system in an arbitrarily short time t_1 by making $|u|$ sufficiently large. *The time-optimal control problem is meaningless if the control is unbounded.* From now on we restrict our attention to **bounded controls** with $-1 \leq u(t) \leq 1$. We will see later how to scale a slightly more realistic problem so that u lies in this convenient range. In general, the components of a bounded control vector satisfy $-1 \leq u_i(t) \leq 1$.

Intuitively, we can see that to push x back towards equilibrium as quickly as possible, we need to push in the appropriate direction as hard as we can by taking $u(t) = -1$, so that (14.3) gives

$$x^0 - \int_0^{t_1} e^{-\tau}\,d\tau = 0,$$

and hence

$$t_1 = -\log(1 - x^0), \tag{14.4}$$

with solution
$$x(t) = 1 + (x^0 - 1)e^t. \qquad (14.5)$$

For $t > t_1$ we take $u = 0$, and the system remains in its equilibrium state, $x = 0$. Equation (14.4) shows that $t_1 \to \infty$ as $x^0 \to 1^-$. If $x^0 \geqslant 1$, the system cannot be controlled back to equilibrium. The tightrope walker cannot push hard enough, and she falls off. Similarly, if $x^0 < 0$, the time-optimal control is $u(t) = 1$, pushing in the other direction. Some time-optimal solutions are shown in Figure 14.1.

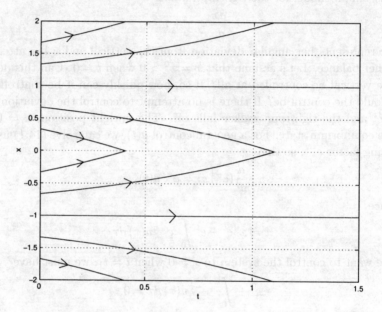

Fig. 14.1. Some time-optimal solutions, (14.5), of the tightrope walker problem.

We can now make a couple of definitions. The **controllable set at time** t_1 is the set of initial states that can be controlled to the origin in time t_1. For the tightrope walker problem, this is the set

$$\mathcal{C}(t_1) = \left\{ x \mid |x| \leqslant 1 - e^{-t_1} \right\}.$$

The **controllable set** is the set of initial states that can be controlled to the origin in some finite time. This is just the union of all the controllable sets at time $t_1 \geqslant 0$. For the tightrope walker problem, the controllable set is

$$\mathcal{C} = \bigcup_{t_1 \geqslant 0} \mathcal{C}(t_1) = \{ x \mid |x| < 1 \}.$$

If $\mathcal{C}$ is the whole real line, $\mathcal{C} = \mathbb{R}$, we say that the system is **completely controllable**. The tightrope walker system is not completely controllable, since $\mathcal{C} \neq \mathbb{R}$ (see Exercise 14.1). This is a good point at which to define the **reachable set**. If

14.2 FIRST ORDER EQUATIONS

$x = x^0$ when $t = 0$, the reachable set in time t_1, $R(t_1, x^0)$, is the set of points x^1 for which there exists a control $u(t)$ such that $x(t_1) = x^1$. From (14.2),

$$x^1 = x^0 e^{t_1} + e^{t_1} \int_0^{t_1} e^{-\tau} u(\tau) \, d\tau,$$

and hence, since $|u(t)| \leqslant 1$,

$$|x^1 e^{-t_1} - x^0| = \left| \int_0^{t_1} e^{-\tau} u(\tau) \, d\tau \right| \leqslant \int_0^{t_1} e^{-\tau} |u(\tau)| \, d\tau \leqslant \int_0^{t_1} e^{-\tau} \, d\tau = 1 - e^{-t_1},$$

so that

$$(x^0 - 1)e^{t_1} + 1 \leqslant x^1 \leqslant (x^0 + 1)e^{t_1} - 1, \qquad (14.6)$$

defines the points in the reachable set, $R(t_1, x^0)$. For $|x^0| \geqslant 1$, the reachable set does not contain the origin, whilst for $|x^0| < 1$ the origin is reachable for $t_1 \geqslant -\log(1 - |x^0|)$, consistent with what we know about the controllable set. The reachable set is shown for two different cases in Figure 14.2. Note that the boundaries of the reachable set are given by the solutions with the controls $u(t) = \pm 1$, a fact that will prove to be important later.

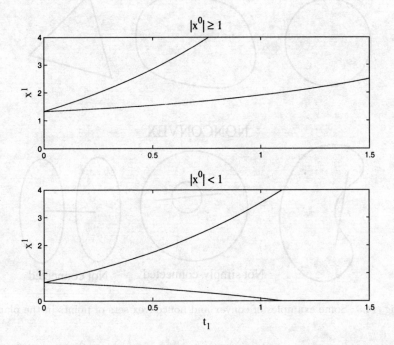

Fig. 14.2. The reachable set for the tightrope walker problem lies between the curved lines.

14.3 Second Order Equations

For second order, linear, constant coefficient systems, we can examine the behaviour of the state variables, x_1 and x_2, in the (x_1, x_2)-phase plane (see Chapter 9). We will show that our intuitive notion that only the extreme values of the bounded control are used in the time-optimal control is correct. This is known as **bang-bang control**. Before we can do this, there are a few mathematical ideas that we must consider. All of the following can be generalized to systems of higher order, and some of it to nonlinear, nonautonomous systems.

14.3.1 Properties of sets of points in the plane

— A **convex set**, S, is one for which the line segment between any two points in S lies entirely within S (see Figure 14.3 for some examples). Note that necessary but not sufficient conditions are that S must be both connected (any two points in S can be joined by a curve lying within S) and simply-connected (S has no holes†). Formally, if S is a convex set and $\mathbf{x}, \mathbf{y} \in S$, $c\mathbf{x} + (1-c)\mathbf{y} \in S$ for all c such that $0 \leqslant c \leqslant 1$.

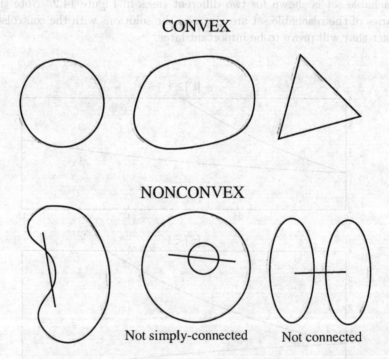

Fig. 14.3. Some examples of convex and nonconvex sets of points in the plane.

---

† More formally, S is connected and any closed loop lying in S can be shrunk continuously to a point without leaving S.

14.3 SECOND ORDER EQUATIONS

— An **interior point** of S is a point $\mathbf{x} \in S$ for which there exists a disc of points centred on $\mathbf{x}$, all of which lie in S (see Figure 14.4).
— The **interior**, $\text{Int}(S)$, of a set S is the set of all the interior points of S.
— An **exterior point** of S is a point in the interior of S^C, the complement of S.
— The **exterior**, $\text{Ext}(S)$, of a set S is the set of all the exterior points of S.
— A **boundary point** of S is a point, not necessarily in S, that lies in neither the interior nor the exterior of S (see Figure 14.4). Note that all discs centred on a boundary point of S contain a point that is not in S.
— The **boundary** of a set S is the set of all boundary points of S, and can therefore be written as

$$\partial S = \{\mathbf{x} \mid (\mathbf{x} \notin \text{Int}(S))\} \cap \{\mathbf{x} \mid (\mathbf{x} \notin \text{Int}(S^C))\}.$$

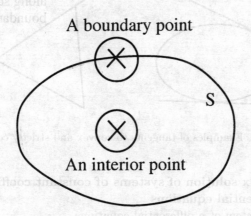

Fig. 14.4. Examples of interior and boundary points of S, and discs centred on them.

— If S does not contain any of its boundary points, it is said to be **open**. For example, the set of points with $|\mathbf{x}| < 1$ (the open unit disc) is open. Note that the boundary of the open unit disc is the unit circle, $|\mathbf{x}| = 1$. An open set has $S = \text{Int}(S)$.
— A **closed set** contains all of its boundary points. For example, the set of points with $|\mathbf{x}| \leqslant 1$ (the closed unit disc) is closed.
— A set S is **strictly convex** if, for each pair of points in S, the line segment joining them is entirely made up of interior points (see Figure 14.5). For a strictly convex set, the tangent to any boundary point does not meet S at any other boundary point. If S is convex, but not strictly convex, some of the tangents to the set will meet the boundary at more than one point along a straight part of the boundary (see Figure 14.6).

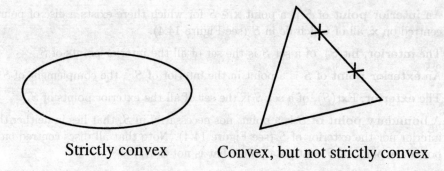

Fig. 14.5. Examples of strictly convex and convex, but not strictly convex sets.

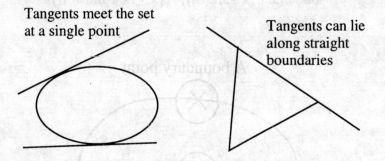

Fig. 14.6. Examples of tangents to convex and strictly convex sets.

14.3.2 Matrix solution of systems of constant coefficient ordinary differential equations

Consider the system of n differential equations

$$\frac{d\mathbf{x}}{dt} = A\mathbf{x} + \mathbf{b}(t), \quad \text{subject to } \mathbf{x}(0) = \mathbf{x}^0, \tag{14.7}$$

with A an $n \times n$ matrix of constants. In order to be able to write the solution of this equation in a compact form, we define the **matrix exponential** of At to be

$$\exp(At) = \sum_{k=0}^{\infty} \frac{A^k t^k}{k!}, \tag{14.8}$$

with $A^0 = I$, the unit matrix. Note that this power series is convergent for all A and t. We can see immediately that

$$\frac{d}{dt}\exp(At) = \sum_{k=1}^{\infty} \frac{A^k t^{k-1}}{(k-1)!} = A\exp(At),$$

a property that the matrix exponential shares with its scalar counterpart, e^{at}.

Now consider the product
$$P(t) = \exp(At)\exp(-At).$$
Firstly, we note that $P(0) = I$. Secondly, using the product rule,
$$\frac{dP}{dt} = A\exp(At)\exp(-At) - \exp(At)A\exp(-At) = 0,$$
using the fact that, from the definition, (14.8),
$$A\exp(At) = \exp(At)A.$$
Similarly, $d^n P/dt^n = 0$, and hence from its Taylor series expansion, $P(t) = I$. In other words,
$$\{\exp(At)\}^{-1} = \exp(-At),$$
again, in line with the result $e^{at}e^{-at} = 1$ for the scalar exponential function. Note that, in general, $\exp(A)\exp(B) \neq \exp(B)\exp(A)$ unless $AB = BA$ (see Exercise 14.4).

These results mean that when $\mathbf{b}(t) = \mathbf{0}$, so that (14.7) is homogeneous, the solution can be written as
$$\mathbf{x} = \exp(At)\mathbf{x}^0.$$
When the equation is inhomogeneous, we can use a matrix integrating factor to write
$$\exp(-At)\frac{d\mathbf{x}}{dt} - \exp(-At)A\mathbf{x} = \frac{d}{dt}\{\exp(-At)\mathbf{x}\} = \exp(-At)\mathbf{b},$$
and hence,
$$\mathbf{x} = \exp(At)\left\{\mathbf{x}^0 + \int_0^t \exp(-A\tau)\mathbf{b}(\tau)\,d\tau\right\}. \tag{14.9}$$
This is the generalization to an nth order system of the first order solution, an example of which is given by (14.2).

Finally, note that we know from the Cayley–Hamilton theorem (Theorem A1.1) which states that every matrix satisfies its own characteristic equation, that, for an $n \times n$ matrix A and $k \geqslant n$, A^k can be written as a linear combination of $I, A, A^2, \ldots, A^{n-1}$, and hence so can $\exp(At)$. In particular, for a 2×2 matrix, $\exp(At)$ is a linear combination of I and A.

Example: Simple harmonic motion

Simple harmonic motion with angular frequency ω is governed by
$$\frac{d^2x}{dt^2} + \omega^2 x = 0.$$
We can write this as a system of two first order equations in the usual way as
$$\dot{x}_1 = x_2, \quad \dot{x}_2 = -\omega^2 x_1,$$

with $x = x_1$. In terms of the generic equation (14.7),

$$A = \begin{pmatrix} 0 & 1 \\ -\omega^2 & 0 \end{pmatrix}.$$

We now note that

$$A^2 = \begin{pmatrix} -\omega^2 & 0 \\ 0 & -\omega^2 \end{pmatrix} = -\omega^2 I.$$

This means that

$$A^{2n} = \omega^{2n}(-1)^n I, \quad A^{2n+1} = \omega^{2n}(-1)^n A,$$

and hence

$$\exp(At) = \sum_{k=0}^{\infty} \frac{A^k t^k}{k!} = \sum_{n=0}^{\infty} \frac{A^{2n} t^{2n}}{(2n)!} + \sum_{n=0}^{\infty} \frac{A^{2n+1} t^{2n+1}}{(2n+1)!}$$

$$= I \sum_{n=0}^{\infty} \frac{(-1)^n \omega^{2n} t^{2n}}{(2n)!} + A \sum_{n=0}^{\infty} \frac{(-1)^n \omega^{2n} t^{2n+1}}{(2n+1)!}$$

$$= \cos \omega t \, I + \frac{1}{\omega} \sin \omega t \, A = \begin{pmatrix} \cos \omega t & \frac{1}{\omega} \sin \omega t \\ -\omega \sin \omega t & \cos \omega t \end{pmatrix}.$$

With the initial condition $x_1(0) = x_1^0$, $x_2(0) = x_2^0$, this means that the solution is

$$\mathbf{x} = \mathbf{x}^0 \exp(At) = \begin{pmatrix} x_1^0 \cos \omega t + \frac{x_2^0}{\omega} \sin \omega t \\ -\omega x_1^0 \sin \omega t + x_2^0 \cos \omega t \end{pmatrix}.$$

14.4 Examples of Second Order Control Problems

Example 1: The positioning problem

Consider the one-dimensional problem of positioning an object of mass m in a frictionless groove using a bounded applied force $F(t)$ such that $-F_{\max} \leq F(t) \leq F_{\max}$. Newton's second law gives

$$F(t) = m \frac{d^2 x}{dt^2}, \quad \text{subject to } x(0) = X, \, \dot{x}(0) = 0, \, x(t_1) = 0, \, \dot{x}(t_1) = 0,$$

minimizing t_1. If we let

$$x_1 = \frac{F_{\max}}{m} x, \quad x_2 = \frac{F_{\max}}{m} \dot{x}, \quad u = \frac{F(t)}{F_{\max}}, \quad x_1^0 = \frac{F_{\max}}{m} X,$$

we have a problem in the standard form, with $|u| \leq 1$,

$$\frac{dx_1}{dt} = x_2, \quad \frac{dx_2}{dt} = u, \quad x_1(0) = x_1^0, \, x_2(0) = 0. \tag{14.10}$$

14.4 EXAMPLES OF SECOND ORDER CONTROL PROBLEMS

In terms of our matrix notation[†]

$$A = \begin{pmatrix} 0 & 1 \\ 0 & 0 \end{pmatrix}, \quad B = \begin{pmatrix} 0 \\ 1 \end{pmatrix}.$$

Let's now think what the optimal control might be. We want to push the particle to the origin as quickly as possible, but it must be at rest when we get it there. Presumably we should push as hard as we can for some period of time, $0 \leqslant t \leqslant T$, then decelerate as strongly as we can by pushing as hard as possible in the opposite direction for $T \leqslant t \leqslant t_1$, so that the particle is at rest at the origin when $t = t_1$. We will prove later that this is indeed the optimal method of control, but for now let's just construct the solution.

Assuming that $x_1^0 > 0$, we take $u = -1$ for $0 \leqslant t \leqslant T$. We can easily integrate (14.10) and find that

$$x_1 = x_1^0 - \frac{1}{2}t^2, \quad x_2 = -t \quad \text{for } 0 \leqslant t \leqslant T. \tag{14.11}$$

Then we take $u = 1$ for $T \leqslant t \leqslant t_1$, and use $x_1 = x_1^0 - \frac{1}{2}T^2$, $x_2 = -T$ when $t = T$ to fix the constants of integration, which gives us

$$x_1 = \frac{1}{2}t^2 - 2Tt + T^2 + x_1^0, \quad x_2 = t - 2T \quad \text{for } T \leqslant t \leqslant t_1. \tag{14.12}$$

Now we just need to find T, the time at which we have to switch from accelerating as hard as possible to decelerating as hard as possible, and t_1, the total time taken to control the particle to the origin. We can obtain this by using $x_1 = x_2 = 0$ when $t = t_1$. This gives $T = \frac{1}{2}t_1$, so that the periods of acceleration and deceleration are equal, and $t_1 = 2\sqrt{x_1^0}$. The full solution is

$$x_1 = \begin{cases} x_1^0 - \frac{1}{2}t^2 & \text{for } 0 \leqslant t \leqslant \sqrt{x_1^0}, \\ \frac{1}{2}t^2 - 2\sqrt{x_1^0}\,t + 2x_1^0 & \text{for } \sqrt{x_1^0} \leqslant t \leqslant 2\sqrt{x_1^0}, \end{cases} \tag{14.13}$$

$$x_2 = \begin{cases} -t & \text{for } 0 \leqslant t \leqslant \sqrt{x_1^0}, \\ t - 2\sqrt{x_1^0} & \text{for } \sqrt{x_1^0} \leqslant t \leqslant 2\sqrt{x_1^0}. \end{cases} \tag{14.14}$$

A typical solution is plotted as a function of t, and various solutions plotted in the (x_1, x_2)-phase plane in Figure 14.7. We will return later to the problem of determining the time-optimal control if the particle is not initially stationary, so that $x_2(0) \neq 0$.

Example 2: The steering problem / The positioning problem with friction

The forced motion of a ship is unstable, and tends to drift off course if it is not controlled. If x_1 represents the deviation of the ship from a straight path, we can

---

[†] Note that, strictly speaking, the matrix B should have another column of zeros, since there is no dependence on a second control function in this problem. We have omitted this for clarity.

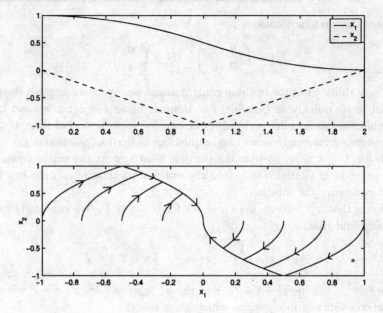

Fig. 14.7. Some time-optimal solutions of the positioning problem when the particle is initially stationary.

write a simple model as

$$\frac{dx_1}{dt} = x_2, \quad \frac{dx_2}{dt} = -qx_2 + u. \tag{14.15}$$

The drag force caused by the resistance of the water to the lateral motion of the ship is represented by the term $-qx_2$, with q a positive constant. This is exactly the same as the positioning problem, but in a groove with friction. In terms of our matrix notation,

$$A = \begin{pmatrix} 0 & 1 \\ 0 & -q \end{pmatrix}, \quad B = \begin{pmatrix} 0 \\ 1 \end{pmatrix}.$$

Example 3: Controlling a linear oscillator

The equation for a linear oscillator (simple harmonic motion) of unit angular frequency, subject to an external force, $u(t)$, can be written as

$$\frac{dx_1}{dt} = x_2, \quad \frac{dx_2}{dt} = -x_1 + u. \tag{14.16}$$

In terms of our matrix notation,

$$A = \begin{pmatrix} 0 & 1 \\ -1 & 0 \end{pmatrix}, \quad B = \begin{pmatrix} 0 \\ 1 \end{pmatrix}.$$

If the oscillator is not initially at rest and we wish to bring it to a halt as quickly as possible, we must solve (14.16) subject to

$$x_1(0) = x_1^0, \quad x_2(0) = x_2^0, \quad x_1(t_1) = x_2(t_1) = 0, \tag{14.17}$$

minimizing t_1. We can think of this as the problem of stopping a child on a swing as quickly as possible.

Example 4: The positioning problem with two controls

Suppose that in the positioning problem we are also able to change the velocity of the particle through an extra control, so that

$$\frac{dx_1}{dt} = x_2 + u_1, \quad \frac{dx_2}{dt} = u_2. \tag{14.18}$$

In terms of our matrix notation,

$$A = \begin{pmatrix} 0 & 1 \\ 0 & 0 \end{pmatrix}, \quad B = \begin{pmatrix} 1 & 0 \\ 0 & 1 \end{pmatrix}.$$

With the exception of the steering problem (see Exercise 14.9), we will solve all of these problems below after we have studied some more of the theory relevant to this type of control problem.

14.5 Properties of the Controllable Set

Apart from determining the time-optimal control, we often want to construct $\mathcal{C}(t_1)$, the controllable set for any time t_1. There are a number of statements that we can make about the geometry of these sets, and of $\mathcal{C}$, the controllable set.

— If $t_1 < t_2$, $\mathcal{C}(t_1) \subset \mathcal{C}(t_2)$. In other words, the controllable set never gets smaller as t increases. Any state controllable to zero in time t_1 is also controllable to zero in time $t_2 > t_1$.
— $\mathcal{C}(t)$ and $\mathcal{C}$ are connected sets.
— $\mathcal{C}$ is open if and only if $\mathbf{0} \in \text{Int}(\mathcal{C})$.

These results hold for nonlinear, nonautonomous systems of ordinary differential equations. For linear, constant coefficient equations, we also have that $\mathcal{C}(t)$ and $\mathcal{C}$ are symmetric about the origin and convex. Note that this also implies that these sets are simply-connected, since all convex sets are simply-connected.

Theorem 14.1 *If $t_1 < t_2$, $\mathcal{C}(t_1) \subset \mathcal{C}(t_2)$.*

Proof Let $\mathbf{x}^0$ be a point in $\mathcal{C}(t_1)$, with control $\mathbf{u} = \mathbf{v}(t)$. If we apply the control

$$\mathbf{u} = \begin{cases} \mathbf{v}(t) & \text{for } 0 \leqslant t \leqslant t_1, \\ \mathbf{0} & \text{for } t_1 < t \leqslant t_2, \end{cases}$$

the trajectory reaches $\mathbf{x} = \mathbf{0}$ when $t = t_1$, then remains there for $t_1 \leqslant t \leqslant t_2$, since $\mathbf{x} = \mathbf{0}$, $\mathbf{u} = \mathbf{0}$ is an equilibrium state. Therefore $\mathbf{x}^0 \in \mathcal{C}(t_2)$, which means that $\mathcal{C}(t_1) \subset \mathcal{C}(t_2)$ (see Figure 14.8). □

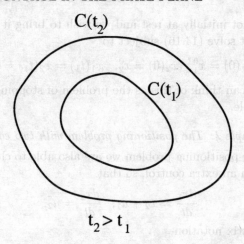

Fig. 14.8. If $t_1 < t_2$, $\mathcal{C}(t_1) \subset \mathcal{C}(t_2)$.

Theorem 14.2 *If $\mathbf{x}^0 \in \mathcal{C}(t_1)$ and $\mathbf{y}$ is a point on the trajectory from $\mathbf{x}^0$ to $\mathbf{0}$, $\mathbf{y} \in \mathcal{C}(t_1)$. In other words, all points on controllable trajectories are controllable.*

Proof Let $\mathbf{x} = \mathbf{X}(t)$ be the trajectory containing both $\mathbf{x}^0$ and $\mathbf{y}$, with control $\mathbf{u}(t)$. When $t = \tau_1$, $\mathbf{X}(\tau_1) = \mathbf{y}$, and when $t = \tau_2$, $\mathbf{X}(\tau_2) = \mathbf{0}$, with $\tau_1 \leqslant \tau_2 \leqslant t_1$ (see Figure 14.9). If we now consider the solution with control $\mathbf{v}(t) = \mathbf{u}(t + \tau_1)$ and

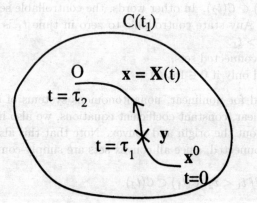

Fig. 14.9. All points on controllable trajectories are controllable.

initial condition $\mathbf{x}(0) = \mathbf{y}$, the system follows the same trajectory, $\mathbf{x} = \mathbf{X}(t + \tau_1)$, and reaches $\mathbf{x} = \mathbf{0}$ when $t = \tau_2 - \tau_1$. Therefore $\mathbf{y} \in \mathcal{C}(\tau_2 - \tau_1)$ and, by Theorem 14.1, $\mathbf{y} \in \mathcal{C}(t_1)$. □

14.5 PROPERTIES OF THE CONTROLLABLE SET

Theorem 14.3 $\mathcal{C}(t_1)$ and $\mathcal{C}$ are connected sets.

Proof If $\mathbf{x}^0 \in \mathcal{C}(t_1)$ and $\mathbf{y}^0 \in \mathcal{C}(t_1)$, there are, by definition, trajectories that connect each point to the origin (see Figure 14.10). By Theorem 14.2, all points on

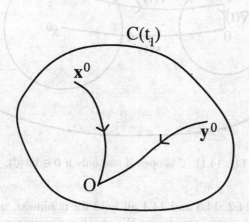

Fig. 14.10. $\mathcal{C}(t_1)$ and $\mathcal{C}$ are connected sets.

each of these trajectories lie in $\mathcal{C}(t_1)$. Therefore the union of these two trajectories is a curve made up of points in $\mathcal{C}(t_1)$ that connects $\mathbf{x}^0$ and $\mathbf{y}^0$, and hence, by definition, $\mathcal{C}(t_1)$ is connected. Since $\mathcal{C} = \bigcup_{t_1 \geqslant 0} \mathcal{C}(t_1)$, $\mathcal{C}$ is also connected. $\square$

Theorem 14.4 $\mathcal{C}$ is open if and only if $\mathbf{0} \in Int(\mathcal{C})$.

Proof If $\mathcal{C}$ is open, all of its points are interior points, so clearly $\mathbf{0} \in \text{Int}(\mathcal{C})$. It is less straightforward to prove that $\mathbf{0} \in \text{Int}(\mathcal{C})$ implies that $\mathcal{C}$ is open.

If $\mathbf{0} \in \text{Int}(\mathcal{C})$, by definition there is a disc of radius r centred on $\mathbf{0}$, which we write as $D(\mathbf{0}, r)$, that lies entirely within $\mathcal{C}$. Now suppose that $\mathbf{u} = \mathbf{v}(t)$ is a control that steers some point $\mathbf{x}^0$ to $\mathbf{0}$ in time t_1. Let $D(\mathbf{x}^0, r^0)$ be a disc of radius r^0 centred on $\mathbf{x}^0$, and let $\mathbf{y}^0$ be another point within this disc, as shown in Figure 14.11. By continuity of the solutions of the underlying differential equations, if r^0 is sufficiently small, the control $\mathbf{v}(t)$ steers $\mathbf{y}^0$ into the disc $D(\mathbf{0}, r)$ on a path $\mathbf{y}(t)$ with $\mathbf{y}(t_1) \in D(\mathbf{0}, r)$ at some time t_1. Since $D(\mathbf{0}, r) \in \mathcal{C}$, we can also find a control $\hat{\mathbf{v}}(t)$ that steers $\mathbf{y}(t_1)$ to $\mathbf{0}$ in some time t_2. Therefore $\mathbf{y}^0$ can be controlled to the origin in time $t_1 + t_2$ using the control

$$\mathbf{u}(t) = \begin{cases} \mathbf{v}(t) & \text{for } 0 \leqslant t \leqslant t_1, \\ \hat{\mathbf{v}}(t) & \text{for } t_1 < t \leqslant t_2. \end{cases}$$

Therefore $\mathbf{y}^0 \in \mathcal{C}(t_1 + t_2) \subset \mathcal{C}$, and hence, for r^0 sufficiently small, $D(\mathbf{x}^0, r^0) \in \mathcal{C}$ for all $\mathbf{x}^0 \in \mathcal{C}$. By definition, $\mathcal{C}$ is therefore open. $\square$

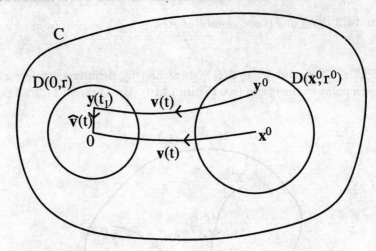

Fig. 14.11. $\mathcal{C}$ is open if and only if $\mathbf{0} \in \text{Int}(\mathcal{C})$.

Theorems 14.1, 14.2, 14.3 and 14.4 all hold for nonlinear, nonautonomous systems of ordinary differential equations. We now focus on linear, constant coefficient equations. Recall that if

$$\frac{d\mathbf{x}}{dt} = A\mathbf{x} + B\mathbf{u},$$

with A and B constant matrices, the solution is

$$\mathbf{x} = \exp(At)\left\{\mathbf{x}^0 + \int_0^t \exp(-A\tau)B\mathbf{u}(\tau)\, d\tau\right\}.$$

This means that $\mathbf{x}^0 \in \mathcal{C}(t_1)$ if and only if there is a control $\mathbf{u}(t)$ such that

$$\mathbf{x}^0 = -\int_0^{t_1} \exp(-A\tau)B\mathbf{u}(\tau)\, d\tau. \tag{14.19}$$

Theorem 14.5 $\mathcal{C}(t_1)$ *is symmetric about the origin and convex.*

Proof If $\mathbf{x}^0 \in \mathcal{C}(t_1)$ with control $\mathbf{u}(t)$, (14.19) shows that $-\mathbf{x}^0 \in \mathcal{C}(t_1)$ with control $-\mathbf{u}(t)$.† Therefore $\mathcal{C}(t_1)$ is symmetric about the origin.

Now note that the set of bounded controls

$$U = \{\mathbf{u}(t) \mid -1 \leqslant u_i(t) \leqslant 1\}$$

is convex, since if $\mathbf{u}^0 \in U$ and $\mathbf{u}^1 \in U$, $c\mathbf{u}^0 + (1-c)\mathbf{u}^1 \in U$ for $0 \leqslant c \leqslant 1$. Therefore, if $\mathbf{u}^0$ and $\mathbf{u}^1$ are controls that steer $\mathbf{x}^0$ and $\mathbf{x}^1$ to the origin in time t_1,

$$c\mathbf{x}^0 + (1-c)\mathbf{x}^1 = -\int_0^{t_1} \exp(-A\tau)B\left\{c\mathbf{u}^0(\tau) + (1-c)\mathbf{u}^1(\tau)\right\} d\tau,$$

† Note that the fact that the control variables are scaled so that $-1 \leqslant u_i(t) \leqslant 1$ is crucial here.

14.6 THE CONTROLLABILITY MATRIX

and hence the control $c\mathbf{u}^0(\tau) + (1-c)\mathbf{u}^1(\tau) \in U$ steers $c\mathbf{x}^0 + (1-c)\mathbf{x}^1$ to the origin in time t_1. The line segment that joins $\mathbf{x}^0$ to $\mathbf{x}^1$ therefore lies entirely within $\mathcal{C}(t_1)$, and hence, by definition, $\mathcal{C}(t_1)$ is convex. □

Theorem 14.6 $\mathcal{C}$ *is symmetric about the origin and convex.*

Proof A union of symmetric sets is symmetric, so $\mathcal{C} = \bigcup_{t_1 \geqslant 0} \mathcal{C}(t_1)$ is symmetric. Although a union of convex sets is not necessarily convex (see Figure 14.12), since Theorem 14.3 tells us that $\mathcal{C}(t_1) \subset \mathcal{C}(t_2)$ for $t_1 < t_2$, $\mathcal{C}$ is a union of nested convex sets and therefore is itself convex. □

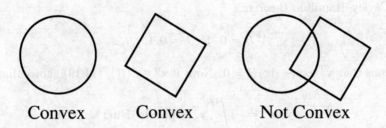

Fig. 14.12. A union of convex sets is not necessarily convex.

14.6 The Controllability Matrix

For a second order system, we define the **controllability matrix** to be

$$M = [B \ AB]. \qquad (14.20)$$

We will show that the system is completely controllable ($\mathcal{C} = \mathbb{R}^2$) if and only if

(i) $\text{rank}(M) = 2$,
(ii) all the eigenvalues of A have zero or negative real part.

Although we will not do so here, it is straightforward to generalize this result to n^{th} order systems.

Example: The positioning problem

For the positioning problem

$$A = \begin{pmatrix} 0 & 1 \\ 0 & 0 \end{pmatrix}, \quad B = \begin{pmatrix} 0 \\ 1 \end{pmatrix},$$

and hence

$$M = \begin{pmatrix} 0 & 1 \\ 1 & 0 \end{pmatrix}.$$

Clearly, $\text{rank}(M) = 2$, since its columns are linearly independent. Also, since

$$\det(A - \lambda I) = \lambda^2,$$

the matrix A has the repeated eigenvalue zero. We conclude that the system is completely controllable. All of the other examples that we described in Section 14.4 are also completely controllable (see Exercise 14.6).

Let's now prove that these properties of the controllability matrix do determine whether or not a second order, linear, constant coefficient system is controllable.

Theorem 14.7 $\mathbf{0} \in \text{Int}(\mathcal{C})$ *if and only if* $\text{rank}(M) = 2$.

Proof Suppose that $\text{rank}(M) < 2$. If $\text{rank}(M) = 1$, there is a single direction, $\mathbf{y}$, orthogonal to every column of M. This means that $\mathbf{y}^T B = \mathbf{y}^T AB = \mathbf{0}$, and hence by the Cayley–Hamilton theorem,

$$\mathbf{y}^T A^k B = \mathbf{0} \text{ for } k = 0, 1, 2, \ldots.$$

This means that $\mathbf{y}^T \exp(-A\tau) B = \mathbf{0}$. Now, if $\mathbf{x}^0 \in \mathcal{C}(t_1)$, (14.19) shows that

$$\mathbf{y}^T \mathbf{x}^0 = \mathbf{y} \cdot \mathbf{x}^0 = -\int_0^{t_1} \mathbf{y}^T \exp(-A\tau) B \mathbf{u}(\tau) d\tau = 0.$$

Therefore if $\text{rank}(M) = 1$, $\mathbf{x}^0$ lies on the straight line through the origin perpendicular to $\mathbf{y}$, which is a closed set, and hence $\mathbf{0} \notin \text{Int}(\mathcal{C})$. If $\text{rank}(M) = 0$, M has only zero entries, and hence so does B. There are therefore no controls, $\mathcal{C} = \{\mathbf{0}\}$, and hence $\mathbf{0} \notin \text{Int}(\mathcal{C})$.

Now suppose that $\mathbf{0} \notin \text{Int}(\mathcal{C})$. Since $\mathcal{C}(t_1) \subset \mathcal{C}$, $\mathbf{0} \notin \text{Int}(\mathcal{C}(t_1))$ at any time t_1. Since $\mathbf{0} \in \mathcal{C}(t_1)$, the origin must be a boundary point of $\mathcal{C}(t_1)$. Since $\mathcal{C}(t_1)$ is convex (Theorem 14.5), there is a tangent to $\mathcal{C}(t_1)$ through $\mathbf{0}$ with outward normal $\mathbf{z}$, and for all $\mathbf{x}^0 \in \mathcal{C}(t_1)$, $\mathbf{z}^T \mathbf{x}^0 = \mathbf{z} \cdot \mathbf{x}^0 \leqslant 0$ (see Figure 14.13). Equation 14.19 then shows

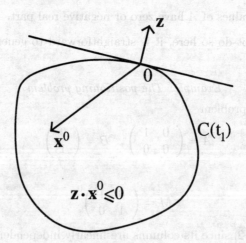

Fig. 14.13. The tangent and outward normal to the set $\mathcal{C}(t_1)$ at the origin.

that
$$\int_0^{t_1} \mathbf{z}^T \exp(-A\tau) B\mathbf{u}(\tau)\, d\tau \geqslant 0$$
for all controls $\mathbf{u}$. However, since $-\mathbf{u}$ is also an admissible control,
$$-\int_0^{t_1} \mathbf{z}^T \exp(-A\tau) B\mathbf{u}(\tau)\, d\tau \geqslant 0,$$
and hence
$$\int_0^{t_1} \mathbf{z}^T \exp(-A\tau) B\mathbf{u}(\tau)\, d\tau = 0 \text{ for } 0 \leqslant t \leqslant t_1.$$
Since this must hold for all controls $\mathbf{u}$,
$$\mathbf{z}^T \exp(-At) B = \mathbf{0} \text{ for } 0 \leqslant t \leqslant t_1.$$
If we now set $t = 0$ we get $\mathbf{z}^T B = \mathbf{0}$, and if we differentiate and put $t = 0$, $\mathbf{z}^T AB = \mathbf{0}$. This means that $\mathbf{z}$ is orthogonal to all of the columns of M, and hence rank$(M) < 2$. □

Since the system can only be completely controllable if $\mathbf{0} \in \text{Int}(\mathcal{C})$, this theorem shows that a necessary condition for the system to be completely controllable is that rank$(M) = 2$.

Theorem 14.8 *If* rank$(M) = 2$ *and* $\text{Re}(\lambda_i) \leqslant 0$ *for each eigenvalue* λ_i *of A, the system is completely controllable.*

Proof We proceed by contradiction. Suppose that rank$(M) = 2$ and A has eigenvalues with zero or negative real parts, but that $\mathcal{C} \neq \mathbb{R}^2$. Consider a point $\mathbf{y} \notin \mathcal{C}$. There is then a tangent to $\mathcal{C}$, with equation $\mathbf{n} \cdot \mathbf{x} = p$, that separates $\mathbf{y}$ from each $\mathbf{x}^0 \in \mathcal{C}$, with $\mathbf{n} \cdot \mathbf{x}^0 \leqslant p$ and $\mathbf{n} \cdot \mathbf{y} \geqslant p$ (see Figure 14.14). Let $\mathbf{z} = \mathbf{n}\{\exp(-At)B\}$. Because rank$(M) = 2$, $\mathbf{z} \neq \mathbf{0}$ for $0 \leqslant t \leqslant t_1$. Now choose a control with components $u_i(t) = -\text{sgn}(z_i(t))$, so that
$$\mathbf{n} \cdot \mathbf{x}^0 = -\int_0^{t_1} \mathbf{n}^T \exp(-A\tau) B\mathbf{u}(\tau)\, d\tau$$
$$= -\int_0^{t_1} \mathbf{z}(\tau)\mathbf{u}(\tau)\, d\tau = \int_0^{t_1} (|z_1| + |z_2|)\, d\tau.$$
By choosing an appropriate coordinate system, we can make each component of $\mathbf{z}$ a sum of terms proportional to $e^{-\lambda_i \tau}$. If any eigenvalue has zero part, the corresponding component of $\mathbf{z}$ will be either a polynomial or a periodic function of t. In each case, $\int_0^{t_1}(|z_1| + |z_2|)\, d\tau \to \infty$ as $t_1 \to \infty$. In particular, for t_1 sufficiently large, $\mathbf{n} \cdot \mathbf{x}^0 > p$. This is a contradiction, and we conclude that $\mathcal{C} = \mathbb{R}^2$. □

Theorem 14.9 *If* rank$(M) = 2$ *and A has at least one eigenvalue with positive real part, the system is not completely controllable.*

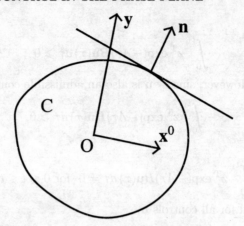

Fig. 14.14. A tangent to the set C that separates each $\mathbf{x}^0 \in C$ from $\mathbf{y} \notin C$.

Proof Let λ be an eigenvalue of A with positive real part, and $\mathbf{e}$ the associated eigenvector, so that $\mathbf{e}^T A = \lambda \mathbf{e}^T$, and $\mathbf{e}^T A^k = \lambda^k \mathbf{e}^T$. This means that $\mathbf{e}^T \exp(-A\tau) = e^{-\lambda \tau} \mathbf{e}^T$, and hence

$$\mathbf{e}^T \mathbf{x}^0 = \mathbf{e} \cdot \mathbf{x}^0 = -\int_0^{t_1} e^{-\lambda \tau} \mathbf{e}^T B \mathbf{u}(\tau)\, d\tau.$$

This integral converges as $t_1 \to \infty$, and is bounded above by some constant c, so that $\mathbf{e} \cdot \mathbf{x}^0 \leqslant c$. The controllable set therefore lies on one side of a line in the plane, and hence $C \neq \mathbb{R}^2$. $\square$

This concludes our proof that the controllability matrix has the properties that we outlined at the start of this section.

14.7 The Time-Optimal Maximum Principle (TOMP)

Consider the set reachable from $\mathbf{x}^0$ in time t, $R(t, \mathbf{x}^0)$. As time increases, this set traces out a volume in (x_1, x_2, t)-space, which we label $RT(t, \mathbf{x}^0)$. If the system can be controlled to the origin, the shortest time in which this can be achieved is t^*, where t^* is the first time when $t^* \in R(t^*, \mathbf{x}^0)$ (see Figure 14.15).

Theorem 14.10 *The time-optimal trajectory lies in the boundary, $\partial RT(t, \mathbf{x}^0)$.*

Proof Let $\mathbf{u} = \mathbf{u}^*(t)$ be the optimal control for $\mathbf{x}(0) = \mathbf{x}^0$, and let $\mathbf{y}(t)$ be a solution with $\mathbf{u} = \mathbf{u}^*$. Now suppose that $\mathbf{y}(t_0) \in \text{Int}(R(t_0, \mathbf{x}^0))$. There must therefore be a disc of sufficiently small radius r, $D(\mathbf{y}(t_0), r)$, that lies entirely within $R(t_0, \mathbf{x}^0)$. If we now apply the optimal control, $\mathbf{u}^*$, to all of the points within $D(\mathbf{y}(t_0), r)$, they will lie in the neighbourhood of $\mathbf{y}(t_1)$ when $t = t_1 > t_0$, and must also lie in $R(t_1, \mathbf{x}^0)$. Therefore $\mathbf{y}(t_1) \in \text{Int}(R(t_1, \mathbf{x}^0))$, so that any trajectory that starts in

14.7 THE TIME-OPTIMAL MAXIMUM PRINCIPLE (TOMP)

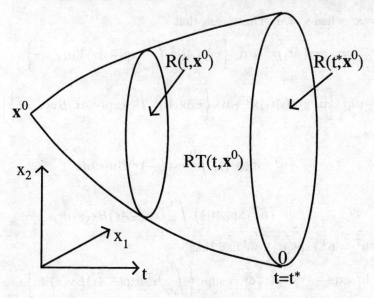

Fig. 14.15. The reachable set, $RT(t, \mathbf{x}^0)$.

$\text{Int}(R(t, \mathbf{x}^0))$ remains there. Since the origin lies in the boundary, $\partial RT(t, \mathbf{x}^0)$, the time-optimal trajectory must lie entirely within this boundary. □

Theorem 14.11 (The time-optimal maximum principle (TOMP)) *The control $\mathbf{u}(t)$ is time-optimal if and only if there exists a nonzero vector, $\mathbf{h}$, such that for $0 \leqslant t \leqslant t_1$,*

$$\mathbf{h}^T \{\exp(-At)B\mathbf{u}(t)\} = \sup_{\mathbf{v}(t)} \mathbf{h}^T \{\exp(-At)B\mathbf{v}(t)\}.$$

The components of the time-optimal control are

$$u_i(t) = \text{sgn}\left[\mathbf{h}^T \{\exp(-At)B\}\right]_i \quad \text{for } i = 1, 2, \ldots, m$$

– *bang-bang control.*

Proof Let $\mathbf{u}(t)$ be a control that steers $\mathbf{x}$ from $\mathbf{x}^0$ when $t = 0$ to $\mathbf{x}^1 \in \partial R(t_1, \mathbf{x}^0)$ when $t = t_1$. The reachable set is convex (this can be proved in the same way that we proved that the controllable set is convex in Theorem 14.5), so there is a tangent at $\mathbf{x}^1$ with normal $\mathbf{n}$ such that

$$\mathbf{n} \cdot \mathbf{x} = \sup_{\mathbf{y}^1 \in R(t_1, \mathbf{x}^0)} \mathbf{n} \cdot \mathbf{y}^1.$$

If $\mathbf{v}(t)$ is an arbitrary control and $\mathbf{y}(t)$ the corresponding solution,

$$\mathbf{y}^1 = \exp(At_1)\left\{\mathbf{x}^0 + \int_0^{t_1} \exp(-A\tau)B\mathbf{v}(\tau)\, d\tau\right\},$$

and $\mathbf{y}^1 = \mathbf{x}^1$ when $\mathbf{v} = \mathbf{u}$. This means that

$$\mathbf{n} \cdot \exp(At_1)\mathbf{x}^0 + \mathbf{n} \cdot \left\{ \exp(At_1) \int_0^{t_1} \exp(-A\tau)B\mathbf{u}(\tau)\, d\tau \right\}$$

$$= \sup_{\mathbf{v}} \left[\mathbf{n} \cdot \exp(At_1)\mathbf{x}^0 + \mathbf{n} \cdot \left\{ \exp(At_1) \int_0^{t_1} \exp(-A\tau)B\mathbf{v}(\tau)\, d\tau \right\} \right],$$

and hence

$$\{\mathbf{n}^T \exp(At_1)\} \int_0^{t_1} \exp(-A\tau)B\mathbf{u}(\tau)\, d\tau$$

$$= \sup_{\mathbf{v}} \, \{\mathbf{n}^T \exp(At_1)\} \int_0^{t_1} \exp(-A\tau)B\mathbf{v}(\tau)\, d\tau.$$

If we let $\mathbf{h}^T = \mathbf{n}^T \exp(At_1)$, we find that

$$\int_0^{t_1} \mathbf{h}^T \exp(-A\tau)B\mathbf{u}(\tau)\, d\tau = \sup_{\mathbf{v}} \left\{ \int_0^{t_1} \mathbf{h}^T \exp(-A\tau)B\mathbf{v}(\tau)\, d\tau \right\}. \quad (14.21)$$

Note that $\mathbf{h}$ is nonzero, because $\exp(At)$ is a nonsingular matrix since it always has an inverse, $\exp(-At)$. Since (14.21) holds for all $t_1 > 0$, we must have

$$\mathbf{h}^T \exp(-A\tau)B\mathbf{u}(\tau) = \sup_{\mathbf{v}} \, \{\mathbf{h}^T \exp(-A\tau)B\mathbf{v}(\tau)\}. \quad (14.22)$$

All of the steps of this argument can be reversed, and we have therefore proved the first part of the theorem.

To obtain the maximum value of the right hand side of (14.22), we must take

$$v_i(t) = \operatorname{sgn}\left[\mathbf{h}^T \exp(-At)B\right]_i \quad \text{for } 0 \leqslant t \leqslant t_1.$$

In other words, each component of the time-optimal control must take one of its extreme values, 1 or -1, and change when $\left[\mathbf{h}^T \exp(-At)B\right]_i$ changes sign – bang-bang control. □

We shall see that this is the main result that we need to solve time-optimal control problems in the phase plane.

Example: The positioning problem

For this problem

$$A = \begin{pmatrix} 0 & 1 \\ 0 & 0 \end{pmatrix}, \quad B = \begin{pmatrix} 0 \\ 1 \end{pmatrix}.$$

Since $A^2 = 0$,

$$\exp(-At) = I - At = \begin{pmatrix} 1 & -t \\ 0 & 1 \end{pmatrix},$$

and

$$\exp(-At)B = \begin{pmatrix} -t \\ 1 \end{pmatrix}.$$

14.7 THE TIME-OPTIMAL MAXIMUM PRINCIPLE (TOMP)

Therefore, if we write $\mathbf{h}^T = (\alpha, \beta)$, we find that

$$\mathbf{h}^T \exp(-At)B = \beta - \alpha t.$$

This means that the time-optimal, bang-bang control changes sign at most once. The TOMP does not tell us when, or even whether, this change in sign occurs. However, with $x_1(0) = x_1^0$ and $x_2(0) = 0$, an initially stationary particle, we have seen that the control must change sign at least once if the solution is to reach the origin. We also constructed the unique solution with bang-bang control that changes sign just once. The TOMP now shows that this is indeed the time-optimal solution.

Let's now consider what happens if the particle is initially moving, with $x_1(0) = x_1^0$ and $x_2(0) = x_2^0$. Rather than just solving the equations, let's think about what the time-optimal solution will look like in the phase plane. We know that the time-optimal control is bang-bang with $u = \pm 1$, so the integral paths are given by

$$\frac{dx_1}{dx_2} = \pm x_2,$$

and hence are parabolas, with

$$x_1 = k \pm \frac{1}{2}x_2^2$$

for some constant of integration k. The only two time-optimal paths that reach the origin are therefore the appropriate branches of $x_1 = \pm \frac{1}{2}x_2^2$, which are labelled as $S_\pm$ in Figure 14.16. On S_+ we need $u = 1$, whilst on S_- we need $u = -1$. Any initial conditions that lie on these curves can be controlled to the origin without changing the sign of u. For any other initial conditions, the system must be controlled onto one of the curves $S_\pm$, when the control must change sign, as illustrated in Figure 14.16.

Example: Controlling a linear oscillator

For a linear oscillator with unit angular frequency,

$$\exp(-At) = \begin{pmatrix} \cos t & -\sin t \\ \sin t & \cos t \end{pmatrix},$$

and

$$\exp(-At)B = \begin{pmatrix} -\sin t \\ \cos t \end{pmatrix}.$$

With $\mathbf{h}^T = (\alpha, \beta)$,

$$\mathbf{h}^T \exp(-At)B = \beta \cos t - \alpha \sin t.$$

The TOMP therefore shows that the time-optimal control is

$$u(t) = \text{sgn}(\beta \cos t - \alpha \sin t) = \text{sgn}\{a \cos(t + b)\},$$

for some constants a and b. Since $\cos(t + b)$ changes sign at intervals of π, so must $u(t)$. The first change of sign occurs when $t = T_0$ with $0 < T_0 \leq \pi$, and must

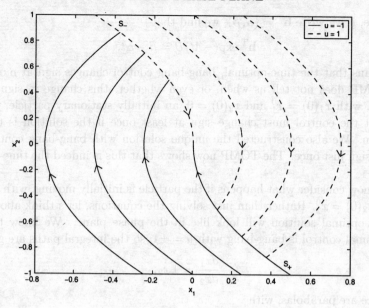

Fig. 14.16. The time-optimal trajectories for the positioning problem.

subsequently change when $t = T_n = T_0 + n\pi$, for $n = 0, 1, 2, \ldots$. Let's consider the solution when $u = \pm 1$ in the phase plane. The governing equations, (14.16), show that

$$(x_1 \mp 1)\frac{dx_1}{dt} + x_2 \frac{dx_2}{dt} = 0,$$

and hence that

$$(x_1 \mp 1)^2 + x_2^2 = k^2,$$

for some constant of integration, k. These trajectories are circles, centred on $(\pm 1, 0)$, as shown in Figure 14.17. Proceeding as we did for the positioning problem, we can see that only the circles marked $S_\pm$ in Figure 14.17, given by $(x \mp 1)^2 + x_2^2 = 1$, enter the origin. We can now construct the time-optimal solution by considering the solutions that meet $S_\pm$, with the control changing sign there. A typical example is shown in Figure 14.18. The sign of the control must change with period π, so the time-optimal solution consists of part of S_+ or S_-, and a succession of semicircles of increasing radius with centres alternating between $(1, 0)$ and $(-1, 0)$. Intuitively, we would perhaps have expected the control to change sign when $x_2 = 0$. Thinking in terms of stopping a child on a swing, we might have expected to push in the opposite direction to the motion. In fact, the time-optimal solution is out of phase with this intuitive solution in order to allow the velocity to be zero precisely when the swing is vertical.

14.7 THE TIME-OPTIMAL MAXIMUM PRINCIPLE (TOMP)

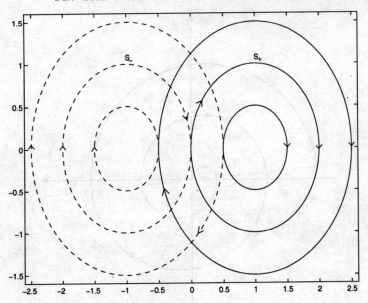

Fig. 14.17. The bang-bang trajectories for the problem of controlling a linear oscillator.

Example: The positioning problem with two controls

For this problem

$$\exp(-At)B = \begin{pmatrix} 1 & -t \\ 0 & 1 \end{pmatrix}.$$

With $\mathbf{h}^T = (\alpha, \beta)$,

$$\mathbf{h}^T \exp(-At)B = (\alpha, \beta - \alpha t).$$

The TOMP therefore says that the time-optimal control $u_2 = \text{sgn}(\beta - \alpha t)$ changes sign at most once, as was the case for the positioning problem with just one control variable. In contrast, $u_1 = \text{sgn}(\alpha)$ does not change sign in the time-optimal control, provided that $\alpha \neq 0$. If, however, $\alpha = 0$, u_2 does not change sign and u_1 is undetermined. Let's begin by considering this case, with $u_2 = -1$. This immediately gives us

$$x_2 = x_2^0 - t,$$

which is therefore only appropriate when $x_2^0 > 0$. The optimal control time is therefore $t_1 = x_2^0$. If we now integrate the equation for x_1, we obtain

$$x_1 = x_1^0 + x_2^0 t - \frac{1}{2}t^2 + \int_0^{t_1} u_1(\tau) \, d\tau.$$

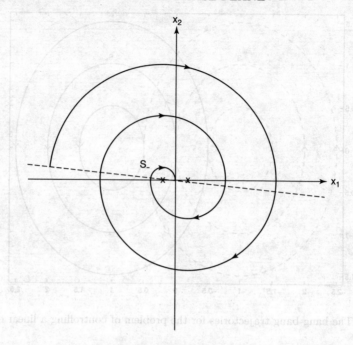

Fig. 14.18. A typical time-optimal trajectory for the problem of controlling a linear oscillator.

Since $x_1(t_1) = 0$, we have a time-optimal trajectory for any control u_1 such that

$$\int_0^{x_2^0} u_1(\tau)\, d\tau = -x_1^0 - \frac{1}{2}(x_2^0)^2.$$

However, since $|u_1| \leqslant 1$, this can only be achieved when

$$\left| -x_1^0 - \frac{1}{2}(x_2^0)^2 \right| \leqslant x_2^0,$$

and hence when

$$-x_2^0 - \frac{1}{2}(x_2^0)^2 \leqslant x_1^0 \leqslant x_2^0 - \frac{1}{2}(x_2^0)^2.$$

This is marked as region B in Figure 14.19. Similarly, when $x_2^0 < 0$ and $u_2 = 1$ we have nonunique time-optimal paths in region A, where

$$x_2^0 + \frac{1}{2}(x_2^0)^2 \leqslant x_1^0 \leqslant -x_2^0 + \frac{1}{2}(x_2^0)^2.$$

Outside these two regions, the time-optimal control is unique, with $u_1 = u_2 = -1$ in region C and $u_1 = u_2 = 1$ in region D, as marked in Figure 14.19. In region C, each time-optimal path meets the boundary of region A, where the sign of u_1 changes, and then follows the boundary of region A to the origin. Similarly in region D.

Note that, since $t_1 = |x_2^0|$ in regions A and B where the time-optimal solution

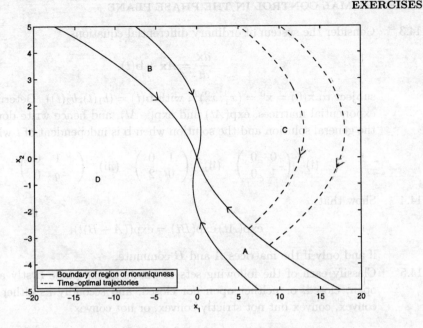

Fig. 14.19. The regions A, B, C and D, and some time-optimal solutions for the positioning problem with two controls.

is not unique, the boundaries of $\mathcal{C}(t_1)$, the reachable set in time t_1, are straight lines in these two regions. This means that $\mathcal{C}(t_1)$ is not strictly convex. Although we will not discuss this further, controllable sets that are not strictly convex are always associated with nonunique time-optimal solutions.

In practice, the difficulty with applying bang-bang control lies in determining when the control needs to change by making measurements of the state of the system. This is known as **measurement–action lag** (see, for example, Marlin, 1995).

Exercises

14.1 We have seen that the tightrope walker system, $\dot{x} = x + u$, is not completely controllable. Show, by solving the governing equations, that the other two qualitatively different, first order, linear, constant coefficient systems, $\dot{x} = -x + u$ and $\dot{x} = u$, are completely controllable.

14.2 For the tightrope walker system, $\dot{x} = x + u(t)$ with $|u| \leqslant 1$, find the reachable set from $x = 2$ in time t_1, $R(t_1, 2)$ and the set controllable to $x = 2$ in time t_1, $\mathcal{C}(t_1, 2)$, and sketch them. Show that $R(t_1, 2) \not\subset R(t_2, 2)$ and $\mathcal{C}(t_1, 2) \not\subset \mathcal{C}(t_2, 2)$ when $t_1 < t_2$.

14.3 Consider the system of ordinary differential equations

$$\frac{d\mathbf{x}}{dt} = A\mathbf{x} + \mathbf{b}(t),$$

subject to $\mathbf{x}(0) = \mathbf{x}^0 = (x_1^0, x_2^0)^T$, with $\mathbf{b}(t) = (b_1(t), b_2(t))$. Determine the exponential matrices, $\exp(At)$ and $\exp(-At)$, and hence write down both the general solution and the solution when $\mathbf{b}$ is independent of t when $A =$

(i) $\begin{pmatrix} 0 & 0 \\ 1 & 0 \end{pmatrix}$ (ii) $\begin{pmatrix} 1 & 0 \\ 0 & 2 \end{pmatrix}$ (iii) $\begin{pmatrix} 1 & 0 \\ -q & 0 \end{pmatrix}$

14.4 Show that

$$\exp(At)\exp(Bt) = \exp((A+B)t),$$

if and only if the matrices A and B commute.

14.5 Classify each of the following sets of points in the plane firstly as either open, closed, or neither open nor closed, and secondly as either strictly convex, convex but not strictly convex, or not convex.

(a) $\{(x_1, x_2) \mid (x_1 - 1)^2 + x_2^2 < 1\} \cup \{(x_1, x_2) \mid x_1^2 + x_2^2 < 1\}$,
(b) $\{(x_1, x_2) \mid (x_1 - 1)^2 + x_2^2 \leqslant 1\} \cap \{(x_1, x_2) \mid x_1^2 + x_2^2 \leqslant 1\}$,
(c) $\{(x_1, x_2) \mid x_1^2 + x_2^2 \leqslant 1\} \cap \{(x_1, x_2) \mid x_1 \geqslant 0\}$,
(d) $\{(x_1, x_2) \mid x_1^2 + x_2^2 \leqslant 1\} \cap \{(x_1, x_2) \mid x_1 > 0\}$.

14.6 Use the controllability matrix to show that Examples 2, 3 and 4 given in Section 14.4 are completely controllable systems.

14.7 If $\dot{\mathbf{x}} = A\mathbf{x} + B\mathbf{u}$ with $\mathbf{u}$ the vector of bounded controls, construct the controllability matrix, M, and hence determine whether the system is completely controllable when

(a) $A = \begin{pmatrix} 0 & 1 \\ 0 & 2 \end{pmatrix}$, $B = \begin{pmatrix} 1 \\ 1 \end{pmatrix}$,

(b) $A = \begin{pmatrix} -1 & 1 \\ 0 & 0 \end{pmatrix}$, $B = \begin{pmatrix} 1 \\ 1 \end{pmatrix}$,

(c) $A = \begin{pmatrix} -1 & 1 \\ 0 & -1 \end{pmatrix}$, $B = \begin{pmatrix} 1 & 0 \\ 0 & 1 \end{pmatrix}$.

14.8 Consider the positioning problem. Show that for initial conditions lying above the curves $S^{\pm}$ the switching time is

$$T = x_2^0 + \sqrt{x_1^0 + \frac{1}{2}(x_2^0)^2},$$

and that the optimal control time is

$$t_1 = x_2^0 + 2\sqrt{x_1^0 + \frac{1}{2}(x_2^0)^2}.$$

Determine $C(t_1)$, the controllable set in time t_1.

14.9 Consider the steering problem. Show that

$$\exp(-At) = \begin{pmatrix} 1 & \frac{1}{q}(1 - e^{qt}) \\ 0 & e^{qt} \end{pmatrix}.$$

Use the TOMP to show that the time-optimal control has at most one change of sign. If $x_1(0) = x_1^0$ and $x_2(0) = 0$, show that the optimal control time is

$$t_1 = \frac{2}{q} \log \left\{ \exp\left(\frac{1}{2}q^2 x_1^0\right) + \sqrt{\exp(q^2 x_1^0) - 1} \right\}.$$

Show that the integral paths for $u = \pm 1$ are

$$x_1 = k - \frac{1}{q^2} \{qx_2 \pm \log(1 \mp qx_2)\},$$

with k a constant. Sketch the time-optimal trajectories and the controllable set in time t_1 in the (x_1, x_2)-phase plane.

14.10 The population of a pest is increasing exponentially. To control the pest, a genetically engineered, sterile, predatory beetle is introduced into the environment. Since the beetles are harmful to crops, it is desirable to remove them as soon as possible.

Let the populations of the pest and the beetle be denoted by $x_1(t)$ and $x_2(t)$ creatures per square metre respectively. Initially, $x_1 = X > 0$ and $x_2 = 0$, and the target is to reduce both populations to zero simultaneously and in the shortest possible time. Model equations for the two populations are

$$\dot{x}_1 = x_1 - x_2, \quad \dot{x}_2 = -x_2 + u(t),$$

with $|u| \leq 1$ representing the rate at which beetles are released into the environment. Use the time-optimal maximum principle to show that the optimal control changes sign just once.

Show that the optimal control time is

$$t_1 = \log\left\{\frac{X + 1 + \sqrt{X^2 + 4X}}{2X - 1}\right\}.$$

What is the upper limit on X below which the system is controllable?

14.11 Consider the system

$$\dot{x}_1 = x_1 + x_2, \quad \dot{x}_2 = -x_2 + u(t),$$

subject to $x_1(0) = x_1^0$, $x_2(0) = x_2^0$, with $|u(t)| \leq 1$. Use the time-optimal maximum principle to show that the sign of the optimal control changes at most once. Show that when x_2 is initially positive, the time-optimal solution on which the control does not change sign has

$$x_1^0 = -\frac{(x_2^0)^2}{2(1 + x_2^0)}.$$

14.12 Determine the matrices $\exp(-At)$ and $\exp(At)$ when

$$A = \frac{1}{\sqrt{2}} \begin{pmatrix} 1 & 1 \\ 1 & -1 \end{pmatrix}.$$

Consider the second order system of ordinary differential equations

$$\frac{d\mathbf{x}}{dt} = A\mathbf{x} + B\mathbf{u},$$

where $\mathbf{x}$ is the vector of state variables, $\mathbf{u}$ the vector of bounded control variables with $|u_i(t)| \leq 1$, and

$$B = \begin{pmatrix} 1 \\ 0 \end{pmatrix}.$$

Use the time-optimal maximum principle to show that the sign of $u(t)$ can change no more than once in the time-optimal control. Show that the time-optimal solutions on which the control does not change sign have

$$x_1^0 = \pm \left(\sinh t_1 - \frac{1}{\sqrt{2}} \cosh t_1 + \frac{1}{\sqrt{2}} \right), \quad x_2^0 = \pm \left(-\frac{1}{\sqrt{2}} \cosh t_1 + \frac{1}{\sqrt{2}} \right),$$

where t_1 is the time taken to control the system to the origin.

CHAPTER FIFTEEN

An Introduction to Chaotic Systems

In order to introduce the idea of a chaotic solution, we will begin by studying three simple chaotic systems that arise in different physical contexts. We then look at some examples of mappings, which are important because ordinary differential equations can be related to mappings through the Poincaré return map. After investigating homoclinic tangles in Poincaré return maps, which contain chaotic solutions, we investigate how their existence can be established by examining the zeros of the Mel'nikov function. Finally, we discuss the computation of the Lyapunov spectrum of a differential equation, from which a quantitative measure of chaos can be obtained.

15.1 Three Simple Chaotic Systems

15.1.1 A Mechanical Oscillator

Consider the mechanical system that consists of two rings of mass m threaded onto two horizontal wires a distance a apart, as shown in Figure 15.1. The rings are joined by a spring of natural length $l > a$ that obeys Hooke's law with elastic constant μ. If we move the upper ring, what happens to the lower ring? We denote the displacement of the upper ring from a fixed vertical line by $\phi(t)$, and that of the lower ring by $y(t)$. On the assumption that a frictional force of magnitude $mk\dot{y}$ opposes the motion of the lower ring, Newton's second law in the horizontal direction shows that

$$-\mu \frac{\left\{\sqrt{a^2 + (y-\phi)^2} - l\right\}(y-\phi)}{\sqrt{a^2 + (y-\phi)^2}} - mk\dot{y} = m\ddot{y}, \qquad (15.1)$$

where a dot denotes d/dt. Let's assume that the relative displacement of the two rings, $Y = y - \phi$, is much less than the distance, a, between the wires, so that

$$\sqrt{a^2 + (y-\phi)^2} = a\sqrt{1 + \frac{Y^2}{a^2}} \sim a + \frac{Y^2}{2a}.$$

In terms of Y, (15.1) becomes

$$\ddot{Y} + k\dot{Y} - \frac{\mu(l-a)}{ma}Y + \frac{\mu l}{2ma^3}Y^3 \sim -\ddot{\phi} - k\dot{\phi}.$$

AN INTRODUCTION TO CHAOTIC SYSTEMS

If we now define the dimensionless variables

$$x = \sqrt{\frac{l}{2a^2(l-a)}} Y, \quad \hat{t} = \sqrt{\frac{\mu(l-a)}{ma}} t,$$

we obtain

$$\ddot{x} + \epsilon \delta \dot{x} - x + x^3 = F(\hat{t}), \tag{15.2}$$

where

$$\epsilon \delta = k \sqrt{\frac{ma}{\mu(l-a)}}, \quad F(\hat{t}) = -\left(\ddot{\phi} + k\dot{\phi}\right)\sqrt{\frac{2m^2 a^4}{\mu^2 l(l-a)}}.$$

The reason for our rather odd choice of dimensionless parameters will become clear later. Equation (15.2) with $F = 0$ is known as **Duffing's equation**, and arises in many other contexts in mechanics. With $F \neq 0$, (15.2) is the **forced Duffing equation**.† We will assume that the forcing takes the form $F = \epsilon \gamma \cos \omega t$ (dropping the hat on the time variable for convenience), so that (15.2) becomes

$$\ddot{x} + \epsilon \delta \dot{x} - x + x^3 = \epsilon \gamma \cos \omega t. \tag{15.3}$$

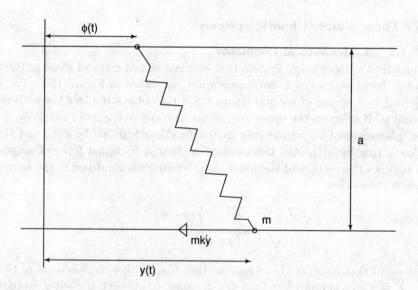

Fig. 15.1. A mechanical system whose small amplitude oscillations are governed by the forced Duffing equation, (15.3).

Figure 15.2 shows the solution, calculated numerically using MATLAB (see Section 9.3.4), when $\epsilon \delta = \frac{1}{2}$, $\epsilon \gamma = \frac{3}{5}$ and $y = dy/dt = 0$ when $t = 0$. This solution

† For more on the dynamics of the forced Duffing equation, see Arrowsmith and Place (1990) and references therein.

behaves erratically. Indeed, it is tempting to think of it as 'random' in some sense. However, we know from Chapter 8 that the solution of (15.3) exists and is unique. Figure 15.3 superimposes another solution, this time with $y = 10^{-4}$, $dy/dt = 0$ when $t = 0$. The two solutions begin close to each other, but, as time increases, drift further apart, and soon diverge completely. We say that the system exhibits **sensitive dependence upon initial conditions**. In practice, we can only know the initial state of a physical system with a finite degree of accuracy. After a sufficient time has elapsed, solutions with different initial conditions, but which are close enough together that, in practice, they are indistinguishable, will diverge in a chaotic system.

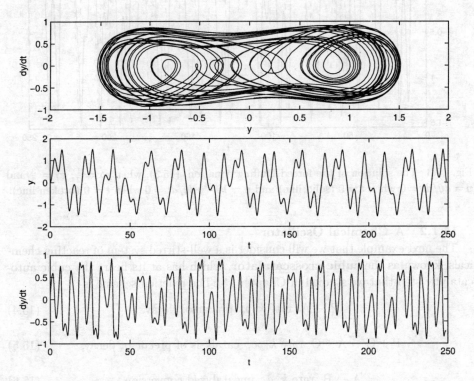

Fig. 15.2. The solution of the forced Duffing equation, (15.3), when $\epsilon\delta = \frac{1}{2}$, $\epsilon\gamma = \frac{3}{5}$ and $y = dy/dt = 0$ when $t = 0$.

We can now give an informal definition of a **chaotic solution** as a bounded, aperiodic, recurrent solution, that has a random aspect due to its sensitive dependence on initial conditions. Adjacent chaotic solutions diverge exponentially fast, a property that we will later measure using the Lyapunov spectrum, and remain in a bounded region, where they undergo repeated folding, and are, in practice, unpredictable in the long term.

450 AN INTRODUCTION TO CHAOTIC SYSTEMS

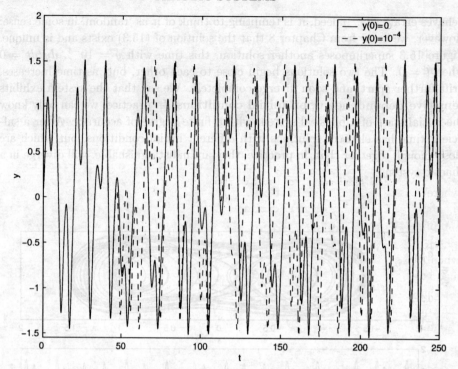

Fig. 15.3. The solution of the forced Duffing equation, (15.3), when $\epsilon\delta = \frac{1}{2}$, $\epsilon\gamma = \frac{3}{5}$ and $y = dy/dt = 0$ when $t = 0$ (solid line) and $y = 10^{-4}$, $dy/dt = 0$ when $t = 0$ (broken line).

15.1.2 A Chemical Oscillator

The next example that we will consider is a well-stirred system of reacting chemicals, known as the **cubic crosscatalator**, which has at its heart the cubic autocatalytic step that we studied in Chapter 13. The reaction scheme is

$$\text{P} \to \text{A} \quad \text{rate } k_0 p, \text{ precursor decay,} \tag{15.4}$$

$$\text{P} + \text{C} \to \text{A} + \text{C} \quad \text{rate } k_1 pc, \text{ catalysis of precursor decay,} \tag{15.5}$$

$$\text{A} \to \text{B} \quad \text{rate } k_u a, \text{ uncatalyzed conversion,} \tag{15.6}$$

$$\text{A} + 2\text{B} \to 3\text{B} \quad \text{rate } k_1 ab^2, \text{ cubic autocatalysis,} \tag{15.7}$$

$$\text{B} \to \text{C} \quad \text{rate } k_2 b, \text{ autocatalyst decay,} \tag{15.8}$$

$$\text{C} \to \text{D} \quad \text{rate } k_3 c, \text{ catalyst decay.} \tag{15.9}$$

In addition to the reactant, A, and autocatalyst, B, there is a **precursor**, P, which decays to produce A, and a **catalyst**, C, which accelerates the decay of the precursor, and is produced by the decay of the autocatalyst. The action of C both at the start of the reaction cascade, catalyzing the decay of the precursor, and as a product at the end of the sequence P → A → B → C is the essential ingredient that leads to complex behaviour. The reactant A also decays spontaneously to produce B, and C is itself unstable, decaying to the inert product D.

Under the assumption that the precursor, P, is in large excess and decays slowly, we can derive the dimensionless governing equations (see Exercise 15.1)

$$\begin{aligned}\dot{\alpha} &= \kappa(1+\eta\gamma) - \alpha\beta^2 - \epsilon\alpha, \\ \dot{\beta} &= \alpha\beta^2 - \beta + \epsilon\alpha, \\ \dot{\gamma} &= \beta - \chi\gamma,\end{aligned} \quad (15.10)$$

where α, β and γ are the dimensionless concentrations of A, B and C, and κ, η, ϵ and χ are dimensionless constants. Although this system possesses a very complicated set of different types of behavior (see Petrov, Scott and Showalter 1992), we will focus on a single chaotic solution. Figure 15.4 shows the solution when $\kappa = 0.71$, $\eta = 0.054$, $\epsilon = 0.005$ and $\chi = 0.25$. The single equilibrium point has a one-dimensional stable manifold and a two-dimensional unstable manifold associated with complex eigenvalue. The solution is continually attracted towards the equilibrium point close to the stable manifold, and then spirals away close to the unstable manifold before the process begins again.

15.1.3 The Lorenz Equations

Our third example is the system of three autonomous differential equations,

$$\begin{aligned}\dot{x} &= -\frac{8}{3}x + yz, \\ \dot{y} &= -10(y-z), \\ \dot{z} &= -xy + 28y - z,\end{aligned} \quad (15.11)$$

known as the **Lorenz equations**. Equations (15.11) were derived by Lorenz (1963) as the leading order approximation to the behaviour of an idealized model of the Earth's atmosphere. To claim that these simple equations model the weather is perhaps going a little too far, but they certainly have very interesting dynamics.

There are three equilibrium points, one at the origin and two at $x = 27$, $y = z = \pm 6\sqrt{2}$, each of which is unstable. The two equilibrium points away from the origin each have a two-dimensional unstable manifold, associated with complex eigenvalues, and hence oscillatory behaviour, and a one-dimensional stable manifold. The system is rather like two copies of the cubic crosscatalator system interacting with each other. Typical solutions bounce back and forth between the two equilibrium points away from the origin, continually being attracted towards an equilibrium point along a trajectory close to the stable manifold, and then spiralling away close to the unstable manifold, as shown in Figure 15.5 for the solution with $x = y = z = 1$ when $t = 0$.

The best way to get a feel for the dynamics of the Lorenz equations, and an indication of their iconic status as the first chaotic system to be discovered, is to type `lorenz` in MATLAB, which runs an animated simulation.

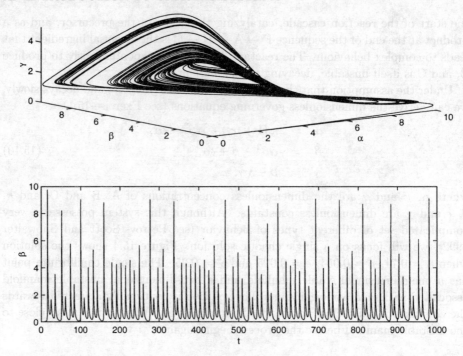

Fig. 15.4. The solution of the cubic crosscatalator equations, (15.10), with $\kappa = 0.71$, $\eta = 0.054$, $\epsilon = 0.005$, $\chi = 0.25$ and $\alpha(0) = \beta(0) = \gamma(0) = 0$.

15.2 Mappings

Although this book is about differential equations, we will see later that it is often helpful to relate the solutions of differential equations to those of **mappings**, sometimes known as **difference equations**. Before we give some basic definitions, let's consider three examples of nonlinear mappings, which illustrate how complicated the solutions of these deceptively simple systems can be.

Example: A shift map

The **shift map** is defined by

$$x \mapsto ax | 1, \tag{15.12}$$

which maps $[0, 1)$ to itself, where $b|c$ means the remainder when b is divided by c. Let's focus on the case $a = 10$. The shift map is then equivalent to shifting the decimal point one place to the right and throwing away the integer part. For example, $\frac{1}{8} = 0.125$ maps to 0.25, which maps to 0.5, which maps to zero, an **equilibrium**, or **fixed point** of the map. Another equilibrium point of the map is $\frac{1}{9} = 0.11111\ldots$. The map also has many **periodic solutions**, for example, $\frac{1}{11} = 0.09090909\ldots \mapsto \frac{10}{11} = 0.909090\ldots \mapsto \frac{1}{11} \mapsto \ldots$, which has period 2.

15.2 MAPPINGS

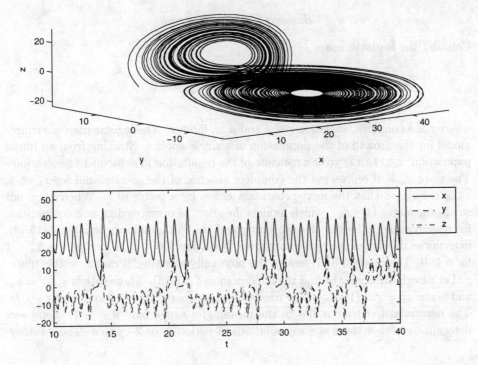

Fig. 15.5. The solution of the Lorenz equations, (15.11), with $x(0) = y(0) = z(0) = 1$.

Every rational number has either a finite decimal expansion, and therefore is eventually mapped to zero, for example $\frac{1}{8}$, or has a repeated decimal expansion, in which case it is part of a periodic solution, for example $\frac{1}{11}$. Irrational numbers, for example $\sqrt{2}$, π and e, have decimal expansions that do not repeat themselves and have no pattern, for example $\sqrt{2} - 1 \approx 0.4142135\ldots \mapsto 0.1421356\ldots \mapsto 0.4213562 \mapsto 0.2135623 \mapsto \ldots$. In addition, two irrational numbers may be arbitrarily close to each other, but the corresponding solutions eventually diverge. For example, consider $x = x_1 = \sqrt{2}-1$ and $x = x_2 = \sqrt{2}-1+10^{-5}\pi \approx 0.4142449\ldots$. These differ by just $10^{-5}\pi$, but after four iterations of (15.12), x_1 maps to $0.135\ldots$ whilst x_2 maps to $0.449\ldots$. This is a simple example of sensitive dependence on initial conditions. Each solution initially at an irrational number therefore satisfies our conditions to be called a chaotic solution, since they behave in an apparently random manner, and initially close solutions diverge. The chaotic solutions, which correspond to the irrational numbers, are dense in $[0, 1)$, as are the periodic solutions, which correspond to rational numbers with no finite decimal representation.

Example: The logistic map

Consider the **logistic map**

$$x_{n+1} = rx_n(1 - x_n), \tag{15.13}$$

where r is a constant, with $0 < r \leqslant 4$ and n an integer. The logistic map is a simple model for the growth of the population of a single species. Starting from an initial population, x_0, (15.13) gives a measure of the population in subsequent generations. The state $x_n = 0$ represents the complete absence of the species, and for $x_n \ll 1$, $x_{n+1} \sim rx_n$, so that the next generation grows by a factor of r. When x_n is not small, the factor $(1 - x_n)$, which models the effect of overcrowding and competition for resources, is no longer close to unity, and the full, nonlinear equation, (15.13), determines the size of the next generation. Note that if $0 \leqslant x_0 \leqslant 1$, then $0 \leqslant x_n \leqslant 1$ for $n \geqslant 0$. The interval $[0, 1]$ is also the physically meaningful range for this map.

Let's begin by trying to find the fixed points of (15.13). These satisfy $x_{n+1} = x_n$, and hence $x_n = rx_n(1 - x_n)$. The fixed points are therefore $x = 0$ and $x = (r-1)/r$. The nontrivial fixed point lies in the meaningful range only if $r > 1$. Let's now determine whether there are any solutions of period 2, or **2-cycles**. These satisfy

$$x_{n+1} = rx_n(1 - x_n),$$
$$x_{n+2} = x_n = rx_{n+1}(1 - x_{n+1}).$$

By eliminating x_{n+1}, we obtain the equation for x_n,

$$x_n \left(x_n - \frac{r-1}{r}\right) \{r^2 x_n^2 - r(1+r)x_n + (1+r)\} = 0.$$

This equation is easy to factorize, since we know that it must also be satisfied by the two fixed points. The discriminant of the quadratic factor is $r^2(r^2 - 2r - 3)$, which is positive provided $r > 3$. This means there are points of period 2 for $r > 3$. In fact, it can be shown that the nontrivial fixed point is stable for $r < 3$, but loses stability in a bifurcation† at $r = 3$, where the points of period 2 emerge and are stable. Similarly, as r increases, the points of period 2 eventually lose stability, and a stable 4-cycle emerges. This process is known as **period doubling**, and, as r increases, eventually leads to chaotic solutions. We will not go into the details of this process, since we want to concentrate on maps relevant to differential equations. The interested reader is referred to Arrowsmith and Place (1990). Figure 15.6 shows the period doubling process as a bifurcation diagram. This was produced using the MATLAB script

† a flip bifurcation

```
for r = 0:0.005:4
    x = rand(1);
    for j = 1:100
        x = r*x*(1-x);
    end
    xout = [];
    for j = 1:400
        x = r*x*(1-x); xout = [xout x];
    end
    plot(r*ones(size(xout)),xout,'.','MarkerSize',3)
    axis([0 4 0 1]), hold on, pause(0.01)
end
```

This iterates the map 100 times, starting from randomly generated initial conditions (x = rand(1)), and then saves the next 400 iterates of the map before plotting them. Figure 15.7 shows 100 iterates of the logistic map with $r = 4$ for two initial conditions separated by just 10^{-16}. The apparently random nature of the solution can be seen, as can the fact that these initially very close solutions have completely diverged after about 50 iterations of the map.

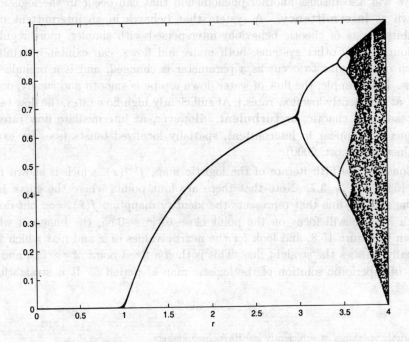

Fig. 15.6. The bifurcation diagram for the logistic map.

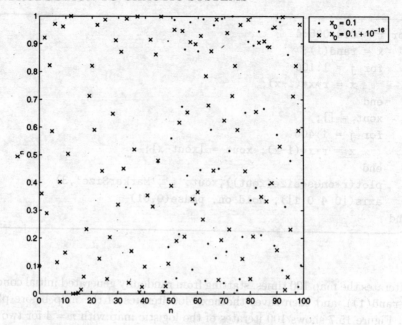

Fig. 15.7. A sequence of 100 iterates of the logistic map with $r = 4$ and $x_0 = 0.1$ and $x_0 = 0.1 + 10^{-16}$

We will also discuss another phenomenon that can occur in the logistic map, known as **intermittency**. A system that behaves in an intermittent manner exhibits bursts of chaotic behaviour interspersed with simpler, more regular behaviour. Many other systems, both maps and flows, can exhibit intermittency, which often begins to occur as a parameter is changed, and is a prelude to full chaos. For example, the flow of water down a pipe is smooth and steady, or **laminar**, at sufficiently low flow rates.† At sufficiently high flow rates, the flow becomes unsteady and chaotic, or **turbulent**. However, at intermediate flow rates, turbulence can appear in intermittent, spatially localized bursts (see, for example, Mathieu and Scott, 2000).

Consider the fifth iterate of the logistic map, $f^{(5)}(x)$, which is shown in Figure 15.8 for $r = 3.7$. Note that there are four points where the curve is close to the straight line that represents the identity mapping, $f(x) = x$, but does not touch it. We will focus on the point close to $x = 0.65$, the image of which is shown in Figure 15.8, and look for the nearby values of x and r at which $f^{(5)}(x)$ actually touches the straight line. This is then a fixed point of $f^{(5)}(x)$, and hence part of a periodic solution of the logistic map of period 5. It is straightforward

---

† Strictly speaking, at sufficiently low Reynolds numbers.

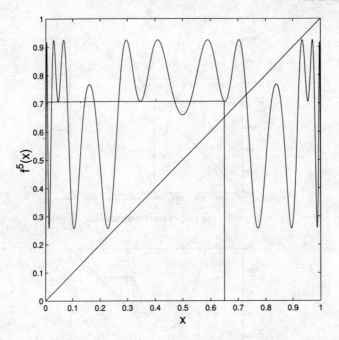

Fig. 15.8. The fifth iterate of the logistic map for $r = 3.7$. Also shown is the image of the point $x = 0.65$.

to show, using MATLAB, that $g(x) = f^{(5)}(x) - x$ and its derivative are zero when $r = r_c \approx 3.73817237526634$ and $x = x_c \approx 0.66045050397608$. Figure 15.9 shows $f^{(5)}(x)$ in the neighbourhood of this point when $r = r_c$ and for two values of r slightly above and below r_c. We can see that for $r > r_c$ there are two fixed points, one stable, one unstable, whilst for $r < r_c$, locally there are no fixed points. Clearly, $r = r_c$ is a bifurcation point of the map $f^{(5)}$, analogous to the saddle–node bifurcation that we discussed in Section 13.3.1, and is known as a **tangent bifurcation**. For values of r slightly greater that r_c, points initially close to x_c are attracted to the fixed point, and therefore the solution of the logistic map is attracted to a stable periodic solution of period 5. For r slightly less than r_c, we can see in Figure 15.9 that $f^{(5)}(x)$ is very close to the straight line, and that iterates of the map can be trapped close to $x = x_c$ before moving away, and also close to the other points where the curve is close to the straight line. The solution therefore exhibits almost steady behaviour before moving away and behaving irregularly, as shown in Figure 15.10. This is intermittency. The equivalent solution of the logistic map displays almost periodic behaviour interrupted by chaotic bursts, which is also shown in Figure 15.10. For further examples of intermittency, see Guckenheimer and Holmes (1983).

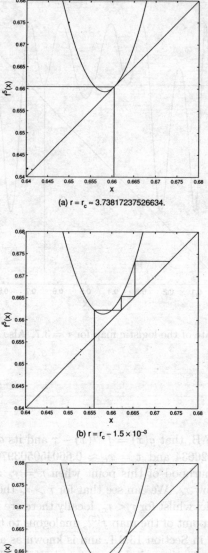
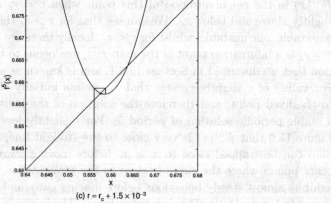

Fig. 15.9. The function $f^5(x)$ at r_c and $r_c \pm 1.5 \times 10^{-3}$. The iterates of a point in this neighbourhood are also shown.

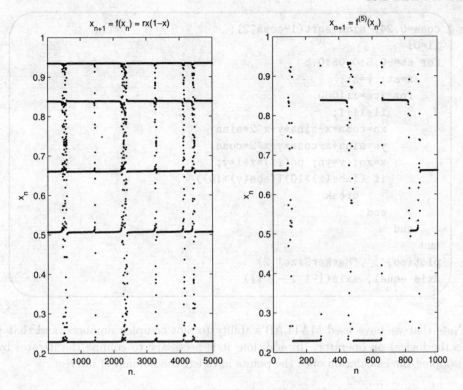

Fig. 15.10. The action of the logistic map and its fifth iterate with $x_1 = x_c$ and $r = r_c - 1 \times 10^{-6}$

Example: The Hénon map

The **Hénon map** is a nonlinear, two-dimensional map, which is defined by

$$x_{n+1} = x_n \cos\theta - y_n \sin\theta + x_n^2 \sin\theta,$$

$$y_{n+1} = x_n \sin\theta + y_n \cos\theta - x_n^2 \cos\theta.$$

This is the most simple area-preserving, quadratically nonlinear map with linear part a rotation through the angle θ about the origin. We can see that it preserves areas, since the determinant of its Jacobian, the definition of which we will discuss in more detail below, is

$$\left| \begin{array}{cc} \frac{\partial x_{n+1}}{\partial x_n} & \frac{\partial x_{n+1}}{\partial y_n} \\ \frac{\partial y_{n+1}}{\partial x_n} & \frac{\partial y_{n+1}}{\partial y_n} \end{array} \right| = \left| \begin{array}{cc} \cos\theta + 2x_n \sin\theta & -\sin\theta \\ \sin\theta - 2x_n \cos\theta & \cos\theta \end{array} \right| = 1.$$

In Figure 15.11 we show the results of an iteration using the MATLAB script

AN INTRODUCTION TO CHAOTIC SYSTEMS

```
cosa=0.24; sina=sqrt(1-cosa^2);
ii=0;
for st=-0.5:0.05:0.5
    x=st; y=st;
    for its=1:1000
        ii=ii+1;
        xn=cosa*x-sina*y+x^2*sina;
        yn=sina*x+cosa*y-x^2*cosa;
        x=xn; y=yn; po(ii)=x+i*y;
        if ((abs(x)>10)|(abs(y)>10))
            break
        end
    end
end
plot(po,'.','MarkerSize',2)
axis equal, axis([-1 1 -1 1])
```

Note that we have used MATLAB's ability to plot complex numbers, and that | is the logical **or** operator. In addition, it is necessary to confine the iterates by stopping the calculation once the points have left the domain of interest.

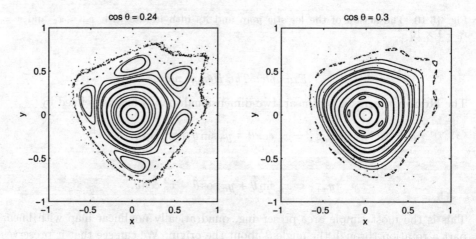

Fig. 15.11. Iterates of the Hénon map for $\cos\theta = 0.24$ and $\cos\theta = 0.3$.

It is clear that this apparently simple map gives rise to extremely complex behaviour. The regions where there are concentric sets of points, or **islands**, owe their appearance to the existence of periodic solutions. In particular, for $\cos\theta = 0.24$ we see evidence of a period 5 structure. In fact, there is a 5-cycle consisting of the five points at the centre of the chain of five islands, separated by five further unstable equilibrium points. Chaotic solutions exist near these points. If we consider other values of $\cos\theta$, other periodic structures become evident (for instance, when $\cos\theta = 0.34$ we find a structure of period 11). We also note that if we consider certain regions in more detail, we can see structures with even longer periods (see, for example, Figure 15.12, which we obtained from Figure 15.11 using MATLAB's ability to zoom in on a small region of an existing figure).

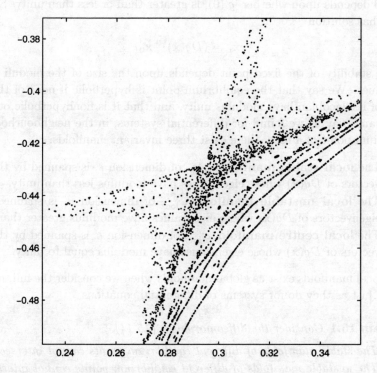

Fig. 15.12. Iterates of the Hénon map for $\cos\theta = 0.24$ in the neighbourhood of one of the hyperbolic fixed points.

15.2.1 Fixed and Periodic Points of Maps
Consider the map
$$\mathbf{x}_{n+1} = \mathbf{g}(\mathbf{x}_n), \tag{15.14}$$

where $\mathbf{g} : \mathbb{R}^n \to \mathbb{R}^n$ is a **diffeomorphism**; that is $\mathbf{g}$ is a bijection and is differentiable with a differentiable inverse†. A **fixed point**, $\bar{\mathbf{x}}$, of the map satisfies $\bar{\mathbf{x}} = \mathbf{g}(\bar{\mathbf{x}})$. A **point of period** k, $\mathbf{x}^*$, satisfies $\mathbf{x}^* = \mathbf{g}^k(\mathbf{x}^*)$, and $\mathbf{x}^* \neq \mathbf{g}^m(\mathbf{x}^*)$ for all positive integer $m < k$.

† Note that, of the maps that we consider in this section, only the Hénon map is a diffeomorphism, since the logistic, tent and horseshoe maps are not bijective and have discontinuous derivatives.

Let's now consider the linearization of (15.14) about a fixed point, $\bar{\mathbf{x}}$. If we let $\mathbf{x}_n = \bar{\mathbf{x}} + \hat{\mathbf{x}}_n$ we find that, at leading order,

$$\hat{\mathbf{x}}_{n+1} = Dg(\bar{\mathbf{x}})\hat{\mathbf{x}}_n, \tag{15.15}$$

where $Dg(\bar{\mathbf{x}})$ is the Jacobian matrix associated with $\mathbf{g}$ at $\mathbf{x} = \bar{\mathbf{x}}$. For one-dimensional maps, the solution of $x_{n+1} = g'(0)x_n$ is $x_n = (g'(0))^n x_0$, and hence the stability of $\mathbf{x} = 0$ depends upon whether $|g'(0)|$ is greater than or less than unity. Similarly, (15.15) has solution

$$\hat{\mathbf{x}}_n = (Dg(\bar{\mathbf{x}}))^n \mathbf{x}_0, \tag{15.16}$$

and the stability of the fixed point depends upon the size of the moduli of the n eigenvalues. We say that the equilibrium point is hyperbolic if none of the eigenvalues of its Jacobian have modulus unity, and that it is nonhyperbolic otherwise. In the same way as we found for differential systems, in the neighbourhood of an equilibrium point of a map, there exist three invariant manifolds.

(i) The **local stable manifold**, $\omega_{\text{loc}}^{\text{s}}$, of dimension s, is spanned by the eigenvectors of $Dg(\bar{\mathbf{x}})$ whose eigenvalues have modulus less than unity.
(ii) The **local unstable manifold**, $\omega_{\text{loc}}^{\text{u}}$, of dimension u, is spanned by the eigenvectors of $Dg(\bar{\mathbf{x}})$ whose eigenvalues have modulus greater than unity.
(iii) The **local centre manifold**, $\omega_{\text{loc}}^{\text{c}}$, of dimension c, is spanned by the eigenvectors of $Dg(\bar{\mathbf{x}})$ whose eigenvalues have modulus equal to unity.

These local manifolds exist as global manifolds when we consider the full, nonlinear system, just as they do for systems of differential equations.

Theorem 15.1 *Consider the diffeomorphism (15.14).*

(i) *The stable manifolds of different equilibrium points cannot intersect.*
(ii) *The unstable manifolds of different equilibrium points cannot intersect.*
(iii) *The stable manifold of an equilibrium point cannot intersect itself.*
(iv) *The unstable manifold of an equilibrium point cannot intersect itself.*

Proof Recall that every point $\mathbf{x} \in \mathbb{R}^n$ has a unique image and preimage, since $\mathbf{g}$ is a diffeomorphism.

(i) Since a point in the stable manifold of an equilibrium point, $\bar{\mathbf{x}}$, must asymptote to $\bar{\mathbf{x}}$ as $n \to \infty$, it cannot also asymptote to a different equilibrium point as $n \to \infty$, which proves (i).
(ii) As for case (i) but with $n \to -\infty$.
(iii) Consider a point, $\mathbf{x}^*$, that is mapped to a point of intersection of the stable manifold with itself, $\mathbf{g}(\mathbf{x}^*)$. Points on the stable manifold in the neighbourhood of $\mathbf{x}^*$ must map to points on the stable manifold in the neighbourhood of $\mathbf{g}(\mathbf{x}^*)$. Since $\mathbf{g}$ is continuous, this is not possible.
(iv) As for case (iii), but on the unstable manifold.

□

Although this theorem eliminates several possibilities, it *is* possible for a stable manifold to intersect an unstable manifold, either of the same equilibrium point or a different one. This will prove to be crucial later.

15.2.2 Tents and Horseshoes

Before we return to consider systems of differential equations, we will examine two more examples of maps, which will prove to be of direct relevance later.

Example: A tent map

A **tent map** is a map $x \mapsto f(x)$ where $f : \mathbb{R} \to \mathbb{R}$ and

$$f(x) = \begin{cases} sx & \text{for } x \leqslant \tfrac{1}{2}, \\ s(1-x) & \text{for } x \geqslant \tfrac{1}{2}. \end{cases}$$

We will concentrate on the case $s = 3$, when the function f is as shown in Figure 15.13. Note that $f(x) = 1$ when $x = \tfrac{1}{3}$ or $x = \tfrac{2}{3}$ and that the only equilibrium

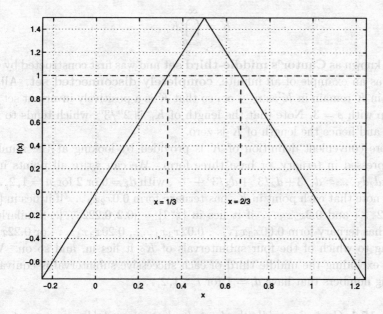

Fig. 15.13. The function $f(x)$ when $s = 3$.

point of the map is zero. Let's now consider where this tent map sends various sets of points.

(i) If $x < 0$, $x \mapsto 3x < x$. Clearly, $f^n(x) \to -\infty$ as $n \to \infty$ for $x \in (-\infty, 0)$.
(ii) If $x > 1$, $x \mapsto 3(1-x) < 0$. After one iteration, all points with $x > 1$ are therefore mapped to a point with $x < 0$, and case (i) applies for subsequent iterations, with $f^n(x) \to -\infty$ as $n \to \infty$ for $x \in (1, \infty)$.

(iii) If $x \in \left(\frac{1}{3}, \frac{2}{3}\right)$, $x \mapsto f(x) > 1$. After one iteration, all points with $\frac{1}{3} < x < \frac{2}{3}$ are therefore mapped to a point with $x > 1$, and case (ii) applies for subsequent iterations, with $f^n(x) \to -\infty$ as $n \to \infty$.

(iv) Now consider points with $x \in [0, \frac{1}{3}] \cup [\frac{2}{3}, 1]$, which are mapped to $[0, 1]$. Only points in the set $[0, \frac{1}{9}] \cup [\frac{2}{9}, \frac{3}{9}] \cup [\frac{6}{9}, \frac{7}{9}] \cup [\frac{8}{9}, 1]$ have images in the set $[0, \frac{1}{3}] \cup [\frac{2}{3}, 1]$, so the remaining points have $f^n(x) \to -\infty$ as $n \to \infty$.

If we continue this process, we can see that the set of points, K, that are *not* expelled from $[0, 1]$ as $n \to \infty$ can be constructed in an iterative way by deleting the middle third from $K_0 = [0, 1]$ to leave $K_1 = [0, \frac{1}{3}] \cup [\frac{2}{3}, 1]$, then deleting the middle third from each of the remaining intervals to leave $K_2 = [0, \frac{1}{9}] \cup [\frac{2}{9}, \frac{3}{9}] \cup [\frac{6}{9}, \frac{7}{9}] \cup [\frac{8}{9}, 1]$, and so on, with K_n the union of the 2^n closed subintervals $[r/3^n, (r+1)/3^n]$, each of which has length 3^{-n}. The index r takes the values in the set R_n, which can be generated iteratively using

$$R_0 = \{0\}, \quad R_n = \{R_{n-1}, 3^n - 1 - R_{n-1}\} \quad \text{for } n \geq 1.$$

We then have

$$K = \bigcap_{n=0}^{\infty} K_n,$$

which is known as **Cantor's middle-third set** and was first constructed by Cantor in 1883 as an example of an infinite, **completely disconnected set**. All points initially in K remain in K as $n \to \infty$, so that K is a positively invariant set for the tent map with $s = 3$. Note that the length of K_n is $2^n/3^n$, which tends to zero as $n \to \infty$, and hence the length of K is zero.

A more convenient definition of K is provided by looking at the numbers in $[0, 1]$ expressed in ternary, or base three form. We can write all points in K as $x = 0.d_1 d_2 d_3 \ldots = d_1/3 + d_2/3^2 + d_3/3^3 + \cdots$, with $d_i = 0$ or 2 for $n = 1, 2, \ldots$. To see this, note that each point in K_1 has ternary form $0.0 x_2 x_3 \ldots$ if it lies in $[0, \frac{1}{3}] = [0, 0.0222\ldots]$ and $0.2 x_2 x_3 \ldots$ if it lies in $[\frac{2}{3}, 1] = [0.2, 0.222\ldots]$. Similarly, each x in K_2 has ternary form $0.00 x_3 x_4 \ldots$, $0.02 x_3 x_4 \ldots$, $0.20 x_3 x_4 \ldots$, or $0.22 x_3 x_4 \ldots$ according to which of the four subintervals of K_2 it lies in, and so on. We can see that excluding the middle third of each successive subinterval is equivalent to excluding numbers that have $d_i = 1$ for $i = 1, 2, \ldots$.

Lemma 15.1 *Cantor's middle-third set, K, has uncountably many points.*

Proof Suppose that K is countable, so that *all* the members of K can be ordered as $x_1 = 0.x_{11} x_{12} x_{13} \ldots < x_2 = 0.x_{21} x_{22} x_{23} \ldots$, in ternary form. We can now define $y = 0.y_1 y_2 y_3 \ldots$, with $y_m = 0$ if $x_{mm} = 2$ and $y_m = 2$ if $x_{mm} = 0$. Then $0 < y < 1$, $y \in K$ and $y \neq x_n$ for all n – a contradiction. The elements of K cannot therefore be ordered, and must be uncountable.† □

† This is just a variation of Cantor's famous diagonalization proof that the real numbers are uncountable.

Nonetheless, by construction, K contains no subinterval, however small, of $[0,1]$, and $[0,1]$ contains infinitely many subintervals that do not intersect K.

Other Cantor sets can be constructed by successively removing different parts of $[0,1]$. An invariant set that is also a Cantor set is, as we shall see, typical of a chaotic system.

Example: The Smale horseshoe map

Consider a two-dimensional map from the unit square,

$$D = \{(x,y) \in \mathbb{R}^2 \mid 0 \leqslant x \leqslant 1, \ 0 \leqslant y \leqslant 1\},$$

to itself, $f : D \to D$. The function f contracts the square in the horizontal direction and expands it in the vertical direction, and then folds the resulting strip back on itself, as shown in Figure 15.14. The map is only defined on the unit square, D, and points that are mapped out of D are discarded. This is the **Smale horseshoe map**. The inverse map, which can be visualized in terms of stretching and folding the unit square in the opposite way, is shown in Figure 15.15.

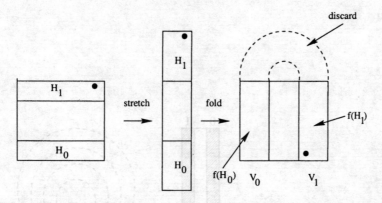

Fig. 15.14. The Smale horseshoe map.

The invariant set, Λ, which is both positively and negatively invariant, is the intersection of the image of D under any number of forward or backward iterations,

$$\Lambda = \quad \cdots \cap f^{-2}(D) \cap f^{-1}(D) \cap D \cap f(D) \cap f^2(D) \cap \cdots.$$

As we can see, each application of the map removes the middle third of each remaining strip in each direction, so we conclude that Λ is a two-dimensional Cantor set given by the set of points (x,y) such that $x \in K_x$ and $y \in K_y$, where K_x and K_y are the Cantor middle-third sets in each direction.

It is now helpful to show that each point in the invariant set can be uniquely labelled by a sequence of 0's and 1's. After each forward iteration of the map, we append a_k to the *right* of the symbol sequence, where

$$a_k = \begin{cases} 0 & \text{for points in the left half of } D, \\ 1 & \text{for points in the right half of } D, \end{cases}$$

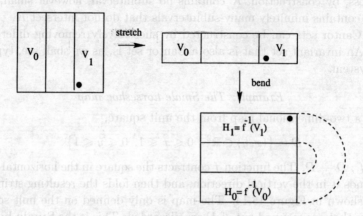

Fig. 15.15. The inverse Smale horseshoe map.

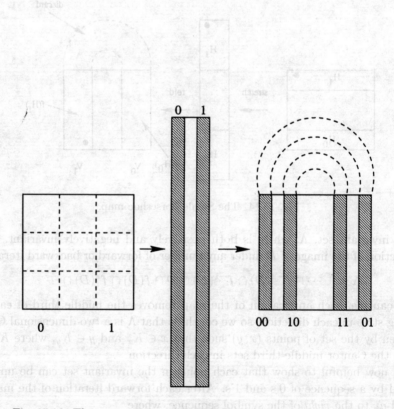

Fig. 15.16. The construction of the left half of the bi-infinite sequence.

as illustrated in Figure 15.16. After each backward iteration of the map we append a_k to the *left* of the symbol sequence, where

$$a_k = \begin{cases} 0 & \text{for points in the bottom half of } D, \\ 1 & \text{for points in the top half of } D, \end{cases}$$

as shown in Figure 15.17. We can then represent any point in Λ as the bi-infinite

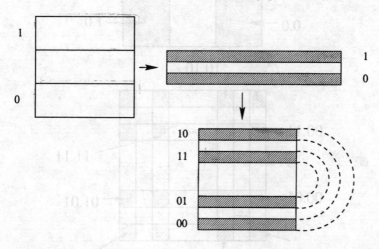

Fig. 15.17. The construction of the right half of the bi-infinite sequence.

sequence of symbols,

$$a = \quad \ldots a_2 a_1 a_0 . a_{-1} a_{-2} \ldots ,$$

as shown in Figure 15.18. The digits to the right of the decimal point reflect the vertical location and to the left the horizontal location. For any point in Λ, the action of the horseshoe map is simply to shift the decimal point in the bi-infinite sequence one place to the right. The map is therefore equivalent to the shift map $x \mapsto 2x|1$, which is known as the **Bernoulli shift map**. For the same reasons that the shift map that we studied as our first example had chaotic solutions, so does the Bernoulli shift map, and hence the horseshoe map.

15.3 The Poincaré Return Map

Now that we have seen several examples of maps, we will demonstrate how the solutions of a system of ordinary differential equations can be related to the solutions of a map, the Poincaré map, that is easier to work with than the original system, and then develop a technique for deciding whether this map is chaotic.

Consider the autonomous system of ordinary differential equations

$$\frac{d\mathbf{x}}{dt} = \mathbf{f}(\mathbf{x}), \tag{15.17}$$

where $\mathbf{f} : \mathbb{R}^n \to \mathbb{R}^n$ and $\mathbf{x} \in \mathbb{R}^n$. We introduce the **flow evolution operator** or **flow operator**, ϕ^{t_1,t_0}, which maps the point $\mathbf{x}_0 \in \mathbb{R}^n$ and an interval $(t_0, t_1) \in \mathbb{R}$

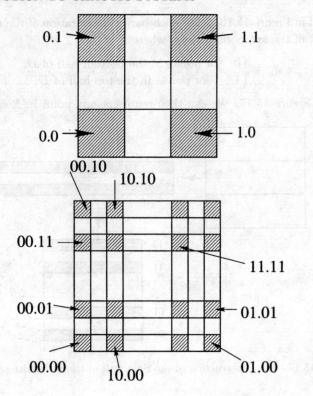

Fig. 15.18. The points in Λ, the invariant set.

to the point $\mathbf{x}_1 \in \mathbb{R}^n$, where $\mathbf{x}_1 = \mathbf{x}(t_1)$ and $\mathbf{x}$ is the solution of (15.17) subject to $\mathbf{x} = \mathbf{x}_0$ when $t = t_0$. That is $\mathbf{x}(t_0) = \mathbf{x}_0$ evolves to $\mathbf{x}(t_1) = \mathbf{x}_1$ in the n-dimensional phase space during the time interval (t_0, t_1). Note that $\phi^{t_0,t_0}(\mathbf{x}_0) = \mathbf{x}_0$ and $\phi^{t_2,t_1} \cdot \phi^{t_1,t_0} \equiv \phi^{t_2,t_0}$ for all $t_1 \in (t_0, t_2)$. In addition, since (15.17) is an autonomous system, we can express the flow evolution operator in terms of the length of the time interval alone, and we write

$$\phi^{t_1,t_0}(\mathbf{x}_0) = \phi^{t_1-t_0}(\mathbf{x}_0).$$

Consequently we have that ϕ^0 is the identity transformation, $\phi^t \cdot \phi^s \equiv \phi^{t+s}$ and $(\phi^t)^{-1} = \phi^{-t}$, so that ϕ is a group operator (see Chapter 10).

Example: The logistic equation

Let's construct the flow operator for the **logistic equation**,

$$\frac{dx}{dt} = x(1-x).$$

This is the continuous version of the logistic map, which we studied earlier. The logistic equation is separable, and we can therefore integrate it subject to the initial

15.3 THE POINCARÉ RETURN MAP

condition that $x = x_0$ when $t = 0$ to obtain the solution, and hence the evolution operator,

$$x(t) = \phi^t(x_0) = \frac{x_0 e^t}{x_0 e^t - x_0 + 1}.$$

Note that solutions of the logistic equation all have $x \to 1$ as $t \to \infty$, in complete contrast to the solutions of the logistic map. This shows that choosing to use a continuous rather than a discrete system can have dramatic consequences for the behaviour of the solutions.

We can now verify that $\phi^0(x_0) = x_0$ and

$$\phi^t \cdot \phi^s(x_0) = \phi^t(\phi^s(x_0)) = \frac{\phi^s(x_0)e^t}{\phi^s(x_0)e^t - \phi^s(x_0) + 1},$$

and, since $\phi^s(x_0) = x_0 e^s/(x_0 e^s - x_0 + 1)$,

$$\phi^t \cdot \phi^s(x_0) = \frac{x_0 e^s e^t}{x_0 e^s e^t - x_0 e^s + (x_0 e^s - x_0 + 1)} = \frac{x_0 e^{s+t}}{x_0 e^{s+t} - x_0 + 1},$$

which is equal to $\phi^{s+t}(x_0)$, as expected.

Equilibrium points of (15.17) satisfy $\phi^t(\mathbf{x}^*) = \mathbf{x}^*$ for all $t \in \mathbb{R}$ (or equivalently $f(\mathbf{x}^*) = \mathbf{0}$). Periodic solutions of (15.17) with period T satisfy $\mathbf{x}^*(t) = \mathbf{x}^*(t+T)$ for all $t \in \mathbb{R}$, and can be written in terms of the flow operator as $\phi^{t+T}(\mathbf{x}^*) = \phi^t(\mathbf{x}^*)$ for all $t \in \mathbb{R}$.

One way of obtaining a map from the system (15.17) is to sample the solution with period τ, so that $\phi^n(\mathbf{x}_0) = \mathbf{x}(t_0 + n\tau)$ for $n \in \mathbb{Z}$. This allows us to track the trajectory of particles in a stroboscopic way. A more useful map, which we can use to analyze the behaviour of integral paths close to a periodic solution, is the **Poincaré return map**. Let γ be a trajectory of (15.17), and consider the intersections of γ with $\Sigma \subset \mathbb{R}^n$ such that

(i) Σ has dimension $n - 1$,
(ii) Σ is transverse (nowhere parallel) to the integral paths,
(iii) all solutions in the neighbourhood of γ pass through Σ.

If γ intersects Σ at $\mathbf{x} = \mathbf{p}$, and its next intersection with Σ is at $\mathbf{x} = \mathbf{q}$, then the Poincaré return map, $P : \Sigma \to \Sigma$, maps the point $\mathbf{p}$ to the point $\mathbf{q}$. This is illustrated in Figure 15.19 for $n = 3$ where Σ is a plane. Note that an equilibrium point of the Poincaré return map corresponds to a limit cycle that intersects Σ once, and a periodic solution of period k corresponds to a limit cycle that intersects Σ k times.

Example

Let's try to construct the Poincaré return map for (15.17) with $n = 2$ when

$$\mathbf{f}(\mathbf{x}) = \begin{pmatrix} 4x + 4y - x(x^2 + y^2) \\ 4y - 4x - y(x^2 + y^2) \end{pmatrix},$$

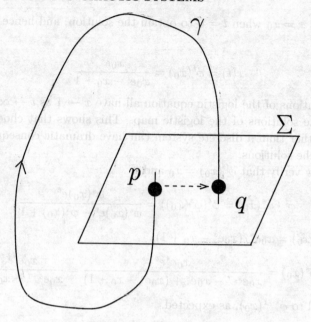

Fig. 15.19. The definition of the Poincaré return map.

and Σ is the positive x axis. It is easier to write this system in terms of plane polar coordinates, and, using (9.22) and (9.23), we find that $\dot{r} = r(4 - r^2)$ and $\dot{\theta} = -4$. These expressions can be integrated with initial conditions in Σ, the positive x-axis, so that $r(0) = x_0$ and $\theta(0) = 0$, to obtain

$$r = \sqrt{\frac{4x_0^2 e^{8t}}{4 - x_0^2 + x_0^2 e^{8t}}}, \quad \theta = -4t.$$

The orbit next returns to this axis when θ is an integer multiple of 2π, that is when $t = \pi/2$, so that

$$x_1 = P(x_0) = \sqrt{\frac{4x_0^2 e^{4\pi}}{4 - x_0^2 + x_0^2 e^{4\pi}}}.$$

The solution for $x_0 = 10$ is shown in Figure 15.20, along with the Poincaré return map, which rapidly asymptotes to $x = 2$.

Example: The Lorenz equations

Solutions of the Lorenz equations repeatedly cross the plane $y = -z$, which separates the two equilibrium points at $x = 27$, $y = z = \pm 6\sqrt{2}$. Indeed the equations are symmetric about this plane, since they are unchanged by the transformation $y \mapsto -y$, $z \mapsto -z$. We can therefore use the plane $y = -z$ as Σ, and hence define a Poincaré return map. In order to investigate this map, we must proceed numerically. In MATLAB, we need to define an **event function**

15.3 THE POINCARÉ RETURN MAP

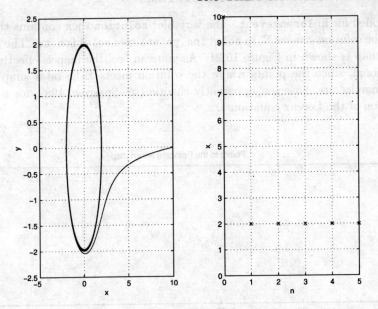

Fig. 15.20. The solution when $x_0 = 10$ and the corresponding Poincaré return map.

```
function [value, isterminal, direction] = lorenzevent(t,y)
value = y(2)+y(3); isterminal = 0; direction = 1;
```

This function returns the value zero when the solution intersects Σ (when $y + z =$ y(2)+y(3) $= 0$), but only as the solution approaches Σ with $y + z$ increasing (direction $= 1$), and allows the integration to continue (isterminal $= 0$). We can then use the commands

```
options = odeset('Events',@lorenzevent);
solution = ode45(@lor,[0 500], [1 1 1], options);
```

This integrates the Lorenz equations, which are in the function lor

```
function dy = lor(t,y)
dy(1) = -8*y(1)/3+y(2)*y(3);
dy(2) = 10*(y(3)-y(2));
dy(3) = -y(2)*y(1)+28*y(2)-y(3);
dy=dy';
```

for $0 \leqslant t \leqslant 500$, starting from $x = y = z = 1$ when $t = 0$. Note that the function odeset allows us to create a variable options that we can pass to ode45 as an argument, which controls its execution. In this case, we tell ode45 to detect the

AN INTRODUCTION TO CHAOTIC SYSTEMS

event coded for in `lorenzevent`. The variable† `solution` then contains the points where the solution crosses Σ, in `solution.ye`, at times `solution.te`. The Poincaré return map is shown in Figure 15.21. As you can see, the map is effectively one-dimensional, since the points where the solution meets Σ lie on a simple curve. The dynamics are, however, apparently chaotic (see Sparrow, 1982, for a detailed discussion of the Lorenz equations).

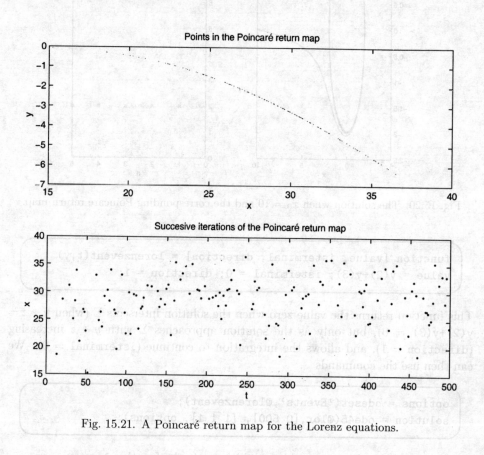

Fig. 15.21. A Poincaré return map for the Lorenz equations.

15.4 Homoclinic Tangles

Let's now consider a two-dimensional Poincaré return map, with $\mathbf{f} : \Sigma \to \Sigma$ and $\Sigma \subset \mathbb{R}^2$, associated with a three-dimensional system, (15.17). Each limit cycle solution of (15.17) is associated with an equilibrium point, $\bar{\mathbf{x}}$, of the map. If this equilibrium point is stable, the limit cycle is stable. Let's assume that a particular equilibrium point, $\bar{\mathbf{x}}$ of the Poincaré return map, which corresponds to a limit cycle solution

---

† The variable `solution`, produced by `ode45`, is a **structure**, and effectively contains more than one type of data in its definition.

15.4 HOMOCLINIC TANGLES

of (15.17), has a one-dimensional stable manifold, $\omega^s(\bar{\mathbf{x}})$, and a one-dimensional unstable manifold, $\omega^u(\bar{\mathbf{x}})$. As we saw earlier, it is possible for these manifolds to intersect. As we shall see, if this intersection is transverse, the manifolds become tangled, intersecting infinitely often, as shown in Figure 15.22. The manifolds then contain embedded horseshoe maps, and thus have chaotic solutions. We will now discuss why this should be so. Although we will proceed through lemmas and theorems, our approach is fairly informal, and should not be read as a rigorous proof.

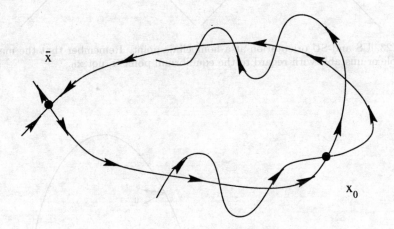

Fig. 15.22. A homoclinic tangle.

A **homoclinic point** is a point $\mathbf{x} \neq \bar{\mathbf{x}}$ that lies in the set $\omega^s(\bar{\mathbf{x}}) \cap \omega^u(\bar{\mathbf{x}})$, the intersection of the stable and unstable manifolds of $\bar{\mathbf{x}}$. Such a point asymptotes to $\bar{\mathbf{x}}$ as $n \to \pm\infty$. Under successive applications of the map and its inverse, a homoclinic point is mapped to a **homoclinic orbit**, a discrete set of points that is a subset of the stable and unstable manifolds.

We will now assume that the map $\mathbf{f}$ is area-preserving, so that, for any set $D \subset \Sigma$, the area of $\mathbf{f}(D)$ is equal to that of D. This assumption greatly simplifies the following discussion, but is not actually necessary (see Wiggins, 1988).

Lemma 15.2 *If $\mathbf{x}_0$ is a transverse homoclinic point, then all positive and negative iterates of $\mathbf{x}_0$ have the same orientation, either US or SU, as shown in Figure 15.23.*

Proof The Poincaré return map is associated with a flow. Figure 15.24 shows the trajectory through the points $\mathbf{x}_0$ and $\mathbf{f}(\mathbf{x}_0)$. Since the stable and unstable manifolds associated with this trajectory vary continuously, their orientations can at most rotate and cannot flip during their passage from one side of Σ to the other. □

474 AN INTRODUCTION TO CHAOTIC SYSTEMS

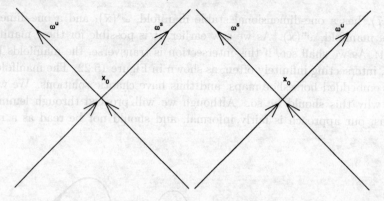

Fig. 15.23. US and SU orientation at a homoclinic point. Remember that the manifolds are stable or unstable with regard to the equilibrium point $\bar{\mathbf{x}}$, not $\mathbf{x}_0$.

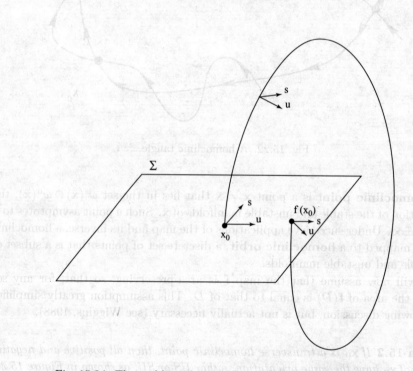

Fig. 15.24. The stable and unstable manifolds of the limit cycle.

Lemma 15.3 *If $\mathbf{x}_0$ is a transverse homoclinic point, then there must be another transverse homoclinic point between $\mathbf{x}_0$ and $\mathbf{f}(\mathbf{x}_0)$. In other words, the next transverse homoclinic point along say ω^s after $\mathbf{x}_0$ cannot be $\mathbf{f}(\mathbf{x}_0)$.*

15.4 HOMOCLINIC TANGLES

Proof If the orientation associated with x_0 is US, then the next homoclinic point along ω^s, z, has SU orientation, and by Lemma 15.2, $z \neq f(x_0)$, as shown in Figure 15.25. □

Note that this means that there are at least two transverse homoclinic orbits, one associated with x_0 and one associated with z.

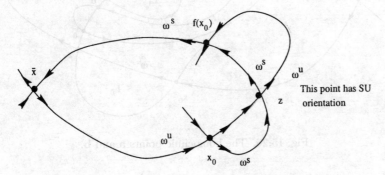

Fig. 15.25. The orientation of successive homoclinic points.

Lemma 15.4 *The existence of a single transverse homoclinic point ensures the existence of an infinite number of transverse homoclinic points.*

Proof We consider the image of the points within the lobe L_0, which is bounded by the stable and unstable manifolds through x_0 and z, as shown in Figure 15.26. The image of L_0 is the lobe L_1, which is bounded by the portions of ω^u and ω^s between $f(x_0)$ and $f(z)$. Similarly if $x \in L_0$ then $f^j(x) \in L_j$ which is bounded by the portions of ω^u and ω^s between $f^j(x_0)$ and $f^j(z)$. Since we know that $f^n(x_0) \to \bar{x}$ and $f^n(z) \to \bar{x}$ along ω^s as $n \to \infty$, the distance between these images must decrease. However, the distance between the points along ω^u must increase. This leads to long thin lobes, bounded by short sections of ω^s and long sections of ω^u. We therefore have a finite area covered by an infinite set of lobes of finite area equal to the area of A_0 (recall that we are assuming that the map preserves areas). Consequently, these lobes must overlap. We can assume, without loss of generality, that they overlap between z and $f(x_0)$. There are, therefore, two further transverse intersection points, which we label **a** and **b**. We can repeat this argument indefinitely, and conclude that there must be an infinite number of homoclinic points. □

Theorem 15.2 (Smale–Birkhoff) *Let* $f : \mathbb{R}^2 \to \mathbb{R}^2$ *be a diffeomorphism with a hyperbolic equilibrium point* $\bar{x}$. *If* $\omega^s(\bar{x})$ *and* $\omega^u(\bar{x})$ *intersect transversally at a point other than* $\bar{x}$, *in other words, if there is a homoclinic tangle, then the map has a horseshoe map embedded within it.*

Note that, since horseshoe maps have chaotic orbits, this theorem shows that the

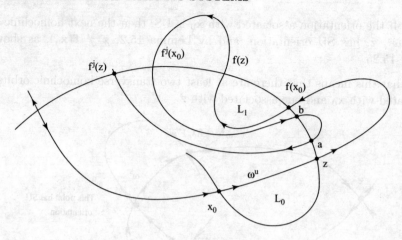

Fig. 15.26. The homoclinic points **a** and **b**.

existence of a homoclinic tangle implies the existence of chaotic orbits. As we shall see later, there is an algebraic test, Mel'nikov's method, that can be used to determine whether a system has a transverse homoclinic point.

Proof We will simply give an outline of the proof here. A more detailed proof, which uses the idea of Markov partitioning, is given in Guckenheimer and Holmes (1983). Our aim here is simply to convince you that the forward and backward maps of the region D, which we define below, intersect.

Consider a region D that contains a transverse homoclinic point, x_0, associated with an equilibrium point, $\bar{x}$ of a map $\mathbf{f} : \mathbb{R}^2 \to \mathbb{R}^2$, as shown in Figure 15.27. Then

(i) $f^k(D)$ contains the iterates of $f^k(x_0)$ for $k \in \mathbb{Z}$.

(ii) For $k < 0$, $f^k(D)$ is stretched along the stable manifold and contracted along the unstable manifold.

(iii) For $k > 0$, $f^k(D)$ is stretched along the unstable manifold and contracted along the stable manifold.

(iv) For some backward iterate $-q$, $f^{-q}(D) = D^-$ will have a horseshoe shape and intersect with $f^p(D) = D^+$ for some forward iterate p.

(v) The map $f^{-(p+q)}(D^+) = D^-$, which maps the region D^+ into a horseshoe shaped region, D^-, with overlap between D^- and D^+, is a horseshoe map.

□

15.4 HOMOCLINIC TANGLES

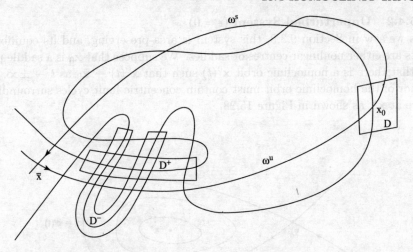

Fig. 15.27. A horseshoe map arising from the dynamics in a homoclinic tangle.

15.4.1 Mel'nikov Theory

We will now describe an algebraic test with which we can determine whether a system has a homoclinic tangle, and hence, by Theorem 15.2, chaotic solutions. We will focus on two-dimensional Hamiltonian systems perturbed by a periodic function of time only, although the method can also be used for perturbed non-Hamiltonian systems (see Wiggins, 1988).

We consider a Hamiltonian system (see Section 9.3.9) with Hamiltonian $H = H(x,y)$ and associated differential equations

$$\dot{\mathbf{x}} = \mathbf{f}(\mathbf{x})$$

where $\mathbf{f} = (-\partial H/\partial y, \partial H/\partial x)^{\mathrm{T}}$ or, written in component form,

$$\frac{dx}{dt} = -\frac{\partial H}{\partial y}, \quad \frac{dy}{dt} = \frac{\partial H}{\partial x}. \tag{15.18}$$

We will consider this system under the influence of a small perturbation that is a function of space and periodic in time, in the form

$$\dot{\mathbf{x}} = \mathbf{f}(\mathbf{x}) + \epsilon \mathbf{g}(\mathbf{x}, t), \tag{15.19}$$

with $\mathbf{g}(\mathbf{x},t) = \mathbf{g}(\mathbf{x},t+T)$ for some period $T > 0$, and $\epsilon \ll 1$. We start by considering the unperturbed system.

15.4.2 Unperturbed System ($\epsilon = 0$)

As we saw in Section 9.3.9, this system is area-preserving, and its equilibrium points are either nonlinear centres or saddles. We suppose that $\bar{\mathbf{x}}_0$ is a saddle point and that there is a homoclinic orbit $\mathbf{x}^0(t)$ such that $\mathbf{x}^0(t) \to \bar{\mathbf{x}}_0$ as $t \to \pm\infty$. The interior of the homoclinic orbit must contain concentric limit cycles surrounding a centre at $\mathbf{x}^*$, as shown in Figure 15.28.

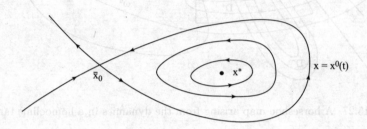

Fig. 15.28. A homoclinic orbit of the unperturbed Hamiltonian system.

15.4.3 Perturbed System ($0 < \epsilon \ll 1$)

We can think of the perturbed system as autonomous in the three-dimensional (x, y, t)-phase space. We can define an associated map by stroboscopically sampling the flow at $t = nT$ for $n \in \mathbb{Z}$. Since $\mathbf{g}(\mathbf{x}, t) = \mathbf{g}(\mathbf{x}, t + nT)$, this is equivalent to a Poincaré return map with Σ the plane $t = 0$. Note that the equilibrium point, $\bar{\mathbf{x}}_0$, of the *unperturbed* system of differential equations is also an equilibrium point of the associated map, and that the stable and unstable manifolds of $\bar{\mathbf{x}}_0$ are the same for the *unperturbed* system of differential equations and the associated map.

We assume that the influence of the perturbation is to modify the equilibrium point of the map to the point $\mathbf{x}^\epsilon$, with, since $\epsilon \ll 1$, $|\mathbf{x}^\epsilon - \mathbf{x}_0| \ll 1$. For the perturbed map, the stable and unstable manifolds of the saddle point do not necessarily connect smoothly, as shown in Figure 15.29. We would now like to find some way of distinguishing between the two cases – either disjoint or tangled manifolds. Let t_0 denote the time at which we wish to consider the fate of the solutions. Since the unperturbed system is autonomous, its orbits are invariant under arbitrary transformations in time, so that the orbits $\mathbf{x}^0(t)$ and $\mathbf{x}^0(t - t_0)$ are the same. This is not true for the solutions of the perturbed system. In Figure 15.30, we show the orbit $\mathbf{x}^0(t)$ as a dashed line and the stable and unstable manifolds of the saddle point of the perturbed system as solid lines. We will now introduce the idea of the distance between points on the stable and unstable manifolds of the saddle point. The idea is to find an expression for this distance and determine whether it can be

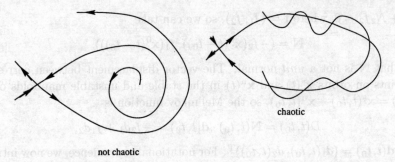

Fig. 15.29. The perturbed stable and unstable manifolds of the associated map.

zero. We can do this by defining the Mel'nikov function, which then provides an algebraic sufficient condition for the existence of transverse homoclinic intersections and hence chaotic dynamics.

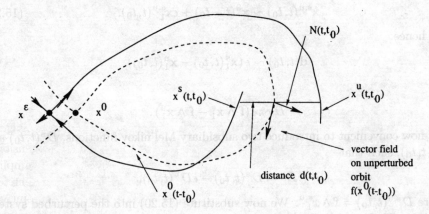

Fig. 15.30. The perturbed (solid lines) and unperturbed (broken lines) stable and unstable manifolds of the saddle point.

The **Mel'nikov function**, $D(t, t_0)$, is proportional to the component of the distance between corresponding points on the stable and unstable manifolds in the direction normal to the unperturbed homoclinic orbit, and is given by

$$D(t, t_0) = \mathbf{N}(t, t_0) \cdot \mathbf{d}(t, t_0).$$

Here $\mathbf{N}(t, t_0)$ is a normal to the unperturbed orbit and $\mathbf{d}(t, t_0)$ connects corresponding points on the stable and unstable manifolds. If $D(t, t_0)$ has simple zeros, then there must be transverse homoclinic intersections, and we can conclude that the system has chaotic solutions.

Firstly, we construct the normal to the unperturbed homoclinic orbit, $\mathbf{x}^0(t)$. The tangent to the orbit is $\mathbf{f}$, and hence $\mathbf{N}(t, t_0) \cdot \mathbf{f}(\mathbf{x}^0(t - t_0)) = 0$. In component form,

$N_1 f_1 + N_2 f_2 = 0$, where $\mathbf{f} = (f_1, f_2)$, so we can take

$$\mathbf{N} = (-f_2(\mathbf{x}^0(t - t_0)), f_1(\mathbf{x}^0(t - t_0))).$$

Note that $\mathbf{N}$ is not a *unit* normal. The vector displacement between corresponding points on orbits $\mathbf{x}^s(t)$ and $\mathbf{x}^u(t)$ in the stable and unstable manifolds of $\mathbf{x}^\epsilon$ is $\mathbf{d}(t, t_0) = \mathbf{x}^s(t, t_0) - \mathbf{x}^u(t, t_0)$, so the Mel'nikov function is

$$D(t, t_0) = \mathbf{N}(t, t_0) \cdot \mathbf{d}(t, t_0) = -f_2 d_1 + f_1 d_2,$$

where $\mathbf{d}(t, t_0) = (d_1(t, t_0), d_2(t, t_0))^T$. For notational convenience, we now introduce the binary operation $\wedge$ such that $\mathbf{u} \wedge \mathbf{v} = u_1 v_2 - v_1 u_2$, so that $D \equiv \mathbf{f} \wedge \mathbf{d}$. We can now use perturbation theory to obtain an expression for $D(t, t_0)$.

We assume that points on the orbits associated with $\mathbf{x}^\epsilon$ remain close to points on the homoclinic orbit $\mathbf{x}^0(t - t_0)$, so that we can write their locations as a position on the homoclinic orbit plus a small perturbation proportional to ϵ. We also introduce the superscript ^{s,u} so that we can discuss the stable and unstable cases simultaneously, writing

$$\mathbf{x}^{s,u}(t, t_0) \sim \mathbf{x}^0(t - t_0) + \epsilon \mathbf{x}_1^{s,u}(t, t_0), \qquad (15.20)$$

and hence

$$\mathbf{d}(t, t_0) \sim \epsilon \left(\mathbf{x}_1^s(t, t_0) - \mathbf{x}_1^u(t, t_0) \right)$$

and

$$D \sim \epsilon \left(\mathbf{f} \wedge \mathbf{x}_1^s - \mathbf{f} \wedge \mathbf{x}_1^u \right).$$

It is now convenient to introduce two subsidiary Mel'nikov functions, $D^s(t, t_0)$ and $D^u(t, t_0)$, such that

$$D \sim \epsilon D^s(t, t_0) - \epsilon D^u(t, t_0),$$

where $D^{s,u}(t, t_0) = \mathbf{f} \wedge x_1^{s,u}$. We now substitute (15.20) into the perturbed system, (15.19), and obtain

$$\dot{\mathbf{x}}^0 + \epsilon \dot{\mathbf{x}}_1^{s,u} = \mathbf{f}(\mathbf{x}^0 + \epsilon \mathbf{x}_1^{s,u}) + \epsilon \mathbf{g}(\mathbf{x}^0 + \epsilon \mathbf{x}_1^{s,u}, t)$$

$$= \mathbf{f}(\mathbf{x}^0) + \epsilon D f(\mathbf{x}^0) \mathbf{x}_1^{s,u} + \epsilon \mathbf{g}(\mathbf{x}^0, t) + O(\epsilon^2),$$

where $Df(\mathbf{x}^0)$ is the Jacobian of the unperturbed system. This equation is automatically satisfied at leading order, since $\mathbf{x}^0$ is a solution of the unperturbed equation, whilst at $O(\epsilon)$,

$$\dot{\mathbf{x}}_1^{s,u}(t, t_0) = Df(\mathbf{x}^0(t - t_0)) \mathbf{x}_1^{s,u}(t, t_0) + \mathbf{g}(\mathbf{x}^0(t - t_0), t). \qquad (15.21)$$

Now, differentiating the functions $D^{s,u}(t, t_0)$, we obtain

$$\dot{D}^{s,u}(t, t_0) = \dot{\mathbf{f}} \wedge \mathbf{x}_1^{s,u} + \mathbf{f} \wedge \dot{\mathbf{x}}_1^{s,u},$$

since there is a product rule associated with the $\wedge$ operator. However, we note that

$$\dot{\mathbf{f}} = \frac{d}{dt} \mathbf{f}(\mathbf{x}^0(t - t_0)) = Df(\mathbf{x}^0(t - t_0)) \dot{\mathbf{x}}^0(t - t_0),$$

15.4 HOMOCLINIC TANGLES

which implies that $\dot{\mathbf{f}} = D\mathbf{f}\mathbf{f}$, and hence

$$\dot{D}^{s,u}(t,t_0) = (D\mathbf{f}\mathbf{f}) \wedge \mathbf{x}_1^{s,u} + \mathbf{f} \wedge \dot{\mathbf{x}}_1^{s,u}.$$

Substituting from (15.21) into this equation gives

$$\dot{D}^{s,u}(t,t_0) = (D\mathbf{f}\mathbf{f}) \wedge \mathbf{x}_1^{s,u} + \mathbf{f} \wedge D\mathbf{f}\mathbf{x}_1^{s,u} + \mathbf{f} \wedge \mathbf{g}.$$

If we now use the identity

$$D\mathbf{f}\mathbf{f} \wedge \mathbf{x} + \mathbf{f} \wedge D\mathbf{f}\mathbf{x} \equiv (\nabla \cdot \mathbf{f})\mathbf{f} \wedge \mathbf{x},$$

and note that $\nabla \cdot \mathbf{f} = 0$ since the unperturbed system is Hamiltonian, we find that

$$\dot{D}^{s,u}(t,t_0) = \mathbf{f} \wedge \mathbf{g}.$$

Considering the unstable manifold first, we integrate from $-\infty$ to t_0 to obtain

$$\int_{-\infty}^{t_0} \dot{D}^u(t,t_0)\, dt = \int_{-\infty}^{t_0} \mathbf{f} \wedge \mathbf{g}\, dt,$$

which gives

$$D^u(t_0, t_0) - D^u(-\infty, t_0) = \int_{-\infty}^{t_0} \mathbf{f} \wedge \mathbf{g}\, dt. \qquad (15.22)$$

Now consider the stable case integrated from t_0 to $+\infty$, which gives

$$\int_{t_0}^{\infty} \dot{D}^s(t,t_0)\, dt = \int_{t_0}^{\infty} \mathbf{f} \wedge \mathbf{g}\, dt,$$

and hence

$$-D^s(t_0, t_0) + D^s(\infty, t_0) = \int_{t_0}^{\infty} \mathbf{f} \wedge \mathbf{g}\, dt. \qquad (15.23)$$

We note that as $t \to -\infty$ along the unstable orbit and as $t \to \infty$ along the stable orbit, both solutions tend to $\mathbf{x}^\epsilon$, so that $D^u(-\infty, t_0) = D^s(\infty, t_0)$. Adding (15.22) and (15.23) and changing notation slightly so that we replace (t_0, t_0) with (t_0), we have

$$D(t_0) \sim \epsilon(D^s(t_0) - D^u(t_0)) = -\epsilon \int_{-\infty}^{\infty} \mathbf{f} \wedge \mathbf{g}\, dt$$

$$= -\epsilon \int_{-\infty}^{\infty} \left[\mathbf{f}(\mathbf{x}^0(t - t_0)) \wedge \mathbf{g}(\mathbf{x}^0(t - t_0), t) \right] dt.$$

Note that

(i) If $D(t_0)$ has *simple zeros* for sufficiently small ϵ, then ω^u and ω^s intersect transversally, and hence by Theorem 15.2, there are chaotic solutions.
(ii) If $D(t_0)$ is bounded away from zero, there are no transverse homoclinic intersections.
(iii) If we can find one zero of a Mel'nikov function, the dynamics of the system ensure that there will be infinitely many zeros.

Example

Consider the perturbed Hamiltonian system given by the forced Duffing equation, (15.3),

$$\ddot{x} = x - x^3 - \epsilon(\delta \dot{x} - \gamma \cos \omega t),$$

when $\epsilon \ll 1$. For what values of δ, γ and ω are there chaotic solutions?

We begin by rewriting the system as

$$\frac{d}{dt}\begin{pmatrix} x \\ y \end{pmatrix} = \begin{pmatrix} y \\ x - x^3 - \epsilon(\delta y - \gamma \cos \omega t) \end{pmatrix}. \qquad (15.24)$$

The unperturbed system is

$$\frac{d}{dt}\begin{pmatrix} x \\ y \end{pmatrix} = \begin{pmatrix} y \\ x - x^3 \end{pmatrix}.$$

This is a Hamiltonian system with

$$\frac{d}{dt}\begin{pmatrix} x \\ y \end{pmatrix} = \begin{pmatrix} \partial H/\partial y \\ -\partial H/\partial x \end{pmatrix}.$$

and Hamiltonian

$$H = \frac{1}{2}y^2 + \frac{1}{4}x^4 - \frac{1}{2}x^2.$$

The unperturbed system has equilibrium points at $(x, y) = (0, 0)$ and $(x, y) = (\pm 1, 0)$, and it is straightforward to show that the origin is a saddle point and that the other two equilibria are centres. The unperturbed phase portrait is shown in Figure 15.31. We now need to determine the equation of the homoclinic orbit. This orbit must pass through the origin, where $H = 0$. However, H is a constant on the orbit, so the homoclinic orbit is given by

$$y^2 = x^2 - \frac{1}{2}x^4.$$

Setting y equal to zero, we find that the homoclinic orbit also meets the x-axis where $x = \pm\sqrt{2}$. We will take $y(0) = 0$ and $x(0) = \sqrt{2}$. We can then solve the differential equation for the homoclinic orbit and find that $x^0(t) = \sqrt{2}\text{sech}\,t$. This gives us

$$x^0(t - t_0) = \sqrt{2}\text{sech}(t - t_0), \quad y^0(t - t_0) = -\sqrt{2}\text{sech}(t - t_0)\tanh(t - t_0).$$

We can now construct the Mel'nikov function. The vectors $\mathbf{f}$ and $\mathbf{g}$ are

$$\mathbf{f}(\mathbf{x}^0(t - t_0)) = \begin{pmatrix} y^0(t - t_0) \\ x^0(t - t_0) - (x^0(t - t_0))^3 \end{pmatrix}$$

and

$$\mathbf{g}(\mathbf{x}^0(t - t_0)) = \begin{pmatrix} 0 \\ \gamma \cos \omega t - \delta y^0(t - t_0) \end{pmatrix},$$

so that their wedge product is

$$\mathbf{f} \wedge \mathbf{g} = y^0(t - t_0)(\gamma \cos \omega t - \delta y^0(t - t_0)).$$

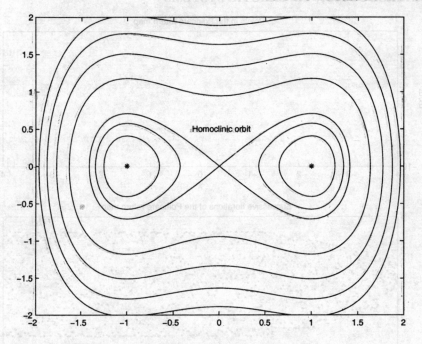

Fig. 15.31. The phase portrait of (15.24) with $\epsilon = 0$.

Now, using the definition of the Mel'nikov function,

$$D(t_0) = -\epsilon \int_{-\infty}^{\infty} -\sqrt{2}\operatorname{sech}(t-t_0)\tanh(t-t_0)$$

$$\times \left[\cos\omega t + \delta\sqrt{2}\operatorname{sech}(t-t_0)\tanh(t-t_0)\right] dt.$$

This can be integrated to give

$$D(t_0) = -\epsilon\left(-\sqrt{2}\pi\gamma\omega\operatorname{sech}\left(\frac{\pi\omega}{2}\right)\sin\omega t_0 + \frac{4\delta}{3}\right).$$

This has simple zeros when

$$\delta = \frac{3\sqrt{2}\pi}{4}\gamma\omega\operatorname{sech}\left(\frac{\pi\omega}{2}\right)\sin\omega t_0,$$

provided that

$$\delta < \frac{3\sqrt{2}\pi}{4}\gamma\omega\operatorname{sech}\left(\frac{\pi\omega}{2}\right).$$

This means that there are transverse homoclinic points when this condition is satisfied, and we can infer that the system is chaotic. Figure 15.32 shows the Poincaré return map when $\epsilon = 0.1$, $\delta = 0$, $\gamma = 1$ and $\omega = \frac{1}{3}$.

484 AN INTRODUCTION TO CHAOTIC SYSTEMS

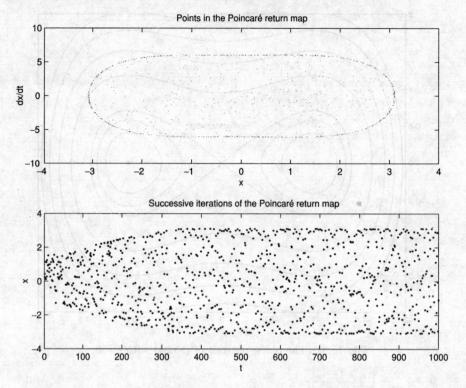

Fig. 15.32. The Poincaré return map for the forced Duffing equation when $\epsilon = 0.1$, $\delta = 0$, $\gamma = 1$ and $\omega = \frac{1}{3}$.

15.5 Quantifying Chaos: Lyapunov Exponents and the Lyapunov Spectrum

We have now seen how to determine analytically when chaotic solutions exist for weakly, periodically perturbed Hamiltonian systems. What can we say about other nonlinear systems of differential equations? If we can solve such a system numerically, and it appears to have chaotic solutions, can we characterize and quantify the chaos? In this final section, we will introduce the ideas of the maximum Lyapunov exponent and the Lyapunov spectrum, which we can use for this purpose.

15.5.1 Lyapunov Exponents of Systems of Ordinary Differential Equations

In a chaotic system, we have seen that neighbouring trajectories diverge. In fact this divergence is usually exponentially fast. If we can quantify this rate of

15.5 LYAPUNOV EXPONENTS AND THE LYAPUNOV SPECTRUM

divergence, we can quantify the chaotic solutions. Consider the system

$$\frac{d\mathbf{x}}{dt} = \mathbf{f}(\mathbf{x}), \tag{15.25}$$

where $\mathbf{f}: \mathbb{R}^n \to \mathbb{R}^n$. Let $\bar{\mathbf{x}}(t)$ be a **reference trajectory**, and consider a neighbouring trajectory, $\mathbf{y}(t)$, that has $\mathbf{y}(0) = \bar{\mathbf{x}}(0) + \Delta \mathbf{x}(0)$. As $t \to \infty$, and the trajectories diverge, we expect that $\Delta \mathbf{x}(t) = \mathbf{y}(t) - \bar{\mathbf{x}}(t) \sim \Delta \mathbf{x}(0) e^{\lambda t}$, as shown in Figure 15.33. If $\lambda < 0$, the trajectories actually converge, whilst if $\lambda > 0$ the trajectories diverge, and λ gives a measure of the rate of divergence.

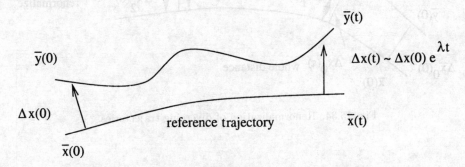

Fig. 15.33. Neighbouring trajectories diverge exponentially fast.

Formally, we define the **maximum Lyapunov exponent** with respect to a reference trajectory of a flow as

$$\lambda_{\max} = \lim_{\substack{||\Delta \mathbf{x}(0)|| \to 0 \\ t \to \infty}} \frac{1}{t} \log \frac{||\Delta \mathbf{x}(t)||}{||\Delta \mathbf{x}(0)||}, \tag{15.26}$$

where $||\mathbf{x}|| = \sqrt{\sum_i x_i^2}$ is the Euclidean norm (see Section A1.2). This gives us the basic numerical recipe for computing $\lambda_{\max}$ from two neighbouring solutions. Recall that the idea which is central to this definition is linearization about the trajectory $\bar{\mathbf{x}}(t)$, and consequently we need to ensure that we can indeed linearize by considering neighbouring trajectories ($\Delta \mathbf{x}(0) \to 0$). In this limit, $\Delta \mathbf{x}(t)$ is governed by the linearized equation

$$\frac{d\Delta \mathbf{x}}{dt} = Df(\bar{\mathbf{x}}(t))\Delta \mathbf{x}. \tag{15.27}$$

Formally $\Delta \mathbf{x}(t) = \phi_L^t(\Delta \mathbf{x}(0))$ where $\phi_L^t(.)$ is a linear evolution operator, so that $\phi_L^t(\alpha \Delta \mathbf{x}) = \alpha \phi_L^t(\Delta \mathbf{x})$. If $\lambda_{\max}$ is positive, numerical integration of (15.27) will lead to exponentially growing solutions. This can be avoided in a nonlinear system by **renormalizing**. The idea is to integrate forward in time until the two trajectories become a given distance apart, and then scale $\Delta \mathbf{x}$, so that the calculation can continue with a trajectory that is within a small distance of the reference orbit, as shown in Figure 15.34. One way of doing this is to renormalize after regular time intervals τ. The maximum Lyapunov exponent is then given by the average of the

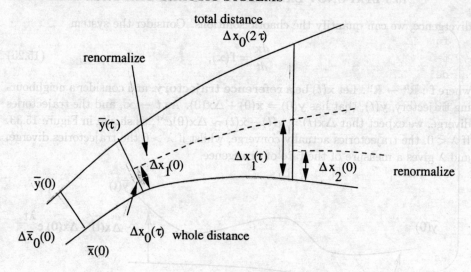

Fig. 15.34. Renormalization of diverging trajectories.

exponent calculated between each renormalization,

$$\lambda_{\max} = \lim_{n \to \infty} \frac{1}{(n+1)\tau} \sum_{j=0}^{n} \log \frac{||\Delta \mathbf{x}_j(\tau)||}{||\Delta \mathbf{x}_j(0)||}, \qquad (15.28)$$

provided that we define our renormalization to be

$$\Delta \mathbf{x}_n(0) = \delta \frac{\Delta \mathbf{x}_{n-1}(\tau)}{||\Delta \mathbf{x}_{n-1}(\tau)||} \quad \text{for } n = 1, 2, \ldots,$$

where $\delta \ll 1$ and we note that $||\mathbf{x}_n(0)|| = \delta$.

Example: The forced Duffing equation

Consider the forced Duffing equation, (15.3), with $\epsilon\delta = \frac{1}{2}$ and $\epsilon\gamma = \frac{3}{5}$, a solution of which is shown in Figure 15.2. We can now construct the maximum Lyapunov exponent using the MATLAB function

15.5 LYAPUNOV EXPONENTS AND THE LYAPUNOV SPECTRUM

```
function avls = lyapunov(n,tau,del0,y,eqn)
t=0; ls=zeros(1,n); avls=ls;
del=del0*[1 1]/sqrt(2);

for i=1:n
   tspan = [t t+tau];
   [tout1 yout1]=ode45(eqn,tspan, y);
   [tout2 yout2]=ode45(eqn,tspan, y+del);
   delxe= [yout1(end,:)-yout2(end,:)];
   nd=norm(delxe);
   ls(i) = log(nd/del0); avls(i) = sum(ls)/i/tau;
   del = del0*delxe/nd;
   y = yout1(end,:); t = t+tau;
end
```

The arguments of the function lyapunov are n, the number of separate evaluations of the maximum Lyapunov exponent, tau = τ, del0 = δ, y, the vector of initial data, and eqn, a handle containing the name of the equation to be integrated. The built-in function norm(delxe) calculates the Euclidean norm of the vector delxe. The command

$$\text{plot(lyapunov(1000,1,0.01,[0 0],@duffing))}$$

produces Figure 15.35, which shows how the average maximum Lyapunov exponent converges to a value of about 0.1. Since this is positive, neighbouring trajectories diverge exponentially fast – an indication that the solution has sensitive dependence upon initial conditions, and hence is chaotic.

One difficulty associated with calculating the maximum Lyapunov exponent in this way is that the choice of the direction of the initial displacement, $\Delta \mathbf{x}_0$, can have an effect. We will see how to overcome this problem in the next section.

15.5.2 The Lyapunov Spectrum

Although we now have a way of characterizing the rate of divergence of neighbouring trajectories, this is rather a blunt tool, and can depend upon the direction of the initial displacement from the reference trajectory. For an n-dimensional system of differential equations, we can overcome these problems by defining n quantities that characterize the growth of line[†], area, volume and hypervolume elements.

Consider the system (15.25). As we saw above, the time evolution of small variations $\Delta \mathbf{x}(t)$ about the reference trajectory, $\bar{\mathbf{x}}(t)$, is governed by

$$\frac{d\Delta \mathbf{x}}{dt} = Df(\bar{\mathbf{x}}(t))\Delta \mathbf{x}.$$

We can write the solution of this system as

$$\Delta \mathbf{x}(t) = M(t)\Delta \mathbf{x}(0),$$

† This is just the maximum Lyapunov exponent.

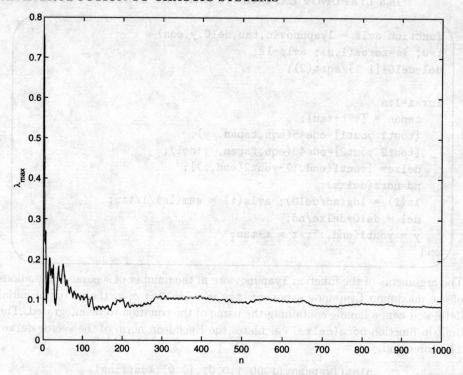

Fig. 15.35. The maximum Lyapunov exponent of the forced Duffing equation, estimated over 1000 iterations.

where the evolution operator $M(t)$ is the **fundamental matrix**. We can construct M numerically. The first column of M is $\Delta \mathbf{x}^{(1)}(t)$, the solution subject to the initial condition $\Delta \mathbf{x}^{(1)}(0) = (1, 0, \ldots, 0)^T$. The jth column is $\Delta \mathbf{x}^{(j)}$ subject to the initial condition $\Delta \mathbf{x}^{(j)}(0) = (0, \ldots, 1, 0, \ldots, 0)^T$ where the one is in the jth position. By using this construction we find that

$$\Delta \mathbf{x}^{(j)}(t) = M(t) \Delta \mathbf{x}^{(j)}(0),$$

with

$$M(t) = (\Delta \mathbf{x}^{(1)}(t), \Delta \mathbf{x}^{(2)}(t), \ldots, \Delta \mathbf{x}^{(n)}(t)).$$

The fundamental matrix $M(t)$ has n eigenvalues $\{m_i(t)\}$ and the **Lyapunov exponents** are defined as

$$\lambda_i = \lim_{t \to \infty} \frac{1}{t} \log |m_i(t)| \text{ for } i = 1, 2, \ldots, n. \tag{15.29}$$

We say that the set of n Lyapunov exponents is the **Lyapunov spectrum**. The exponents can be ordered as $\lambda_1 \geq \lambda_2 \geq \cdots \geq \lambda_n$, and $\lambda_1 = \lambda_{\max}$, the maximum Lyapunov exponent. As before, if one of the exponents is positive, this is a sign of

15.5 LYAPUNOV EXPONENTS AND THE LYAPUNOV SPECTRUM

sensitive dependence upon initial conditions, and we need to use a renormalization scheme to calculate the Lyapunov spectrum.

Example: The cubic crosscatalator

We can calculate the Lyapunov spectrum of the cubic crosscatalator equations, (15.10) using the MATLAB function

```
function avlambda = lyapunovspectrum(n,tau,del0,y,eqn)
t=0; avlambda = zeros(3,n); lambda = avlambda;
ex = [1 0 0]; ey = [0 1 0]; ez = [0 0 1];

for i=1:n
   tspan = [t t+tau];
   [tout yout]=ode45(eqn,tspan, y);
   [toutx youtx]=ode45(eqn,tspan, y+del0*ex);
   [touty youty]=ode45(eqn,tspan, y+del0*ey);
   [toutz youtz]=ode45(eqn,tspan, y+del0*ez);

   delx= [youtx(end,:)-yout(end,:)]/del0;
   dely= [youty(end,:)-yout(end,:)]/del0;
   delz= [youtz(end,:)-yout(end,:)]/del0;

   m = eig([delx; dely; delz]);
   lambda(:,i) = log(abs(m));
   avlambda(:,i) = sum(lambda,2)/i/tau;
   y = yout(end,:); t = t+tau;
end
```

The built-in function `eig(A)` calculates the eigenvalues of the matrix **A**. Figure 15.36 shows the estimates of the three Lyapunov exponents converging over 1000 iterations. One of these is positive, which indicates that the solution depends sensitively upon initial conditions. Since the other two elements of the Lyapunov spectrum are negative, this indicates that neighbouring solutions diverge in one direction and approach each other in the two perpendicular directions.

We should note that there are some difficulties associated with calculating the Lyapunov exponents numerically. There may be different sets of exponents in different regions of phase space. This means that the Lyapunov spectrum may depend upon the starting condition. It is also notoriously hard to obtain convergence for Lyapunov exponents, especially where there are regions characterized by rotation, such as in the neighbourhood of centres.

There are several further points that we can make about the Lyapunov spectrum.

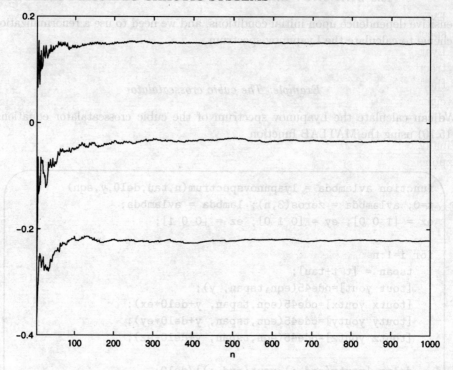

Fig. 15.36. The Lyapunov spectrum of the cubic crosscatalator equations.

We can also define the N-**dimensional Lyapunov exponent**,

$$\Lambda_N = \sum_{j=1}^{N} \lambda_j$$

This allows us to consider the growth of various elements, so that line elements grow like $e^{\Lambda_1 t}$, area elements grow like $e^{\Lambda_2 t}$, and so on. For the cubic crosscatalator, we can see that Λ_1 and Λ_2 are positive, so that line and area elements expand, but that Λ_3 is negative, so that volume elements contract.

The Lyapunov exponents of an equilibrium point are just the real parts of its eigenvalues. To see this, note that, in the neighbourhood of an equilibrium point, $\mathbf{x} = \mathbf{x}^*$,

$$\frac{d\Delta \mathbf{x}}{dt} = Df(\mathbf{x}^*)\Delta \mathbf{x},$$

with $Df(\mathbf{x}^*)$ a constant matrix. This means that the fundamental matrix is $M = \exp(Df(\mathbf{x}^*)t)$ (see Section 14.3.2). If μ_i are the eigenvalues of $Df(\mathbf{x}^*)$, then the eigenvalues of M are $e^{\mu_i t}$. On substituting these into the definition (15.29), we obtain $\lambda_i = \text{Re}(\mu_i)$.

— Every point in the basin of attraction of an attractor has the same Lyapunov spectrum.
— Lyapunov exponents characterize average rates of expansion ($\lambda_i > 0$) and contraction ($\lambda_i < 0$) in phase space. For conservative systems, $\sum_{i=1}^{n} \lambda_i = 0$, since the determinant of the Jacobian is unity, which means that the product of its eigenvalues is unity, and hence the sum of the Lyapunov exponents is zero. A **dissipative system**, for which volume elements contract, has $\sum_{i=1}^{n} \lambda_i < 0$.

Finally, we note that if a dynamical system is chaotic then at least one Lyapunov exponent is positive.

Exercises

15.1 Write down the equations that govern the concentrations of the chemicals involved in the cubic crosscatalator scheme, (15.4) to (15.9). After defining suitable dimensionless variables, derive (15.10), noting carefully the conditions on the initial concentration of P and the rate at which P decays under which the equations are a good approximation.

15.2 Determine the period 2 points of the map
$$f : x \mapsto 4x|1.$$

15.3 Consider the map
$$H(x) = \begin{cases} 3x, & \text{for } x \in [0, \tfrac{1}{2}], \\ -2 + 3x, & \text{for } x \in (\tfrac{1}{2}, 1]. \end{cases}$$

Prove that the only points that remain in $[0, 1]$ have the form
$$x = \sum_{n=1}^{\infty} \frac{a_n}{3^n}, \quad \text{with } a_n \in \{0, 2\}.$$

If $\mathbf{a}(x) = a_1 a_2 a_3 \ldots$, with $a_n \in \{0, 2\}$, show that
$$\mathbf{a}(H(x)) = \sigma(\mathbf{a}(x)),$$
where $\sigma \mathbf{a} = \mathbf{b}$ if $b_n = a_{n+1}$. For what value of x is $\mathbf{a}(x) = 002002002002\ldots$? Show that this point is periodic of period 3 under H.

15.4 In order to determine the nth roots of a, we can try to determine the zeros of $f(x, a) = x^n - a$ using the Newton–Raphson iteration. This is given by
$$x_{i+1} = x_i - \frac{f(x_i)}{f'(x_i)}.$$

(a) Write down the map for the iterates in the determination of the roots, and show for $n = 2$ that
$$x_{i+1} = \frac{1}{2}\left(x_i + \frac{a}{x_i}\right).$$

(b) Show that, for general n, the fixed points of the map are $a^{1/n}$, and determine the stability of these points by considering $x_i + \epsilon$ for $\epsilon \ll 1$. Do you think that this is a good way of determining the nth roots of a?

15.5 Determine the fixed point of the map

$$y_{n+1} = \frac{1}{1+y_n} \quad \text{for } y_n > 0,$$

and show that with $y_n = x_n/x_{n+1}$, x_n are the Fibonacci numbers.

15.6 Consider two maps $\mathbf{f}$ and $\mathbf{g}$ such that $\mathbf{f}(\mathbf{x}) : \mathbf{x} \mapsto A\mathbf{x}$ and $\mathbf{g}(\mathbf{x}) : \mathbf{x} \mapsto B\mathbf{x}$, where $\mathbf{x} = (x, y)^T$, and A, B are real 2×2 matrices.

(a) Determine a condition for $\mathbf{f}(\mathbf{x})$ to have a fixed point other than the trivial one. Comment on the fixed points of the composition of $\mathbf{f}(\mathbf{g}(\mathbf{x}))$ and $\mathbf{g}(\mathbf{f}(\mathbf{x}))$, stating a condition for which these are the same set of points.

(b) If $\det(A) = \det(B) = 1$, and hence the corresponding maps are area-preserving, comment on the properties of the composition $\mathbf{f}(\mathbf{g}(\mathbf{x}))$. Discuss the different options for $\mathbf{f}$ and $\mathbf{g}$ given that the maps are area-preserving.

(c) Consider

$$A = \begin{pmatrix} 1 & 0 \\ 0 & -1 \end{pmatrix}, \quad B = \begin{pmatrix} 1 & 1 \\ 0 & 1 \end{pmatrix},$$

and determine the unstable manifolds, where applicable, of the maps $\mathbf{f}$, $\mathbf{g}$ and $\mathbf{f}(\mathbf{g})$.

15.7 Express the system

$$\dot{x} = x^3 + xy^2 - x - 2y, \quad \dot{y} = yx^2 + y^3 - y + 2x$$

in terms of the polar coordinates (r, θ) and hence calculate the Poincaré return map $P : \mathbb{R} \to \mathbb{R}$ as the map of successive intersections of the orbit $\mathbf{x}(t)$ with the positive y-axis. Show that orbits starting on the y-axis outside the circle $r = \sqrt{e^{2\pi}/(e^{2\pi}-1)}$ never return to the y-axis.

15.8 After defining a suitable plane Σ, use MATLAB to calculate a Poincaré return map for (i) the forced Duffing equation and (ii) the cubic crosscatalator equations.

15.9 The equation of motion of a forced simple pendulum is

$$\frac{d^2\theta}{dt^2} + \sin\theta = \epsilon(\alpha + \gamma \cos \omega t),$$

where α, γ and ϵ are positive constants.

Show that if $\epsilon = 0$ then there is a pair of heteroclinic orbits connecting saddle points of $(\pm\pi, 0)$ in the (θ, ϕ)-plane, where $\phi = d\theta/dt$. Deduce that one of the orbits is given by

$$\theta_0(t) = 2\tan^{-1}(\sinh t), \quad \phi_0(t) = 2\text{sech}(t).$$

Show that the Mel'nikov function for the perturbation problem of intersection near $(0,2)$ of the unstable manifold from $(-\pi, 0)$ and the stable manifold to $(\pi, 0)$ can be expressed as

$$M(t_0) = \int_{-\infty}^{\infty} \phi_0(t - t_0)(\alpha + \gamma \cos \omega t)\, dt,$$

and deduce that

$$M(t_0) = 2\pi \left\{ \alpha + \gamma \operatorname{sech}\left(\frac{\pi \omega}{2}\right) \cos(\omega t_0) \right\}.$$

Hence show that there is chaos for small ϵ if $\gamma > \alpha \cosh(\tfrac{1}{2}\pi\omega)$.

15.10 **Arnol'd's cat map** maps the torus $\mathbb{T}^2 = \mathbb{R}^2/\mathbb{Z}^2$ to itself, and is given by $\mathbf{x}_{n+1} = \mathbf{f}(\mathbf{x}_n)$, where $\mathbf{x}_n = (x_n, y_n)$ and

$$\mathbf{f}(x,y) = (x + y \text{ modulo } 1, x + 2y \text{ modulo } 1).$$

Show that this map is area-preserving. Find its Lyapunov exponents.

15.11 Calculate the Lyapunov spectrum of the Lorenz equations.

15.12 **Project** The two-dimensional motion of a particle in a flow with stream function $\psi = \psi(x, y, t)$ is governed by the equations

$$\dot{x} = \frac{\partial \psi}{\partial y}, \quad \dot{y} = -\frac{\partial \psi}{\partial x}. \tag{E15.1}$$

Consider the motion of a particle under the influence of an impulsive Stokes flow, for which momentum is negligibly small, and $\psi = \psi^*(x,y)\delta(t-t_0)$. We can integrate (E15.1) to show that

$$x_1 - x_0 = \left.\frac{\partial \psi^*}{\partial y}\right|_{(x_0,y_0)}, \quad y_1 - y_0 = -\left.\frac{\partial \psi^*}{\partial x}\right|_{(x_0,y_0)}, \tag{E15.2}$$

where (x_0, y_0) is the position of the particle before the impulse and (x_1, y_1) its position after the impulse.

(a) Show that the map defined in (E15.2) is not area-preserving, but that the map

$$x_1 - x_0 = \left.\frac{\partial \psi^*}{\partial y}\right|_{(x_0,y_1)}, \quad y_1 - y_0 = -\left.\frac{\partial \psi^*}{\partial x}\right|_{(x_0,y_1)}, \tag{E15.3}$$

is area-preserving.

(b) Write a MATLAB script that iterates points forward under the influence of the pulsed flow

$$\psi(x, y, t) = \sum_{n=0}^{\infty} \psi^*(x, y)\delta(t - n),$$

where the stream function is that associated with a point force at

$(x, y) = (d, h)$ above a solid plane at $y = 0$. This is known as a **Stokeslet**, and its stream function is given by

$$\psi_{d,h}^*(x, y) =$$

$$\alpha(x - d) \left[\frac{1}{2} \log \left\{ \frac{(x-d)^2 + (y+h)^2}{(x-d)^2 + (y-h)^2} \right\} - \frac{2hy}{(x-d)^2 + (y+h)^2} \right].$$

Here α is the strength parameter (see Otto, Yannacopolous and Blake, 2001, for more details). Use the area-preserving map (E15.3). This will give a stroboscopic plot of the trajectory of the point.

(c) Consider the effect of alternating Stokeslets at $(0, \frac{1}{2})$ and $(0, \frac{3}{2})$ with the same strengths. Show that this leads to chaotic dynamics. This is similar to Aref's blinking vortex which leads to chaotic advection (Aref, 1984).

(d) Now consider the flow associated with Stokeslets that are not on a vertical line, for instance $(-\frac{1}{2}, \frac{1}{2})$ and $(\frac{1}{2}, \frac{1}{2})$ or $(-\frac{1}{2}, \frac{1}{2})$ and $(\frac{1}{2}, \frac{3}{2})$.

(e) By constructing the Jacobian associated with the flow, determine the nature of any fixed or periodic points, and determine the Lyapunov exponents associated with the flow.

(f) This model can be extended to other flows (see Ottino, 1989). Investigate the combination of other fundamental solutions of Stokes flow.

APPENDIX 1

Linear Algebra

A1.1 Vector Spaces Over the Real Numbers

Let V be a set of objects called **vectors**, of the form $V = \{\ldots, \mathbf{x}, \mathbf{y}, \mathbf{z}, \ldots\}$, and let $\mathbb{R}$ denote the real numbers, or **scalars**. The set V forms a **vector space** over $\mathbb{R}$ if, for all $\mathbf{x}, \mathbf{y}, \mathbf{z} \in V$ and $\alpha, \beta \in \mathbb{R}$,

(i) $\mathbf{x} + \mathbf{y} = \mathbf{y} + \mathbf{x}$,
(ii) $(\mathbf{x} + \mathbf{y}) + \mathbf{z} = \mathbf{x} + (\mathbf{y} + \mathbf{z})$,
(iii) $\mathbf{x} + \mathbf{0} = \mathbf{x}$,
(iv) $\mathbf{x} + (-\mathbf{x}) = \mathbf{0}$,
(v) $\alpha(\mathbf{x} + \mathbf{y}) = \alpha\mathbf{x} + \alpha\mathbf{y}$,
(vi) $(\alpha + \beta)\mathbf{x} = \alpha\mathbf{x} + \beta\mathbf{x}$,
(vii) $(\alpha\beta)\mathbf{x} = \alpha(\beta\mathbf{x})$,
(viii) $1\mathbf{x} = \mathbf{x}$.

These conditions are the familiar laws of commutativity, associativity and distributivity for the vectors and scalars, together with the existence of inverses and identities for the scalars and vectors.

Examples

(i) $V = \mathbb{R}^n$, so that

$$\mathbf{x} = \begin{pmatrix} x_1 \\ x_2 \\ \vdots \\ x_n \end{pmatrix}, \quad \mathbf{y} = \begin{pmatrix} y_1 \\ y_2 \\ \vdots \\ y_n \end{pmatrix},$$

are n-dimensional vectors that can be written in terms of their coordinates. If we then define vector addition and scalar multiplication by

$$\mathbf{x} + \mathbf{y} = \begin{pmatrix} x_1 + y_1 \\ x_2 + y_2 \\ \vdots \\ x_n + y_n \end{pmatrix}, \quad \alpha\mathbf{x} = \begin{pmatrix} \alpha x_1 \\ \alpha x_2 \\ \vdots \\ \alpha x_n \end{pmatrix},$$

it is straightforward to verify that $\mathbb{R}^n$ is a vector space over $\mathbb{R}$.

(ii) $V = P_n = \{a_n x^n + a_{n-1} x^{n-1} + \cdots + a_1 x + a_0 \mid a_n \in \mathbb{R}, \, x \in [\alpha, \beta]\}$, so that V is the set of polynomials of degree n with domain $x \in [\alpha, \beta]$. A typical member of V would be an object of the form $6x^3 - 2x + 1$. If we define vector addition and scalar multiplication by

$$(\mathbf{f} + \mathbf{g})(x) = \mathbf{f}(x) + \mathbf{g}(x), \quad (\alpha \mathbf{f})(x) = \alpha \mathbf{f}(x),$$

for $x \in [\alpha, \beta]$, and the zero function by $\mathbf{0}(x) = 0$, it is again easy to verify that V is a vector space over $\mathbb{R}$.

A subset $B = \{\mathbf{b}_1, \mathbf{b}_2, \ldots, \mathbf{b}_n\}$ of a vector space V is said to be **linearly independent** if $\alpha_1 \mathbf{b}_1 + \alpha_2 \mathbf{b}_2 + \cdots + \alpha_n \mathbf{b}_n = \mathbf{0}$ implies that $\alpha_1 = \alpha_2 = \cdots = \alpha_n = 0$. If $\alpha_1 \mathbf{x}_1 + \alpha_2 \mathbf{x}_2 + \cdots + \alpha_n \mathbf{x}_n = \mathbf{0}$, and the α_i are *not* all zero, we say that the set of vectors $\mathbf{x}_1, \mathbf{x}_2, \ldots, \mathbf{x}_n$ is **linearly dependent**.

The subset B forms a **basis** for V if, for every $\mathbf{x} \in V$, we can write $\mathbf{x}$ as a **linear combination** of the elements of B, so that $\mathbf{x} = \alpha_1 \mathbf{b}_1 + \alpha_2 \mathbf{b}_2 + \cdots + \alpha_n \mathbf{b}_n$, for some $\alpha_i \in \mathbb{R}$. The set of *all* linear combinations of $\mathbf{b}_1, \ldots, \mathbf{b}_n$ is called the **span** of these vectors. If span $(\mathbf{b}_1, \ldots, \mathbf{b}_n) = V$, then $\mathbf{b}_1, \ldots, \mathbf{b}_n$ form a basis for V.

A vector space V is **finite dimensional** if it has a basis with a finite number of elements. If it is not finite dimensional, and it has a basis with an infinite number of elements, it is said to be **infinite dimensional**.

Examples

(i) Consider $V = \mathbb{R}^n$. The subset

$$B = \left\{ \begin{pmatrix} 1 \\ 0 \\ \vdots \\ 0 \end{pmatrix}, \begin{pmatrix} 0 \\ 1 \\ \vdots \\ 0 \end{pmatrix}, \ldots, \begin{pmatrix} 0 \\ 0 \\ \vdots \\ 1 \end{pmatrix} \right\} = \{\mathbf{b}_1, \mathbf{b}_2, \ldots, \mathbf{b}_n\},$$

forms a basis, since

$$\begin{pmatrix} x_1 \\ x_2 \\ \vdots \\ x_n \end{pmatrix} = x_1 \mathbf{b}_1 + x_2 \mathbf{b}_2 + \cdots + x_n \mathbf{b}_n.$$

There are other bases for $\mathbb{R}^n$, but this is the simplest. All other bases also have n elements.

(ii) Consider the vector space

$$V = \left\{ \sum_{n=0}^{\infty} a_n \cos nx \,\middle|\, a_n \in \mathbb{R}, \, x \in [-\pi, \pi] \right\},$$

which consists of convergent Fourier series defined on $-\pi \leqslant x \leqslant \pi$. A basis for V is $B = \{1, \cos x, \cos 2x, \ldots\}$, which contains an infinite number of elements. This shows that V is infinite dimensional.

A1.2 Inner Product Spaces

The **inner** or **dot product** of two elements, $\mathbf{x} = (x_1, x_2, \ldots, x_n)$ and $\mathbf{y} = (y_1, y_2, \ldots, y_n)$, of $\mathbb{R}^n$ is defined to be

$$\langle \mathbf{x}, \mathbf{y} \rangle = x_1 y_1 + x_2 y_2 + \cdots + x_n y_n.$$

For a general vector space, V over $\mathbb{R}$, the inner product is a mapping, $\langle ., . \rangle : V \times V \to \mathbb{R}$, with the three properties

(i) $\langle \mathbf{x}, \mathbf{x} \rangle \geqslant 0$,
(ii) $\langle \mathbf{x}, \mathbf{x} \rangle = 0$ if and only if $\mathbf{x} = \mathbf{0}$,
(iii) $\langle \alpha \mathbf{x} + \beta \mathbf{y}, \mathbf{z} \rangle = \alpha \langle \mathbf{x}, \mathbf{z} \rangle + \beta \langle \mathbf{y}, \mathbf{z} \rangle$,

for $\mathbf{x}, \mathbf{y}, \mathbf{z} \in V$, $\alpha, \beta \in \mathbb{R}$.

A vector space with an inner product defined on it is called an **inner product space**. An important example is the space of all real-valued functions, $C(I)$, defined on an interval $I = [a, b]$. It is straightforward to confirm, using the properties of the Riemann integral, that

$$\langle f, g \rangle = \int_a^b f(x) g(x) \, dx$$

is an inner product. For example, if $I = [-1, 1]$, $f(x) = x$ and $g(x) = x^3$, then $\langle f, g \rangle = \int_{-1}^1 x^4 \, dx = \frac{2}{5}$.

Two nonzero vectors, $\mathbf{x}$ and $\mathbf{y}$, in an inner product space are said to be **orthogonal** if $\langle \mathbf{x}, \mathbf{y} \rangle = 0$. A set of nonzero vectors in an inner product space, $\{\mathbf{x}_i\}$ for $i \geqslant 1$, whose members are mutually orthogonal is necessarily linearly independent. To see this, suppose that $\alpha_1 \mathbf{x}_1 + \alpha_2 \mathbf{x}_2 + \cdots + \alpha_n \mathbf{x}_n = \mathbf{0}$ for $\alpha_1, \alpha_2, \ldots, \alpha_n \in \mathbb{R}$. This means that $\alpha_1 \langle \mathbf{x}_1, \mathbf{x}_j \rangle + \alpha_2 \langle \mathbf{x}_2, \mathbf{x}_j \rangle + \cdots + \alpha_n \langle \mathbf{x}_n, \mathbf{x}_j \rangle = \alpha_j \langle \mathbf{x}_j, \mathbf{x}_j \rangle = 0$, and hence, by property (ii) above, $\alpha_j = 0$ for $j \geqslant 0$.

A **norm**, $||\mathbf{x}||$, of a vector $\mathbf{x}$ must have the four properties

(i) $||\mathbf{x}||$ is a non-negative real number,
(ii) $||\mathbf{x}|| = 0$ if and only if $\mathbf{x} = \mathbf{0}$,
(iii) $||k \mathbf{x}|| = |k| \, ||\mathbf{x}||$ for all real k,
(iv) $||\mathbf{x} + \mathbf{y}|| \leqslant ||\mathbf{x}|| + ||\mathbf{y}||$, the **triangle inequality**.

The **Euclidean norm** of a vector is defined to be $||\mathbf{x}|| = \sqrt{\langle \mathbf{x}, \mathbf{x} \rangle}$, and gives the size or length of $\mathbf{x}$. This is familiar in $\mathbb{R}^3$, with $\langle \mathbf{x}, \mathbf{y} \rangle = x_1 y_1 + x_2 y_2 + x_3 y_3$, so that $||\mathbf{x}|| = \sqrt{x_1^2 + x_2^2 + x_3^2}$. In $C(I)$, there are different types of norm. Using the inner product discussed above, we can define a norm

$$||f|| = \sqrt{\int_a^b f^2(x) \, dx}.$$

A useful relationship between the inner product and this norm is the **Cauchy–Schwartz inequality**,

$$|\langle \mathbf{x}, \mathbf{y} \rangle| \leqslant ||\mathbf{x}|| \, ||\mathbf{y}||.$$

We can also define the **sup norm** through

$$||f|| = \sup\{f(x) \mid x \in I\}.$$

The function space $C(I)$ is **complete** under the sup norm.† This can be useful when proving rigorous results about differential equations.

A1.3 Linear Transformations and Matrices

A transformation, T, from a vector space, V, into itself, denoted by $T : V \to V$, is **linear** if

(i) $T(\mathbf{x} + \mathbf{y}) = T(\mathbf{x}) + T(\mathbf{y}) \;\; \forall \mathbf{x}, \mathbf{y} \in V$,
(ii) $T(\lambda \mathbf{x}) = \lambda T(\mathbf{x}) \;\; \forall \lambda \in \mathbb{R},\, \mathbf{x} \in V$.

It follows immediately from this definition that $T(\lambda \mathbf{x} + \mu \mathbf{y}) = \lambda T(\mathbf{x}) + \mu T(\mathbf{y})$ and $T(\mathbf{0}) = \mathbf{0}$.

Examples

(i) If $V = \mathbb{R}^3$ and $T : \mathbb{R}^3 \to \mathbb{R}^3$ is defined by

$$T\begin{pmatrix} x_1 \\ x_2 \\ x_3 \end{pmatrix} = \begin{pmatrix} x_2 \\ x_3 \\ x_1 \end{pmatrix},$$

then

$$T(\mathbf{x} + \mathbf{y}) = T\left(\begin{pmatrix} x_1 + y_1 \\ x_2 + y_2 \\ x_3 + y_3 \end{pmatrix}\right) = \begin{pmatrix} x_2 + y_2 \\ x_3 + y_3 \\ x_1 + y_1 \end{pmatrix} = T(\mathbf{x}) + T(\mathbf{y}),$$

$$T(\lambda \mathbf{x}) = T\left(\begin{pmatrix} \lambda x_1 \\ \lambda x_2 \\ \lambda x_3 \end{pmatrix}\right) = \begin{pmatrix} \lambda x_2 \\ \lambda x_3 \\ \lambda x_1 \end{pmatrix} = \lambda T(\mathbf{x}),$$

so that T is a linear transformation.

(ii) If $V = P_n$, the vector space of polynomials of degree n, and

$$T(a_n x^n + \cdots + a_1 x + a_0) = n a_n x^{n-1} + \cdots + a_1,$$

T is a linear transformation, and can be identified with the operation of differentiation.

Linear transformations can be represented by considering their effect on the basis vectors. If $T(\mathbf{b}_j) = \alpha_{1j}\mathbf{b}_1 + \alpha_{2j}\mathbf{b}_2 + \cdots + \alpha_{nj}\mathbf{b}_n$, and we take a general vector $\mathbf{x} = \lambda_1 \mathbf{b}_1 + \lambda_2 \mathbf{b}_2 + \cdots \lambda_n \mathbf{b}_n$, then

$$T(\mathbf{x}) = \lambda_1 T(\mathbf{b}_1) + \lambda_2 T(\mathbf{b}_2) + \cdots + \lambda_n T(\mathbf{b}_n),$$

$$= \lambda_1(\alpha_{11}\mathbf{b}_1 + \alpha_{21}\mathbf{b}_2 + \cdots + \alpha_{n1}\mathbf{b}_n) + \cdots + \lambda_n(\alpha_{1n}\mathbf{b}_1 + \alpha_{2n}\mathbf{b}_2 + \cdots + \alpha_{nn}\mathbf{b}_n),$$

† See Kreider, Kuller, Ostberg and Perkins (1966) for a discussion of completeness.

which, using the standard definition of matrix multiplication, we can write as

$$T \begin{pmatrix} \lambda_1 \\ \lambda_2 \\ \vdots \\ \lambda_n \end{pmatrix} = \begin{pmatrix} \alpha_{11} & \alpha_{12} & \cdots & \alpha_{1n} \\ \alpha_{21} & \alpha_{22} & \cdots & \alpha_{2n} \\ \vdots & \vdots & & \vdots \\ \alpha_{n1} & \alpha_{n2} & \cdots & \alpha_{nn} \end{pmatrix} \begin{pmatrix} \lambda_1 \\ \lambda_2 \\ \vdots \\ \lambda_n \end{pmatrix}.$$

We say that the matrix

$$A = \begin{pmatrix} \alpha_{11} & \alpha_{12} & \cdots & \alpha_{1n} \\ \alpha_{21} & \alpha_{22} & \cdots & \alpha_{2n} \\ \vdots & \vdots & & \vdots \\ \alpha_{n1} & \alpha_{n2} & \cdots & \alpha_{nn} \end{pmatrix}$$

is a **representation** of the transformation. In example (i) above,

$$T \begin{pmatrix} x_1 \\ x_2 \\ x_3 \end{pmatrix} = \begin{pmatrix} x_2 \\ x_3 \\ x_1 \end{pmatrix},$$

so that

$$A = \begin{pmatrix} 0 & 1 & 0 \\ 0 & 0 & 1 \\ 1 & 0 & 0 \end{pmatrix}.$$

A1.4 The Eigenvalues and Eigenvectors of a Matrix

If A is an $n \times n$ matrix and $\mathbf{x}$ is an $n \times 1$ column vector, the eigenvalues of A are defined as those values of λ for which the equation $A\mathbf{x} = \lambda\mathbf{x}$ has a nontrivial solution. For each of these values of λ, the eigenvectors of A are the corresponding values of $\mathbf{x}$. This defining equation can be rearranged into the form $(A - \lambda I)\mathbf{x} = \mathbf{0}$, where I is the $n \times n$ identity matrix, for which the condition for nontrivial solutions is $\det(A - \lambda I) = 0$. This is known as the **characteristic equation** associated with the matrix A.

Example

Find the eigenvalues and eigenvectors of the matrix

$$A = \begin{pmatrix} 4 & -1 \\ 2 & 1 \end{pmatrix}.$$

The eigenvalues satisfy

$$\det \begin{pmatrix} 4 - \lambda & -1 \\ 2 & 1 - \lambda \end{pmatrix} = 0.$$

This gives $\lambda^2 - 5\lambda + 6 = 0$ so that $\lambda = 2$ or 3. The eigenvectors satisfy

$$\begin{pmatrix} 4 & -1 \\ 2 & 1 \end{pmatrix} \begin{pmatrix} x \\ y \end{pmatrix} = \lambda \begin{pmatrix} x \\ y \end{pmatrix}.$$

For $\lambda = 2$, $2x - y = 0$, so that the eigenvector is $\alpha(1,2)^T$ for any nonzero real α. Here, the superscript T denotes the transpose of the vector. For $\lambda = 3$, $x - y = 0$ so that the eigenvector is $\beta(1,1)^T$ for any nonzero real β. The eigenvectors are linearly independent and span $\mathbb{R}^2$.

Example

Find the eigenvalues and eigenvectors of the matrix

$$A = \begin{pmatrix} 0 & 0 & 1 \\ 0 & -1 & 0 \\ 2 & 2 & 1 \end{pmatrix}.$$

The eigenvalues satisfy

$$\det \begin{pmatrix} -\lambda & 0 & 1 \\ 0 & -1-\lambda & 0 \\ 2 & 2 & 1-\lambda \end{pmatrix} = 0,$$

so that $(\lambda - 2)(\lambda + 1)^2 = 0$, and hence $\lambda = 2$ or -1. The associated eigenvectors are $\alpha(1, 0, 2)^T$ corresponding to $\lambda = 2$, and $\beta(1, 0, -1)^T$ corresponding to $\lambda = -1$. These eigenvectors span a two-dimensional subspace of $\mathbb{R}^3$.

A particularly useful result concerning eigenvalues is that, if an $n \times n$ matrix A has n distinct eigenvalues $\lambda_1, \lambda_2, \ldots, \lambda_n$, there exists a nonsingular matrix B, whose columns are the eigenvectors of A, such that

$$B^{-1}AB = \text{diag}(\lambda_1, \lambda_2, \ldots, \lambda_n) = \begin{pmatrix} \lambda_1 & 0 & \cdots & 0 \\ 0 & \lambda_2 & 0 & \cdots \\ \vdots & & & \\ 0 & 0 & \cdots & \lambda_n \end{pmatrix}.$$

The only nonzero entries of this matrix are on its diagonal, and are the eigenvalues of A.

Consider the system of differential equations $\dot{\mathbf{x}} = A\mathbf{x}$, where $\mathbf{x} = (x_1, x_2, \ldots, x_n)^T$ and a dot denotes differentiation. If we write $\mathbf{x} = B\mathbf{y}$, the system changes to $\dot{\mathbf{y}} = B^{-1}AB\mathbf{y}$, which is considerably easier to analyze if $B^{-1}AB$ is diagonal, since the differential equations are all decoupled.

Example

Let's try to find a simplification of the system of differential equations

$$\begin{aligned} \dot{x}_1 &= x_3, \\ \dot{x}_2 &= x_1, \\ \dot{x}_3 &= 2x_1 + x_2. \end{aligned}$$

We can write this in matrix form as $\dot{\mathbf{x}} = A\mathbf{x}$ with

$$A = \begin{pmatrix} 0 & 0 & 1 \\ 1 & 0 & 0 \\ 2 & 1 & 0 \end{pmatrix}.$$

A1.4 THE EIGENVALUES AND EIGENVECTORS OF A MATRIX

We need to simplify the structure of the matrix A. To do this we firstly find the eigenvalues, $\lambda = -1, \frac{1}{2}(1+\sqrt{5})$ and $\frac{1}{2}(1-\sqrt{5})$, and the corresponding eigenvectors,

$$\begin{pmatrix} 1 \\ -1 \\ -1 \end{pmatrix}, \quad \begin{pmatrix} 1 \\ -\frac{1}{2}(1-\sqrt{5}) \\ \frac{1}{2}(1+\sqrt{5}) \end{pmatrix}, \quad \begin{pmatrix} 1 \\ -\frac{1}{2}(1+\sqrt{5}) \\ \frac{1}{2}(1-\sqrt{5}) \end{pmatrix}.$$

These are orthogonal, and hence linearly independent, and form a basis for $\mathbb{R}^3$. After choosing our matrix B to be

$$B = \begin{pmatrix} 1 & 1 & 1 \\ -1 & -\frac{1}{2}(1-\sqrt{5}) & -\frac{1}{2}(1+\sqrt{5}) \\ -1 & \frac{1}{2}(1+\sqrt{5}) & \frac{1}{2}(1-\sqrt{5}) \end{pmatrix},$$

with inverse

$$B^{-1} = \begin{pmatrix} 1 & 1 & -1 \\ \frac{1}{\sqrt{5}} & \frac{1}{2\sqrt{5}}(3-\sqrt{5}) & -\frac{1}{2\sqrt{5}}(1-\sqrt{5}) \\ \frac{1}{\sqrt{5}} & -\frac{1}{2\sqrt{5}}(3+\sqrt{5}) & \frac{1}{2\sqrt{5}}(1+\sqrt{5}) \end{pmatrix},$$

we find that

$$B^{-1}AB = \begin{pmatrix} -1 & 0 & 0 \\ 0 & \frac{1}{2}(1+\sqrt{5}) & 0 \\ 0 & 0 & \frac{1}{2}(1-\sqrt{5}) \end{pmatrix} = \mathrm{diag}(-1, \frac{1}{2}(1+\sqrt{5}), \frac{1}{2}(1-\sqrt{5}))$$

is the representation of A with respect to the basis of eigenvectors. The transformed system of differential equations therefore takes the form

$$\dot{y}_1 = -y_1, \quad \dot{y}_2 = \frac{1}{2}(1+\sqrt{5})y_2, \quad \dot{y}_3 = \frac{1}{2}(1-\sqrt{5})y_3.$$

These have the simple solutions $y_1 = c_1 \exp(-t)$, $y_2 = c_2 \exp\{\frac{1}{2}(1+\sqrt{5})t\}$, $y_3 = c_3 \exp\{\frac{1}{2}(1-\sqrt{5})t\}$. Finally, the solution is $x_1 = y_1 + y_2 + y_3$, $x_2 = -y_1 - \frac{1}{2}(1-\sqrt{5})y_2 - \frac{1}{2}(1+\sqrt{5})y_3$ and $x_3 = -y_1 + \frac{1}{2}(1+\sqrt{5})y_2 + \frac{1}{2}(1-\sqrt{5})y_3$, in terms of the original variables.

Finally, we will often make use of the **Cayley–Hamilton** theorem.

Theorem A1.1 (Cayley–Hamilton) *Every square matrix A satisfies its own characteristic equation, $\det(A - \lambda I) = 0$.*

The proof of this theorem is rather involved, and we will not consider it here (see Morris, 1982). The Cayley–Hamilton theorem shows that A^k, with $k \geq n$, can be written as a linear combination of $I, A, A^2, \ldots, A^{n-1}$.

APPENDIX 2

Continuity and Differentiability

Let f be a real-valued function defined on some open interval that contains the point $x = c$. We say that

$$f \text{ is \textbf{continuous} at } x = c \text{ if and only if } \lim_{x \to c} f(x) = f(c).$$

The function f is continuous on the interval (a, b) if and only if it is continuous for all $x \in (a, b)$. The set of all continuous functions $f : (a, b) \to \mathbb{R}$ ($\mathbb{R}$ denotes the set of real numbers) forms a vector space denoted by $C(a, b)$ or $C^0(a, b)$.

A function f is continuous on $[a, b]$ if it is continuous on (a, b) and

$$\lim_{x \to a^+} f(x) = f(a), \quad \lim_{x \to b^-} f(x) = f(b).$$

In other words, there is no singular behaviour at the ends of the interval. We will use $C[a, b]$ to denote the vector space of continuous functions $f : [a, b] \to \mathbb{R}$.

An alternative, but equivalent, definition of continuity is that f is continuous at a point $x = x_0$ if, for every $\epsilon > 0$, there exists a function $\delta(x_0, \epsilon)$ such that $|f(x) - f(x_0)| < \epsilon$ when $|x - x_0| < \delta$. If δ depends upon ϵ but *not* upon x_0 in some interval I, we say that f is **uniformly continuous** on I. It can be shown that, if a function is continuous on a *closed* interval, then it is also uniformly continuous there.

If $\lim_{x \to c^+} f(x) \neq \lim_{x \to c^-} f(x)$, $f(x)$ is *not* continuous at $x = c$. There is an important class of functions of this form, for which there are a finite number of discontinuities, at each of which $f(x)$ jumps by a finite amount. These are known as **piecewise continuous functions**, and, like the continuous functions, form a vector space, which we denote by $PC(a, b)$.

If the limit of $\{f(x) - f(c)\}/(x - c)$ exists as $x \to c$, we say that f is **differentiable** at $x = c$, and denote its derivative by

$$f'(c) = \lim_{x \to c} \left\{ \frac{f(x) - f(c)}{x - c} \right\}.$$

The set of all once-differentiable functions on (a, b) forms a vector space which we denote by $C^1(a, b)$. We can make similar definitions of $C^2(a, b), C^3(a, b), \ldots, C^n(a, b)$ $\ldots, C^\infty(a, b)$, in which the higher derivatives are defined in terms of limits of derivatives of one order lower.

Examples of functions that live in these spaces are:

(i) $f : (0, 2\pi) \to \mathbb{R} : f(x) = \sin x$. Since $f'(x) = \cos x$ and $f''(x) = -\sin x =$

$-f(x)$, this function can be differentiated as many times as we like and the result will be a continuous function. We conclude that $f \in C^\infty(0, 2\pi)$.

(ii) $f : (-1, 1) \to \mathbb{R} : f(x) = x^2$ for $x \geqslant 0$, $f(x) = 0$ for $x < 0$. This function can be differentiated twice, but, since $f''(x) = 0$ for $x < 0$ and $f''(x) = 2$ for $x \geqslant 0$, no more than twice. We conclude that $f \in C^2(-1, 1)$.

(iii) $f : \mathbb{R} \to \mathbb{R} : f(x) = \sum_{n=1}^{\infty} \frac{1}{n!} \sin(n!)^2 x$. This function, which is illustrated in Figure A2.1, is continuous on $\mathbb{R}$ but nowhere differentiable, so that $f \in C^0(\mathbb{R})$, but $f \notin C^n(\mathbb{R})$ for any $n > 0$.

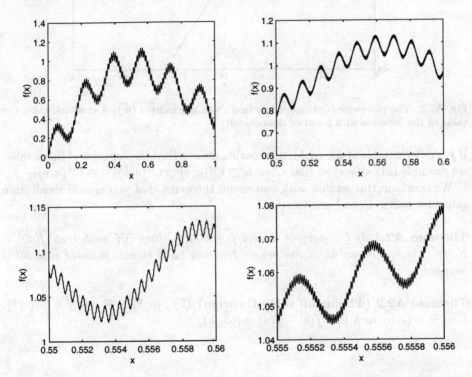

Fig. A2.1. The continuous, nowhere differentiable function $f(x) = \sum_{n=1}^{\infty} \frac{1}{n!} \sin(n!)^2 x$.

(iv) $f : [0, 3] \to \mathbb{R}$:

$$f(x) = \begin{cases} x^2 & \text{for } 0 \leqslant x < 1, \\ \cos x & \text{for } 1 \leqslant x \leqslant 2, \\ e^{-x} & \text{for } 2 < x \leqslant 3, \end{cases} \quad \text{(A2.1)}$$

which is plotted in Figure A2.2. This function is discontinuous at $x = 1$ and $x = 2$, and therefore $f \in PC[0, 3]$.

504 CONTINUITY AND DIFFERENTIABILITY

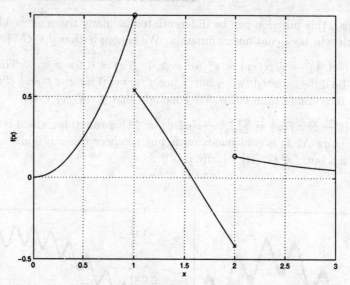

Fig. A2.2. The piecewise continuous function given as example (iv). A cross indicates the value of the function at a point of discontinuity.

If f is differentiable at $c \in (a,b)$, it is continuous at c (but the converse of this is false, see example (iii) above) so that $C^0(a,b) \supset C^1(a,b) \supset C^2(a,b) \cdots \supset C^n(a,b) \cdots$.

We conclude this section with two useful theorems that you should recall from your first course in real analysis.

Theorem A2.1 *If $f : [a,b] \to \mathbb{R}$ and $f \in C[a,b]$, then $\exists K$ such that $|f(x)| < K \; \forall x \in [a,b]$. In words, a continuous function on a closed, bounded interval is bounded.*

Theorem A2.2 (The mean value theorem) *If $f : (a,b) \to \mathbb{R}$ and $f \in C^1(a,b)$, then $\exists c \in (a,b)$ such that $f(b) - f(a) = (b-a)f'(c)$.*

APPENDIX 3

Power Series

Many commonly encountered functions can be expressed as power series, for example,

$$\sin x = x - \frac{x^3}{3!} + \frac{x^5}{5!} - \cdots = \sum_{n=0}^{\infty} (-1)^n \frac{x^{2n+1}}{(2n+1)!},$$

$$\cos x = 1 - \frac{x^2}{2!} + \frac{x^4}{4!} - \cdots = \sum_{n=0}^{\infty} (-1)^n \frac{x^{2n}}{(2n)!}. \tag{A3.1}$$

In general, we can develop a power series representation for a sufficiently differentiable function, provided that it involves no singularities, fractional powers of x or logarithms.

A3.1 Maclaurin Series

The **Maclaurin series** is a power series about $x = 0$. If $f \in C^{\infty}(a, b)$ for some $a < 0 < b$, the series takes the form

$$f(x) = f(0) + xf'(0) + \frac{x^2}{2!}f''(0) + \cdots + \frac{x^n}{n!}f^{(n)}(0) + \cdots.$$

Here, a prime denotes a derivative with respect to x and $f^{(n)}(x)$ is the nth derivative.

Example

Let's determine the Maclaurin series expansion for the function $f(x) = 1/(1+x)$. Since

$$f(x) = \frac{1}{1+x}, \qquad f(0) = 1,$$

$$f'(x) = -\frac{1}{(1+x)^2}, \qquad f'(0) = -1,$$

$$f''(x) = \frac{2}{(1+x)^3}, \qquad f''(0) = 2,$$

$$f^{(3)}(x) = \frac{-3 \cdot 2}{(1+x)^4}, \qquad f^{(3)}(0) = -3!,$$

$$\vdots$$

$$f^{(n)}(x) = \frac{(-1)^n n!}{(1+x)^{n+1}}, \qquad f^{(n)}(0) = (-1)^n n!,$$

the Maclaurin series is

$$f(x) = 1 + (-1)x + 2\frac{x^2}{2!} + (-1)3!\frac{x^3}{3!} + \cdots + (-1)^n n!\frac{x^n}{n!} + \cdots$$

$$= 1 - x + x^2 - x^3 + \cdots + (-1)^n x^n + \cdots = \sum_{n=0}^{\infty} (-1)^n x^n.$$

A3.2 Taylor Series

The Maclaurin series is a special case of the **Taylor series**, which is a power series about $x = x_0$. Its existence also requires reasonable behaviour of the function at $x = x_0$, and for a C^∞ function takes the form,

$$f(x) = f(x_0) + (x - x_0)f'(x_0) + \frac{(x - x_0)^2}{2!}f''(x_0) + \cdots + \frac{(x - x_0)^n}{n!}f^{(n)}(x_0) + \cdots.$$

Example

Let's determine the Taylor series of the function $f(x) = 1/(1 + x)$ at the point $x_0 = 1$. Since

$$f(x) = \frac{1}{1 + x}, \qquad f(x_0) = \frac{1}{2},$$

$$f'(x) = -\frac{1}{(1 + x)^2}, \qquad f'(x_0) = -\frac{1}{2^2},$$

$$f''(x) = \frac{2}{(1 + x)^3}, \qquad f''(x_0) = \frac{2}{2^3},$$

$$f^{(3)}(x) = \frac{-3 \cdot 2}{(1 + x)^4}, \qquad f^{(3)}(x_0) = -\frac{3!}{2^4},$$

$$\vdots$$

$$f^{(n)}(x) = \frac{(-1)^n n!}{(1 + x)^{n+1}}, \qquad f^{(n)}(x_0) = (-1)^n \frac{n!}{2^{n+1}},$$

the Taylor series is

$$f(x) = \sum_{n=0}^{\infty} \frac{(x - 1)^n}{n!} \frac{(-1)^n n!}{2^{n+1}} = \frac{1}{2} \sum_{n=0}^{\infty} \frac{(x - 1)^n (-1)^n}{2^n}. \qquad (A3.2)$$

A3.3 Convergence of Power Series

In order to determine for what values of x a power series is convergent, we can usually use the **ratio test**, which says that

the series $\sum_{n=0}^{\infty} a_n$ $\begin{cases} \text{converges if } \lim_{n \to \infty} |a_{n+1}/a_n| < 1, \\ \text{diverges if } \lim_{n \to \infty} |a_{n+1}/a_n| > 1. \end{cases}$

In the previous example, (A3.2),

$$a_n = \frac{(x-1)^n(-1)^n}{2^{n+1}}, \quad a_{n+1} = \frac{(x-1)^{n+1}(-1)^{n+1}}{2^{n+2}},$$

so that, for convergence, we need

$$\lim_{n\to\infty}\left|\frac{a_{n+1}}{a_n}\right| = \left|\frac{x-1}{2}\right| < 1.$$

This means that the series converges when $-1 < x < 3$. This defines the **domain of convergence** of the power series (A3.2). This is often written as $|x-1| < 2$, which defines the **radius of convergence** of the series as two units from the point $x = 1$. Notice that the function $f(x)$ is singular at $x = -1$, and will not have a Taylor series about this point. It is worth pointing out that the Taylor series of a polynomial is a terminating series and hence has an infinite radius of convergence.

A useful composite result from the theory of Taylor series is that if $f(x) = \sum_{n=0}^{\infty} a_n(x-x_0)^n$ is a Taylor series with radius of convergence R (it converges for $|x-x_0| < R$), f is differentiable for all x such that $|x-x_0| < R$, and $f'(x) = \sum_{n=1}^{\infty} n a_n (x-x_0)^{n-1}$ has the same radius for convergence as the series for $f(x)$. This 'term by term' differentiation result allows us to develop a method for solving differential equations using power series – the method of Frobenius, which we discuss in Chapter 1.

Note that another useful test for the convergence of a series is the **comparison test**, which says that

if $\sum_{n=0}^{\infty} a_n$ and $\sum_{n=0}^{\infty} b_n$ are series of positive terms then

(i) if $\sum_{n=0}^{\infty} a_n$ converges and $b_k \leqslant a_k$ for $k > k_0$, $\sum_{n=0}^{\infty} b_n$ also converges,

(ii) if $\sum_{n=0}^{\infty} a_n$ diverges and $b_k \geqslant a_k$ for $k > k_0$, $\sum_{n=0}^{\infty} b_n$ also diverges.

A3.4 Taylor Series for Functions of Two Variables

If $f = f(x,y)$ is a scalar field that has sufficient partial derivatives at the point (x_0, y_0), the Taylor series about this point is

$$f(x,y) = f(x_0, y_0) + (x-x_0)\left.\frac{\partial f}{\partial x}\right|_{(x_0,y_0)} + (y-y_0)\left.\frac{\partial f}{\partial y}\right|_{(x_0,y_0)}$$

$$+ \frac{(x-x_0)^2}{2!}\left.\frac{\partial^2 f}{\partial x^2}\right|_{(x_0,y_0)} + (x-x_0)(y-y_0)\left.\frac{\partial^2 f}{\partial x \partial y}\right|_{(x_0,y_0)} + \frac{(y-y_0)^2}{2!}\left.\frac{\partial^2 f}{\partial y^2}\right|_{(x_0,y_0)} + \cdots$$

Example

Let's find the first two terms in the Taylor series of $e^{-(x^2+y^2)}$ about the point $(0,0)$.

$$\left.\frac{\partial f}{\partial x}\right|_{(0,0)} = \left.\frac{\partial f}{\partial y}\right|_{(0,0)} = 0,$$

$$\left.\frac{\partial^2 f}{\partial x^2}\right|_{(0,0)} = \left.\frac{\partial^2 f}{\partial y^2}\right|_{(0,0)} = -2, \quad \left.\frac{\partial^2 f}{\partial x \partial y}\right|_{(0,0)} = 0,$$

so that $e^{-(x^2+y^2)} = 1 - x^2 - y^2 + \cdots$.

We can also generalize the Taylor series from scalar- to vector-valued functions of two variables. Let's define a vector field,

$$\mathbf{f}(\mathbf{x}) = \begin{pmatrix} f(\mathbf{x}) \\ g(\mathbf{x}) \end{pmatrix}, \quad \mathbf{x} = \begin{pmatrix} x \\ y \end{pmatrix}.$$

Using the Taylor series for each component of this vector we can write

$$\mathbf{f}(\mathbf{x}) = \mathbf{f}(\mathbf{x}_0) + J(\mathbf{f})(\mathbf{x}_0)(\mathbf{x} - \mathbf{x}_0) + \cdots,$$

where

$$J(\mathbf{f})(\mathbf{x}_0) = \begin{pmatrix} \dfrac{\partial f}{\partial x}(\mathbf{x}_0) & \dfrac{\partial f}{\partial y}(\mathbf{x}_0) \\ \dfrac{\partial g}{\partial x}(\mathbf{x}_0) & \dfrac{\partial g}{\partial y}(\mathbf{x}_0) \end{pmatrix}$$

is called the **Jacobian matrix**. This result proves to be very useful in Chapter 9, where we study nonlinear differential equations. Note that the definition of the Jacobian can easily be generalized to higher dimensional vector fields.

Example

Let's find the linear approximation to the vector field

$$\mathbf{f} = \begin{pmatrix} e^x - 1 \\ (1-y)e^x \end{pmatrix}$$

about the point $\mathbf{x}_0 = \begin{pmatrix} 0 \\ 1 \end{pmatrix}$. Since $\mathbf{f}(\mathbf{x}_0) = \mathbf{0}$, the Taylor series takes the form

$$\mathbf{f}(\mathbf{x}) = J(\mathbf{f})\begin{pmatrix} 0 \\ 1 \end{pmatrix}\left\{\mathbf{x} - \begin{pmatrix} 0 \\ 1 \end{pmatrix}\right\} + \cdots.$$

After calculating the Jacobian, this gives

$$\mathbf{f}(\mathbf{x}) = \begin{pmatrix} 1 & 0 \\ 0 & -1 \end{pmatrix}\begin{pmatrix} x \\ y-1 \end{pmatrix} + \cdots = \begin{pmatrix} x \\ 1-y \end{pmatrix} + \cdots.$$

APPENDIX 4

Sequences of Functions

The concepts that we briefly describe in this section are used in Chapter 5, where we discuss the convergence of Fourier series. Consider a **sequence of functions**, $\{f_k(x)\}$ for $k = 1, 2, \ldots$, defined on some interval of the real line, I. We say that $\{f_k(x)\}$ **converges pointwise** to $f(x)$ on I if $\lim_{k\to\infty} f_k(x_0)$ exists for all $x_0 \in I$, and $f_k(x_0) \to f(x_0)$. For example, consider the sequence $\{f_k(x)\} = \{x + 1/k\}$ on $\mathbb{R}$. This sequence converges pointwise to $f(x) = x$ as $k \to \infty$. The functions $f_k(x)$ and their limit, $f(x) = x$, are continuous. In contrast, consider the sequence $\{f_k(x)\} = \{x^k\}$ on $[0, 1]$. Although each of these functions is continuous, as $k \to \infty$, $f_k(x)$ converges pointwise to

$$f(x) = \begin{cases} 0 & \text{for } 0 \leqslant x < 1, \\ 1 & \text{for } x = 1, \end{cases}$$

which is not continuous. As a final example, consider the sequence $f_k(x)$, defined on $x \geqslant 0$, with

$$f_k(x) = \begin{cases} k^2 x(1 - kx) & \text{for } 0 \leqslant x \leqslant 1/k, \\ 0 & \text{for } x \geqslant 1/k, \end{cases} \quad \text{(A4.1)}$$

which is illustrated in Figure A4.1. Since $f_k(x) = 0$ for $x \geqslant 1/k$, $f_k(x_0) \to 0$ as $k \to \infty$ for all $x_0 > 0$. Moreover, $f_k(0) = 0$ for all k, so we conclude that $\{f_k(x)\}$ converges pointwise to $f(x) = 0$, a continuous function. However, the individual members of the sequence don't really provide a good approximation to the pointwise limit, firstly because the maximum value of $f_k(x)$ is $k/4$, which is unbounded as $k \to \infty$, and secondly because

$$\int_0^\infty f_k(x)\, dx = 1/6, \quad \int_0^\infty f(x)\, dx = 0.$$

In order to eliminate this sort of behaviour, we need to introduce the concept of uniform convergence. A sequence of functions $\{f_k(x)\}$ is **uniformly convergent** to a function $f(x)$ on an interval of the real line, I, as $k \to \infty$ if for every $\epsilon > 0$ there exists a positive integer, $K(\epsilon)$, which is *not* a function of x, such that $|f_k(x) - f(x)| < \epsilon$ for all $k \geqslant K(\epsilon)$ and $x \in I$. This means that, for sufficiently large k, we can make $f_k(x)$ arbitrarily close to $f(x)$ over the entire interval I. For example, to see that the sequence defined by (A4.1) does *not* converge uniformly to zero, note that, for a given value of ϵ, we can only guarantee that $|f_k(x) - f(x)| = f_k(x) < \epsilon$ for $k \geqslant 1/x$, which is *not* independent of x.

510 SEQUENCES OF FUNCTIONS

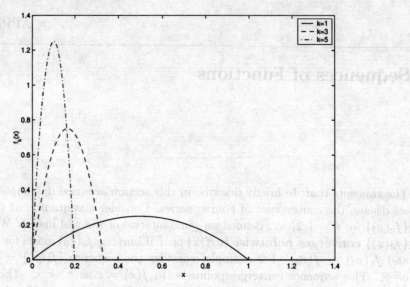

Fig. A4.1. The sequence of functions given by (A4.1).

Theorem A4.1 *If $\{f_k(x)\}$ is a sequence of continuous functions defined on an interval of the real line, $I = [a,b]$, that converges uniformly to $f(x)$, then*
  (i) *$f(x)$ is continuous*
  (ii) *for every $x \in I$*

$$\lim_{k \to \infty} \int_a^x f_k(s)\, ds = \int_a^x f(s)\, ds.$$

A proof of this theorem can be found in any textbook on analysis.

APPENDIX 5

Ordinary Differential Equations

Chapters 1 to 4 are concerned with the solution of linear second order differential equations with variable coefficients. In this appendix, we will review simple methods for solving first order differential equations, and second order differential equations with constant coefficients.

A5.1 Variables Separable

We will start by looking at a class of equations that can be rewritten in the form

$$g(y)\frac{dy}{dt} = f(t). \tag{A5.1}$$

These equations can be integrated with respect to t to give

$$\int g(y)\frac{dy}{dt}\, dt = \int f(t)\, dt,$$

and, using the chain rule,

$$\int g(y)\, dy = \int f(t)\, dt.$$

Example

Determine the solution of the differential equation

$$\frac{dy}{dt} + e^y \cos t = 0,$$

subject to the condition that $y(0) = 1$. Firstly, we convert the equation to the form (A5.1), so that

$$e^{-y}\frac{dy}{dt} = -\cos t.$$

Integrating with respect to t we have

$$-e^{-y} = -\sin t + C.$$

Using the boundary condition we find that $C = -e^{-1}$, and rearranging gives the solution

$$y(t) = \log\left(\frac{1}{e^{-1} + \sin t}\right).$$

A5.2 Integrating Factors

We now consider the linear, first order differential equation

$$\frac{dy}{dx} + P(x)y = Q(x).$$

You should recognize this as a type of equation that can be solved by finding an integrating factor. If we multiply through by $\exp\left\{\int^x P(t)dt\right\}$, where t is a dummy variable for the integration, we have

$$\exp\left\{\int^x P(t)dt\right\}\frac{dy}{dx} + \exp\left\{\int^x P(t)dt\right\}P(x)y = Q(x)\exp\left\{\int^x P(t)dt\right\}.$$

We can immediately see that the left hand side of this is $\dfrac{d}{dx}\left\{\exp\left\{\int^x P(t)dt\right\}y\right\}$, so we have

$$\frac{d}{dx}\left\{\exp\left\{\int^x P(t)dt\right\}y\right\} = Q(x)\exp\left\{\int^x P(t)dt\right\}.$$

This can be directly integrated to give

$$y = A\exp\left\{-\int^x P(t)dt\right\} + \exp\left\{-\int^x P(t)dt\right\}\int^x Q(s)\exp\left\{\int^s P(t)dt\right\}ds,$$

where A is a constant of integration. Here it is important to notice the structure of the solution as $y = y_h + y_p$ where

$$y_h = A\exp\left\{-\int^x P(t)dt\right\}$$

is the solution of the **homogeneous differential equation**,

$$\frac{dy_h}{dx} + P(x)y_h = 0,$$

and

$$y_p = \exp\left\{-\int^x P(t)dt\right\}\int^x Q(s)\exp\left\{\int^s P(t)dt\right\}ds$$

is the **particular integral solution** of the **inhomogeneous differential equation**

$$\frac{dy_p}{dx} + P(x)y_p = Q(x).$$

The key idea of an integrating factor used here may seem rather like pulling a rabbit out of a hat. In fact, the derivation can be performed systematically using the idea of a Lie group (see Chapter 10).

Example

Let's find the solution of the differential equation

$$\frac{dy}{dx} - y = e^x,$$

subject to the condition that $y(0) = 0$. The integrating factor is

$$\exp\left(\int^x -1\,dx\right) = e^{-x},$$

so that we have

$$e^{-x}\frac{dy}{dx} - e^{-x}y = \frac{d}{dx}\left(e^{-x}y\right) = 1.$$

We can integrate this to obtain $e^{-x}y = x + c$, where c is a constant. Since $y(0) = 0$, we must have $c = 0$ and hence $y = xe^x$.

A5.3 Second Order Equations with Constant Coefficients

We will now remind you how to solve second order ordinary differential equations with constant coefficients through a series of examples.

Example

Solve the second order ordinary differential equation

$$\frac{d^2y}{dx^2} + 3\frac{dy}{dx} + 2y = 0,$$

subject to the boundary conditions that $y(0) = 0$ and $y'(0) = 1$.

We can solve constant coefficient equations by seeking a solution of the form $y = e^{mx}$, which we substitute into the equation in order to determine m. This gives

$$\frac{d^2y}{dx^2} + 3\frac{dy}{dx} + 2y = \left(m^2 + 3m + 2\right)e^{mx} = 0.$$

Since e^{mx} is never zero, we require $m^2 + 3m + 2 = (m+2)(m+1) = 0$. This gives $m = -1$ or $m = -2$. The general solution of the equation is therefore

$$y(x) = Ae^{-x} + Be^{-2x},$$

with A and B constants. The first boundary condition, $y(0) = 0$, yields

$$A + B = 0, \tag{A5.2}$$

whilst the second, $y'(0) = 1$, gives

$$-A - 2B = 1. \tag{A5.3}$$

We now need to solve (A5.2) and (A5.3), which gives us $A = 1$ and $B = -1$, so that the solution is

$$y(x) = e^{-x} - e^{-2x}.$$

It is possible to solve this equation by direct integration, but the method we have presented above is far simpler. Notice that it is not necessary that both boundary conditions are imposed at the same place. Boundary conditions at different values of x will just lead to slightly more complicated simultaneous equations.

Example: Equal roots

Solve the differential equation

$$\frac{d^2y}{dx^2} + 2\frac{dy}{dx} + y = 0,$$

subject to the boundary conditions $y(0) = 1$ and $y'(0) = 0$.

Again we can look for a solution of the form $y = e^{mx}$. We find that m satisfies the quadratic equation $m^2 + 2m + 1 = (m+1)^2 = 0$, so that $m = -1$ is a repeated root, and we have only determined one solution. In fact, the full solution has the form

$$y(x) = Ae^{-x} + Bxe^{-x}. \tag{A5.4}$$

The boundary condition $y(0) = 1$ gives $A = 1$. Since

$$y'(x) = -Ae^{-x} + B(-x+1)e^{-x},$$

$y'(0) = 0$ gives $-A + B = 0$, and hence $B = 1$. The solution is therefore

$$y(x) = e^{-x} + xe^{-x} = (x+1)e^{-x}.$$

Example: Imaginary roots

Solve the differential equation

$$\frac{d^2y}{dx^2} + 4y = 0,$$

subject to the boundary conditions $y(0) = 1$ and $y'(0) = 3$.

As usual, we seek a solution of the form $y = e^{mx}$, and find that m satisfies $m^2 + 4 = 0$. This means that $m = \pm 2i$. The general solution is therefore

$$y(x) = Ae^{2ix} + Be^{-2ix}. \tag{A5.5}$$

We could proceed with the solution in this form. However, it may be better to use $e^{i\alpha} = \cos\alpha + i\sin\alpha$. Substituting this into (A5.5) gives

$$y(x) = A(\cos 2x + i\sin 2x) + B(\cos 2x - i\sin 2x)$$

$$= (A+B)\cos 2x + i(A-B)\sin 2x = \tilde{A}\cos 2x + \tilde{B}\sin 2x.$$

We have introduced new constants $\tilde{A}$ and $\tilde{B}$, which we determine from the boundary conditions to be $\tilde{A} = 1$ and $\tilde{B} = 3/2$. The solution is therefore

$$y(x) = \cos 2x + \frac{3}{2}\sin 2x.$$

Example: An inhomogeneous equation

Solve the differential equation

$$\frac{d^2y}{dx^2} + 3\frac{dy}{dx} + 2y = e^x,$$

subject to the boundary conditions $y(0) = y'(0) = 1$.

A5.3 SECOND ORDER EQUATIONS WITH CONSTANT COEFFICIENTS

Initially we consider just the homogeneous part of the equation and solve

$$\frac{d^2 y_h}{dx^2} + 3\frac{dy_h}{dx} + 2y_h = 0.$$

As we saw earlier, the general solution of this is

$$y_h(x) = Ae^{-x} + Be^{-2x}.$$

We now have to find a function $f(x)$, the particular integral, which, when substituted into the inhomogeneous differential equation, yields the right hand side. We notice that when we differentiate e^x we get it back again, so we postulate that $f(x)$ takes the form αe^x. Substituting $y = f(x) = \alpha e^x$ into the differential equation gives

$$\frac{d^2 f}{dx^2} + 3\frac{df}{dx} + 2f = \alpha e^x + 3\alpha e^x + 2\alpha e^x = e^x.$$

From this expression we find that we need $\alpha = 1/6$. The solution of the inhomogeneous equation is therefore

$$y(x) = Ae^{-x} + Be^{-2x} + \frac{1}{6}e^x.$$

It is at this point that we impose the boundary conditions, which show that $A = 5/2$ and $B = -5/3$, and hence the solution is

$$y(x) = \frac{5}{2}e^{-x} - \frac{5}{3}e^{-2x} + \frac{1}{6}e^x.$$

Example: Right hand side a solution of the homogeneous equation

Solve the differential equation equation

$$\frac{d^2 y}{dx^2} + 3\frac{dy}{dx} + 2y = e^{-x},$$

subject to the boundary conditions $y(0) = y'(0) = 1$. Following the previous example, we need to find a function which, when substituted into the left hand side of the equation, yields the right hand side. We could try $y = \alpha e^{-x}$, but, since e^{-x} is a solution of the equation, substituting it into the left hand side just gives us zero. Taking a lead from the case of repeated roots, we try a solution of the form $y = f(x) = \alpha x e^{-x}$. Since

$$f(x) = \alpha x e^{-x}, \quad f'(x) = \alpha(e^{-x} - xe^{-x}) = \alpha(1-x)e^{-x},$$

$$f''(x) = \alpha(-1 - (1-x))e^{-x} = \alpha(x-2)e^{-x},$$

on substituting into the differential equation we obtain

$$\alpha(x-2)e^{-x} + 3\alpha(1-x)e^{-x} + 2\alpha x e^{-x} = 3\alpha e^{-x} = e^{-x},$$

which gives $\alpha = 1/3$. The solution is therefore

$$y(x) = Ae^{-x} + Be^{-2x} + \frac{1}{3}xe^{-x}.$$

We now need to satisfy the boundary conditions, which give $A = 8/3$ and $B = -5/3$, and hence

$$y(x) = \frac{8}{3}e^{-x} - \frac{5}{3}e^{-2x} + \frac{1}{3}xe^{-x}.$$

This technique is usually referred to as the trial solution method. Clearly it is not completely satisfactory, as it requires an element of guesswork. We present a more systematic method of solution in Chapter 1.

APPENDIX 6

Complex Variables

The aim of this brief appendix is to provide a reminder of the results that are needed in order to be able to invert Laplace transforms using the Bromwich inversion integral (6.5), and to understand the material on the asymptotic evaluation of complex integrals in Section 11.2. We have had to be selective in what we have presented, and note that this appendix is no substitute for a proper course on complex variables! A good textbook is Ablowitz and Fokas (1997).

A6.1 Analyticity and the Cauchy–Riemann Equations

Consider a complex-valued function of a complex variable, $f(s)$, with $s = x + iy$. The natural way to define the derivative of f is

$$\frac{df}{ds} = \lim_{\Delta s \to 0} \frac{f(s + \Delta s) - f(s)}{\Delta s},$$

provided that this limit is independent of the way that $\Delta s = \Delta x + i\Delta y$ tends to zero. A function $f(s)$ is said to be **analytic** in some region, R, of the complex s-plane if df/ds exists and is unique in R.

Theorem A6.1 *If the complex-valued function $f(s) = u(x,y) + iv(x,y)$, where $s = x + iy$ and u, v, x and y are real, is analytic in a region, R, of the complex s-plane, then*

$$\frac{\partial u}{\partial x} = \frac{\partial v}{\partial y}, \quad \frac{\partial v}{\partial x} = -\frac{\partial u}{\partial y}. \tag{A6.1}$$

*These are known as the **Cauchy–Riemann equations**.*

Proof By definition,

$$\frac{df}{ds} = \lim_{\Delta s \to 0} \frac{f(s + \Delta s) - f(s)}{\Delta s}$$

$$= \lim_{\substack{\Delta x \to 0 \\ \Delta y \to 0}} \frac{u(x + \Delta x, y + \Delta y) + iv(x + \Delta x, y + \Delta y) - u(x,y) - iv(x,y)}{\Delta x + i\Delta y}.$$

If $\Delta y = 0$,

$$\frac{df}{ds} = \lim_{\Delta x \to 0} \frac{u(x + \Delta x, y) - u(x,y) + i\{v(x + \Delta x, y) - v(x,y)\}}{\Delta x} = \frac{\partial u}{\partial x} + i\frac{\partial v}{\partial x},$$

whilst if $\Delta x = 0$,

$$\frac{df}{ds} = \lim_{\Delta y \to 0} \frac{u(x, y + \Delta y) - u(x,y) + i\{v(x, y + \Delta y) - v(x,y)\}}{i\Delta y} = \frac{\partial v}{\partial y} - i\frac{\partial u}{\partial y}.$$

These are clearly not equal unless the Cauchy–Riemann equations hold. □

Theorem A6.2 *If the Cauchy–Riemann equations, (A6.1), hold in a region, R, of the complex s-plane for some complex-valued function $f(s) = u(x,y) + iv(x,y)$, where $s = x + iy$ and u, v, x and y are real, then, provided that all of the partial derivatives in (A6.1) are continuous in R, $f(s)$ is analytic in R.*

We will not give a proof of this here.

A6.2 Cauchy's Theorem, Cauchy's Integral Formula and Taylor's Theorem

The **contour integral** of a complex-valued function $f(s)$ along some contour C in the complex s-plane is defined to be

$$\int_C f(s)\,ds = \int_a^b f(s(t))\frac{ds}{dt}(t)\,dt,$$

where $s(t)$ for $a \leq t \leq b$ is a parametric representation of the contour C. By convention, the integral around a closed contour is taken in the anticlockwise direction.

Theorem A6.3 (Cauchy's theorem) *If $f(s)$ is a single-valued, analytic function in a simply-connected domain D in the complex s-plane, then, along any simple closed contour, C, in D,*

$$\int_C f(s)\,ds = 0.$$

Proof If $f(s) = u + iv$ and $ds = dx + i\,dy$,

$$\int_C f(s)\,ds = \int_C (u\,dx - v\,dy) + i\int_C (v\,dx + u\,dy).$$

If df/ds is continuous in D, then u and v have continuous partial derivatives there. We can therefore use Green's theorem in the plane to write

$$\int_C f(s)\,ds = -\int\int_D \left(\frac{\partial v}{\partial x} + \frac{\partial u}{\partial y}\right) dx\,dy + i\int\int_D \left(\frac{\partial u}{\partial x} - \frac{\partial v}{\partial y}\right) dx\,dy.$$

Since f is analytic in D, the Cauchy–Riemann equations hold, and the result is proved. If df/ds is not continuous in D, the proof is rather more technical, and we will not give it here. □

A6.2 CAUCHY'S THEOREM AND TAYLOR'S THEOREM

Theorem A6.4 (Cauchy's integral formula) *If $f(s)$ is a single-valued, analytic function in a simply-connected domain D in the complex s-plane, then*

$$f(z) = \frac{1}{2\pi i} \int_C \frac{f(s)}{(s-z)} \, ds,$$

where C is any simple closed contour in D that encloses the point $s = z$.

Proof Let C_δ be a small circle of radius δ, centred on the point $s = z$. Using Cauchy's theorem,

$$\int_C \frac{f(s)}{(s-z)} \, ds = \int_{C_\delta} \frac{f(s)}{(s-z)} \, ds,$$

which we can rewrite as

$$\int_C \frac{f(s)}{(s-z)} \, ds = f(z) \int_{C_\delta} \frac{ds}{(s-z)} + \int_{C_\delta} \frac{f(s) - f(z)}{(s-z)} \, ds.$$

By writing the first of these two integrals in terms of polar coordinates, $s = z + \delta e^{i\theta}$, we find that

$$\int_{C_\delta} \frac{ds}{(s-z)} = \int_0^{2\pi} \frac{i\delta e^{i\theta}}{\delta e^{i\theta}} \, d\theta = 2\pi i.$$

We can deal with the second integral by noting that, since $f(s)$ is continuous, $|f(s) - f(z)| < \epsilon$ for sufficiently small $|s - z| = \delta$. This means that

$$\left| \int_{C_\delta} \frac{f(s) - f(z)}{(s-z)} \, ds \right| \leq \int_{C_\delta} \frac{|f(s) - f(z)|}{|s - z|} \, |ds| \leq \frac{\epsilon}{\delta} \int_{C_\delta} |ds| = 2\pi \epsilon.$$

Since $\epsilon \to 0$ as $\delta \to 0$, the result is proved. □

An extension of Cauchy's integral formula is

Theorem A6.5 *If $f(s)$ is a single-valued, analytic function in a simply-connected domain D in the complex s-plane, then all the derivatives of f exist in D, and are given by*

$$f^{(n)}(z) = \frac{n!}{2\pi i} \int_C \frac{f(s)}{(s-z)^{n+1}} \, ds, \qquad (A6.2)$$

where C is any simple closed contour in D that encloses the point $s = z$.

The proof of this is similar to that of Cauchy's integral formula.

Theorem A6.6 (Taylor's theorem) *If $f(z)$ is analytic and single-valued inside and on a simple closed contour, C, and z_0 is a point inside C, then*

$$f(z) = f(z_0) + (z - z_0)f'(z_0) + \frac{1}{2!}(z - z_0)^2 f''(z_0) + \cdots = \sum_{n=0}^\infty \frac{1}{n!}(z - z_0)^n f^{(n)}(z_0).$$

$$(A6.3)$$

Proof Let δ be the minimum distance from z_0 to the curve C, and let γ be any circle centred on z_0 with radius $\rho < \delta$, so that, for any point z inside γ, $|z - z_0| < \rho < \delta$. By Cauchy's integral formula,

$$f(z) = \frac{1}{2\pi i} \int_C \frac{f(w)}{w - z} dw = \frac{1}{2\pi i} \int_\gamma \frac{f(w)}{w - z} dw.$$

We also have

$$\frac{1}{w - z} = \frac{1}{w - z_0 - z + z_0} = \frac{1}{w - z_0} \frac{1}{1 - \frac{z - z_0}{w - z_0}}$$

$$= \frac{1}{w - z_0} \left\{ 1 + \frac{z - z_0}{w - z_0} + \frac{(z - z_0)^2}{(w - z_0)^2} + \cdots \right\}$$

$$= \frac{1}{w - z_0} + \frac{z - z_0}{(w - z_0)^2} + \cdots + \frac{(z - z_0)^n}{(w - z_0)^{n+1}} + \cdots,$$

and this series is uniformly convergent inside γ, since $|z - z_0| / |w - z_0| < 1$. This means that

$$f(z) = \frac{1}{2\pi i} \int_\gamma \frac{f(w)}{w - z} dw = \frac{1}{2\pi i} \int_\gamma \frac{f(w)}{w - z_0} dw + (z - z_0) \frac{1}{2\pi i} \int_\gamma \frac{f(w)}{(w - z_0)^2} dw + \cdots$$

$$+ (z - z_0)^n \frac{1}{2\pi i} \int_\gamma \frac{f(w)}{(w - z_0)^{n+1}} dw + \cdots.$$

Equation (A6.2) then gives us (A6.3). This series converges in any circle centred on z_0 with radius less than δ. If $z_0 = 0$ we get the simpler form

$$f(z) = f(0) + z f'(0) + \frac{1}{2!} z^2 f''(z) + \cdots.$$

□

Theorem A6.7 (Cauchy's inequality) *If $f(z)$ is analytic for $|z| < R$ with $f(z) = \sum_{n=0}^\infty a_n z^n$, then, if M_r is the supremum of $|f(z)|$ on the circle $|z| = r < R$, we have $|a_n| < M_r / r^n$.*

Proof By Taylor's theorem,

$$a_n = \frac{1}{2\pi i} \int_{|z|=r} \frac{f(z)}{z^{n+1}} dz,$$

-so that

$$|a_n| \leq \frac{1}{2\pi} \int_{|z|=r} \frac{|f(z)|}{r^{n+1}} |dz| \leq \frac{1}{2\pi} \frac{M_r}{r^{n+1}} 2\pi r = \frac{M_r}{r^n}.$$

□

A6.3 The Laurent Series and Residue Calculus

Theorem A6.8 *Any function $f(s)$ that is single-valued and analytic in an annulus $r_1 \leq |s - s_0| \leq r_2$ for $r_1 < r_2$ real and positive has a series expansion*

$$f(s) = \sum_{n=-\infty}^{\infty} c_n (s - s_0)^n,$$

*which is convergent for $r_1 < |s - s_0| < r_2$. This is known as the **Laurent series**, and its coefficients are given by*

$$c_n = \frac{1}{2\pi i} \int_C \frac{f(s)}{(s - s_0)^{n+1}} ds$$

for any simple closed contour that encloses $s = s_0$ and lies in the annulus $r_1 \leq |s - s_0| \leq r_2$.

Although we quote this important theorem, which underlies the techniques of residue calculus, we will not give a proof here. We do note however that

$$\int_C f(s) \, ds = 2\pi i c_{-1}.$$

For an analytic function, $c_{-1} = 0$, and we have Cauchy's theorem. If f is not analytic within C, this result shows that we can calculate the integral by determining the coefficient of $1/(s - s_0)$ in the Laurent series. The coefficient c_{-1} is called the **residue** of $f(s)$ at $s = s_0$.

Theorem A6.9 (The residue theorem) *Let C be a simple, closed contour, and let $f(s)$ be a complex-valued function that is single-valued and analytic on and within C, except at n isolated singular points, $s = s_1, s_2, \ldots, s_n$. Then*

$$\int_C f(s) \, ds = 2\pi i \sum_{j=1}^{n} a_j,$$

where a_j is the residue of $f(s)$ at $s = s_j$.

Proof Consider the closed contour $\hat{C}$, shown in Figure A6.1, which is the contour C deformed into small circles around each singular point and joined to the original position of C by straight contours. Since f is analytic within $\hat{C}$,

$$\int_{\hat{C}} f(s) \, ds = 0,$$

by Cauchy's theorem. The integrals along the straight parts of $\hat{C}$ cancel out in the limit as they approach each other, so we conclude that

$$\int_{\hat{C}} f(s) \, ds = \int_C f(s) \, ds - \sum_{j=1}^{n} \int_{C_j} f(s) \, ds = 0,$$

and hence

$$\int_C f(s)\,ds = 2\pi i \sum_{j=1}^{n} a_j.$$

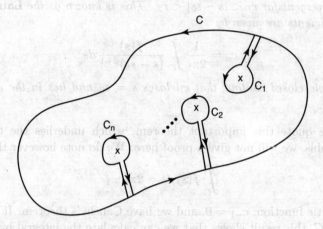

Fig. A6.1. The contours C and C_j.

The residue theorem simply states that the integral around a simple, closed contour is equal to $2\pi i$ times the sum of the residues enclosed by C. This can be used in a variety of ways, in particular to evaluate the integral in the Laplace inversion formula, (6.5).

Note that, for a function with a **pole of order** m at $s = s_0$, and Laurent series expansion

$$f(s) = \frac{a_{-m}}{(s-s_0)^m} + \frac{a_{-(m-1)}}{(s-s_0)^{m-1}} + \cdots,$$

we can write

$$f(s) = \frac{1}{(s-s_0)^m} g(s),$$

where

$$g(s) = a_{-m} + a_{-(m-1)}(s-s_0) + \cdots + a_{-1}(s-s_0)^{m-1} + \cdots,$$

and hence

$$\frac{d^{m-1}g}{ds^{m-1}} = (m-1)!\,a_{-1} + m!\,a_0(s-s_0) + \cdots,$$

which shows that the residue of $f(s)$ at $s = s_0$ is

$$a_{-1} = \frac{1}{(m-1)!} \frac{d^{m-1}}{ds^{m-1}} \{(s-s_0)^m f(s)\}\Big|_{s=s_0}.$$

In particular, at a simple pole, for which $m = 1$, $a_{-1} = (s - s_0)f(s_0)$, and at a double pole, for which $m = 2$,

$$a_{-1} = \frac{d}{ds}\{(s-s_0)^2 f(s)\}\Big|_{s=s_0}.$$

A6.4   Jordan's Lemma

Jordan's lemma allows us to neglect an integral that often arises when evaluating inverse Laplace transforms of functions $F(s)$, provided that $|F(s)| \to 0$ uniformly as $|s| \to \infty$ in the left half plane.

Lemma A6.1 (Jordan) *Let C_J be a semi-circular arc of radius R centred at $s = s_0$ in the left half of the complex s-plane. If $F(s)$ is a complex-valued function of s and $|F(s)| \to 0$ uniformly on C_J as $R \to \infty$, then*

$$\lim_{R \to \infty} \int_{C_J} e^{st} F(s)\, ds = 0,$$

for t a positive constant.

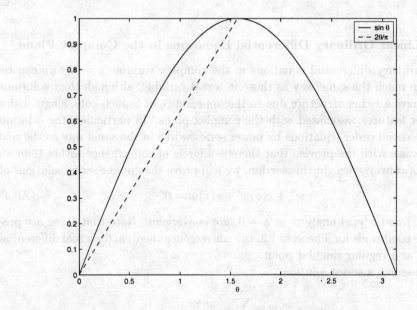

Fig. A6.2. $\sin\theta \geqslant 2\theta/\pi$.

524 COMPLEX VARIABLES

Proof On C_J, $s = s_0 + Re^{i\theta}$, so

$$I = \int_{C_J} e^{st} F(s)\, ds = \int_{\pi/2}^{3\pi/2} e^{ts_0 + tRe^{i\theta}} F(s_0 + Re^{i\theta}) iRe^{i\theta}\, d\theta,$$

and hence

$$|I| \leq R \int_{\pi/2}^{3\pi/2} \left| e^{ts_0 + tRe^{i\theta}} F(s_0 + Re^{i\theta}) \right| d\theta$$

$$= R \int_{\pi/2}^{3\pi/2} |e^{ts_0}| \left| e^{tR\cos\theta + itR\sin\theta} F(s_0 + Re^{i\theta}) \right| d\theta$$

$$= R|e^{ts_0}| \int_{\pi/2}^{3\pi/2} e^{tR\cos\theta} |F(s_0 + Re^{i\theta})|\, d\theta$$

$$= R|e^{ts_0}| \int_0^{\pi} e^{-tR\sin\theta'} \left| F\left(s_0 + Re^{i(\theta' + \frac{\pi}{2})}\right) \right| d\theta',$$

after making the change of variable $\theta = \theta' + \pi/2$. Since $|F(s_0 + Re^{i\theta})| \to 0$ as $R \to \infty$, $|F(s_0 + Re^{i\theta})| < K(R)$ for some real-valued, positive function $K(R)$, such that $K(R) \to 0$ as $R \to \infty$. In addition, since $\sin\theta \geq 2\theta/\pi$, as shown in Figure A6.2,

$$|I| \leq R|e^{ts_0}| K(R) \int_0^{\pi} e^{-2tR\theta'/\pi} d\theta' = \frac{\pi |e^{ts_0}| K(R)}{2t} \left(1 - e^{-2tR}\right) \to 0 \text{ as } R \to \infty.$$

$\square$

A6.5 Linear Ordinary Differential Equations in the Complex Plane

Linear ordinary differential equations in the complex variable $z = x + iy$ can be studied in much the same way as those in a real variable, although their solutions tend to have a richer structure due to the appearance of branch cuts, singularities and other features associated with the complex plane. In particular, the solution of linear second order equations by power series works in the same way as the real variable case, with the proviso that there is a circle of convergence rather than an interval of convergence. In this section, we will prove that power series solutions of

$$w'' + q(z)w' + r(z)w = 0, \tag{A6.4}$$

with $zq(z)$ and $z^2 r(z)$ analytic at $z = 0$, are convergent. Note that these are precisely the conditions for Theorem 1.3, the convergence theorem for a real differential equation at a regular singular point:

If we assume a series solution,

$$w = \sum_{n=0}^{\infty} a_n z^{n+c},$$

A6.5 ORDINARY DIFFERENTIAL EQUATIONS IN THE COMPLEX PLANE

and expansions
$$zq(z) = q_0 + q_1 z + \cdots, \quad z^2 r(z) = r_0 + r_1 z + \cdots,$$
then the indicial equation is
$$a_0 \{c(c-1) + cq_0 + r_0\} = 0,$$
and, in general,
$$a_n f(c+n) + \sum_{s=0}^{n-1} a_s \{(c+s)q_{n-s} + r_{n-s}\} = 0, \tag{A6.5}$$
where $f(c) = c(c-1) + cq_0 + r_0$. The indicial equation always has one root that will produce a well-defined series solution. We will proceed assuming that we are not dealing with a difficult case in which $f(c+n)$ vanishes, when special care would be needed.

We can rewrite (A6.5) in the form
$$a_n n (n + c_1 - c_2) = -\sum_{s=0}^{n-1} a_s \{(c_1 + s) q_{n-s} + r_{n-s}\}, \tag{A6.6}$$
where c_1 and c_2 are the solutions of the indicial equation, and we have used the fact that $c_1 + c_2 = 1 - q_0$. Since $zq(z)$ and $z^2 r(z)$ are analytic at $z = 0$, they each have Taylor expansions that converge in some disc centred on $z = 0$. Let R be the smaller of these two radii. Then, by Cauchy's inequality (Theorem A6.7), there exist r and $K = K(r)$ such that
$$|q_n| \leqslant \frac{K}{r^n}, \quad |r_n| \leqslant \frac{K}{r^n} \quad \text{for } r < R \text{ and } n = 0, 1, 2, \ldots.$$
If we take the modulus of (A6.6), we can now see that
$$n|a_n||n + c_1 - c_2| \leqslant K \sum_{s=0}^{n-1} |a_s| \frac{|c_1| + s + 1}{r^{n-s}}.$$
Writing $|c_1 - c_2| = \lambda$ and $|c_1| = \mu$, we can define a sequence of coefficients A_n with $|a_n| \leqslant A_n$ using
$$\begin{aligned} A_n &= |a_n| & \text{for } 0 \leqslant n < \lambda, \\ n(n-\lambda) A_n &= K \sum_{s=0}^{n-1} A_n (\mu + s + 1)/r^{n-s} & \text{for } n \geqslant \lambda. \end{aligned}$$
The definitions of A_{n-1} and A_n for $n \geqslant \lambda$ show that
$$n(n-\lambda) A_n - (n-1)(n-1-\lambda) \frac{A_{n-1}}{r} = K(\mu + n) \frac{A_{n-1}}{r}.$$
If we now divide through by $n(n-\lambda) A_{n-1}$ and let $n \to \infty$, we find that $A_n/A_{n-1} \to 1/r$. This shows that the radius of convergence of $\sum_{n=0}^{\infty} A_n z^n$ is r. But, from our definition, $|a_n| \leqslant A_n$, so that the radius of convergence of $\sum_{n=0}^{\infty} a_n z^n$ is at least r for all $r < R$. Hence $\sum_{n=0}^{\infty} a_n z^n$ converges for $|z| < R$ as we claimed.

APPENDIX 7

A Short Introduction to MATLAB

MATLAB† is a programming language and environment that is both powerful and easy to use. It contains **built-in functions** that perform most of the tasks that we need in order to illustrate the material in this book, for example, the numerical integration of ordinary differential equations‡, the graphical display of data, and much more. In particular, MATLAB was originally developed to provide an easy way to access programs that act on matrices, and contains extremely efficient routines, for example, to solve systems of linear equations and to extract the eigenvalues of a matrix§.

In this appendix, our aim is to show complete newcomers to MATLAB enough for them to be able understand the material in the book that uses MATLAB. We introduce several other MATLAB functions in the main part of the text. For a more extensive guide to the power of MATLAB, see Higham and Higham (2000).

A7.1 Getting Started

MATLAB can be used in two ways. As we shall see, we can save **functions**, which have arguments and return values, and **scripts**, which are just sequences of MATLAB commands, as files, and call them from MATLAB. Alternatively, we can simply type commands into MATLAB at the command prompt (>>), and obtain results immediately. For example, MATLAB has all the functionality of a scientific calculator.

```
>> (4+5-6)*14/5, exp(0.4), gamma(0.5)/sqrt(pi)
ans = 8.4000
ans = 1.4918
ans = 1.0000
```

Note that `pi` is a built-in approximation to π and `gamma` is the gamma function, $\Gamma(x)$. The three expressions separated by commas are evaluated successively, and the calculated answers displayed. The default is to display five significant digits of the answer. This can be changed using the `format` command. For example,

---
† MATrix LABoratory
‡ We will often use the built-in subroutine `ode45` to solve ordinary differential equations. An explanation of this is given in Section 9.3.4.
§ See Trefethen and Bau (1997) for an introduction to numerical linear algebra that uses MATLAB extensively to illustrate the material.

```
>> format long
>> (4+5-6)*14/5, exp(0.4), gamma(0.5)/sqrt(pi)
ans = 8.40000000000000
ans = 1.49182469764127
ans = 1.00000000000000
```

MATLAB has comprehensive online documentation, which can be accessed with the command help. Typing help format at the command prompt lists the many other formats that are available.

It should be noted that MATLAB, like any programming language, can only perform calculations with a finite precision, since numbers can only be represented digitally with a finite precision. In particular, each calculation is subject to an error of the order of eps, the **machine precision**.

```
>> eps,sin(pi)
ans = 2.2204e-016
ans = 1.2246e-016
```

This tells us that the machine precision is about 10^{-16}[†], which corresponds to double precision arithmetic. A calculation of $\sin \pi = 0$ leads to an error comparable in magnitude to eps.

A7.2 Variables, Vectors and Matrices

As well as being able to perform calculations directly, MATLAB allows the use of **variables**. Most programming languages require the user to declare the type of each variable, for example a double precision scalar or a single precision complex array. This is not necessary in MATLAB, as storage is allocated as it is needed. For example,

```
>> A = 1, B = 1:3, C = 0:0.1:0.45, D = linspace(1,2,5)'...
   E = [1 2 3; 4 5 6], F = [1+i; 2+2*i]
A = 1
B = 1      2      3
C = 0      0.1000    0.2000    0.3000    0.4000
D = 1.0000
    1.2500
    1.5000
    1.7500
    2.0000
E = 1      2      3
    4      5      6
F = 1.0000 + 1.0000i
    2.0000 + 2.0000i
```

[†] The precise value of eps varies depending upon the system on which MATLAB is installed.

Note that ... allows the current command to overflow onto a new line. We also need to remember that MATLAB variables are **case-sensitive**. For example, X and x denote distinct variables.

— A is a 1×1 matrix with the single, real entry, 1. Of course, we can just treat this as a scalar.
— B is a 1×3 matrix (a row vector). The colon notation, a:b, just denotes a vector with entries running from a to b at unit intervals.
— C is another row vector. In the notation a:c:b, c denotes the spacing as the vector runs from a to b.
— D is a 5×1 matrix (a column vector). The function linspace(a,b,n) generates a row vector running from a to b with n evenly spaced points. Note that A' is the transpose of A when A is real, and its adjoint if A is complex.
— E is a 2×3 matrix. The semicolon is used to denote a new row.
— F is a complex column vector. Note that i is the square root of -1.

MATLAB has a rich variety of functions that operate upon matrices (type help elmat), which we do not really need for the purposes of this book. However, as an example, matrix multiplication takes the obvious form,

```
>> E'*F
ans = 9.0000 +  9.0000i
     12.0000 + 12.0000i
     15.0000 + 15.0000i
>> E*F
??? Error using ==> *
Inner matrix dimensions must agree.
```

Remember, an $m_1 \times n_1$ matrix can only be multiplied by an $m_2 \times n_2$ matrix if $n_1 = m_2$.

It is also possible to operate on matrices element by element, for example

```
>> B.*C
ans = 0     0.2000    0.6000    1.2000    2.0000
```

This takes each element of B and multiplies it by the corresponding element of C. The matrices B and C must be of the same size. The use of a full stop in, as another example, B./C, always denotes element by element calculation. Matrix addition and subtraction take the obvious form. For example,

```
>> C + D'
ans = 1.0000    1.3500    1.7000    2.0500    2.4000
>> C - D
??? Error using ==> -
Matrix dimensions must agree.
```

The one exception to the rule that matrices must have the same dimensions comes when we want to add a scalar to each element of a matrix. For example,

```
>> C + 1
ans = 1.0000    1.1000    1.2000    1.3000    1.4000
```

This adds 1 to each element of C, even though 1 is a scalar.

It is often useful to be able to extract rows or columns of a matrix. For example,

```
>> E(:,3), E(2,:)
ans = 3
      6
ans = 4    5    6
```

extracts the third column and second row of E.

Most of MATLAB's built-in functions can also take matrix arguments. For example

```
>> sin(C)
ans = 0    0.0998    0.1987    0.2955    0.3894
>> E.^2
ans = 1     4     9
      16   25    36
```

A7.3 User-Defined Functions

We can add to MATLAB's set of built-in functions by defining our own functions. For example, if we save the commands

```
function a = x2(x)
a = x.^2;
```

in a file named x2.m, we will be able to access this function, which simply squares each element of x, from the command prompt. Note that the semicolon suppresses the output of a. For example,

```
>> x2(0:-0.1:-0.5)
ans = 0    0.0100    0.0400    0.0900    0.1600    0.2500
```

It is always prudent to write functions in such a way that they can take a matrix argument. For example, if we had written x2 using a = x^2, MATLAB would return an error unless x were a square matrix, when the result would be the matrix x*x.

A7.4 Graphics

One of the most useful features of MATLAB is its wide range of different methods of displaying data graphically (type help graphics). The most useful of these is the plot command. The basic idea is that plot(x,y) plots the data contained in the vector y against that contained in the vector x, which must, of course have the same dimensions. For example,

```
>> x = linspace(1,10,500); y = log(gamma(x)); plot(x,y)
```

produces a plot of the logarithm of the gamma function with $1 \leqslant x \leqslant 10$. We can add axis labels and titles with

```
>> xlabel('x'), ylabel('log \Gamma(x)')
>> title('The logarithm of the gamma function')
```

This produces the plot shown in Figure A7.1. Note that MATLAB automatically picks an appropriate range for the y axis. We can override this using, for example, `YLim([a b])`, to reset the limits of the y axis. There are many other ways of controlling the axes (type `help axis`).

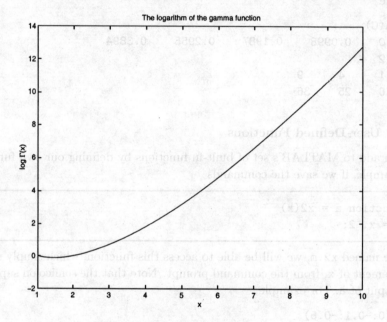

Fig. A7.1. A plot of $\log \Gamma(x)$, produced by MATLAB.

The command `plot` can produce graphs of more than one function. For example, `plot(x1,y1,x2,y2,x3,y3)` produces three lines in the obvious way. By default, the lines are produced in different colours to distinguish between them, and the command `legend('label 1', 'label 2', 'label 3')` adds a legend that names each line. The style of each line can be controlled, and individual points plotted if necessary. For example, `plot(x1,y1,'--',x2,y2,'x',x3,y3,'o-')` plots the first data set as a dashed line, the second as discrete crosses, and the third as a solid line with circles at the discrete data points. For a comprehensive list of options, type `help plot`.

As we shall see in the main part of the book, there is also an easy way of plotting functions to get an idea of what they look like. For example, `ezplot(@sin)` produces a plot of $\sin x$. We will also use the command `ezmesh`, which produces a mesh plot of a function of two variables (see Section 2.6.1 for an explanation). The quantity `@sin` is called a **function handle**. These allow functions to take other functions as arguments, by passing the function handle as a parameter. This includes functions supplied by the user, so that, for example, `ezplot(@x2)` plots the function x2 that we defined earlier.

A7.5 Programming in MATLAB

All of the control structures that you would expect a programming language to have are available in MATLAB. The most commonly used of these are FOR loops and the IF, ELSEIF structure.

FOR Loops

There are many examples of FOR loops in the main text. The basic syntax is

```
for c = x
    statements
end
```

The statements are executed with c taking as its value the successive columns of x. For example, if we save the script

```
x = linspace(0,2*pi,500);
for k = 1:4
    plot(x,sin(k*x))
    pause
end
```

to a file `sinkx.m`, and then type `sinkx` at the command prompt, the commands in `sinkx.m` are executed. This is a MATLAB script. Note that the indentation of the lines is not necessary, but makes the script more readable. The editor supplied with MATLAB formats scripts in this way automatically. The command `pause` waits for the user to hit any key before continuing, so this script successively plots $\sin x$, $\sin 2x$, $\sin 3x$ and $\sin 4x$ for $0 \leqslant x \leqslant 2\pi$.

As a general rule, if a script needs to execute quickly, FOR loops should be avoided where possible, in favour of matrix and vector operations. For example,

```
>> tic, A = (1:50000).^2; toc
elapsed_time = 0.0100
>> tic, for k = 1:50000, A(k) = k^2; end, toc
elapsed_time = 0.4010
```

The command `tic` sets an internal clock to zero, and the variable `toc` contains the time elapsed, in seconds, since the last `tic`. We can see that allocating k^2 to $A(k)$, the kth entry of A, using element by element squaring of 1:50000, known as **vectorizing** the function, executes about forty times faster than performing the same task using a FOR loop. For an in-depth discussion of vectorization, see Van Loan (1997).

Another way of speeding up a MATLAB script is to preallocate the array. In the above sequence, the array A already exists, and is of the correct size when the FOR loop executes. Consider

```
>>A = [];
>>tic, for k = 1:50000, A(k) = k^2; end,toc
elapsed_time = 94.1160
```

If we set A to be the empty array, and then allocate k^2 to A(k), expecting MATLAB to successively increase the size of A at each iteration of the loop, we can see that the execution time becomes huge compared with that when A is already known to be a vector of length 50000.

The IF, ELSEIF Structure

The basic syntax for an IF, ELSEIF statement is

```
if expression
    statements
elseif expression
    statements
elseif expression
    statements
    .
    .
    .
else
    statements
end
```

For example, consider the function

```
function f = f1(x)
f = zeros(size(x));
for k = 1:length(x)
    if (x(k)<0)|(x(k)>3)
        disp('x out of range')
    elseif x(k)<1
        f(k) = x(k)^2;
    elseif x(k)<2
        f(k) = cos(x(k));
    else
        f(k) = exp(-x(k));
    end
end
```

A7.5 PROGRAMMING IN MATLAB

This evaluates the function (A2.1), which is plotted in Figure A2.2, at the points in the vector x, and displays a warning if any of the elements of x lie outside the range $[0,3]$, where the function is defined. The logical function **or** is denoted by | in MATLAB. The function length(x) calculates the length of a vector x. Note that we initialize f as a vector of zeros with the same dimensions as x using f = zeros(size(x)). As we have already discussed, this preallocation of the matrix significantly speeds up the execution of the function. In fact, although this is a transparent way of programming the function, it is still not as efficient as it could be. Consider the function

```
function f = f2(x)
if ((x<0)|(x>3)) == zeros(size(x))
    f = ((x>=0)&(x<1)).*x.^2 + ((x>=1)&(x<2)).*cos(x)...
      + ((x>=2)&(x<=3)).*exp(-x);
else
    disp('x out of range')
end
```

Note that ==, not =, is used to test the equality of two matrices. The vector (x>=0) is the same size as x, and contains ones where the corresponding element of x is positive or zero, and zeros elsewhere. The operator & acts in the obvious way as the logical function **and**. The function f2 evaluates $f(x)$ in a vectorized manner, avoiding the use of a FOR loop. Now consider

```
>> tic, y = f1(x); toc
elapsed_time = 1.4320
>> tic, y = f2(x); toc
elapsed_time = 0.1300
```

It is clear that the vectorized function f2 evaluates (A2.1) far more efficiently than f1, which uses a FOR loop and an IF, ELSEIF statement.

Bibliography

Ablowitz, M.J. and Fokas, A.S., 1997, *Complex Variables*, Cambridge University Press.
Acheson, D.J., 1990, *Elementary Fluid Dynamics*, Oxford University Press.
Aref, H., 1984, 'Stirring by chaotic advection', *J. Fluid Mech.*, **143**, 1–21.
Arrowsmith, D.K. and Place, C.M., 1990, *An Introduction to Dynamical Systems*, Cambridge University Press.
Bakhvalov, N. and Panasenko, G., 1989, *Homogenization: Averaging Processes in Periodic Media*, Kluwer.
Billingham, J., 2000, 'Steady state solutions for strongly exothermic thermal ignition in symmetric geometries', *IMA J. of Appl. Math.*, **65**, 283–313.
Billingham, J. and King, A.C., 1995, 'The interaction of a fluid/fluid interface with a flat plate', *J. Fluid Mech.*, **296**, 325–351.
Billingham, J. and King, A.C., 2001, *Wave Motion*, Cambridge University Press.
Billingham, J. and Needham, D.J., 1991, 'A note on the properties of a family of travelling wave solutions arising in cubic autocatalysis', *Dynamics and Stability of Systems*, **6**, 1, 33–49.
Bluman, G.W. and Cole, J.D., 1974, *Similarity Methods for Differential Equations*, Springer-Verlag.
Bourland, F.J. and Haberman, R., 1988, 'The modulated phase shift for strongly nonlinear, slowly varying, and weakly damped oscillators', *SIAM J. Appl. Math.*, **48**, 737–748.
Brauer, F. and Noble, J.A., 1969, *Qualitative Theory of Ordinary Differential Equations*, Benjamin.
Buckmaster, J.D. and Ludford, G.S.S., 1982, *Theory of Laminar Flames*, Cambridge University Press.
Byrd, P.F., 1971, *Handbook of Elliptic Integrals for Engineers and Scientists*, Springer-Verlag.
Carrier, G.F. and Pearson, C.E., 1988, *Partial Differential Equations: Theory and Technique*, Academic Press.
Coddington, E.A. and Levinson, N., 1955, *Theory of Ordinary Differential Equations*, McGraw-Hill.
Courant, R. and Hilbert, D., 1937, *Methods of Mathematical Physics*, Interscience.
Dunlop, J. and Smith, D.G., 1977, *Telecomunications Engineering*, Chapman and Hall.
Flowers, B.H. and Mendoza, E., 1970, *Properties of Matter*, John Wiley.
Garabedian, P.R., 1964, *Partial Differential Equations*, John Wiley.
Gel'fand, I.M., 1963, 'Some problems in the theory of quasilinear equations', *Amer. Math. Soc. Translations Series 2*, **29**, 295–381.
Glendinning, P., 1994, *Stability, Instability and Chaos: An Introduction to the Theory of Nonlinear Differential Equations*, Cambridge University Press.
Grindrod, P., 1991, *Patterns and Waves: The Theory and Applications of Reaction-Diffusion Equations*, Oxford University Press.
Grobman, D.M., 1959, "Homeomorphisms of systems of differential equations', *Dokl. Akad. Nauk SSSR*, **128**, 880.
Guckenheimer, J. and Holmes, P.J., 1983, *Nonlinear Oscillations, Dynamical Systems*

and Bifurcations of Vector Fields, Springer-Verlag.

Hartman, P., 1960, 'A lemma in the theory of structural stability of differential equations', *Proc. Amer. Math. Soc.*, **11**, 610–620.

Higham, D.J. and Higham, N.J., 2000, *MATLAB Guide*, SIAM.

Hydon, P.E., 2000, *Symmetry Methods for Differential Equations*, Cambridge University Press.

Ince, E.L., 1956, *Ordinary Differential Equations*, Dover.

Kevorkian, J., 1990, *Partial Differential Equations: Analytical Solution Techniques*, Wadsworth and Brooks.

King, A.C., 1988, 'Periodic approximations to an elliptic function', *Applicable Analysis*, **27**, 271–278.

King, A.C. and Needham, D.J., 1994, 'The effects of variable diffusivity on the development of travelling waves in a class of reaction-diffusion equations', *Phil. Trans. R. Soc. Lond. A*, **348**, 229–260.

Körner, T.W., 1988, *Fourier Analysis*, Cambridge University Press.

Kreider, D.L., Kuller, R.G., Ostberg, D.R. and Perkins, F.W., 1966, *An Introduction to Linear Analysis*, Addison-Wesley.

Kuzmak, G.E., 1959, 'Asymptotic solutions of nonlinear second order differential equations with variable coefficients', *J. Appl. Math. Mech.*, **23**, 730–744.

Landau, L.D. and Lifschitz, E.M., 1959, *Theory of Elasticity*, Pergamon.

Lighthill, M.J., 1958, *Introduction to Fourier Analysis and Generalised Functions*, Cambridge University Press.

Lorenz, E.N., 1963 'Deterministic non-periodic flows', *J. Atmos. Sci.*, **20**, 130–141.

Lunn, M., 1990, *A First Course in Mechanics*, Oxford University Press.

Marlin, T.E., 1995, *Process Control: Designing Processes and Control Systems for Dynamic Performance*, McGraw-Hill.

Mathieu, J. and Scott, J., 2000, *An Introduction to Turbulent Flow*, Cambridge University Press.

Milne-Thompson, L.M., 1952, *Theoretical Aerodynamics*, Macmillan.

Milne-Thompson, L.M., 1960, *Theoretical Hydrodynamics*, Macmillan.

Morris, A.O., 1982, *Linear Algebra: An Introduction*, Van Nostrand Reinhold.

Ottino, J.M., 1989, *The Kinematics of Mixing: Stretching, Chaos and Transport*, Cambridge University Press.

Otto, S.R., Yannacopoulos, A. and Blake, J.R., 2001, 'Transport and mixing in Stokes flow: the effect of chaotic dynamics on the blinking stokeslet', *J. Fluid Mech.*, **430**, 1–26.

Pedley, T.J., 1980, *The Fluid Mechanics of Large Blood Vessels*, Cambridge University Press.

Petrov, V., Scott, S.K. and Showalter, K., 1992, 'Mixed-mode oscillations in chemical systems', *J. Chem. Phys.*, **97**, 6191–6198.

Schiff, L.I., 1968, *Quantum Mechanics*, McGraw-Hill.

Sobey, I., 2000, *An Introduction to Interactive Boundary Layer Theory*, Oxford University Press.

Sparrow, C.T., 1982, *The Lorentz Equations: Bifurcations, Chaos and Strange Attractors*, Springer-Verlag.

Trefethen, L.N. and Bau, D., III, 1997, *Numerical Linear Algebra*, SIAM.

Van Dyke, M., 1964, *Perturbation Methods in Fluid Mechanics*, Academic Press.

Van Loan, C.F., 1997, *Introduction to Scientific Computing*, Prentice Hall.

Watson, G.N., 1922, *A Treatise on the Theory of Bessel Functions*, Cambridge University Press.

Wiggins, S., 1988, *Global Bifurcations and Chaos*, Springer-Verlag.

Wiggins, S., 1990, *Introduction to Applied Nonlinear Dynamical Systems and Chaos*, Springer-Verlag.

Index

Abel's formula, 10, 34, 77, 99
aerofoils, 195, 274
Airy's equation, 79, 214, 350
analyticity, 517
Arrhenius law, 319
asymptotic balance, 305
autocatalysis, 374, 409

bang-bang control, 422, 437
basin of attraction, 223
Bendixson's negative criterion, 240
    Dulac's extension, 240, 253E
Bernoulli shift map, 467
Bernoulli's equation, 190
Bessel functions, 42, 58, 293
    Fourier–Bessel series, 71, 74
    generating function, 64
    modified Bessel functions of first and second kind, 71
    orthogonality, 71
    recurrence relations, 69
    Weber's, 62
Bessel's equation, 28, 58, 108
    $\nu = 1$, 10
    $\nu = \frac{1}{2}$, 12, 28E
    inhomogeneity, 77
    Lommel's functions, 79
Bessel's inequality, 114
Bessel–Lommel theorem, 75
bifurcation
    codimension two, 397
    diagram, 390
    flip, 454
    global, 408
    homoclinic, 408
    Hopf, 402
    pitchfork, 399, 402
    saddle–node, 390, 392, 402, 457
    subcritical Hopf, 402
    subcritical pitchfork, 399
    supercritical Hopf, 402
    supercritical pitchfork, 399
    tangent, 457
    transcritical, 398, 402
bifurcation theory, 388
boundary value problem, 46, 54E, 93, 121E, 183, 370, 376
    inhomogeneous, 96

bounded controls, 419

Cantor
    diagonalization proof, 464
cantor
    middle-third set, 464
Cauchy problem, 180
Cauchy's integral formula, 519
Cauchy's theorem, 287, 298, 518
Cauchy–Kowalewski theorem, 180
Cauchy–Riemann equations, 186
Cauchy–Schwartz inequality, 497
Cayley–Hamilton theorem, 425, 434, 501
centre manifold
    local, 373, 379, 462
centre manifold theorem, 372
chain rule, 511
chaotic solutions, 449
    islands, 460
characteristic equation, 499
characteristic variables, 358
chemical oscillator, 450
comparison theorems, 212
complete elliptic integral
    first kind, 334
    second kind, 336
completely controllable, 420, 444E
completeness, 42, 114, 498
complex analysis
    branch cut, 165
    branch point, 164
    Cauchy–Riemann equations, 186
    keyhole contour, 165
    poles and residues, 163, 521
composite expansions, 312
conformal mapping, 191
    Joukowski transformation, 195
connection problems, 348
conservation law, 271
conservative systems, 218
continuity, 502
    piecewise, 502
    uniform, 502
contour integral, 518
control
    bang-bang, 422
    variables, 418
    vector, 418

control problem
  optimal, 418
  time-optimal, 418
controllability matrix, 433
controllable set, 420
convolution, 130, 160, 352
CSTR (continuous flow, stirred tank reactor), 393, 400, 415
cubic autocatalysis, 374
cubic crosscatalator, 450, 489, 492E

D'Alembert's solution, 180, 357
diffeomorphism, 379, 461, 475
  smooth, 379
differentiable, 502
diffusion equation
  heat, 45, 123
  point source solution, 271
diffusion problem, 352
dimensionless
  groups, 274
  variables, 123
Dirac delta function, 135, 271
Dirichlet kernel, 128
Dirichlet problem, 55, 143, 183, 193
domain of attraction, 223

eigenfunctions, 95
  expansions, 104
eigensolutions, 369E
eigenvalues, 95, 103, 372, 499
eigenvectors, 103, 499
elastic membrane, 80
electrostatics, 39, 148
equation
  Airy, 79, 122E, 214, 280, 350
  Bernoulli, 190
  Cauchy–Riemann, 186, 290, 517
  characteristic (matrices), 425
  diffusion, 123, 175, 184, 270
  Duffing's, 448
  Euler, 413E
  forced Duffing, 448, 482, 486, 492E
  forced wave, 151
  Fourier, 124
  Helmholtz, 351
  Laplace, 143, 147, 175, 179, 182, 186, 193
  logistic, 468
  Lorenz, 451, 470, 493E
  modified Helmholtz, 151, 352
  Navier–Stokes, 170
  porous medium, 273
  reaction–diffusion, 185, 370, 375
  Ricatti's, 269
  Schrödinger's, 119
  Tricomi's, 177
  Volterra integral, 162
  wave, 148, 175, 288, 357
equilibrium point
  centre, 230
  hyperbolic, 224, 475
  nonhyperbolic, 224, 384

nonlinear centre, 401
saddle, 227, 294, 378
stable, 220
stable node, 227
stable spiral, focus, 229
unstable, 220
unstable node, 226
unstable spiral, focus, 228
equilibrium solutions, 220
Euclidean norm, 485, 487, 497
exchange of stabilities, 398
existence, 418
  local, 204
exponential integral, 276

Floquet theory, 335
flow evolution operator, 467
fluid dynamics, 48
  Bernoulli's equation, 51, 186
  boundary conditions, 49
  boundary layer, 274
  circle theorem, 189
  complex potential, 186
  flow past a flat plate, 194
  flow past an aerofoil, 195
  inviscid, irrotational, 186
  Kutta condition, 195
  Navier-Stokes equations, 170
  Reynolds number, 274, 371, 456
  stream function, 186
  triple deck, 370
  turbulence, 456
  velocity potential, 49
  viscosity, 167
Fourier integral, 133
Fourier series, 68, 125, 496, 509
  Bessel functions, 71, 74
  cosine, 127
  generalised coefficients, 113
  Gibbs' phenomenon, 75, 128
  Legendre polynomials, 43
  sine, 127
  uniform convergence, 132
Fourier theorem, 131
Fourier transform, 133, 280, 289, 352
  convolution integral, 140
  higher dimensions, 145
  inverse, 138
  inversion formula, 133
  linearity, 138
  shifting, 150
Fourier's equation, 108, 124
Fourier's law, 44
Fourier–Bessel series, 71, 74, 90E, 126
Fourier–Legendre series, 126
Fredholm alternative, 96
frequency modulation, 87
Friedrichs boundary conditions, 109
functions
  $\phi(n)$, 23
  Airy, 79, 296
  Bessel, 280

INDEX

functions (*cont.*)
  Bessel (large argument), 346
  Bessel of order zero, 293
  complementary error, 167, 270, 357
  cost (control), 418
  Dirac delta, 135
  error, 167
  gamma, 58, 153, 281
  gauge, 275
  good, 134
  Heaviside, 135
  Jacobian elliptic, 251, 314, 331, 333, 368E
  Kronecker delta, 74
  Laguerre polynomials, 122
  Mel'nikov, 479
  modified Bessel, 71, 281, 300E, 302E
  negative definite, 381
  of exponential order, 154
  parabolic cylinder, 302
  positive definite, 381
  sequences of, 509
  sign function, 135
  unit function, 135

gamma function, 153, 294
gauge functions, 275
generalized function, 134
  Dirac delta, 135
  Heaviside, 135
  multi-dimensional delta function, 145
  sign function, 135
  unit function, 135
generating functions
  Bessel functions, 64
  Legendre polynomials, 35
Gibbs' phenomenon, 75, 128
good functions, 134
Green's formula, 111
Green's function, 98, 121E, 141
  free space, 142
Gronwall's inequality, 210, 212, 216E
group
  extended, 262
  infinitesimal generators, 259
  isomorphic, 258
  magnification, 260
  one-parameter, 257, 266
  PDEs, 270
  rotation, 261
  two-parameter, 270
group invariants, 256, 261
group theoretic methods, 322, 512

Hamiltonian system, 247, 477
Hartman-Grobman theorem, 230
Heaviside function, 135, 156
Hermite's equation, 101, 112
Hermitian, 103
heteroclinic
  path, 221
  solutions, 221

homeomorphism, 379
homoclinic
  path, 221
homoclinic point
  transverse, 474
homoclinic tangles, 447, 472
Hopf bifurcation theorem, 403, 415E
hypergeometric equation, 29
hysteresis, 397

initial value problem, 3, 181, 288, 357
inner product, 102, 497
  complex functions, 111
integral equations, 172E
  Volterra, 162
integral path, 219, 225, 373
integrating factors, 512
intermittency, 456
Inverse
  Fourier transform, 138

Jacobian, 234, 248, 250, 322, 377, 380, 480, 508
Jordan curve theorem, 243
Jordan's lemma, 163, 287, 523

Kronecker delta, 113
Kuzmak's method, 332

Lagrange's identity, 101
Laguerre polynomials, 122
Laplace transform, 152, 352
  Bromwich inversion integral, 163
  convolution, 160
  first shifting theorem, 155, 164
  inverse, 155, 280
  inversion, 162
  linearity, 154
  ODEs, 158
  of a derivative, 157
  second shifting theorem, 156
Laplace's equation, 45, 54E, 143, 193
Laplace's method, 281, 300E
Laurent series, 521
leading order solutions, 303
Legendre equation
  $n = 1$, 6
Legendre polynomials, 31, 61, 106
  $P_0(x)$, $P_1(x)$, $P_2(x)$, $P_3(x)$ and $P_4(x)$, 33
  associated Legendre polynomials, 52
  Fourier-Legendre series, 43, 54E
  generating function, 35, 54E
  Laplace's representation, 40
  orthogonality, 41
  recurrence relations, 38
  Rodrigues' formula, 39
  Schläfli's representation, 40
  special values, 37
Legendre's equation, 31, 108
  associated, 52, 120
  general solution, 32

INDEX   539

generalized, 121
  order, 31
Lerch's theorem, 154
Lie group, 257
Lie series, 259
limit cycles, 221
limit point
  ω, 242
linear dependence, 9, 496
linear independence, 9, 496
linear operator
  Hermitian, 103
  self-adjacency, 103
  self-adjoint, 42, 100
linear oscillator
  controlling, 428, 439
linear transformations, 498
linearity
  Fourier transform, 138
  Laplace transform, 154
linearization of nonlinear systems, 221
Liouville's theorem, 249
Lipschitz condition, 205, 212
lobe, 475
Lommel's functions, 79
Lyapunov exponents, 484
  maximum, 485
  spectrum, 487
Lyapunov function, 372, 381
Lyapunov stable, 381

Maclaurin series, 505
manifold, 379
maps, 452
  Bernoulli shift, 467
  Hénon, 459
  horseshoe, 475
  logistic, 454
  Poincaré return, 467
  shift, 452
  Smale horseshoe, 465
  tent, 463
matched asymptotic expansions, 310
MATLAB functions
  airy, 80
  besselj, 63
  ceil, 125
  contour, 50
  eig, 489
  ellipj, 314, 318
  ellipke, 337
  erf, 169
  erfc, 169
  ezmesh, 48
  fzero, 76, 318
  gamma, 60
  legend, 33
  legendre, 33
  linspace, 33
  lorenz, 451
  meshgrid, 50

  ode45, 237, 489
  plot, 33, 529
  polyval, 17
  subplot, 63
matrix exponential, 424, 444
mean value theorem, 504
Mel'nikov function, 493E
Mel'nikov theory, 477
method of Frobenius, 11–24, 31, 54E, 60, 507
  General Rule I, 15, 27, 32
  General Rule II, 19, 61
  General Rule III, 23, 62
  indicial equation, 13
method of matched asymptotic expansions, 307, 366E
method of multiple scales, 325, 332, 333, 359, 367E
method of reduction of order, 5, 33, 54E, 269
  formula, 6
method of stationary phase, 285–290, 301E
method of steepest descents, 290, 350
Method of variation of parameters, 7, 34, 77, 96, 351
  formula, 8, 98
modified Bessel functions, 71

negatively invariant set, 241
Neumann problem, 144, 184
Newton's
  first law, 94
  second law, 50, 80, 218, 231, 255, 426, 447
nonautonomous, 269, 322, 429, 432
normal form, 390
  pitchfork bifurcation, 402, 415E
  transcritical bifurcation, 402, 414E
nullclines, 234

ODEs, 511
  autonomous, 222
  complementary function, 4
  coupled systems, 158
  existence, 203, 204
  integral paths, 225
  integrating factors, 512
  irregular singular point, 25
  nonautonomous, 222
  nonsingular, 3, 10
  ordinary points, 225
  particular integral, 4, 8
  regular point, 3
  second order equations (constant coefficients), 513
  separable variables, 511
  singular point, 3, 24
  successive approximations, 204
  uniqueness, 203, 210
one-parameter group
  horizontal translation, 257
  magnification, 257
  rotation, 257
  vertical translation, 257

INDEX

orthogonality, 98, 103, 111, 497
   Bessel functions, 71
   Legendre polynomials, 41
orthonormality, 113

particular integral, 512
PDEs
   asymptotic methods, 351
   canonical form, 175, 177
   Cauchy problem, 180
   characteristic variables, 83, 177, 358
   classification, 175
   d'Alembert's solution, 180
   elliptic, 175
   Fourier transforms, 143
   group theoretic methods, 270
   hyperbolic, 175
   parabolic, 175
   separable solution, 124
   similarity variable, 270
Peano uniqueness theorem, 216E
period doubling, 454
periodic solutions, 221
perturbation problem
   regular, 274, 303
   singular, 274, 304
phase plane, 219
phase portrait, 245
plane polar coordinates, 146
planetary motion, 66
Pockhammer symbol, 60
Poincaré return map, 467, 492E
Poincaré index, 237, 253E
Poincaré projection, 245
Poincaré–Bendixson theorem, 241, 387
population dynamics, 233
positioning problem, 426, 433, 438, 444E
   with friction, 427
   with two controls, 429, 441
positively invariant set, 241
power series
   comparison test, 507
   convergence, 506
   radius of convergence, 507
   ratio test, 506
Prüfer substitution, 115
predator–prey system, 234, 253E, 445E
product rule, 5, 425

quantum mechanics, 118
   hydrogen atom, 120

Rayleigh's problem, 168
reachable set, 420, 421
residue theorem, 521
resonance, 97
Riemann integral, 497

Riemann–Lebesgue lemma, 130
Rodrigues' formula, 39

saddle point, 227
Schrödinger's equation, 119
secular terms, 327
separatrices, 227, 252E, 323, 378, 408
set
   Cantor's middle-third, 464
   closed, 423
   completely disconnected, 464
   connected, 431
   controllable, 429
   convex, 422
   open, 423
   reachable, 421
   strictly convex, 423
simple harmonic motion, 217, 233, 331, 425
Smale-Birkhoff theorem, 475
solubility, 97
spherical harmonics, 54
spherical polar coordinates, 39, 53, 146
stability
   asymptotic, 381
   exchange of, 398
   Lyapunov, 381
   structurally unstable, 389
stable manifold
   global, 380
   local, 372, 378, 379, 462
stable node, 227
stable spiral, focus, 229
state variables, 418
state vector, 418
steering problem, 427
Stirling's formula, 282, 301E
structurally unstable, 389
Sturm comparison theorem, 30
Sturm separation theorem, 11
Sturm–Liouville problems, 42, 107
   regular endpoint, 109
   singular endpoint, 109

target state, 418
Taylor series, 506
   functions of two variables, 507
thermal ignition, 318
tightrope walker, 419, 443E
time-optimal control problem, 418
Time-optimal maximum principle, 436
transformations
   identity, 257
   infinitesimal, 258
   inverse, 257
   magnification, 260
   rotation, 261
transversal, 242
travelling wave solution, 375, 409
triangle inequality, 208, 497

uniqueness, 210, 418
unstable manifold
  global, 380
  local, 379, 462
unstable node, 227
unstable spiral, focus, 228

van der Pol Oscillator, 329
Van Dyke's matching principle, 311, 366E

Vector space, 9, 495
  basis, 496

Watson's lemma, 283, 292, 300E
wave equation, 82
  d'Alembert's solution, 83
WKB approximation, 344, 369E
Wronskian, 8, 9, 28E, 54E, 77, 90E, 98